Contributors

R. E. Baier, Calspan Corporation, Advanced Technology Center, Buffalo, New York

G. Bitton, Department of Environmental Engineering Sciences, University of Florida, Gainesville, Florida

William A. Corpe, Department of Biological Sciences, Columbia University, New York, New York

J. W. Costerton, Biology Department, University of Calgary, Calgary, Alberta, Canada

Stacy L. Daniels, Environmental Sciences Research Laboratory, Dow Chemical U.S.A., Midland, Michigan

Frank B. Dazzo, Departments of Microbiology and Public Health, and Crop and Soil Sciences, Michigan State University, East Lansing, Michigan

Adrian Lee, School of Microbiology, The University of New South Wales, Kensington, N.S.W., Australia

K. C. Marshall, School of Microbiology, The University of New South Wales, Kensington, N.S.W., Australia

Hubert N. Newman, Institute of Dental Surgery, Eastman Dental Hospital (British Postgraduate Medical Federation), University of London, London, U.K.

Hans W. Paerl, Institute of Marine Sciences, University of North Carolina, Morehead City, North Carolina

Dwayne C. Savage, Department of Microbiology, University of Illinois, Urbana, Illinois

Preface

In an age when the activities of scientists must be accountable in terms of relevance to world needs, microbiologists are showing a renewed interest in ecological and pollution-related aspects of their discipline. A direct consequence of this trend is an increasing awareness of the extent of associations between microorganisms and the variety of surfaces found in all microbial ecosystems. Adhesion of microorganisms to such surfaces has enormous consequences in both the beneficial and detrimental functions of microorganisms in natural habitats. In compiling this book, the editors selected authors whose interest range from mechanisms of microbial adhesion, to the role of surface properties, to aspects of methodology in the study of adhesion processes, to health-related aspects, and to the significance of microbial adhesion in a range of natural ecosystems. We feel that this book provides an excellent introduction to and source of concepts and references on microbial adhesion for research workers and students in most areas of microbiology.

The editors wish to thank Associate Professor P. Schofield, The University of New South Wales, for providing the frontispiece cartoon.

GABRIEL BITTON
K. C. MARSHALL

Gainesville, Florida
Kensington, N.S.W., Australia
October 1979

Adsorption of Microorganisms to Surfaces

MICROBIAL ADHESION

Adsorption of Microorganisms to Surfaces

Edited by

Gabriel Bitton
University of Florida

and

Kevin C. Marshall
The University of New South Wales

A Wiley-Interscience Publication

JOHN WILEY & SONS

New York • Chichester • Brisbane • Toronto

Library of Congress Cataloging in Publication Data
Main entry under title:

Adsorption of microorganisms to surfaces.

"A Wiley-Interscience publication."
Includes index.
1. Micro-organisms—Physiology. 2. Adsorption (Biology) 3. Bacterial cell walls.
4. Surface chemistry. I. Bitton, Gabriel.
II. Marshall, Kevin C.
QR84.A42 576'.11 79-19482
ISBN 0-471-03157-7

Printed in the United States of America

10 9 8 7 6 5 4 3 2 1

Contents

Adsorption of Microorganisms to Surfaces

CHAPTER 1

Microbial Adhesion in Perspective

K. C. MARSHALL

School of Microbiology, The University of New South Wales, Kensington, N. S. W. Australia

G. BITTON

Department of Environmental Engineering Sciences, University of Florida, Gainesville

CONTENTS

1.1 INTRODUCTION

Research on the association between microorganisms and solid surfaces following the pioneering work of ZoBell [1,2] continued in a spasmodic fashion until the present decade. Rapid progress achieved in this field recently has prompted us to provide a definitive review of the physicochemical basis of microbial adhesion and its significance in a variety of ecological situations. We have invited specialists to discuss the properties of surfaces; the modes of attachment of viruses and microorganisms to such surfaces; and the significance of adhesion processes in oral, intestinal, plant, soil, and aquatic ecosystems. It is our intention in this prefatory chapter to emphasize and summarize some significant aspects of the microbial adhesion process.

1.2 PHYSICOCHEMICAL BASIS OF MICROBIAL ADHESION

Viruses and most bacteria are so small that they behave as colloidal particles in aqueous systems and, at least in the case of bacteria, can be considered as *living colloids* [3]. Although the biological properties of the organisms cannot be ignored, many aspects of microbial attraction to and attachment at surfaces can be explained only in terms of physicochemical principles defined by colloid scientists. As a result, we have adopted the conventional colloid-science terms of *adhesion* and *adsorption* in describing associations between microorganisms and solid surfaces. There appears to be some reluctance in accepting such terminology in the fields of oral and medical microbiology, where use of the less accurate term adherence has become entrenched. We have rejected the term *adherence* because it is not consistent with colloid science terminology and because it is best used in the context of adherence to a concept or a principle.

1.2.1 Nutrient Accumulation at Surfaces

The nutrient status of many microbial habitats is too low to support active growth of microorganisms. Because nutrient materials (ions, macromolecules, and colloids) tend to concentrate at the solid–liquid interfaces, the availability of colonizable surfaces plays a major role in determining the growth and distribution of microorganisms in such habitats. Another phenomenon that can affect the behavior of microorganisms in the vicinity of the solid–liquid interfaces is an alteration to the surface free energy of immersed solids by the spontaneous adsorption of macromolecular conditioning films (see Chapter 3). There is an urgent need for a better understanding of the role of surface properties and conditioning films in the microbial adhesion process if we are to evolve suitable methods for controlling microbial fouling of ship and other artificially immersed surfaces.

1.2.2 Transport and Attraction to Surfaces

The movement of microorganisms in the aqueous phase of any ecosystem may be either active or passive. Some microorganisms are motile, depending on flagella or cilia for propulsion through water, and are capable of chemotactic responses toward the nutrient gradient established near solid–liquid interfaces [4]. Although *Rhizobium* is attracted to legume root exudates, chemotaxis does not, however, appear to be a major determinant of host specifity in the *Rhizobium*–legume association (see Chapter 8). Transport of nonmotile microorganisms to surfaces, on the other hand, depends on currents, wave motion, or capillary flow. Microorganisms arriving near surfaces then become subject to short-range attraction forces (such as hydrophobic, coulombic, and van der Waals forces) capable of holding the organisms at the solid–liquid interface (see Chapter 2). The organisms are readily removed from the surfaces at this stage, and the term *reversible sorption* is used to describe this situation [5]. An example of this phenomenon is the reversibility of sorption of bacteria and bacteriophage particles to sediment particulates following a decrease in the sediment electrolyte concentration [6].

1.2.3 The Adhesion Process

Contact between bacteria and surfaces is established by means of polymer bridging [3], but several types of adhesion processes have been identified.

Permanent Adhesion. This phenomenon refers to the permanent binding of a microorganism to one site on a surface (see Chapter 2), and various microbial surface components (e.g., capsules, slime layers, microfibrils, and fimbriae) have been implicated in the attachment of different organisms to surfaces (see Chapters 4, 6, and 8). Some microorganisms are capable of permanently adhering to many types of surfaces; this ability is referred to as *nonspecific adhesion*. The extent and strength of this nonspecific adhesion by a microorganism to different surfaces may be dependent on the initial surface properties of the solids involved (see Chapter 3). *Specific adhesion* of certain microorganisms to particular surfaces, on the other hand, must involve interactions between complementary molecular configurations on the solid and the bacterial surfaces (see Chapters 6 and 8).

Temporary Adhesion. Gliding bacteria move only when attached to a solid surface. Consequently, they require a mechanism for adhesion to a surface that will allow movement across that surface. This is achieved by the production of a viscous slime that increases adhesiveness (the force preventing separation) but allows the organism to move across the surface (a relatively low horizontal drag). A slime possessing these properties is referred to as a *temporary* (or Stefan) *adhesive* [7].

1.3 ECOLOGICAL AND PRACTICAL SIGNIFICANCE OF MICROBIAL ADHESION

Sorptive phenomena are of major importance to microorganisms in diverse environments ranging from soil to teeth surfaces. The attachment of a microorganism at a solid surface enables the organism to benefit from the enriched nutrient status at the solid–liquid interface; and in very nutrient deficient ecosystems, such as aquatic (see Chapter 11) and soil (see Chapter 9) habitats, surfaces often provide the only sites where nutrient levels adequate for microbial growth are found. In flowing systems, such as the oral cavity (see Chapter 7), the intestinal tract (see Chapters 5 and 6) and ship surfaces, attachment to surfaces ensures that the microorganisms are not eliminated from the particular ecosystem and can function effectively in a protective role (normal intestinal bacteria) or in a less desirable role (intestinal pathogens or marine fouling microorganisms). Attachment may be a prerequisite for mutualistic relationships between micro- and macroorganisms (see Chapters 5 and 8), or it may be essential for the growth or survival of potentially pathogenic microorganisms (see Chapters 6 and 10).

Understanding the mechanism of bacterial adhesion may help in finding appropriate means for preventing or reducing the establishment of dental plaques (see Chapter 7), the biofouling of heat exchangers in marine waters or the clogging of pipes (see Chapter 3). A thorough understanding of *Rhizobium*–root association may aid in the improvement of biological nitrogen fixation (see Chapter 8). With respect to viruses, the study of their adsorption to surfaces has contributed significantly to the development of detection methods and to the improvement of the virus-removal capacity of water and wastewater treatment plants (see Chapter 10).

Rapid progress has been or is being made in studies on the mechanisms and significance of microbial adhesion. This was made possible through the development of techniques for the recovery and enumeration of microorganisms adsorbed to surfaces (see Chapter 12). It is obvious that microbial adhesion is of such fundamental significance in the function of and interaction between microorganisms that interest and research in this area must continue to accelerate.

1.4 REFERENCES

1. C. E. ZoBell and E. C. Anderson, *J. Bacteriol.*, **29,** 239 (1935).
2. C. E. ZoBell, *J. Bacteriol.*, **46,** 39 (1943).
3. K. C. Marshall, *Interfaces in Microbial Ecology*, Harvard U. P., Cambridge, Mass., 1976, pp. 5, 44.
4. L. Y. Young and R. Mitchell, in R. F. Acker, B. F. Brown, J. R. dePalma, and W. P. Iverson, Eds., *Proceedings of Third International Congress of Marine Corrosion and Fouling*, Northwestern U. P., Evanston, Ill., 1973, p. 617.
5. K. C. Marshall, R. Stout, and R. Mitchell, *J. Gen. Microbiol.*, **68,** 337 (1971).
6. M. M. Roper and K. C. Marshall, *Microbial Ecol.*, **1,** 1 (1974).
7. B. A. Humphrey, M. R. Dickson, and K. C. Marshall, *Arch. Microbiol.* **120**, 231 (1979).

CHAPTER 2

Mechanisms Involved in Sorption of Microorganisms to Solid Surfaces

STACY L. DANIELS

Environmental Sciences Research Laboratory, Dow Chemical U.S.A., Midland, Michigan

CONTENTS

2.1 INTRODUCTION

Sorptive mechanisms are involved in biological systems in the separation and concentration of soluble compounds and ionized species during chemical analysis [165], biochemical purification [135], and wastewater treatment [127]. Sorptive mechanisms are operative in the nonselective concentration of microorganisms at interfaces occurring in natural systems [121,209], such as soil [120] and aqueous [34,218] environments. Similar mechanisms are involved in the direct contact and attachment of suspended microorganisms onto selective solid surfaces and provide a means for isolating cellular components [22] and purifying mixtures of dissimilar cells [40]. Interactions between the surfaces of living cells and other substances dissolved or suspended in the cellular environment have been reported in an extensive and heterogeneous literature involving diverse scientific disciplines of microbiology, biochemistry, engineering, and dentistry [38,40].

Surface-transport phenomena involving microorganisms can be divided into two overlapping areas shown in Fig. 2.1. Nonsorptive charge phenomena involve various electrical, thermal, hydrodynamic, and chemical effects on cellular movement. Sorptive phenomena can involve sorption of soluble and colloidal materials to individual microbial cell surfaces and sorption by the cells themselves to larger surfaces. Sorption to cells involves the movement of soluble or colloidal materials through a fluid to the surface of a freely suspended cell or that of a cell itself adsorbed to a larger surface. Sorption by cells involves the movement of the cells themselves through a fluid to particulate materials, extended natural surfaces, or engineered surfaces having specifically designed purposes.

In this discussion adsorption is interpreted to mean a concentration on the surface of the sorbent. Any subsequent adsorption, that is, penetration or diffusion of an adsorbed material into the interior of the sorbent, is not addressed. All these phenomena must be considered since they relate to the elucidation of the mechanisms and adsorption by microorganisms onto and desorption from solid surfaces.

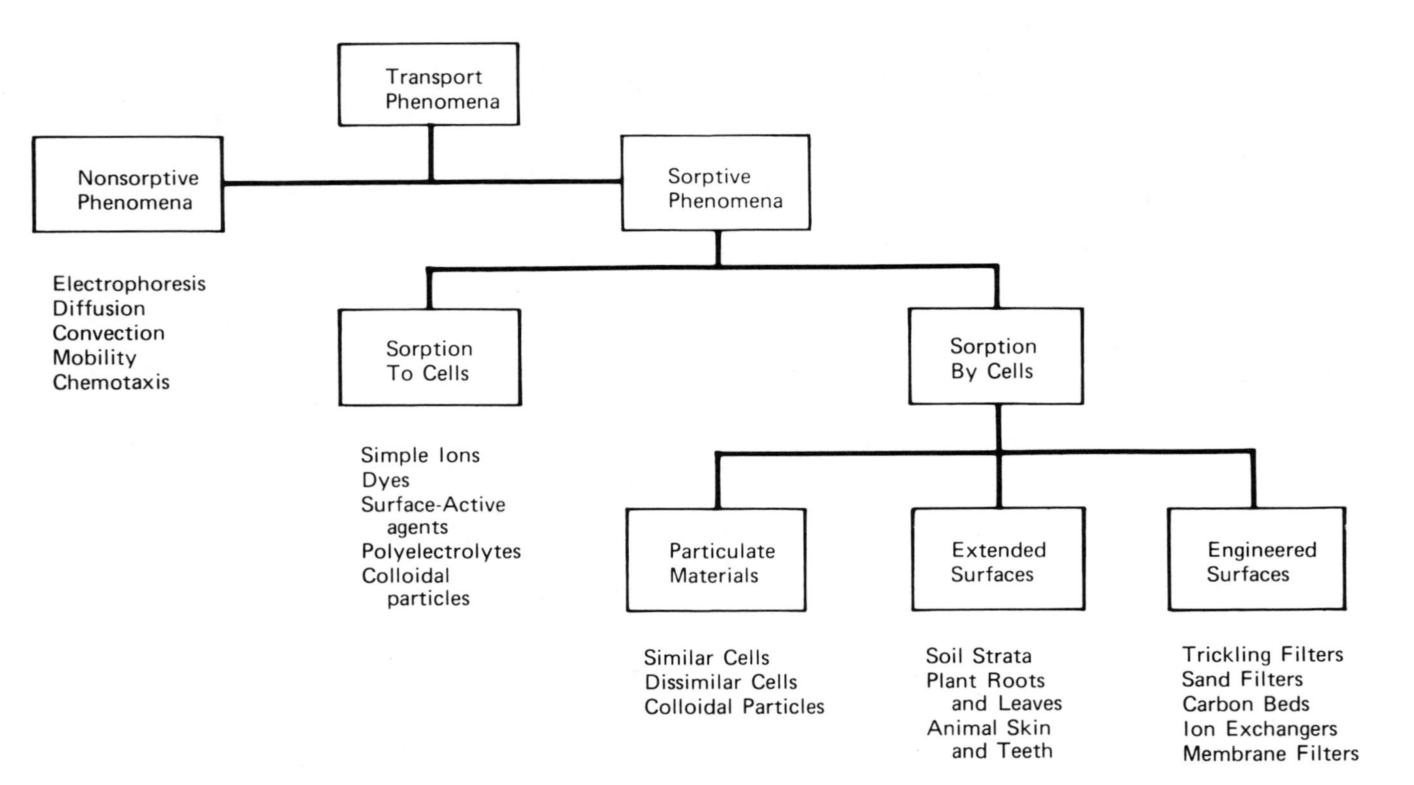

Figure 2.1 Surface transport phenomena involving microorganisms.

2.1.1 Nonsorptive Charge Phenomena

The transfer of cells between two liquid phases [3], or between gas and liquid phases [17,136] and the influence of applied electrical forces on cells suspended in a conducting medium [20] are associated with nonsorptive charge phenomena. Chemotaxis involves a positive or negative response (movement toward or away) from a chemical stimulus that may be a dissolved substance or a chemically active surface [26]. The transport of dissolved material to or from suspended cells [27] and the movement of cells toward or away from surfaces by tactic response [200] may involve diffusional processes that can operate simultaneously with locomotive and sorptive phenomena.

Microbial cells under the influence of an electric field will migrate with respect to the suspending fluid toward electrodes of opposite charge (electrophoresis). This behavior is useful for quantitatively resolving complex mixtures of particles and/or cells [110]. Microorganisms generally travel through a liquid or gas by random diffusion (Brownian motion) [37]. Movement of suspended cells also may occur simultaneously as a result of agitation, convection, or self-propulsion [200]. Cellular movement of all types are influenced by attractive or repulsive forces of neighboring suspended particles and extended surfaces, and by electromagnetic force fields [88].

2.1.2 Sorption to Microorganisms

Cells in bulk suspension may sorb various soluble chemical species to their surfaces [38]. Mass transfer of specific substances may be essential, detrimental, or inconsequential to microbial cells. Flocculation, dispersion, osmotic shock, lysis, or death of the cells may be enhanced. Interactions between charged species existing on cell surfaces and those in bulk solution can be divided arbitrarily into four groups: simple ions, dyes, surface-active agents, and soluble polyelectrolytes [40].

Many dissolved ions are associated with normal cellular growth and development [38]. Simple inorganic and organic ions establish both pH and ionic strength of a given system. More complex ions of amino acids, exoenzymes, and other organic compounds stimulate or inhibit microbial activity by affecting respiration or synthesis. The overall metabolic functioning of these ions can be complex, but certain effects are distinguished by the abilities of such compounds to permeate cell walls, regulate osmotic pressure, facilitate transport, and alter surface electrical charges [27]. These effects can be observed microscopically through changes in the shape and size of individual cells, or by interaction with other cells and surfaces.

Certain multivalent ions can combine with cells of opposite charge [174]. Turbidity changes and coagulation can be observed macroscopically. The sorption of metabolically active ions can often be observed by monitoring changes in the pH, or in the concentration of cells or of specific compounds.

The combinations of dyes with functional chemical groups on microbial cell surfaces have two desirable features: (1) differential staining of specific cellular structures for microscopic observation and (2) visual or optical monitoring of adsorption (and absorption) rates of tracer dyes by cells. The reactions of acidic and basic dyes with cells [130] and the mode of action of the differential Gram stain [10] have been intensively investigated. The bindings of some dyes by microorganisms have been described by Freundlich or Langmuir adsorption isotherms [19,61]. Dyes and cells react stoichiometrically, and ion-exchange capacities of whole cells can be determined through the exchange of dye ions with charged groups on cellular surfaces [78,79,128,129].

Synthetic surface-active agents are not required for normal growth and can interfere with life processes or cause noticeable effects on cellular surfaces. These compounds alter the environment immediately surrounding a cell by reducing surface tension or increasing viscosity and can partially agglomerate cells or otherwise completely disrupt cellular activity [38]. The surface charges of microorganisms can be significantly altered on the adsorption of anionic, cationic, and even nonionic surface-active agents [72].

Polyelectrolytes and hydrophilic colloids are orders of magnitude larger in molecular size than most surface-active agents but are still water soluble or water swellable, respectively. Natural and synthetic polyelectrolytes adsorb strongly to surfaces and may flocculate cells by: (1) reduction of the net surface charge of the cells, and (2) the formation of polymeric bridges among the cells [153]. Flocculation reactions are useful in determining isoelectric points and ion-combining capacities of microbial cells, enumerating bacterial populations, concentrating dilute suspensions, and treating water and wastewater. Such reactions are also important in understanding the mechanisms of sorption of microorganisms to surfaces and are discussed in detail in Section 2.3.4.

2.1.3 Sorption by Microorganisms

The adsorption of microorganisms onto particulate materials, extended surfaces, and engineered surfaces involves many of the same phenomena that cause the suspended cells themselves to adsorb or absorb dissolved substances or other colloids. Cells of dissimilar species having surfaces of opposite or complementing charge can interact mutually. The mechanisms of interaction between bacteria and bacteriophages have been investigated extensively [155,204]. Cells of similar microorganisms also interact, and this results in agglomeration of the entire population [180]. This process is promoted in the biological treatment of wastewater by activated sludge to improve the separation of cellular masses resulting from growth on soluble biodegradable compounds.

Reactions between microorganisms and noncellular extended surfaces significantly affect cellular movement [14]. Microorganisms suspended in a fluid contacting soil strata, plant roots and leaves, and animal skin and teeth are influenced by electrical charges existing on the solid surfaces [38]. These charge reactions are of practical interest in the design and operation of surfaces engineered for water and wastewater treatment. Examples are: eliminating blockages of flow in sand or membrane filters, preventing undesirable growths in beds of activated carbon or ion-exchange resins, and stimulating growth for biological treatment on trickling filters. The metabolic rates of microorganisms adsorbed to surfaces can also be affected since nutrients, cells, and metabolites can all be adsorbed and concentrated at common sites [58,59,86]. The metabolic rates of adsorbed cells can be increased over those of free cells due to the improved availability of nutrients or decreased by the proximity of toxic materials [84,85].

The adsorption of microorganisms onto surfaces can be viewed in three ways as shown in Fig. 2.2: (1) a number of microbial cells can become attached to a single larger surfaces; (2) adsorbent particles and microbial cells of equal size can mutually interact; or (3) several adsorbent particles of dimensions smaller than the microbial cells can adhere to a single cell. The orientations may be end-on or side-on or a combination in some cases. The mechanisms of sorption, rates of transport, and equilibrium capacities can differ in all three cases. The only assumptions are mutually attractive surfaces and a liquid medium through which the contacting surfaces can move together.

2.2 VARIABLES AFFECTING SORPTION

Many of the conclusions put forth in the early literature regarding the sorption of microorganisms to solid surfaces remain irrefutable. Several questions must be reviewed continually, however, to account for new data obtained using better-characterized microorganisms, more efficient adsorbents, and more refined techniques. What is the mechanism by which microorganisms are sorbed at the surface of a solid? Can this process be

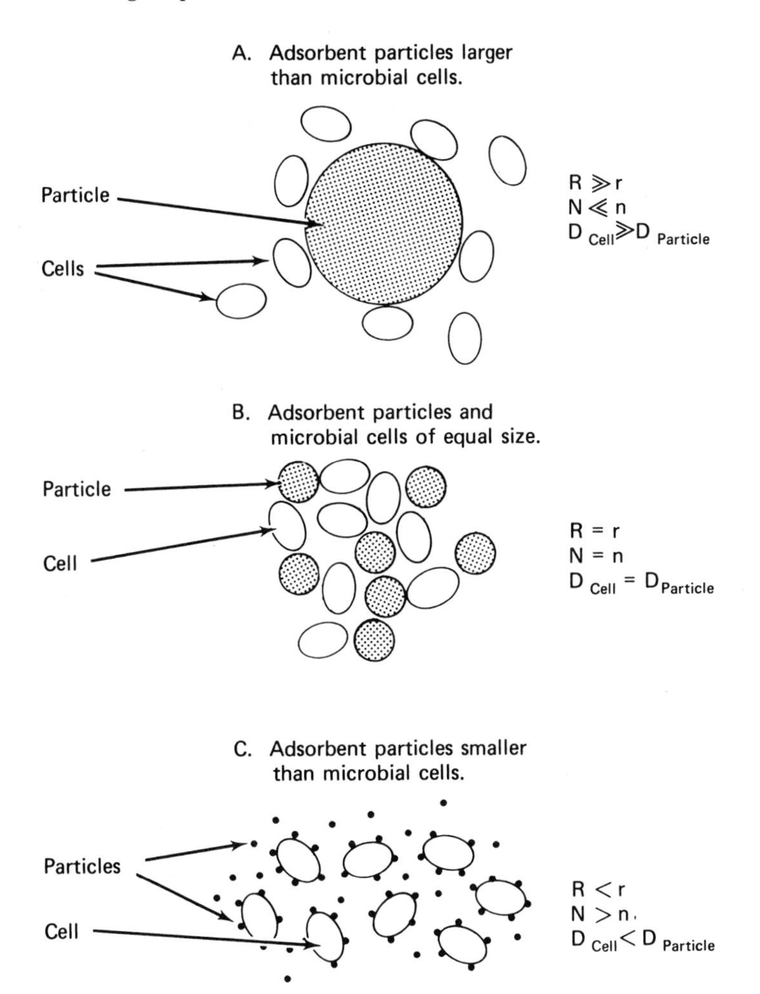

Figure 2.2 Schematic representations of microbial adsorption.

predicted on the basis of specific characteristics of a particular microorganism? Are some materials better or more selective adsorbents?

There have been many studies where the microorganism was well characterized but the adsorbent chosen was a nondescript soil. There have been other studies where the adsorbent was a highly specific ion-exchange resin but the microorganisms were used as a mixed culture. Much of the lack of simultaneous control over both reactants is due to the natural interest of the investigator being directed toward studying the specific activity of a single strain of microorganisms, or removing microorganisms of all types from a specific fluid in real-world situations. Factors related to the sorption of microorganisms are dependent on the surface chemistries of the microbial cells and the adsorbent particles, and on the environmental conditions that affect these related surfaces. These factors are outlined in Table 2.1. The most significant factors influencing selectivity and extent of adsorption are microbial species and adsorbent type. The sorption environment is affected most by pH and the presence of various inorganic and organic compounds that alter surface charges.

2.2.1 General Sorptive Behavior

The bulk of the literature dealing with the sorptive behavior of microorganisms is about equally divided between bacteria and viruses. A few published reports deal with the adsorptions of other microorganisms. General listings of more than 1000 citations of individual species of microorganisms adsorbing onto different solid surfaces have been tabulated in an earlier review [40]. Reviews of sorptive phenomena include those dealing with viruses [12,32,47,67,111,139,154,175,204] and bacteria [38,40,48, 58,70,120,138,160,188,208,209]. Several features of sorption of microorganisms that make understanding of the underlying mechanisms desirable are summarized in Table 2.1.

Investigations of the mechanisms of adsorption of algal cells onto solid surfaces have resulted largely from the sanitary significance of algal cells, present in surface water treated for drinkng purposes, that block filter beds and, subsequently, produce undesirable tastes and odors. Algal cells have been adsorbed onto glass [143], calcium phosphate [2], and both cation-exchange [73] and anion-exchange [112] resins. Adsorbed algal cells can be eluted from surfaces using buffers or other salt solutions. Adsorption is involved in the flocculation of negatively charged algae by positively charged iron and aluminum hydroxides, and by cationic polyelectrolytes [94]. Filter media coated with cationic polyelectrolyes can also remove algae [75].

The mechanisms of adsorption of fungi, protozoa, and rickettsiae are largely unstudied [40]. The adsorptive capacity of various soils, clays, and charcoal for fungi is influenced by pH [145]. Fungal cells also have been removed from aqueous suspensions by adsorption onto cellulose fibers [96] and cation-exchange resins [8]. The adsorptive capacities of sand, soil, and clay for various species of amoebas and flagellates, unlike smaller microorganisms, are less dependent on cell concentration and contact time [36]. Protozoa have also been removed by adsorption onto filter media

Table 2.1 Variables affecting sorption of microorganisms on solid surfaces

1. Character of microorganism
A. Species
B. Culture medium
C. Culture age
D. Concentration
2. Character of adsorbent
A. Type
B. Ionic form (ion-exchange resin)
C. Particle size
D. Cross-linkage (ion-exchange resin)
E. Concentration
3. Character of environment
A. Hydrogen-ion concentration
B. Inorganic salt concentration
C. Organic compounds
D. Agitation
E. Time of contact
F. Temperature

As modified from Daniels [40].

covered with cationic polyelectrolytes [75]. Suspensions of rickettsiae have been separated from debris by nonselective adsorption onto and selective elution from both anion- [92] and cation-exchange [77,203] resins.

The mechanisms of adsorption of animal viruses onto solid surfaces have been studied primarily for purposes of optimizing concentration and detection methods. This topic is covered in greater detail in Chapter 10. Sorptive processes are proven for purification and separation of virus particles from contaminating debris and sundry soluble biochemical species. Charcoal, aluminum hydroxide, and sand were reported [40] as adsorbents for animal viruses in early work. More recently, animal viruses have been reported to be adsorbed onto antibody-coated solids [67]; particulate clays [111], including illite, kaolinite, and montmorillonite; sand; and various insoluble salts, such as aluminum and iron hydroxides, oxides, and phosphates [40]. Both anion- and cation-exchange resins have been reported to be used for the purification of animal viruses [40]. Virus particles thus can be concentrated by adsorption from large volumes of fluid onto adsorbents and then eluted with solutions of differing pH or ionic strength [196].

The adsorption of bacterial viruses onto solid surfaces has been studied mechanistically as a model system to aid in the understanding of virus–host cell reaction [155]. Interactions between bacterial viruses and bacterial cells have been simulated using virus particles in combination with ion-exchange resins [204]. Multivalent cations are necessary for certain bacteriophages to attach to cation exchangers but not for attachment to anion exchangers. This dependency on cation-mediated sorption has been reported for many virus–cell pairs [38]. Other adsorbents for bacteriophages have included kaolin, aluminum oxide, and activated carbon [40]. The majority of the investigations on the adsorption of plant viruses pertain to the adsorption of tobacco mosaic virus (TMV) onto inorganic surfaces, and most are not oriented toward the understanding of mechanisms. Ion-exchange resins [177], calcium phosphate, cellulose, and bentonite have been used to purify plant viruses [154,175]. Solid matrices coupled with antibodies also have been used as specific adsorbents [67].

The possible mechanisms of adsorption of bacterial cells have been discussed frequently in the literature [38]. The reported adsorption of more than 90 bacterial species previously summarized [40] have now expanded to include approximately 100 different species. The adsorption of bacteria onto a variety of soils and other naturally occurring solid surfaces was first reported in detail in the early German and Russian literature [38]. Differences in the adsorption of typhoid and coliform bacteria onto charcoal and clay were of particular interest in studying waterborne diseases [133]. Distinctive differences in adsorption were also evident during the chromatography of cells of various bacterial species in paper strips dipped into cellular suspensions [66,91,102,156].

The metabolic activities of unadsorbed versus adsorbed bacterial cells on soils, sediments, and other inert materials have been studied in detail [46,47,76,98,101,145,151,161,205]. Metabolic activities of bacteria adsorbed on clays have also been detailed [58,59,120,176,188,189]. Studies of the adsorption of bacteria onto synthetic ion-exchange resins of high specificity and capacity were begun following their discovery and general commercial availability. Both respiration and multiplication of bacterial cells have been evaluated on resin surfaces [85,86,207]. Selective adsorption to resolve mixed suspensions into separated cells of their component species using various inorganic materials and ion-exchange resins have also been summarized [40–42].

2.2.2 Microbial Species

The extent and specificity of sorption are influenced mostly by the microbial species undergoing the adsorption [38,40]. Diffusion to the adsorbent is governed more specifically by the surface chemistry of both microbe and adsorbent [120,121]. Several criteria applied to explain variations in the sorption of the cells of different species may be useful in the understanding of mechanisms. These have included the initial conditions of culture, duration of culture, and the methods of purification of suspensions prior to adsorption. Unfortunately, these variables have not been standardized or controlled in many instances. Conditions of growth have been reported to have significant effects on cellular surface charge and any concurrent or subsequent adsorption [40].

In adsorption between species, differences not directly associated with growth have been reported to correlate with staining, viability, motility, and cellular concentration [38]. Correlations of the sorptive behavior of cells of a particular bacterial species with their Gram-staining behavior have been conflicting. The conclusion that all Gram-positive species are preferentially adsorbed by all adsorbents over Gram-negative species is debatable. There are numerous exceptions noted in the resolutions of mixed cellular suspensions [40].

Loss or increase of viability on adsorption, resistance toward desorption, or specificity by a given adsorbent can lead to conflicting results. Living bacteria may [132,161] or may not [98,152] adsorb more strongly than dead bacteria. Bacterial cells may lyse on death, however, and be difficult to detect. The adsorption of motile and nonmotile bacteria may or may not differ, depending on whether the cells are traveling through rigidly confined porous media [14,66,76,186] or interacting with freely suspended solid particles [1,48,98,101,125].

The initial concentration of bacterial cells in suspension, as expected for any sorbable species, affects the extent of adsorption [38,46,54,201]. Saturation of surfaces of adsorbents with bacteria have been observed microscopically [86,168,189,207,211]. The capacity of a given adsorbent may vary from less than 10^{10} bacteria g^{-1} to more than 10^{12} viruses g^{-1} depending on particle size. Adsorption capacities are discussed in more detail in Section 2.5.2. The cells of one species may displace those of another species already adsorbed [41,161,210] or adsorption can occur subsequent to the adsorption of the first species without apparent displacement [42,207].

The marked differences in sorptive behavior among the cells of various bacterial species form the basis for their separation [1,42,104,133,168]. These differences are manifested among different bacterial strains [119,166], and between cells and spores of the same species [106], smooth and rough forms [1], and mother and daughter cells [88,126].

2.2.3 Adsorbent Type

The second most important variable involved in the sorption of cells onto a solid surface is the adsorbent type in contact with the cellular suspension. A general list of major classes of solid adsorbents for microorganisms is given in Table 2.2. A previous review [40] contains detailed citations of specific microorganisms and adsorbents. The more significant adsorbents are summarized in the following discussion.

Early investigations were limited to poorly defined adsorbents that had low capacities and poor selectivity [54,105]. Much of the early work was done using nonspecific soils and sediments [46,98,101,133,145,164]. Common adsorbents, such as clays [4,9,29,54,57,58,76,104,105,108a,120,146], carbon [1,54,105,109,146,164], insoluble calcium and aluminum salts [1,54,76,106,125], filter paper [66,91,102,109,134,156], and glass [34,45,

Table 2.2 Major classes of solid adsorbents for microorganisms

Compound
Organic compounds
Anion-exchange resins
Cation-exchange resins
Cellulose
Cellulose acetate
Cellulose nitrate
Inorganic compounds
Aluminum hydroxide
Aluminum oxide
Aluminum phosphate
Calcium phosphate
Iron oxide
Sodium aluminum silicate
Silica
Soils and minerals
Bentonite
Illite
Kaolinite
Montmorillonite
Living surfaces
Plant tissue
Animal skin and teeth
Intestinal mucosa
Miscellaneous
Charcoal
Filter paper
Glass

Detailed citations are reported elsewhere [38].

123,132,152], have been tested primarily because of their limited water solubility and general availability. Adsorption of microbial cells by any adsorbent is dependent on the quantity available [9,38,46,54,105, 125,161,188]. Smaller particles and those having greater surface areas than smooth spheres are better adsorbents [46,98,101,134,145]. The relative proportions of active and inactive sorbent fractions in a mixed solid further complicate the situation. Capacities of various adsorbents are summarized in Table 2.9.

Synthetic ion-exchange resins have been used in more recent investigations of microbial adsorption. The applications of specific anion- and cation-exchange resins for adsorbing bacterial cells have been summarized [40]. These resins possess specific chemical groupings that dictate the ionic character of their surfaces and their intertices. Ion-exchange resins have been used to study a wide variety of interactions between bacterial cells and their environment, including metabolism [85,86,207] and separation of the cells of individual species [40,210].

The selectivity exhibited by a particular ion-exchange resin is a function of the type of exchangeable counterion [38]. Anion-exchange resins in the chloride form have been the popular and most effective choice. Other anion forms have included phosphate, acetate, nitrate, and hydroxide [40]. Cation-exchange resins in the sodium form have been used most commonly. Other cation forms have included ammonium, potassium, hydrogen, calcium, magnesium, and the divalent forms of iron, copper, and manganese [40].

2.2.4 Environmental Character

The interactions between bacterial cells and adsorbent particles suspended in a fluid are dependent on the pH and ionic strength of the surrounding medium. Adsorption of bacteria by soils [151] clays [108a,145], hydrous metal oxides [151], and ion-exchange resins [38,210] is greatly affected by pH. In many cases adsorption can be reversed simply by altering the pH, suggesting the possibility of a complete charge reversal [9,42,48]. The pH values for optimum adsorption depend on the relative isoelectric points of the microbial cells and the adsorbent [38]. The strongest adsorption of bacterial cells generally occurs at pH 3–6 [38,59,76,188,189].

Different electrolytes can also be bound to the adsorption complex. The adsorption of cells may be enhanced by the addition of multivalent cations to the suspension. The presence of electrolytes, such as calcium and magnesium, can influence the association of microbial cells and adsorbent particles [124]. Multivalent cations can reverse the charge of bacteria suspended in acid media and increase coagulation [162,181] and adsorption

[137,151,156,162,166,189]. This is also true in neutral and basic media where cations such as Fe^{3+}, Al^{3+}, and Ca^{2+} will coagulate suspended solids, including bacterial cells that normally possess negative surface charges. There is also the possibility of adsorption of cells onto hydroxide or carbonate precipitates of these cations.

Additions of various inorganic salts to a suspension after adsorption to solids has occurred can also promote desorption [38,42,71,108a,210]. The relative effectiveness of various anions to elute adsorbed cells has been reported [38,107]. In this case desorption is promoted by salts such as sodium chloride, provided that adsorption has not become irreversible. If very high salt levels are necessary for desorption, the cells may become denatured. Multivalent ions will also promote desorption by exchanging as counterions on ion-exchange resins.

The effects of the duration and intensity of agitation on microbial adsorption to and desorption from solid surfaces have been uncertain. Most researchers have terminated their experiments after no further adsorption of cells was assumed to occur. Equilibrium has generally been reached within 15 min [76,113], although complete equilibrium may require as much as an hour in some cases [38]. The degree of adsorption of bacteria usually is improved up to an optimum point by increased duration and/or intensity of agitation, which increases the probability of contact between the microbial cells and adsorbent particles [38,39]. It is possible to agitate too violently or for too long a period and cause desorption [38].

The affinity of an adsorbent for microbial cells can be greatly altered by chemical and physical means. Heating has been shown to increase the sorptive capacity of soils [145] but reduce that of ion-exchange resins [108]. The capacity of various adsorbents has been both increased and decreased on treatment with acids or bases, oxidizing agents, or certain salts [108,161,171]. It is difficult to predict the actual behaviors of specific microorganism–adsorbent pairs except under carefully controlled environmental conditions.

2.3 MECHANISMS OF SORPTION

A discussion of the mechanisms of sorption begins with division of the attractive and repulsive forces associated with the behavior of microorganisms at interfaces, specific and nonspecific concentration at surfaces, capillary and chromatographic effects, flocculation, and ion exchange. A simplistic and mechanistic division of microbial sorption is proposed in Table 2.3. It should be stressed that these divisions are somewhat arbitrary.

Table 2.3 Mechanistic divisions of microbial sorption

Sorption division	Sorptive interaction	Sorption energy	Forces	Sorption sites	Example
Chemical sorption	Specific; irreversible	High	Multiple; covalent bonding	Fixed	Bacterium-phage interaction; whole cell immobilization
Ion exchange	Specific; reversible	Variable	Electrostatic	Fixed	Cell concentration on charged resins
Flocculation	Nonspecific; reversible and irreversible	Variable	Electrostatic; London–van der Waals	Variable	Polymer-mediated cell aggregation
Physical sorption	Nonspecific; reversible	Low	London–van der Waals; interfacial tension	Variable	Cell concentration on inert solids and bubbles

It is difficult to distinguish between them in actuality, and more than one are probably involved in all microbial sorptions. Some of these divisions are discussed in more detail in the subsequent sections of this chapter.

2.3.1 Attractive and Repulsive Forces

The types of forces leading to the destabilization of microbial suspensions and attachment to surfaces are influenced strongly by the physico-chemical properties of the surfaces of the adsorbent and those of the microbial cells. These properties are discussed in Chapters 3 and 4. A listing of possible attractive and repulsive forces [153] is presented in expanded form in Table 2.4. Those forces of relatively short range and high specificity are grouped at the top of the list. The same forces of interaction between solid surfaces and microbial cells can be involved in the cell–cell and solid–solid interactions.

Table 2.4 Forces of attraction and repulsion between microbial cells and adsorbent surfaces

Forces of attraction

1. Chemical bonding (hydrogen, thio, amide, and ester bonds)
2. Ion-pair formation ($-NH_3^+ \cdots {}^-OOC-$)
3. Ion-triplet formation ($-COO^- \cdots Ca^{2+} \cdots {}^-OOC-$)
4. Interparticle bridging (polyelectrolytes)
5. Charge fluctuations
6. Charge mosaics
7. Charge attraction of opposite signs
8. Electrostatic attraction between surfaces of similar charge
9. Electrostatic attraction due to image forces
10. Surface tension
11. van der Waals forces of attraction
12. Electromagnetic forces
13. Hydrodynamic forces
14. Diffusional forces
15. Gravitational forces
16. Positive chemotaxis (cellular mobility)

Forces of repulsion

1. Charge repulsion between surfaces of similar charge
2. van der Waals forces of repulsion
3. Steric exclusion (hindrance)
4. Negative chemotaxis (cellular mobility)

As expanded from Pethica [153].

The relative importance of each force depends on the types of microbial and adsorbent surfaces. Adhesion forces have been determined for several microorganisms adsorbed to glass [212]. In the case of ion-exchange resins, electrostatic forces and the formation of ion pairs and triplets may be most significant [87]. Adsorption onto activated carbon surfaces may involve predominantly hydrogen bonding and London–van der Waals forces [199]. The balance between London and van der Waals forces (i.e., charge repulsion vs. dispersive attraction) is the basis for the Derjaguin–Landau–Verwey–Overbeek (DLVO) theory of stability for lyophobic colloids [44,93,123,194]. This theory has been extended for the adsorption of virus to oxide surfaces [140] and modified to include steric exclusion for the adhesion of fibroblasts to synthetic substrates [116].

Adhesion of cells to nonspecific surfaces can involve interparticle bridging by natural or synthetic polyelectrolytes [180], or partial charge neutralization by inorganic coagulants [162]. The initial attachment of microbial cells to surfaces may be enhanced [13,35,45,62a,123,124,186] or inhibited [26,62,81] by polymeric materials secreted by the cells or artificially added to the medium. Similar enhancement and inhibition of nonmicrobial cells by polymers have been reported and reviewed [144].

It is impossible to explain the preferential adsorption of the cells of a particular species based on a single force or a simple balance of surface energies or charges. Particles of opposite charge attract each other, but so do other particles possessing the same net charge but different surface potentials. Some hydrodynamic or diffusional forces are required to move the particles close enough to surfaces so that more specific short-range attractive forces can come into play. The physical chemistry of bacterial agglutination itself is related to colloidal theory and antigen–antibody reaction [137]. Viruses have been adsorbed to antibodies coupled to a solid matrix [67]. The entire picture is further complicated by the effects of pH and ionic strength on the electrical double layers of both cells and adsorbent particles.

There are also significant recent studies of the metabolic rates of living microorganisms immobilized on solid surfaces by coupling agents such as polyisocyanates, γ-aminopropyltriethoxysilane [215,216], and 1-ethyl-3-(3′-dimethyl-aminopropyl) carbodiimide [214]. Such immobilization, unlike reversible adsorption, involves covalent bonding and is not sensitive to pH or ionic changes. It is also distinct from immobilization in gels [219].

The possible orientations of microbial cells on adsorbent surfaces are shown in Fig. 2.3. The sorption of certain bacteria to surfaces in an end-on orientation has been reported [43,122]. This may be due in part to contact at points of low radii of curvature [153], where the potential barrier lies

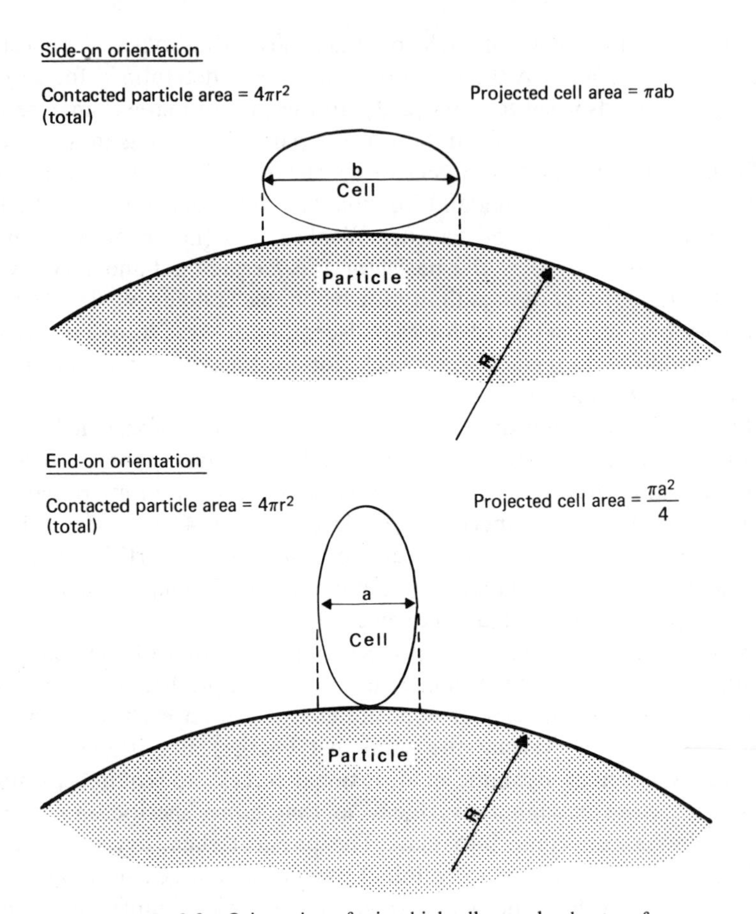

Figure 2.3 Orientation of microbial cells on adsorbent surfaces.

closer to the cellular surface. Adsorption in a side-on orientation [18,38,122] may be thermodynamically more stable if surface charge is considered uniformly distributed. Mosaics of different surface charge [118,119] or unequal distribution of hydrophobic groups [122] at the ends and sides of the cells may also explain the occasional end-on preference. The mobility of the cells also can play a part in their orientation [123].

The energies of interaction between *Achromobacter R8* and a glass surface as computed for varying electrolyte concentrations [123] are plotted in Fig. 2.4. Similar studies of interfacial phenomena have been made for algae [114,143]. The magnitude of repulsion and the particle separation distance both increase with decreasing electrolyte concentration. Reversible sorption may occur with attraction at separation distances up to the secondary minimum. Irreversible sorption may then require the sorbed cell to overcome the repulsive barrier and move still closer to the adsorbent surface. Desorption may be promoted in certain cases during the reversible phase by altering the ionic environment or by increasing the degree of shear. Desorption after irreversible adsorption may result in physical or chemical rupture of the cells.

ζ potentials, streaming currents, surface tensions, and particle-size distributions of colloidal suspensions can be measured by commercially available instruments. The surface-charge density can be calculated from the ζ potential if certain other constants are known and the electrolyte character and concentration are prescribed carefully. The surface charge densities of bacteria typically range 2000–5000 esu/cm^2 (electrostatic units per square centimeter) at ionic strengths of approximately 0.01 [120,167]. There are inherent errors, however, due to surface conductance.

An alternative and somewhat simplistic approach is to consider adsorption to be predominantly electrostatic in nature. Certain physical and chemical phenomena observed with suspensions of bacterial cells have been associated with the apparent isoelectric points (pI) of the cells. Several of these are summarized in Table 2.5. The isoelectric point, as defined according to the zwitterion hypothesis, is the pH at which ionization of the amphoteric bacterial cell is maximum. There exists a considerable volume of literature on electrophoretic mobilities of microbial species and the effects of various ions and differing strengths. A listing of 70 values for pI

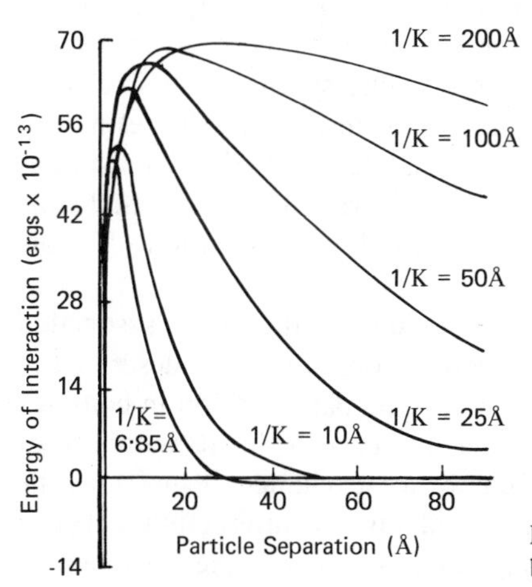

Figure 2.4 Computed energies of interaction between solid and bacterial surfaces.

Table 2.5 Cellular Phenomena associated with the isoelectric point

Minimum phenomena
1. Net surface energy
2. Adsorption of other ions
3. Dispersion
4. Permeability to other ions
5. Growth and germination
6. Viscosity of the suspension
7. Osmotic pressure of the suspension
Maximum phenomena
8. Total surface charge
9. Flocculation
10. Adsorption to anion-exchange resins
11. Refractive index of the suspension
12. Surface tension of the suspension

for 48 species has been compiled [38]. The apparent isoelectric points of various bacteria as determined by electrophoresis, light absorbance, and anion exchange are similar [38]. The pH range of maximum adsorption onto an anion-exchange resin, however, is quite narrow for the cells of *Pseudomonus* (pH 3.9) and much broader for the cells of *Bacillus subtilis* (pH 1.5–4) [38]. The electrophoretic mobility of solid adsorbents is also pH dependent [99].

The microbial cell has been described as a biological ion exchanger in its own right. This is represented schematically [37] in Fig. 2.5. Most of the ion-exchange capacity of the cell resides in the surface polyelectrolytes, which constitute a matrix of fixed charge at ion-exchange sites associated with mobile counterions. This picture is much like synthetic ion-exchange resins, except that the charge distribution is uniform through the resins and not restricted predominantly to the surface, as is the case with cells. There is active transport of soluble ions through the cell wall to interior exchange sites, but such sites are not significant in considering the sorption of whole cells that do not penetrate the surface.

Assuming that the microbial cell is a colloidal particle with associated surface active properties and polymeric surface components as described in the preceding paragraphs, it is easy to visualize its participation in complex sorption phenomena. Specific and generally reversible attachment of microbial cells to specific surfaces such as ion-exchange resins may be electrostatic. Such specific attachment could involve interaction between highly complementary groups opposing each other on the surfaces of microbial cell and solid. Nonspecific and often irreversible adsorption may involve a combination of attractive forces. Nonspecific attachment can result from the nonspecific interaction of uncharged polymeric materials with relatively inert nonionogenic surfaces.

2.3.2 Concentration at Solid Surfaces

The phenomenon of concentration of microorganisms at the surface of a solid phase has been portrayed through the progression of literature under a variety of labels. "Adsorption" has been accepted reluctantly as a term to describe the accumulation of suspended particles of microbial size onto a solid surface. In many cases, adsorption of whole cells has been accompanied and occasionally disguised by other phenomena, such as altered metabolism, multiplication, simultaneous desorption, flocculation, and lysis. Adsorption has become the preferred term, however, in most recent studies of the interaction between microbial cells and well-defined surfaces [1,12,40,71,107,108,120,152,154,160,175]. Sorptive interactions generally involve mechanisms whereby actual transfers and accumulations of cells have occurred at solid surfaces. Evidence supporting this conclusion includes the observed dependence of this interaction on both the pH and the concentrations of electrolytes in the suspension, and the vivid photomicrographs in which bacterial cells can be seen firmly attached to the surfaces of various adsorbents.

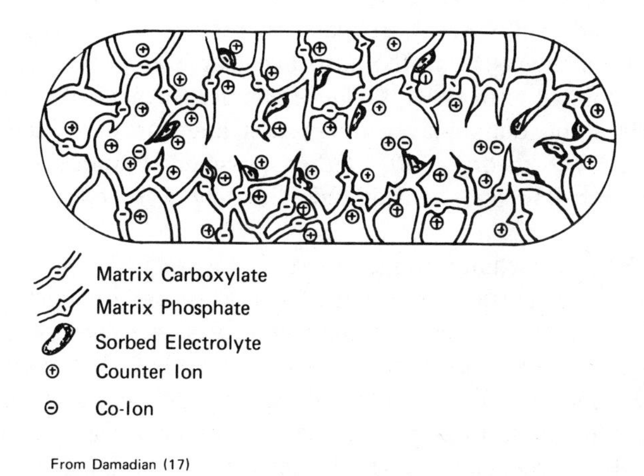

From Damadian (17)
Courtesy of the Biophysical Journal

Figure 2.5 Schematic structure of a biological ion-exchange resin.

The terms *adsorption*, *adsorbent*, and *adsorbate* mean, respectively: (1) the process by which microbial cells and other colloidal particles are concentrated at a solid surface, (2) the solid material promoting this process, and (3) the cellular material undergoing this process. These are no longer simple descriptive conveniences but valid representations. For the purposes of this discussion, the surface area of the adsorbent is considered to be of much greater extent than the contacting area, that is, the projected area of the adsorbate (microorganisms).

The actual mechanism by which bacteria are adsorbed onto surfaces has been explained under many guises. The decrease in the number of bacteria present in a suspension shaken with various powdery materials has been attributed to the "mechanical settling" of the cells with the solid particles [105], or to a purely "mechanical adhesion" of the bacteria to the adsorbent [133]. The adsorption of bacterial cells has also been related to the magnitude of their electrical charges [164] or to the physical rather than the chemical nature of the adsorbent [32]. The relative merits of electrochemical versus mechanical adsorption have been discussed [66,91,102,156]. The dependence of adsorption on electrical effects has been justified, but additional surface factors have also been suggested [46,48,98,101,161].

Adsorption of bacteria by simple reaction of unlike electrical charges, that is, between a negatively charged bacterial cell and a positively charged adsorbent has been concluded [151] and doubted [25]. Reaction with surface ionogenic groups may be involved during the adsorption of bacteria onto ion-exchange resins since counterions are displaced [38,210]. Variations in the degree of adsorption may reflect differences in the electrokinetic potentials of both the cells and the adsorbent particles [9,59,146,201]. The adsorption of poliovirus to oxide surfaces has been explained as a balance between electrostatic double-layer interaction and van der Waals forces, that is, the DLVO theory [140]. Other possible components, such as covalent-ionic bonding and hydrophobic bonding were considered to be insignificant.

The roles of extracellular biopolymers [7,13,34,45,62a, 121–123, 172, 205,211] and various surface appendages in sorptive interactions have been described [205,211,217]. Localized ionogenic groups were not found on either of two bacterial species. It was concluded that electrostatic phenomena were not involved in establishing perpendicular surface orientation [122]. A localized hydrophobic portion of cells was suggested on observation of extracellular adhesive material. Instantaneous reversible sorption of bacterial cells with movement over the adsorbing surface, and then a time-dependent irreversible phase [124] may be indicative of a physical adsorption followed by a more permanent chemical adsorption. Cells of some bacterial species washed free of any surface polymers before adsorption have been desorbed by hydrodynamic forces, whereas those of other species required severe changes in pH or ionic strength for desorption to occur [42].

Microscopic analyses to elucidate the mechanisms of microbial cell adsorption onto colloidal particles can be ambiguous because of artifacts due to drying, staining, convection currents in hanging drops, or adsorption of cells onto container walls. Microscopic [34,38,42,98,101,108a,109, 151,208] and electronmicroscopic [62a,86,122,211] photographs show bacterial cells adsorbed onto surfaces of larger particles primarily in monolayers forming "pincushion" patterns with side-on [38] or occasionally end-on orientations [122]. Similar patterns are observed for the adsorption of smaller particles onto bacteria. Electrophoretic measurements [108a, 117–119] and electronic particle enumerations [166] have also been conducted for cell–particle complexes.

Representative sampling of mixed suspensions to determine the relative proportions of freely suspended and firmly adsorbed cells is necessary in establishing both rate and degree of adsorption. The conglomeration of many cells adsorbing onto ion-exchange resin particles can be separated easily from the bulk of the surrounding medium, which contains any dispersed, unadsorbed cells. Techniques reported in the literature include: sedimentation, filtration, centrifugation, and direct microscopic enumeration.

2.3.3 Capillary and Chromatographic Effects

The movements of microbial cells in porous media are governed by adsorptive mechanisms [220]. The relative penetration of cells in soil strata depends on whether the cells are strongly attracted to the neighboring solid surfaces or restricted in their flow through small channels [74,100,136]. The mobility of bacteria in porous media is strongly influenced by multivalent inorganic cations, particularly ferric iron and aluminum [193]. Attenuation of flow by hydrodynamic chromatography probably occurs simultaneously in the manner observed with suspended colloids in beds of fixed spherical particles.

Differential movements of microbial species have been observed specifically in columns of porous media such as sand filters [18,51,115], ion-exchange beds [38,210], and soil [14,193]. Bacteria rise in paper strips immersed into cellular suspensions. This has been interpreted to be due largely to the capillary action of the carrier fluid. Differences in capillary rising ability among species [66,91,156], however, require that some attendent adsorptive processes also occur to preferentially retard the movement of some cells but not others. The attachment of microorganisms to solid surfaces may also involve chemical factors. The attachment of marine and freshwater bacteria to rocks and other solids, wastewater bacteria to trickling filters, and biological fouling of submerged surfaces may require the synthesis of extracellular adhesives [14,29,34,35,45,62,62a,117, 123,124,205].

2.3.4 Flocculation

The mechanisms of bacterial attachment to surfaces are related to those observed during the flocculation of suspended cells as initiated by high-molecular-weight, water-soluble polyelectrolytes of biological or synthetic origin. An understanding of the mechanisms of flocculation, therefore, has engineering significance in naturally or artificially induced aggregations of cellular suspensions that occur in: (1) aerobic wastewater treatment by trickling filters and activated sludge, (2) anaerobic sludge digestion, (3) stabilization in surface impoundments, (4) nutrient removal or other bioconcentration, (5) sludge dewatering, (6) filtration of surface water, and (7) specific food and antibiotic fermentations.

Microbial cells can be flocculated by a variety of water-soluble polyelectrolytes, as summarized in Table 2.6. Bacterial cells can be flocculated by cationic polyelectrolytes, further verifying the predominantly negative surface charge of cells. Viruses and algae also have been aggregated by cationic flocculants. Nonionic and anionic flocculants usually are not as effective unless aggregation is initiated first by multivalent inorganic cations. Flocculation of negatively charged colloids by anionic polyelectrolytes, however, can be effected in certain cases, demonstrating the reduction of surface potential to zero (isoelectric point) is not a prerequisite for flocculation. Although surface charges of cells are significant in many flocculation mechanisms, they are not an absolute predictor of adsorptive behavior.

Bacterial cells are coagulated by inorganic salts in a manner similar to that of other hydrophilic colloids [162]. Common coagulants are: ferric chloride, aluminum sulfate, various calcium and magnesium salts, and other multivalent inorganic cations. The mechanism of coagulation is primarily a charge neutralization of individual cells since the coagulants are too small to bridge among cells. There are probably some interactions between positively charged or neutralized cells and those still retaining some degree of negativity. Bacterial cells have predominantly negative charges at physiological pH. On adsorption of inorganic cations (cationic polyelectrolytes),

Table 2.6 Flocculation of microorganisms by water-soluble polyelectrolytes

Microbial Species	Polyelectrolyte[a]	Origin[b]	Reference
Algae			
Chlorella sp.	A, C	S	30
Chlorella sp.	A	S	141
Scenedesmus sp.	,,	,,	,,
Scenedesmus sp.	C	S	73
Chlorella sp.	,,	,,	,,
Algae	A, C, N	S	195
Chlorella	C	S	56
Algae	A, C. N	S	178
Chlorella ellipsoida	A, C, N	S	185
Chlorella pyrenoidosa	,,	,,	,,
Green algae	C	S	53
Algae	A, C, N	S	131
Algae	A, C, N	S	213
Bacteria			
15 Gram +/− bacteria	A, C	S	131a
Pasteurella tularensis	A, N	B, S	90
Escherichia freundii	,,	,,	,,
Serratia marcescens	,,	,,	,,
Proteus morganii	,,	,,	,,
Bacillus subtilis	A	S	141
Lactobacillus sp.	,,	,,	,,
Escherichia coli	C	S	75
Staphylococcus albus	,,	,,	,,
Streptococcus sp.	,,	,,	,,
Aerobacter aerogenes	,,	,,	,,
Bacillus subtilis	,,	,,	,,
Bacteria	N	S	179
Escherichia coli	A, C, N	B, S	21, 21a
Aerobacter aerogenes	,,	,,	,,
Escherichia coli	A, C, N	S	50
Pseudomonas fluorescens	C	S	131a
Lactobacillus delbrueckii	,,	,,	,,
Bacteria	A, C, N	B	149, 150
Aerobic, heterotropic bacteria; algae	C	B, S	180
Pullularia pullulans	N	N, S	221
Leuconostoc mesenteroides	N	B	82
Escherichia coli	C	S	187
Virus			
Virus	C	S	75

Table 2.6 (Continued)

Microbial Species	Polyelectrolyte[a]	Origin[b]	Reference
Virus	A, C, N	S	55
Escherichia coli bacteriophage T_4, MS_2	A, C	S	24
Escherichia coli bacteriophage MS_2	C	S	66
Escherichia coli bacteriophage MS_2	A, C	N, S	6
Escherichia coli bacteriophage f_2	A, C, N	S	170
Escherichia coli bacteriophage T_2	A, C, N	S	184
Poliovirus type 1	"	"	"
Poliovirus type 1	C	S	103
Coxsackie virus type A9, B2	"	"	"
Echovirus type 6	"	"	"
Poliovirus type 1	A, C, N	S	63
Yeast			
Saccharomyces sp.	A	S	141
Torulopsis sp.	"	"	"
Saccharomyces sp.	N	B	142
Other microorganisms (31 total)	"	"	"

[a] Abbreviations: A—anionic; C—cationic; N—nonionic.
[b] Abbreviations: B—biological; S—synthetic.

the surface charges become neutralized, and if adsorption continues, the net charge is reversed and becomes positive.

The possible flocculation reactions for colloidal particles destablized by a polyelectrolyte flocculant are shown in Fig. 2.6. If an optimum level of flocculant is added, initial adsorption (reaction 1) with a single particle is followed by visible floc formation (reaction 2) with multiple particles. At insufficient levels of flocculant (reaction 3), the suspension is stable and most individual particles remain dispersed. Aggregation becomes more pronounced as the flocculant concentration is increased (on a logarithmic scale), up to an optimum level of flocculant where turbidity is reduced to the greatest extent. At progressively high levels of flocculant (reaction 4), the aggregation becomes less apparent, turbidity increases, and the suspension becomes redispersed into smaller clumps and individual particles. A secondary adsorption of flocculant can occur (reaction 5). This aggregation and redispersal may be due to a net reversal of charge from negative to positive for the total population of particles. It can also occur if agitation is extreme and/or prolonged, whereby intercellular bridges of flocculant are physically broken.

The shape and location of a flocculation response curve also is dependent on flocculant concentration and aggregate particle size. Collision frequencies among cells in homogeneous suspension are influenced by Brownian motion (diffusion) and agitation (shear fields). Cells in heterogeneous suspension with other particles of grossly different size (e.g., clay particles) are influenced by differential settling as well.

The bell-shaped curve of dispersion–flocculation–redispersion for colloidal particles [181] is similar to those for the flocculation and coagulation of bacterial cells [21a,50,81,150,179,180,185]. It is similar in form to the curve describing the adsorption of bacteriophage to host cells, as a function of salt or cation concentration that mediate the adsorption [155].

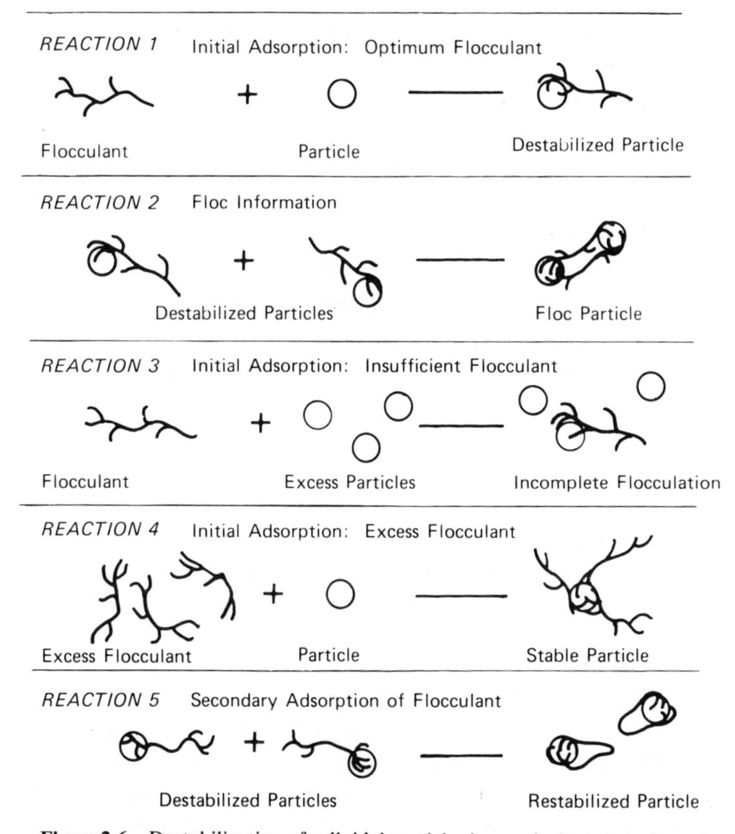

Figure 2.6 Destabilization of colloidal particles by a polyelectrolyte flocculant.

Charged biopolymers are of importance in biological structures. Flocculation and stabilization reactions involve similar natural polymers [65], such as complex polysaccharides and polyamino acids, which are excreted or exposed at cellular surfaces [163]. These polymeric molecules are of sufficient length to form bridges between microbial particles [81]. The interpretation of chemical flocculation of microorganisms, therefore, should be guided by bioflocculation [180]. Aggregation among microorganisms with interparticle bridging by biopolymers has been intensely investigated [21a,35,49,150] and the subject of several doctoral theses [21,80,149].

The mechanisms of cell coagulation and flocculation are obviously complex. They are not simply explained by charge neutralization of surface charges of single cells or by bridging of polyelectrolytes between cells. Nevertheless, several generalizations can be made [81]:

1. Negatively charged bacteria can be aggregated by cationic, anionic, and nonionic polyelectrolytes.
2. The presence of multivalent cations generally is required for aggregation by nonionic or anionic polyelectrolytes.
3. Aggregations of bacteria with synthetic polyelectrolytes can be interpreted in terms of bridging and sorptive interactions between polyelectrolytes and bacterial surfaces.
4. The optimal polyelectrolyte concentration for maximum aggregation is proportional to the concentration of bacterial surface area.
5. The stability of bacterial–polyelectrolyte complexes depends critically on pH, polyelectrolyte molecular weight, bacterial species, agitation, and ionic composition.
6. A theoretical basis exists for bacterial aggregation by natural polyelectrolytes.

Many of the same generalizations summarized here for water-soluble polyelectrolytes can be extrapolated to ion-exchange resins that are water-insoluble polyelectrolytes.

2.3.5 Ion Exchange

Ion-exchange resins are large, insoluble polymeric networks that assume either a positive (anionic) or negative (cationic) charge in association with small, dissociable counterions of opposite charge. Bacterial cells behave as macroscopic ions, as described in Section 2.3.1, and react with charged surfaces, such as those of ion-exchange resins [38,40]. Bacterial cells have been desorbed from anion- and cation-exchange resins by washing the adsorption complex with solutions of different pH or salt content.

Reversible adsorption of bacterial cells to the surface of an ion-exchange resin is predominantly an electrostatic charge mechanism. A cell–resin combination should result only when the reacting particles are of opposing (attracting) electrical charge. This idealization can be complicated by: (1) hydrogen bonding and London–van der Waals forces contributing to overall attraction or repulsion, and (2) the variety of different ionizable groups of the bacterial surface that are available for specific interactions at different pH values.

The isoelectric point is the pH at which maximum ionizations of the acidic and basic groups of a dipolar compound occur. This does not necessarily assume equal numbers of both groups but only a zero net charge. The addition of an acid, HA, or a base, MOH, to a bacterial cell that is behaving as a dipolar ion in suspension can be illustrated in Fig. 2.7, where H and OH are hydrogen and hydroxyl, respectively, and A and M are generalized representations for any anion and cation, respectively. The radical, **R**, is used to represent the composite complex surface of a bacterial cell; the radicals, **R**′ and **R**″, are used subsequently to represent the polymeric backbones of an anion- and a cation-exchange resin, respectively.

Ionizable hydrogen ions can be produced if the pH of a suspension containing bacterial cells is above the isoelectric point (pH > pI) of the carboxyl groups in the bacterial cell or on the cellular surface. These hydrogen ions can conceivably be replaced by any other cation. Each bacterial cell exhibits a net surface charge, and the entire cell behaves as a large anion and is capable of combination with any cation. Alternatively, at pH values below the isoelectric point of the amino groups in the bacterial cell or on the cellular surface (pH < pI), the bacterial cell can assimilate additional hydrogen ions, each cell exhibits a net positive surface charge, and the entire cell behaves as a large cation and is capable of combination with any anion. The charge reversal of some bacteria may not be observed except at extreme pH. The pH values also may be below those encountered in normal

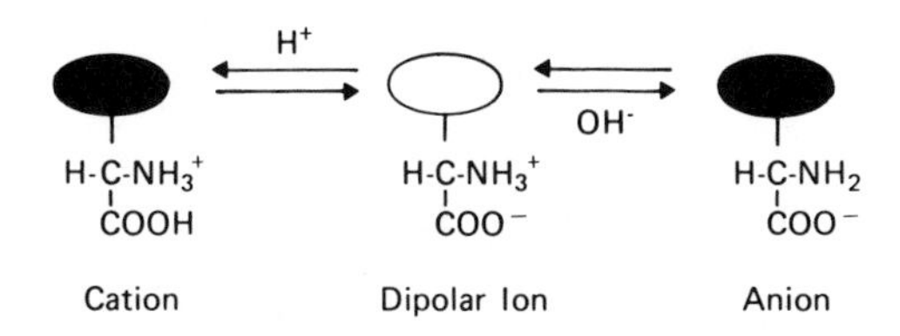

Figure 2.7 Dipolar character of hypothetical microbial cell.

growth, and consequently adsorption may be less than that observed at lower pH. The idealized terminal amino acids depicted in Fig. 2.7 are greatly simplified representations. Such active sites can function as cationic or anionic adsorption complexes in living cells for some bacterial species, however, allowing them to participate in ion-exchange reactions. This idealization is not true in a strict sense for all species where the pH of charge reversal may not be apparent or be a value beyond the normal range of growth.

A formal reaction mechanism now can be proposed to describe the exchange adsorption of bacterial cells from liquid suspensions by synthetic ion-exchange resins. The large complex structure of the bacterial cell can behave as either a cation or an anion and react, respectively, with the charged groups of either a cation- (anionic) or an anion- (cationic) exchange resin. The exchange resin can be represented as a large polymeric network that assumes either a positive or negative charge in association with small, dissociable counterions of opposite charge.

The exchange between a negatively charged bacterial cell and a positively charged ion-exchange resin is shown in Eq. (2.1).

$$\begin{matrix} R \\ | \\ H-C-NH_2 \\ | \\ COO^{\ominus} \end{matrix} + Cl^{\ominus}(H_3C)_3N^{\oplus}-R' \rightleftharpoons \begin{matrix} R \\ | \\ H-C-NH_2 \\ | \\ COO^{\ominus}(H_3C)_3N^{\oplus}-R' \end{matrix} + Cl^{\ominus} \quad (2.1)$$

The corresponding exchange between a postively charged bacterial cell and a negatively charged (cation) ion-exchange resin can be depicted in Eq. (2.2).

$$\begin{matrix} R \\ | \\ H-C-N^{\oplus}H_3 \\ | \\ COOH \end{matrix} + H^{\oplus\ominus}O_3S-R'' \rightleftharpoons \begin{matrix} R \\ | \\ H-C-N^{\oplus}H_3{}^{\ominus}O_3S-R'' \\ | \\ COOH \end{matrix} + H^{\oplus} \quad (2.2)$$

Two additional reactions require mediation by either multivalent cations or anions from simple electrolytes or polyelectrolytes. Cation exchange between a negatively charged bacterial cell and a negatively charged (cation) ion-exchange resin is possible if a multivalent cation, M^{+2}, can convert the cell charge to a cation as shown in Eq. (2.3).

$$\begin{matrix} R \\ | \\ H-C-NH_2 \\ | \\ COO^{\ominus} \end{matrix} + M^{(+2)} + H^{\oplus\ominus}O_3S-R'' \rightleftharpoons \begin{matrix} R \\ | \\ H-C-NH_2 \\ | \\ COO^{\ominus\oplus}M^{\oplus\ominus}O_3S-R'' \end{matrix} + H^{\oplus} \quad (2.3)$$

Anion exchange between a positively charged bacterial cell and a positively charged (anion) ion-exchange resin is also possible if a multivalent anion, A^{-2}, can convert the cell to an anion as shown in Eq. (2.4.).

$$\begin{array}{c} R \\ | \\ H-C-N^{\oplus}H_3 \\ | \\ COOH \end{array} + A^{\ominus 2} + Cl^{\ominus}(H_3C)_3N^{\oplus}-R' \rightleftharpoons \begin{array}{c} R \\ | \\ H-C-N^{\oplus}H_3{}^{\ominus}A^{\ominus}(H_3C)_3N^{\oplus} \\ | \\ COOH \end{array}-R' + Cl^{\ominus} \tag{2.4}$$

Anionic polymers of natural or synthetic origin conceivably could act in this manner. The adsorption to an anionic surface could involve an electrostatic mechanism. Alternatively, adsorption to a more inert surface could involve London–van der Waals forces of attraction.

The four reactions proposed in Eqs. (2.1)–(2.4) are simplified versions of actual exchanges. For adsorption of a single cell, the individual reactions may be multiplied by a factor of 10^8 to account for the approximate number of active exchange sites per bacterial cell [128]. Thus the adsorption of bacterial cells onto ion-exchange resins can be represented by assuming an electrostatic mechanism. Electrostatic attraction is not limited, however, to carboxyl and amino groups, but can involve other charged groups on the bacterial cell surface. Contributions of London–van der Waals forces and hydrogen bonding should not be neglected. No attempt has been made here to explain selective exchange on the basis of subtle differences in surface composition of the cells themselves that are susceptible to changes with age and alterations in their environment.

2.4 SORPTION KINETICS

The nature of most chemical and physicochemical processes involving heterogeneous transformations in liquid and gases are determined largely by hydrodynamic factors. Three steps are involved in heterogeneous reactions: (1) a transfer of reacting particles to the surface at which the reaction occurs, (2) the heterogeneous reaction itself (chemical transformation, adsorption or desorption, the discharge and formation of ions, etc.), and (3) the reverse transfer of reaction products from the reaction site. The reaction is considered to be diffusion controlled if either the introduction of reactants or the removal of product is rate determining.

Processes involving diffusion in the liquid phase and heterogeneous reactions at the liquid–solid interface include: the coagulation of colloidal particles, quenching of fluorescence in solution, process of condensation polymerization, kinetics of enzyme–substrate reactions, adsorption of viruses onto host cells, coagulation of aerosols, and nucleation and growth of crystals. The kinetics of these processes have been reviewed elsewhere [38]. The present discussion is restricted to descriptions of the kinetics of the diffusion-controlled adsorption and desorption of bacterial cells to and from ion-exchange resins.

2.4.1 Kinetics of Diffusional Processes

Several applications of the equations of diffusion to vegetative cells and cellular aggregates have been presented [158]. The adsorption of viruses onto host cells and the adsorption of other particles of virus size onto nonbiological surfaces have been discussed [4,5,192] in developing a rate equation, assuming that adsorption is diffusion controlled and follows the Smoluchowski coagulation theory [147,171]. Diffusion-controlled adsorption-rate models have also been developed for virus-sized particles [60]. The interaction of viruses with host cells has been described by first-order kinetics [89] and several solutions made to the differential equations [28] representing the kinetics of reversible and irreversible attachment of bacteriophages onto host cells [113,148,155]. A power-law equation has been developed to describe the adsorption of viruses [198]. The adsorption of viruses onto activated carbon has been described in terms of reversible, second-order kinetics [33,148].

Criteria for determining film-diffusion control or particle-diffusion control have been applied to bacteria [37]. The adsorption of bacteria to glass surfaces has been reported to involve an instantaneous reversible phase and a time-dependent irreversible phase [123]. A number of different solutions to the diffusion problem have been summarized and applied to the adsorption–desorption of bacterial cells with ion-exchange resin particles [38,39]. The kinetics of diffusional processes have thus been applied to a diversity of natural phenomena. Prediction of reaction rates have been made on the basis of analytical solutions to the partial differential equation describing Fick's second law of diffusion [87]. This law relates the changing concentration of the diffusing species with time, and incorporates a proportionality "constant," called the *diffusion coefficient*, D. Extensions of the specific solutions to the diffusion equation on the molecular level to encompass much larger particles are evidenced in the successes obtained with systems of colloids, aerosols, crystals, and viruses. The literature of these systems has been reviewed elsewhere [38]. The kinetics of adsorption and desorption of bacterial cells with particles of an ion-exchange resin assuming diffusion-controlled reactions is considered in the present development.

The mathematical representation of such a system can be made as complex as desired to account for all possible contributions to the mass transfer. In cases involving mass transfer by ionic exchange, a form more general than Fick's law, known as the *Nernst–Planck equation* [87], should be applied. An external force, equal to the product of the ionic charge and the local electric field strength, is included in this more general form. Although such a force is recognized, this refinement is not warranted for the present state of the art of bacterial sorption with ion-exchange resins.

The general procedure followed has been to solve for the concentration of the adsorbable particles assuming a constant coefficient of diffusion, D, and spherical coordinates. Various boundary conditions have been assumed by several authors [23,31,60,159,171]. The flux of adsorbable particles normal to the boundary surrounding the adsorbent particles of radius, R, is next determined knowing the concentration of adsorbable particles as a function of radial distance, $r'(r' \geq R)$, and time, t. The cumulative transfer of adsorbable particles, M_t of initial mass, M_0, is finally determined on the integration of the flux equation to give a form of the type shown in Eq. (2.5).

$$\ln\left(\frac{M_t}{M_0}\right) = -\frac{4\pi RD}{V}\left[t + \frac{2R}{\sqrt{\pi D}}\sqrt{t}\right] \tag{2.5}$$

This solution [192] is valid for diffusion of adsorbable particles, from a fluid of finite volume initially containing a uniform but slowly falling concentation of adsorbable particles, to a spherical adsorbent surface at which a "zero" concentration of adsorbable particles is maintained. The volume of the adsorbent particles is considered much smaller than that of the fluid volume.

A modified solution [31] has been made assuming the concentration of adsorbable particles at the surface of the adsorbent to be proportional to the gradient in concentration. This recognizes the case of imperfect adsorption; in other words, not every adsorbable particle approaching the adsorbent is adsorbed [169]. The two solutions converge if the adsorbent particles are much larger than the adsorbable particles. The second term in Eq. (2.5) is often ignored for small colloidal particles. This is not a valid assumption, however, for particles as large as bacteria and the larger viruses, for which both terms are significant.

The adsorbent particles (ion-exchange resins) are much larger than the adsorbable particles (bacterial cells) ($R \gg r$), although the latter may greatly outnumber the former ($N \ll n$). Although the bacterium–resin particle system is polydisperse and multiple aggregates are formed, the rate of adsorption is a function of the "disappearance" of the smaller particles, that is, the bacteria. The bacteria are "removed" from suspension when they come into contact with the surfaces of the resin particles. The surface concentration of freely dispersed cells, therefore, is considered to be "zero." The cells do not "disappear" nor diffuse into the interiors of the resin particles. A short distance away from the surface, however, a uniform concentration of freely dispersed cells is maintained throughout the well-stirred fluid.

Diffusion of cells through a thin stagnant film surrounding each resin particle is considered rate limiting. The diffusion rate of the bacterial cells through this film is much greater than that of the resin particles through the bulk fluid. The resin particles are more susceptible to sedimentation than are bacterial cells. The average diffusion coefficient of both particles and cells, therefore, is almost entirely due to the diffusion of the bacterial cells alone (i.e., $D_{\text{bacteria}} \gg D_{\text{resin}}$).

2.4.2 Rate Equations for Bacterial Sorption

Diffusion coefficients for adsorbable particles (bacteria) suspended in a well-stirred fluid can be determined easily by observing the rate of change of the concentration of adsorbable particles remaining unadsorbed in the agitated fluid. The accumulation of adsorbable particles on the adsorbing surface is equal to their disappearance from the bulk suspension. By approximating adsorbable mass, M, with cell number, n, and number in turn with optical absorbance, A, Eq. (2.5) becomes Eq. (2.6).

$$\log\left(\frac{A}{A_0}\right) = kt + k'\sqrt{t} \qquad (2.6)$$

where $k = -4\pi\beta_0 RND$, $k' = \geq 2kR/\sqrt{\pi}D)$, N is the number of resin particles, and β_0 is an empirical constant [38] relating n and A. The constants, k and k', are different for adsorption and desorption, respectively. They can be determined experimentally from data of A/A_0 versus $\sqrt{t}$. The diffusion coefficient, D, can then be determined on substitution.

The diffusion constants for a few widely varying systems are compared in Table 2.7. Constants for adsorption are of the same order of magnitude. The constant for desorption of the bacterium, however, is an order of magnitude smaller than the constant for adsorption. This is perhaps indicative of the greater difficulty of the cells escaping from the sorbing surface than in approaching it. The influence of attractive forces complementing the purely diffusional force is shown in comparison of the apparent diffusion

Table 2.7 Diffusion coefficients for sorbing microorganisms

Microorganism	Sorbent	Diffusion coefficient (10^{-7} cm^2/sec)	Reference
Virus MVL-1	*Acholeplasma* (mycoplasma)	1.2–1.8	64
Escherichia coli	Activated carbon	0.8	33
Bacteriophage T_4	" "	"	"
Bacillus subtilis	Anion-exchange resin	5.97 (adsorption)	38
		0.580 (desorption)	"
		0.049 (Einstein–Stokes)	"

constant for adsorption being two orders of magnitude greater than the value calculated by the Einstein–Stokes equation.

A typical plot of adsorption [38] is shown in Fig. 2.8. These data were fitted by Eq. (2.7).

$$\log\left(\frac{A}{A_0}\right) = -0.035t - 0.090\sqrt{t} \tag{2.7}$$

A typical plot of desorption [38] is shown in Fig. 2.9. These data were fitted by Eq. (2.8).

$$\log\left(\frac{1 - A}{A_0}\right) = -0.0056t - 0.0080\sqrt{t} \tag{2.8}$$

The kinetics of such adsorptions and desorptions need to be understood so that equilibrium conditions are properly recognized in establishing sorption isotherms and calculating adsorption capacities. Such combined relations of optical absorbance as a function of both time and the square root of time [39] have been observed for various conditions of bacterial species, cell concentration, resin type, resin concentration, resin size, resin ionic form, pH, and ionic strength [38,41]. Selective adsorptions and desorptions as means of separating mixed cultures have been defined for given binary mixtures [40] and postulated for a quaternary mixture [42].

An interesting interpretation of the rate of attachment of marine bacteria to various immersed solid surfaces has been reported [45]. The rate of attachment was described by a series of straight-line segments of the form

$$Y = Kt^B \tag{2.9}$$

where Y is number of attached bacteria per unit area, t is the elapsed time of exposure, and B,K are empirical constants. The slow kinetics (days) relative

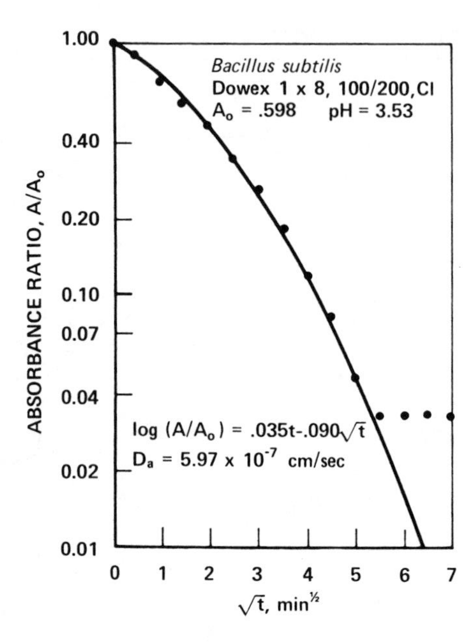

Figure 2.8 Adsorption of bacterial cells onto anion-exchange resin.

to ion exchange (minutes) can be explained in part since the sample panels were immersed in natural seawater with dilute concentrations of microbial cells that probably were actively multiplying once colonized on the panels, as compared to laboratory samples with intense mixing and a static cell count.

2.5 SORPTION EQUILIBRIUM

The initial attachment of microorganisms onto surfaces is considered to be a diffusion-controlled process requiring several minutes [38,76,133] or several hours [45,123,189] to reach equilibrium. It is important to note that adsorption capacities (i.e., loadings) of various solids is a function of the equilibrium concentration of microbial cells remaining in free suspension. The relationship between capacity and equilibrium concentration is defined by an adsorption isotherm. Adsorption isotherms and capacities, determined for the cells of various microbial species and solid adsorbent, are cited in Sections 2.5.1 and 2.5.2, respectively. A few determinations of adsorption energies and other thermodynamics functions have been reported.

2.5.1 Adsorption Isotherms

Several of the classical adsorption isotherms [199] for soluble species applied to the adsorption of suspended microbial cells are described in the paragraphs that follow. Agreements with any of the various isotherms are dependent on the relative size of the microorganism being adsorbed, the type of adsorbent, and the extent to which the adsorption is monolayered.

The initial reversible reaction between microbial cells and a solid adsorbent is depicted in Eq. (2.10):

$$\text{Cell} + \text{adsorbent} \underset{k_2}{\overset{k_1}{\rightleftharpoons}} \text{Cell–adsorbent} \tag{2.10}$$

where k_1 and k_2 are the rate constants for the forward and reverse reactions, respectively. The formations of various cell–adsorbent complexes or aggregates have been postulated by several investigators [83,108a]. Dissociations of such complexes have also been studied [84,177]. Formation and dissociation constants for cell–adsorbent complexes are indicative of their relative stabilities and ease of reversibility.

The subsequent irreversible reaction between microbial cells and solid

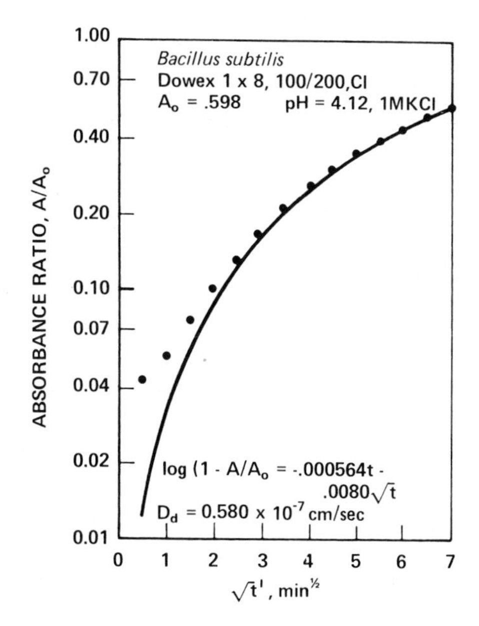

Figure 2.9 Desorption of bacterial cells from anion-exchange resin.

adsorbent is depicted in Eq. (2.11):

$$\text{Cell–adsorbent} \xrightarrow{k_3} \text{[Cell–adsorbent]} \tag{2.11}$$

where k_3 is the rate constant. The initial, loosely bound cell–adsorbent complex is converted into a permanently bound complex. The significances of the reversible and irreversible reactions were discussed in Sections 2.3.1 and 2.4.1, respectively.

The Langmuir isotherm has been applied to the adsorption of viruses [33,148] and bacteria [52,106]. The mathematical form for noncompetitive adsorption is given by Eq. (2.12):

$$\frac{c_e}{q} = \frac{1}{KZ} + \frac{c_e}{Z} \tag{2.12}$$

where c_e = cellular concentration (number of cells) at equilibrium (cells/m^3), q = number of cells adsorbed per unit weight of adsorbent (cells/mg), Z = capacity of adsorbent (sites/mg), and K = equilibrium constant = (k_1/k_2).

A similar development can be applied to competitive adsorption [100,104,115,173] involving more than one adsorbing species competing for the same adsorption site as shown in Eq. (2.13).

$$\text{Cell} + \text{competitor} + \text{adsorbent} \rightarrow \text{Cell–adsorbent} + \text{competitor–adsorbent} \tag{2.13}$$

Assuming there is no interaction between microbial cells and the competing adsorbate, the Langmuir isotherm for two sorbates is given by Eq. (2.14):

$$\frac{c_e}{q} = \frac{1}{KZ} + \frac{K's_e}{KZ} + \frac{c_e}{Z} \tag{2.14}$$

where s_e denotes the equilibrium concentration of the competing adsorbate, which may be a soluble material or cells of another microorganism. Detailed discussion of other competitive adsorption models is beyond the scope of this chapter.

Another alternative for the adsorption of viruses [16,51,68,198], virus-sized particles [60], and bacteria [183] is the empirical Freundlich isotherm given by Eq. (2.15):

$$\frac{c_e}{q} = k''{c_e}^{n''} \tag{2.15}$$

where k'', n'' are empirical constants. Other linear isotherms have been applied to sorption of bacteria on apatite [57] and sand [189]. A specialized adsorption isotherm for bacteria has been found to be S-shaped [104], with the mathematical form given by Eq. (2.16):

$$q = \frac{K''}{1 + \alpha \exp(-\beta c_e)} \tag{2.16}$$

where α, β, and K'' are empirical constants. An unrelated S-shaped function has been developed for plant virus infections [97]. However, the type of adsorption isotherm is not critical to the rate of adsorption, which is a diffusion-controlled process [60].

2.5.2 Adsorption Capacities

Determinations have been made of the cation-exchange capacities (CEC) of the cells of several species of bacteria for hydrogen ions, crystal violet and methylene blue dyes, and macramine (a cationic polyelectrolyte). Values from a few selected references are reported in Table 2.8. Total amounts of free amino groups and *C*-terminal residues of the cell wall are also presented in Table 2.8. The CEC values are approximately 10 times that predicted from the amino-acid analyses of the cell wall. The calculated area covered by macramine is also about 10 times the total surface area of a typical bacterium [181].

There are many confounding factors such as penetration of cell walls by the smaller ions, differential staining of cellular interior structures, steric hindrance by polyelectrolytes, unevenness of the bacterial cell surface, adsorption onto surface appendages, and multilayer adsorption. It is difficult to assume that bacterial cells behave as polyelectrolytes, or to infer any specific conclusions pertaining to surface structures, based solely on chemical-combining reactions. Nevertheless, the exchange capacity of the cells of a particular species grown under specified conditions to a certain

Table 2.8 Exchange capacities and available surface groups of bacterial cells

Compound	Capacity (μmol/g dry cells)	Reference
Hydrogen ion	79–750	79,128,129
Crystal violet	460–720	78
Methylene blue	270	128
Macramine	300–1130	182
Free amino groups	0.4–48.4	163
C-Terminal residues	2.8–43.2	163

Table 2.9 Adsorption capacities of various solids for microbial cells

Microorganism	Adsorbent	Capacity (Cells/g)	Capacity (Cells/cm²)	Coverage (%)	Reference
Bacillus subtilis	Anion resin	6.1×10^9	1.9×10^4	31 (1^a)	38
Staphylococcus aureus	Anion resin			(0.3^a)	210
Bacillus cereus	" "				
Escherichia coli	Anion resin	10^{10}			85
Rhizobium sp.,	Anion paper		10^6–10^7	2–20	202
Escherichia coli	" "				
Bacillus subtilis	" "				
Serratia marcescens	Anion resin		6×10^8		207
Bacillus mycoides	Anion resin		8×10^7		
Bacillus subtilis	Kaolin	50×10^{10}		17	104
Bacillus subtilis	Charcoal	50×10^{10}			
Escherichia coli	Charcoal	50×10^{10}			
Streptococcus mutans	Hydroxyapatite	60×10^6			57
Streptococcus mutans	Hydroxyapatite		3.2×10^6	3	29
Bacillus subtilis	Bentonite	10^{11}			108a
Rhizobium lupini	Montmorillonite	2×10^5			120
Rhizobium trifolii spp.	Montmorillonite	4–7×10^5			
Aeromonas liquiefaceins	Glass		12×10^8		132
Azotobacter sp.	Sand	5.5×10^6			189
Azotobacter chroococcum	Silt loam	6.5×10^8			151
Serratia marcescens	Marine mud	6–14×10^7			161
Marine bacteria	Glass		2.8×10^4 viable 18.2×10^5 total		34
Marine bacteria	Glass, others		1.2×10^9		45

Marine bacteria	Coated metal		10^6 treated		26
			5×10^{12} untreated		
Marine bacteria	Glass slides		40×10^3		123
Bacteria	Marine pebbles			0.3–2	11
Heterotrophic bacteria	Lake sediment		$3–15 \times 10^5$	0.2	190
Bacteria	Soil		$4–6 \times 10^5$		206
Streptococcus sp.	Bovine enamel				191
Actinomyces sp.					
Oral bacteria	Teeth		10^6	1	69
Escherichia coli bacteriophage T_4	Activated carbon	1.6×10^{12}		18	33
Escherichia coli bacteriophage T_2	Activated carbon	3×10^8			59
Escherichia coli bacteriophage T_7	Iron oxide	2×10^7			197
Escherichia coli bacteriophage MS_2	Bituminous coal	4×10^6			148
Escherichia coli bacteriophage f_2	Activated carbon	$2–46 \times 10^5$			198
	Anion resin	2×10^5			
Coxsackie virus type A	Iron oxide	8×10^6			157
Influenza virus type PR8	Hematite	10^5			15
Poliovirus type 1	Activated carbon	2×10^8			68
Poliovirus type 1	Activated carbon	6.5×10^6			115
Poliovirus type 1	Silicates	$7–9 \times 10^5$	$7–14 \times 10^3$		59
Poliovirus type 1	Kaolinite	10^9			95
Saccharomyces cereviseae	Anion resin		5×10^6		207
Chlorella vulgaris	Glass		$10^4–10^5$		186

[a] Percentage of ionic exchange capacity.

age is a reasonably reproducible value. This may imply that an apparent surface charge density could be determined from exchange data.

The capacities of various adsorbents for microbial cells also range according to species and particle size as shown in Table 2.9. Adsorption of cells onto surfaces is usually a fraction of a monolayer depending on whether a side-on or end-on orientation of the cells is assumed or observed as previously illustrated in idealized fashion in Fig. 2.3.

The adsorption of cells of *B. subtilis* onto kaolin and those of *E. coli* onto charcoal reach approximately 50×10^{10} cells g^{-1} adsorbent [104]. In these cases the particles of both adsorbents were approximately the same size as the bacterial cells themselves. About 17% of the total available adsorbent surface was covered. The surface coverage of clay particles by bacteria was relatively constant for particles of 10–500 μm [207] and was estimated to be about 2–5%.

The capacity of an anion-exchange resin for cells of *B. subtilis* was also determined [38]. Approximately 0.61×10^{10} cells g^{-1} resin were adsorbed covering about 31% of the available area of resin particles, which were about two orders of magnitude larger in size (97 μm) than the cells. The theoretical capacity is about eight times greater assuming end-on orientation than for side-on orientation of adsorbed cells (see Fig. 2.3). A bacterial cell is incapable of penetration into an ion-exchange resin particle, and vice versa. Consequently, only the outer surface groups are available for exchange adsorption. The assumption of monolayer adsorption probably is valid if the adsorbing cells are reasonably monodisperse in bulk suspension before adsorption and thus do not adsorb as clumps.

With low surface coverage (31%) assumed, only about 1% of the total exchange capacity of the resin is exhausted [38]. This is corroborated by other research [210] where only 0.3% of exchange capacity was exhausted by adsorbing bacterial cells. One percent of capacity corresponds to the capacity contained within the outer ⅓ of 1% of a particle with a radius of 48.5 μm. A theoretical 100% side-on coverage (1.98×10^{10} cells g^{-1}) or even 100% end-on coverage (183×10^{10} cells g^{-1}) would exhaust only 3.5% or 11%, respectively, of the total exchange capacity. Exchangeable anions (chloride and hydroxide) are detectable on the absorption of bacterial cells [38,210].

The wide range of adsorption capacities observed for the nonspecific soils may be due in part to the uneven distribution of partial monolayers of adsorbed cells as observed microscopically. The capacities of the various soil fractions may also vary, with sand very low and organic matter much higher. The relative sizes of the adsorbing cells and the adsorbent are also important. The three ranges of relative size for categorization are: $R \gg r$, R

$\simeq r$, and $R \ll r$, where R and r are the radii of the adsorbent and the adsorbing cell, respectively (see Fig. 2.2.).

2.5.3 Thermodynamics of Sorption

The thermodynamics of sorption of bacterial cells [100] and virus [140] have been described. Certain thermodynamic functions can be calculated from adsorption equilibrium data. The equilibrium constant is related to the standard-state Gibbs free energy, ΔF^0, in Eq. (2.17):

$$\Delta F^0 = -RT \ln K \tag{2.17}$$

where T is the absolute temperature, K is the equilibrium constant, and R is the gas constant. The enthaphy, ΔH^0, and entropy, ΔS^0, at standard state are related through Eq. (2.18):

$$\ln K = -\frac{\Delta H^0}{RT} + \frac{\Delta S^0}{R} \tag{2.18}$$

If values of K are determined at selected temperatures, all three thermodynamic functions can be calculated. An analysis of thermodynamic functions is not sufficient, however, to explain the physicochemical aspects of adsorptive processes.

Positive ΔH^0 values are indicative of endothermic adsorption. The ΔH^0 for adsorption of bacteria onto soil ranged from +8.5 kcal/mol for noncompetitive adsorption to +23–24 kcal/mol for competitive adsorption with peptone and sodium chloride [100]. Positive values of ΔS^0 observed for all systems are indicative of greater disorder in the adsorbed versus the unadsorbed phases. Negative values of ΔF^0 observed for all systems (−19 to −21 kcal mol^{-1}) are indicative of spontaneous adsorption.

Estimates have not yet been made of the energy required to cause desorption. This could be attempted for the case of reversible sorption based on surface-charge considerations. For the case of irreversible sorption, it could be attempted on the basis of hydrodynamic considerations. The force necessary to rupture bacterial cells of several species adsorbed on glass already has been determined [212].

2.6 SUMMARY

The mechanisms by which microorganisms adsorb to and desorb from solid surfaces involve many overlapping phenomena. The initial movement of microbial cells through a suspending fluid toward and from a sorbent surface can involve nonsorptive charge behavior and diffusional, gravitational, and convective transports. The subsequent attachment and/or release of cells when they are near solid surfaces may involve a balance between London–van der Waals forces and electrostatic forces of electrical double layers.

The typical microbial cell can be depicted as a macroscopic ion having a muliplicity of electrostatically charged surface sites or as a nonionogenic surface with hydrophobic areas. An individual cell can participate in specific exchange reactions with synthetic resins, form bonds with water-soluble polyelectrolytes, and attach nonspecifically to other sorbent surfaces. All three of these activities may be manifested during sorption.

Application of the kinetics of diffusional processes to microbial sorption has been successful in developing rate equations for both adsorption and desorption. Adsorption isotherms used for soluble species can also describe the equilibrium adsorption of microorganisms. Langmiur, Freundlich, and other specialized isotherms have been applied. Various thermodynamic functions have also been derived under equilibrium conditions of sorption.

Sorptive behavior is seen in the restricted movement of cells in: (1) particulate materials, such as other cell and colloidal particles, (2) extended surfaces of soils, plants, and animals, and (3) engineered surfaces designed for water and wastewater treatment. The selectivity and degree of reversibility of these sorptive processes are affected by the microbial species, the adsorbent type, and the character of the environment. The pH and ionic strength of the surrounding fluid affect the rate and extent of microbial sorption.

There is a broad ecological significance and applied practicality to microbial sorption. Much attention has been given by biological researchers and engineering practitioners to microbial sorption, but understanding of the underlying mechanisms, and prediction and control of such behavior, have been limited to incremental advances in diverse fields of microbiology, biochemistry, engineering, medicine, and dentistry. A continued interdisciplinary approach and cross-fertilization of information is a requisite for future advances.

2.7 REFERENCES

1. A. K. Adamov, *Microbiology*, **30** (12), Pt. 2, 5 (1959).
2. P. Albertsson, *Nature*, **177**, 771 (1956).
3. P. Albertsson, *Partition of Cells, Particles, and Macromolecules*, Wiley, New York, 1960.
4. A. C. Allison and R. C. Valentine, *Biochim. Biophys. Acta*, **40**, 393 (1960).

5. A. C. Allison and R. C. Valentine, *Biochim. Biophys. Acta*, **40,** 400 (1960).
6. P. Amirhor and R. S. Engelbrecht, *J. Am. Water Works Assoc.*, **67,** 187 (1975).
7. R. E. Baier, Surface Properties Influencing Biological Adhesion, in R. S. Manley, Ed., *Adhesion in Biological Systems*, Academic, New York, 1970, pp. 15–48.
8. W. J. Bair and J. N. Stannard, *J. Gen. Physiol.*, **38,** 505 (1955).
9. M. Barr, *J. Am. Pharmacol. Assoc.*, **46,** 490 (1957).
10. J. W. Bartholomew and T. Mittwer, *Bacteriol. Rev.*, **16,** 1 (1952).
11. E. Batoosingh and E. H. Anthony, *Can. J. Microbiol.*, **17,** 655 (1971).
12. G. Bitton, *Water Res.*, **9,** 473 (1975).
13. G. Bitton, Y. Henis, and N. Lahav, *Plant Soil*, **44,** 37 (1976).
14. G. Bitton, N. Lahav, and Y. Henis, *Plant Soil*, **40,** 373 (1974).
15. G. Bitton and R. Mitchell, *Water Res.*, **8,** 549 (1974).
16. G. Bitton, O. Pancorbo, and G. E. Gifford, *Water Res.*, **10,** 973 (1976).
17. D. C. Blanchard and L. Syzdek, *Science*, **170,** 626 (1970).
18. J. W. Boyd, T. Yoshida, L. E. Vereen, R. L. Cada, and S. M. Morrison, Sanitary Engineering Papers No. 5, Colorado State University, Fort Collins, June 1969.
19. W. Bozani and M. L. R. Vairo, *J. Bacteriol.*, **77,** 757 (1959).
20. C. C. Brinton, Jr. and M. A. Lauffer, "The Electrophoresis of Viruses, Bacteria, and Cells, and the Microscope Method of Electrophoresis," in M. Bier, *Electrophoresis—Theory, Methods, and Applications*, Academic, New York, 1959, pp. 427–492.
21. P. L. Busch, "Chemical Interactions in the Aggregation of Bacteria," Ph.D. thesis, Harvard University, Cambridge, Mass., 1966.
21a. P. L. Busch and W. Stumm, *Environ. Sci. Technol.*, **2,** 49 (1968).
22. C. Calmon and T. R. E. Kressman, *Ion Exchangers in Organic and Biochemistry*, Interscience, New York, 1957.
23. P. C. Carman and R. A. W. Haul, *Proc. Roy. Soc.(Lond.) Ser. A*, **222,** 109 (1954).
24. M. Chaudhuri and R. S. Engelbrecht, *J. Am. Water Works Assoc.*, **62,** 563 (1972).
25. I. Chet, P. Asketh, and R. Mitchell, *Appl. Microbiol.*, **30,** 1043 (1975).
26. I. Chet and R. Mitchell, *Ann. Rev. Microbiol.*, **30,** 221 (1976).
27. H. N. Christensen, *Biological Transport*, Benjamin, New York, 1961.
28. J. R. Christenson, *Virology*, **27,** 727 (1965).
29. W. B. Clark and R. J. Gibbons, *Infect. Immunol.*, **18,** 514 (1977).
30. J. M. Cohen, G. A. Rourke, and R. L. Woodward, *J. Am. Water Works Assoc.*, **50,** 463 (1958).
31. F. C. Collins, *J. Colloid Interface Sci.*, **5,** 499 (1950).
32. J. T. Cookson, Jr., "Kinetics and Mechanisms of Adsorption of *Escherichia coli* Bacteriophage T_4 to Activated Carbon," Ph.D. thesis, California Institute of Technology, Pasadena, 1966; *Diss. Abstr.*, **26,** 5965 (1966).
33. J. T. Cookson, Jr. and W. J. North, *Environ. Sci. Technol.*, **1,** 26 (1967).
34. W. A. Corpe, "Periphytic Marine Bacteria and the Formation of Microbial Films on Solid Surfaces," in R. R. Colwell and R. Y. Morita, Eds., *Effect of the Ocean Environment on Microbial Activities*, University Park Press, Baltimore, 1974, pp. 387–417.
35. W. A. Corpe, *Dev. Ind. Microbiol.*, **15,** 281 (1974).

36. D. W. Cutler, *J. Agric. Sci.*, **9,** 430 (1919).
37. R. Damadian, *Biophys. J.*, **11,** 773 (1971).
38. S. L. Daniels, "Separation of Bacteria by Adsorption onto Ion Exchange Resins, Ph.D. thesis, University of Michigan, Ann Arbor, 1967; *Diss. Abstr.*, **29,** 1336B (1968).
39. S. L. Daniels, paper presented at the Symposium on Colloid and Surface-Chemical Methods of Separation, 162nd National Meeting of the American Chemical Society, Washington, D. C., September 14, 1971.
40. S. L. Daniels, "The Adsorption of Microorganisms onto Surfaces: A Review," *Dev. Ind. Microbial.*, **13,** 211 (1972).
41. S. L. Daniels and L. L. Kempe, *Chem. Eng. Progr. Symp. Ser.*, **62** (69), 142 (1966).
42. S. L. Daniels and L. L. Kempe, in D. Hershey, Ed., *Chemical Engineering in Biology and Medicine*, Plenum, New York, 1967, pp. 391–415.
43. F. B. Dazzo, C. A. Napoli, and D. H. Hubbell, *Appl. Environ. Microbiol.*, **32,** 166 (1976).
44. B. V. Derjaguin and L. Landau, *Acta Physicochim. USSR*, **14,** 633 (1941).
45. S. C. Dexter, J. D. Sullivan, Jr., J. Williams, III, and S. W. Watson, *Appl. Microbial.*, **30.,** 298 (1975).
46. E. W. Dianova and A. A. Voroschilova, *Nauch. Agron. Zh.*, **2,** 520 (1925).
47. B. H. Dieterich, *A Study of Adsorption Phenomenon in the Removal of Bacterial Virus by Sand Filtration*, M.S. thesis, Harvard University, Cambridge, Mass., 1953; as quoted by Cookson [32].
48. M. M. Dikusar, *Mikrobiol.*, **9,** 895 (1949).
49. J. K. Dixon, R. C. Tilton, and M. W. Zielyk, National Technical Information Service Publication No. PE 195983, 1970.
50. J. K. Dixon and M. W. Zielyk, *Environ. Sci. Technol.*, **3,** 511 (1969).
51. W. A. Drewery and R. Eliassen, *J. Water Pollut. Control Fed.*, **40,** Pt. 2, R257 (1968).
52. F. A. Van Duuren, *Water Treat. Exam.*, *1969*, **18,** Pt. 2, 128; *Chem. Abstr.*, **71,** 94647 (1969).
53. W. F. Echelberger, Jr., J. L. Pavoni, P. C. Singer, and M. W. Tenney, *Proc. ASCE, J. San Eng. Div.*, **97,** (SA5), 721 (1971).
54. P. Eisenberg, *Zentr. Bakteriol., Parasitenk., Abstr. I, Orig.*, **81,** 72 (1918); *Deutsch Med. Wochensch.*, **44,** 634 (1918).
55. R. Eliassen, W. Drewry, O. Chen, and G. Tschobanoglous, "Studies on the Movement of Viruses in Groundwater," Annual Progress Report, Water Quality Control Research Laboratory, Department of Civil Engineering, Stanford, Calif., 1966.
56. B. A. F. Erchul and D. L. Isenberg, *J. Nutr.*, **95,** 374 (1968).
57. T. Ericson, J. Sandham, and I. Magnusson, *Caries Res.*, **9,** 325 (1975).
58. E. F. Estermann, "Digestion of Protein Adsorbed on Clay Minerals by Enzymes and Microbes," Ph.D. thesis, University of California, Berkeley, 1957.
59. E. F. Estermann and A. D. McLaren, *J. Soil Sci.*, **10,** 64 (1959).
60. R. W. Filmer, F. Felton, Jr., and T. Yamamoto, *Univ. Ill. Bull.*, **69** (1), 75 (August 2, 1971).
61. H. Finkelstein and J. W. Bartholomew, *J. Bacteriol.*, **80,** 14 (1960).
62. M. Fletcher, *J. Gen. Microbiol.*, **94,** 400 (1976).

62a. M. Fletcher and G. D. Floodgate, *J. Gen. Microbiol.*, **74,** 235 (1973).
63. J. M. Foliquet and F. Doncoeur, *Water Res.*, **9,** 953 (1975).
64. D. Fraser and C. Fleischmann, *J. Virol.*, **13,** 1067 (1974).
65. R. R. Freeman, *Biotechnol. Bioeng.*, **6,** 87 (1964).
66. E. Friedberger, *Muenchen. Med. Wochenschr.*, **66** (48), 1372 (1919).
67. G. E. Galvez, "Specific Virus Adsorption by Antibodies Coupled to a Solid Matrix," Ph.D. thesis, University of Nebraska, Lincoln, 1964; *Diss. Abstr.*, **25,** 2698 (1964).
68. C. P. Gerba, M. D. Sobsey, C. Wallis, and J. L. Melnick, *Environ. Sci. Technol.*, **9,** 727 (1975).
69. R. J. Gibbons, and J. van Houte, *J. Periodontol.*, **44,** 347 (1973).
70. R. J. Gibbons and J. van Houte, *Annu. Rev. Microbiol.*, **29,** 19 (1975).
71. G. Gillissen, H. Scholz, and C. Dehnert, *Arch. Hyg. Bakteriol.*, **145,** 145 (1961).
72. H. N. Glassman, *Bacteriol. Rev.*, **12,** 105 (1948).
73. C. G. Golueke and W. J. Oswald, *J. Water Pollut. Control Fed.*, **73,** 471 (1965).
74. D. M. Griffin and G. Quail, *Austral. J. Biol. Sci.*, **21,** 579 (1968).
75. K. W. Guebert and J. D. Laman, U.S. Patent 3,242,073, (March 22, 1966).
76. J. B. Gunnison and M. S. Marshall, *J. Bacteriol.*, **33,** 401 (1937).
77. H. Hara, *Jap. J. Microbiol.*, **2,** (1958).
78. J. O. Harris, *J. Bacteriol.*, **65,** 518 (1953).
79. J. O. Harris and T. M. McCalla, *J. Bacteriol.*, **61,** 57 (1951).
80. R. H. Harris, "The Effect of Extracellular Polymers and Inorganic Colloids on the Colloidal Stability of Bacteria," Ph.D. thesis, Harvard University, Cambridge, Mass., 1971.
81. R. H. Harris and R. Mitchell, *Annu. Rev., Microbiol.*, **27,** 27 (1973).
82. R. H. Harris and R. Mitchell, *Water Res.*, **9,** 993 (1975).
83. T. Hattori, Paper presented at the Society of Soil and Manure, Japan, 1968, as quoted by Marshall [120].
84. R. Hattori, *J. Gen. Appl. Microbiol.*, **18,** 319 (1972).
85. R. Hattori, T. Hattori, and C. Furusaka, *J. Gen. Appl. Microbiol.*, **18,** 271 (1972).
86. R. Hattori, T. Hattori, and C. Furusaka, *J. Gen. Appl. Microbiol.*, **18,** 285 (1972).
87. F. Helfferich, *Ion Exchange*, McGraw-Hill, New York, 1962, pp. 267–276.
88. C. E. Helmstetter and D. J. Cummings, *Proc. Nat. Acad. Sci.*, **50,** 767 (1963).
89. C. W. Hiatt, *Bacteriol. Rev.*, **28,** 150 (1964).
90. H. M. Hodge and S. N. Metcalfe, Jr., *J. Bacteriol.*, **75,** 258 (1958).
91. A. Hofman, *Muenchen. Med. Wochenschr.*, **68,** 71 (1921).
92. B. H. Hoyer, E. T. Bolton, R. A. Ormsbee, G. LeBouvier, D. B. Ritter, and C. L. Larson, *Science*, **127,** 859 (1958).
93. J. N. Israelachvili and B. W. Ninham, *J. Colloid Interface Sci.*, **58,** 14 (1977).
94. K. J. Ives, *J. Biochem. Microbiol. Eng.*, **1,** 37 (1959).
95. W. Jakubowski, *Bacteriol. Proc.*, **1969** (V198), 179 (1969).
96. M. Ya, Kalyuzhnii, *Continuous Cultiv. Microorganisms*, **2,** 263 (1964); *Chem. Abstr.*, **63,** 49076 (1965).

97. A. Kammen, D. van Noordam, and T. H. Thung, *Virol*, **14,** 100 (1961).

98. N. S. Karpinskaya, *Nauch. Agron. Zh.*, **3,** 587 (1926).

99. M. A. Kessick and F. A. Wagner, *Water Res.*, **12,** 263 (1978).

100. D. R. Khairnar, "The Effect of Chemical Competition on Thermodynamics of Bacterial Adsorption," Ph.D. thesis, Utah State University, Logan, 1971; *Diss. Abstr. Int. B*, **32,** (7), 3732 (1972); *Chem. Abstr.*, **76,** 125724 (1972).

101. N. N. Khudiakov, *Pochvovedenie*, **21,** 46 (1963); see also *Zentrabl. Bakteriol. Parasitenk., Abstr. II*, **68,** 345 (1926).

102. R. Klinger, *Muenchen. Med. Wochenschr.*, **67,** 74 (1920).

103. K. D. Kostenbader, Jr., and D. O. Cliver, *Appl. Microbiol.*, **24,** 540 (1972).

104. K. Krishnamurti and S. V. Soman, *Proc. Indian Acad. Sci.*, **B34,** 81 (1951).

105. B. Kruger, *Z. Hyg. Infekt.*, **7,** 86 (1889).

106. R. Kunin and R. J. Meyers, *Ion Exchange Resins*, 1st ed., Wiley, New York, 1947, p. 134.

107. T. Kurozumi, M. Itoh, and K. Shibata, *Arch. Biochem. Biophys.*, **109,** 241 (1965).

108. Y. Kuwajima, T. Matsui, and M. Kishigami, *Jap. J. Microbiol.*, **1,** 375 (1957).

108a. N. Lahav, *Plant Soil*, **17,** 191 (1962).

109. P. Lasseur, P. Dombray, and W. Palgen, *Trav. Lab. Microbiol. Fac. Pharm.* (*Nancy*), **7,** 117 (1934).

110. C. Lerche, *Acta Pathol. Microbiol. Scand. Suppl.*, **98,** 1 (1953).

111. A. M. Lerner, *Bacteriol. Rev.*, **28,** 391 (1964).

112. M. R. Lewis and H. B. Andervont, *Am. J. Hyg.*, **7,** 505 (1927).

113. A. A. Lindberg, *Annu. Rev. Microbiol.*, **27,** 205 (1973).

114. H. F. Linskens, *Planta*, **68** (2), 99 (1966); *Biol. Abstr.*, **49,** 20969 (1968).

115. S. H. Lo and O. J. Sproul, *Water Res.*, **11,** 653 (1977).

116. N. G. Maraudas, *Nature* (*London*), **254,** 695 (1975).

117. K. C. Marshall, *Biochim. Biophys. Acta*, **156,** 179 (1968).

118. K. C. Marshall, *J. Gen. Microbiol.*, **56,** 301 (1969).

119. K. C. Marshall, *Biochim. Biophys. Acta*, **193,** 472 (1969).

120. K. C. Marshall, "Sorptive Interactions Between Soil Particles and Microorganisms," in A. D. McLaren and J. J. Skujins, Eds., *Soil Biochemistry*, Vol. II, Marcel Dekker, New York, 1971, pp. 409–445.

121. K. C. Marshall, *Interfaces in Microbiol Ecology*, Harvard U. P., Cambrige, Mass., 1976, 156 pp.

122. K. C. Marshall and R. H. Cruickshank, *Arch. Mikrobiol.*, **91,** 29 (1973).

123. K. C. Marshall, R. Stout, and R. Mitchell, *J. Gen. Microbiol.*, **68,** 337 (1971).

124. K. C. Marshall, R. Stout, and R. Mitchell, *Can. J. Microbiol.*, **17,** 1413 (1971).

125. G. J. Martin, Ion Exchange and Adsorption Agents in Medicine, Little Brown, Boston, 1955, pp. 54–57, 67–69.

126. U. Maruyama and T. Yanigita, *J. Bacteriol.*, **71,** 542 (1956).

127. J. McCabe and W. W. Eckenfelder, Jr., Eds., *Biological Treatment of Sewage and Industrial Waste*, Vol. I, Reinhold, New York, 1956.

128. T. M. McCalla, *J. Bacteriol.*, **40,** 23 (1940).

129. T. M. McCalla, *J. Bacteriol.*, **41,** 775 (1941).
130. T. M. McCalla, *Stain Technol.*, **16,** 27 (1941).
131. M. G. McGarry and C. Tongkasame, *J. Water Pollut. Control Fed.*, **43,** 824 (1971).
131a. W. C. McGregor and R. K. Finn, *Biotech. Bioeng.*, **11,** 127 (1969).
132. P. A. Meadows, *Arch. Mikrobiol.*, **75,** 374 (1971).
133. L. Michaelis, *Berlin. Klin. Wochenschr.*, **55,** 710 (1918).
134. A. R. Minenkov, *Zentrabl. Bakteriol. Parasitenk., Abstr. II*, **78,** 109 (1929).
135. C. J. O. R. Morris and P. Morris, *Separation Methods in Biochemistry*, Interscience, New York, 1963.
136. S. Mudd and E. B. H. Mudd, *J. Exp. Med.*, **40,** 633 (1924).
137. S. Mudd, R. L. Nugent, and L. T. Bullock, *J. Phys. Chem.*, **36,** 229 (1932).
138. G. Muller and B. Hickisch, *Zkentrabl. Bakteriol. Parasitenk., Abstr. II, Orig.*, **124,** 271 (1970).
139. R. H. Muller, "Virology," in C. Calmon and T. R. Kressman, Eds., *Ion Exchangers in Organic and Biochemistry*, Interscience, New York, 1957, pp. 248–254.
140. J. P. Murray, G. A. Parks, and C. E. Schwerdt, paper presented before the Division of Environmental Chemistry, American Chemical Society, Anaheim, Calif., March 12–17, 1978;
141. H. Nakamura, *J. Biochem. Microbiol. Technol. Eng.*, **3,** 395 (1961).
142. J. Nakamura, S. Miyashiro, and Y. Hirose, *Agric. Biol. Chem.*, **40,** 377 (1976).
143. J. S. Nordin, H. M. Tsuchiya, and A. G. Fredreickson, *Biotechnol. Bioeng.*, **9,** 545 (1967).
144. S. Nordling, A. Vaheri, E. Saxen, and K. Penttinen, *Exp. Cell Res.*, **37,** 406 (1965).
145. D. M. Novogrudskii, *Mikrobiol.*, **5,** 623 (1936).
146. V. G. Oksentian, *Mikrobiol.*, **9,** 3 (1940).
147. J. T. G. Overbeek, "Kinetics of Flocculation," in H. R. Kruyt, Ed., *Colloid Science* (Vol. I.,) *Irreversible Systems*, Elsevier, New York, 1952, pp. 278–301.
148. P. P. Oza, and M. Chaudhur, *J. Env. Eng. Div., ASCE*, **102,** 1244 (1976).
149. J. L. Pavoni, "Fractional Composition of Microbially Produced Exocellular Polymers and Their Relationship to Biological Flocculation," Ph.D. thesis, University of Notre Dame, South Bend, Ind., 1970.
150. J. L. Pavoni, M. W. Tenney, and W. F. Echelberger, Jr., *J. Water Pollut. Control. Fed.*, **44,** 414 (1972).
151. T. C. Peele, *N. Y. State Agric. Exp. Sta. (Geneva, N. Y.), Memoir.*, **197,** 3 (1936).
152. A. F. Pertsovskaya and D. G. Zvyagintsev, *Biol. Nauk*, **14,** 100 (1971); *Biol. Abstr.*, **52,** 13430 (1972).
153. B. A. Pethica, *Exp. Cell Res., Suppl.*, **8,** 123 (1961).
154. L. Philipson, K. Maramorosch, and H. Koprowski, Eds., *Methods in Virology*, Vol. II, Academic, New York, 1967, pp. 179–233.
155. T. T. Puck and B. Sagik, *J. Exp. Med.*, **97,** 807 (1953).
156. E. Putter, *Arch. Hyg. Bakteriol.*, **89,** 71 (1920); Dissertation, Griefswald, 1919; *Kolloid Z.*, **27,** 204 (1920).
157. N. V. Rao and N. A. Labzoffsky, *Can. J. Microbiol.*, **15,** 399, (1969).

158. N. Rashevsky, *Mathematical Biophysics: Physico-Mathematical Foundations of Biology*, Vol I, 3rd ed., Dover, New York, 1960, pp. 7–142.

159. A. T. Reid, *Arch. Biochem. Biophys.*, **43,** 416 (1953).

160. B. Rotman, *Bacteriol. Rev.*, **24,** 251 (1960).

161. L. I. Rubentschik, M. B. Roisin, and F. M. Bieljansky, *Mikrobiol.*, **3,** 16 (1934); *J. Bacteriol.*, **32,** 11 (1936).

162. A. J. Rubin, P. L. Hayden, and G. P. Hanna, Jr., *Water Res.*, **3,** 843 (1969).

163. M. R. J. Salton, *The Bacterial Cell Wall*, Elsevier, New York, 1964.

164. G. Salus, *Biochem. Z.*, **84,** 378 (1917).

165. O. Samuelson, *Ion Exchange Separations in Analytical Chemistry*, Wiley, New York, 1963.

166. T. Santoro and G. Stotzky, *Can. J. Microbiol.*, **14,** 299 (1968).

167. H. Schott and C. Y. Young, *J. Pharmacol. Sci.* **61,** 182 (1972).

168. E. Schwartz and J. Mayer, *Zentrabl. Bakteriol. Parasitenk., Abstr. I., Orig.*, **189,** 489 (1963).

169. M. Schwartz, *J. Molec. Biol.*, **103,** 521 (1976).

170. S. P. Shelton and W. A. Drewry, *J. Am. Water Works Assoc.*, **65,** 627 (1973).

171. M. von Smoluchowski, *Z. Physik Chem.* (*Leipzig*), **92,** 129 (1917).

172. Society for General Microbiology, Microbial Cell Surfaces and Membranes Group Symposium, "Adhesion of Microorganisms to Surfaces," *Proc. Soc. Gen. Microbiol.*, **5** (1), xiii (1977).

173. O. J. Sproul, J. Warner, D. R. Brunner, and S. H. Jenkins, Eds. *Advances in Water Pollution Research*, Proceedings of Fourth International Conference, Prague, 1969, Pergamon, New York, 1969, pp. 541–547.

174. A. E. Stearn and E. W. Stearn, *Univ. Mo. Stud.*, **3** (2), 1 (1928).

175. R. L. Steere, *Adv. Virus Res.*, **6,** 1 (1959).

176. G. Stotzky and L. T. Rem, *Can. J. Microbiol.*, **12,** 547 (1966).

177. I. Taniguchi, *Virology*, **18,** 646 (1962).

178. M. W. Tenny, W. F. Echelberger, Jr., R. G. Schuessler, and J. L. Pavoni, *Appl. Microbiol.*, **18,** 965 (1966).

179. M. W. Tenny and W. Stumm, *J. Water Pollut. Control Fed.*, **37,** 1370 (1965).

180. M. W. Tenny and F. H. Verhoff, *Biotechnol. Bioeng.*, **15,** 1045 (1973).

181. A. S. Teot and S. L. Daniels, *Environ. Sci. Technol.*, **3,** 825 (1969).

182. H. Terayama, *Arch. Biochem. Biophys.*, **50,** 55 (1954).

183. C. J. Thomas, T. A. McMeeken, and C. Balis, *Appl. Environ. Microbiol.*, **34,** 456 (1977).

184. R. T. Thorup, F. P. Nixon, D. F. Wentworth, and O. J. Sproul, *J. Am. Water Works Assoc.*, **62** 97 (1969).

185. R. C. Tilton, J. Murphy, and J. K. Dixon, *Water Res.*, **6,** 155–64 (1972).

186. T. R. Tosteson and W. A. Corpe, *Can. J. Microbiol.*, **21,** 1025 (1975).

187. G. P. Treweek and J. J. Morgan, Paper presented at the 50th Annual Conference of Water Pollution Control Federation, October 2–7, 1977, Philadelphia.

188. M. Tschapek, *Cienc. Invest.*, **18,** 310 (1962).

189. M. Tschapek and A. J. Garbosky, *Transact. 4th Internat. Conf. Soil Sci.* (*Amsterdam*), **3,** 102 (1950).

190. D. Tsernoglou and E. H. Anthony, *Can. J. Microbiol.*, **17,** 217 (1971).

191. D. G. Tustian and R. P. Ellen, *J. Periodontal Res.*, **12,** 323 (1977); *Biol. Abstr.*, **65,** 13817 (1978).

192. R. C. Valentine and A. C. Allison, *Biochim. Biophys. Acta*, **34,** 10 (1959).

193. L. E. Vereen, "Bacterial Mobility in Porous Media," Ph.D. thesis, Colorado State University, Fort Collins, 1968; *Dist. Abstr.*, **29,** 1775B (1968).

194. E. J. W. Verwey and J. T. G. Overbeek, *Theory of the Stability of Lyophobic Colloids*, Elsevier, New York, 1948.

195. L. R. Van Vuuren and F. A. Van Duuren, *J. Water Pollut. Control Fed.*, **37,** 1256 (1965).

196. C. Wallis and J. L. Melnick, in G. Berg, Ed., *Transmission of Viruses by the Water Route*, Interscience, New York, 1967, pp. 129–138.

197. J. Warren, A. Neal, and D. Rennels, *Proc. Soc. Exp. Biol. Med.*, **12,** 1250 (1966).

198. J. T. Watson and W. A. Drewry, Department of Civil Engineering research series No. 11 and Water Resources Research Center report No. 14, University of Tennessee, Knoxville, August 1971.

199. W. J. Weber, Jr., *Physicochemical Processes for Water Quality Control*, Wiley-Interscience, New York, 1972.

200. C. Weibull, "Movement," in I. C. Gunsalus and R. Y. Stanier, Eds., *The Bacteria*, Vol. I, *Structure*, Academic, New York, 1960, pp. 153–205.

201. C. M. Weiss, *J. Water Pollut. Control Fed.*, **23,** 227 (1951).

202. D. Werner, J. Wilcockson, and E. Zimmerman, *Arch. Microbiol.*, **105,** 27 (1975).

203. T. Yamamoto, A. Kawamura, Jr., H. Hara, and K. Aikawa, *J. Exp. Med.*, **28,** 329 (1958).

204. F. Zago, "Studies on the Mechanism of Adsorption of Neutropic Viruses," Ph.D. thesis, University of Michigan, Ann Arbor, 1956.

205. C. E. ZoBell, *J. Bacteriol.*, **46,** 39 (1943).

206. D. G. Zvyagintsev, *Sov. Soil Sci.* (*Engl. transl.*), **1962,** 140; as quoted by Marshall [120].

207. D. G. Zvyagintsev, *Microbiology*, **31,** 275 (1962).

208. D. G. Zvyagintsev, *Nauch. Dokl. Vyssh. Shk.*, *Biol. Nauk*, **1967,** (3), 97; *Chem. Abstr.*, **67,** 8814 (1967).

209. D. G. Zvyagintsev, *Interactions of Microorganisms with Solid Surfaces*, University of Moscow Press, 1973.

210. D. G. Zvyagintsev and V. S. Guzev, *Mikrobiology*, **40,** 139 (1971).

211. D. G. Zyvagintsev, A. F. Pertsovskaya, V. I. Duda, and D. I. Nikitin, *Microbiology*, **38,** (1969).

212. D. G. Zyvagintsev, A. F. Pertsovskaya, E. D. Yukhnin, and E. I. Averback, *Mikrobiology*, **40,** 889 (1971).

213. M. T. Bond and B. A. Mowry, *Removal of Algae from Waste Stabilization Pond Effluent*, Water Resources Research Institute, Mississippi State University, State College, July 1976; NTIS PB-256 512.

214. T. R. Jack and J. E. Zajic, *Biotech. Bioeng.*, **19,** 631 (1977).

215. R. A. Messing and R. A. Oppermann, *Biotech. Bioeng.*, **21,** 49 (1979).

216. R. A. Messing, R. A. Oppermann, and F. B. Kolot, *Biotech. Bioeng.*, **21,** 59 (1979).

217. A. F. Pertsovskaya, V. I. Duda, and D. G. Zvyagintsev, *Sov. Soil Sci.*, **4,** 684 (1973).

218. G. E. Sechler, "On the Microbiology of Slime Layers Formed on Immersed Materials in a Marine Environment," Ph.D. Thesis, University of Hawaii, Honolulu, 1972, Diss. Abstr., **33,** 4414B (1973).

219. T. Tosa, T. Sato, Y. Nishida, and I. Chibata, *Biochim. Biophys. Acta*, **483,** 193–202 (1977).

220. V. L. Vilker, L. H. Frommhagen, R. Kamdar, and S. Sandaram, AIChE Symp. Ser. **178,** 84 (1978).

221. J. E. Zajic and A. LeDuy, *Appl. Microbiol.*, **25,** 628 (1973).

CHAPTER 3

Substrata Influences on Adhesion of Microorganisms and Their Resultant New Surface Properties

R. E. BAIER
Calspan Corporation, Advanced Technology Center, Buffalo, New York

CONTENTS

3.1 INTRODUCTION

Microbiological colonization of solid surfaces is very common in both healthy and diseased states and in circumstances that are both environmentally desirable and undesirable. Practical municipal systems for wastewater treatment depend on bacterial adhesion and growth on various substrata, whereas in modern solar-energy extraction designs, which require enormous heat-exchange surfaces placed in the tropical oceans, microorganism–substratum associations are undesirable.

To understand, control, and beneficially modify such substratum–organism relationships, we must learn the details of the influence of the original substratum surface properties on the initial events of microbial colonization. Further, we must learn the new surface qualities imparted to various natural, synthetic, and engineering substrata before we can successfully approach issues of improving colonization success, preventing unwanted biofouling, or decontaminating materials bearing microorganisms of undesired character.

3.2 BACKGROUND INFORMATION

3.2.1 Relations between Cell Adhesion and Surface Energy Parameters

Most prior work on bioadhesion phenomena has dealt with influences of substrata surface properties on the relative adhesion, growth, and behavior of mammalian cells because of the obvious importance of these phenomena in cellular adhesive events that apparently change as the carcinogenic or malignant cellular state is assumed. Little has been done in similarly required studies of microbial interactions with well-defined materials. Nevertheless, surveys of the abundant literature from fields of study as diverse as experimental pathology, the evaluation of synthetic vascular grafts, and the analysis of adhesion between teeth and the gingival cuff that surrounds them suggest that at least a qualitative relationship exists between the relative surface free energy of various synthetic (usually plastic or polymeric) materials and the relative biological response (e.g., cell adhesion or fibrous encapsulation) they evoke. Looking closely at such data, it is found that simple linear relationships do not hold between relative surface free energies and measures of cell–substrata interactions. Rather, a discrete intermediate range of relative surface energies seems to correlate best with minimal biological response, as measured by the number of adsorbed cells, strength of cell adhesion, spread area of individual adsorbed cells, tendency of cells to grow to confluent monolayers over those substrata, and other phenomena [1–3]. Figure 3.1 presents a typical correlation of this general type, ranking the relative binding of particular mammalian cells [4] in contact with various substrata versus the nominal critical surface tension, an

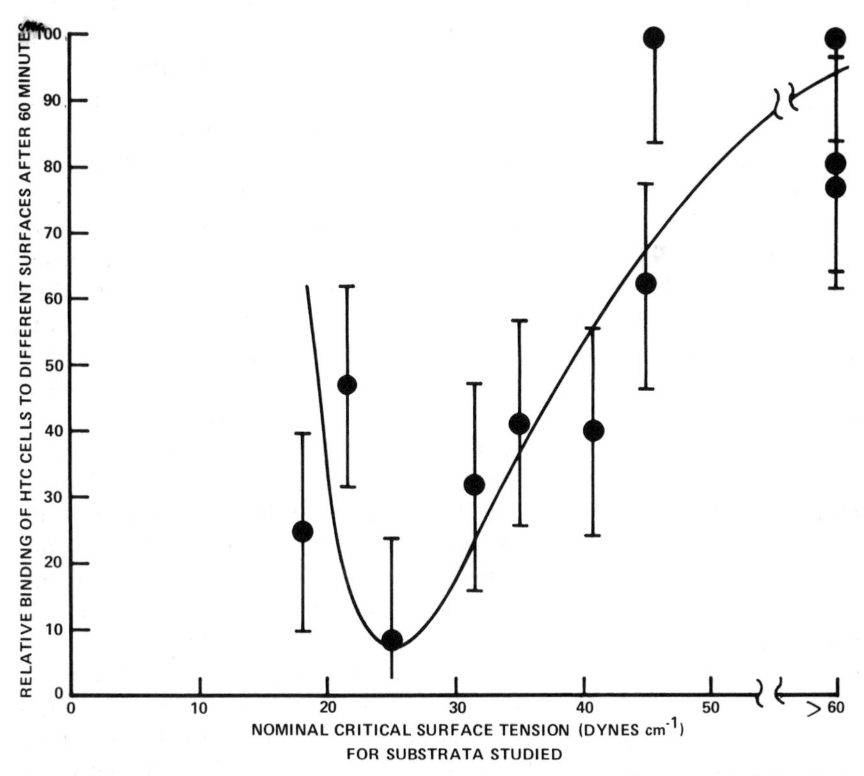

Figure 3.1 Typical correlation of surface properties of various substrata with their relative support for biological adhesion. Replotted from data of F. Grinnell et al. [4]

empirical parameter showing a 1:1 mapping with surface free energy [5–7]. Note particularly the apparent minimum in biological adhesive strength obtained with substrata having critical surface tensions of 20–30 dynes cm^{-1}. A similar correlation was proposed for adhesive interactions with engineering and synthetic materials that must contact flowing blood in artificial internal organs [8]. In fact, successful blood handling devices ranging from artificial hearts [9] to modified human umbilical cord veins [10] have been put into clinical use, as deliberately fabricated to exhibit surface properties in this apparently biocompatible (20–30 dynes cm^{-1}) critical surface tension range.

3.2.2 Complicating Factors

It is much too early to accept or depend on a single criterion such as critical surface tension when selecting materials to which either strong or weak biological adhesion is desired. The great complexity of cellular and microbial adhesion in saline, biochemically active, dynamic systems that can modify most synthetic materials even in the absence of potentially adherent cells has yet to be deciphered. Other data, drawn from a series of experiments with various malignant cell lines [11], do not as clearly support the proposed relationship between critical surface tension of various test substrata and a different parameter of cell adhesion, the relative spread areas of cells gravitationally settled to the substrata in protein-supplemented tissue-culture media (Fig. 3.2). The wide range of spread areas found in these experiments does not allow easy acceptance of the hypothetical curve drawn in Fig. 3.2, again suggesting a minimum in the region of 20–30 dynes cm^{-1}. In contrast, the relative spread areas of adsorbed cells did show a much clearer conformation of the minimal bioadhesive tendencies of rat gingival cells to the proposed "minimal" critical surface tension range [1,12].

We must be aware of and, in fact, look at other complications in these tests that may cloud the interpretation of the resulting data. One such complication in most prior work arises from the use of different substrata from a variety of commercial or laboratory sources. Even when scrupulous care was taken to otherwise clean, prepare, or stabilize the surfaces, clear differences in texture (e.g., surface roughness and molding or machining marks) persisted. Our own studies with variously textured materials of otherwise identical composition or surface treatment, implanted into the cardiovascular system of test animals, clearly showed surface textural variations to be major determinants of cell-adhesion phenomena [13]. Therefore, the opportunity was sought to examine, in precisely defined systems, the influence of changes in relative surface chemistry (i.e., interfacial free

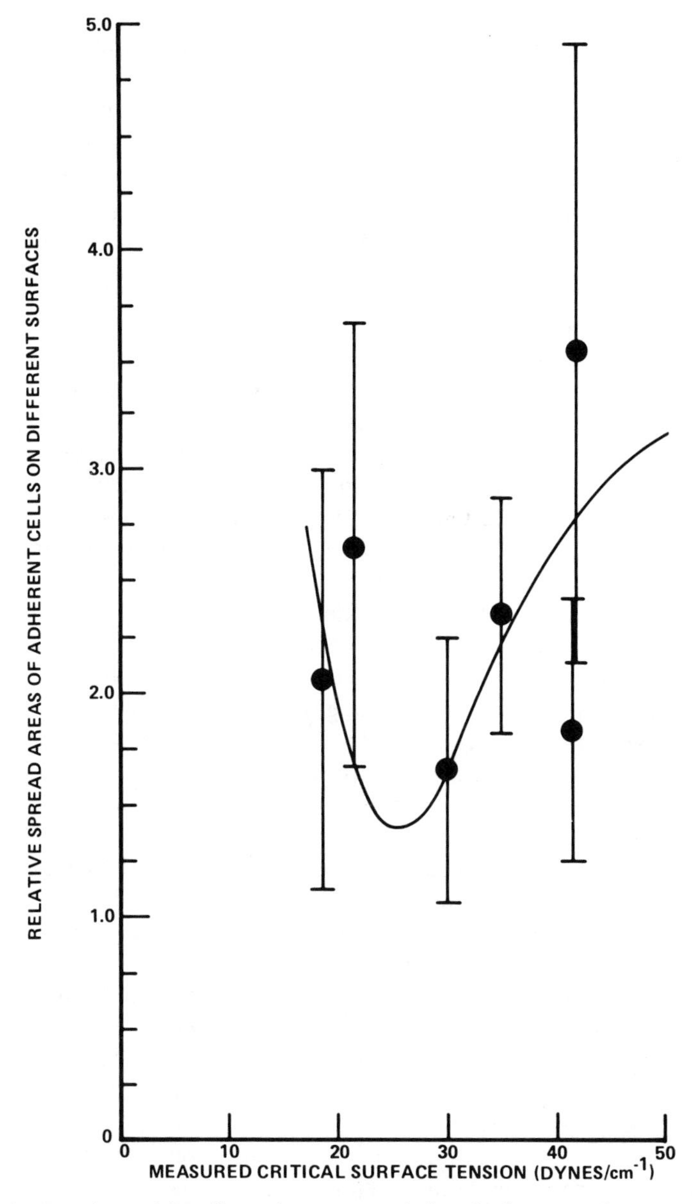

Figure 3.2 Experimental data illustrating poor correlation of substrata surface properties with spread areas of biological cells adsorbed to their surfaces. Ehrlich ascites tumor cells adsorbed in presence of supplemental calf serum.

energies) on cell adhesion to substrata of otherwise identical (very smooth) surface architecture [14]. To further aid in interpretation of results, the surface charge distribution on these substrata also was experimentally determined. Although those data are not reviewed here, differing surface charge distributions were shown to affect the phenomena studied. With the permission of the author [15], a series of graphical summaries of the experimental data bearing on the relationship between cell adhesion and substratum surface chemistry (alternatively, surface energies) is presented here, redrawn from the original figures.

Role of Adsorbed Films. One of the most important observations about biological adhesion in relevant natural circumstances is that it never occurs except through the mediation of preadsorbed films of biological macromolecules, primarily glycoproteins, proteoglycans, or their end-product humic residues. These films are spontaneously adsorbed from the aqueous biological media in which the potentially adhesive cells are suspended [3,16]. The findings in typical tissue-culture–cell-adhesion media containing supplemental calf serum, which mammalian cell lines often require for their successful maintenance and propagation, are documented in Fig. 3.3. Four substrata of equivalent surface texture were incubated simultaneously for increasing time intervals. After taking the test plates from the tissue-culture media, being careful to avoid transfer of any denatured proteinaceous films from the gas–liquid interface, the substrata themselves were used as infrared (IR) transmitting supports for the detection and analysis of adsorbed proteins acquired on surfaces of clearly different chemistry. It is obvious in Fig. 3.3 that all surface energy states imparted to the same bulk material allow spontaneous adsorption and continuing buildup of proteinaceous material from the culture medium, in the same relative amounts and at the same rates. The immediate supposition could be that differences in the surface chemistries of these variously treated test plates are effectively masked or obscured by their overcoating with essentially the same biological macromolecules, accumulated in the same amounts over the same incubation times. If this supposition is correct, subsequent cell contacts, obviously limited to the new exterior face of each protein-coated substratum, no longer could exhibit differential degrees of cell adhesion or cellular spreading. Yet the empirical observation, confirmed again in these studies, is that striking differences in cell adhesion are noted, even with cells from the same cell line gravitationally arriving at these apparently identically "conditioned" test substrata in the same tissue-culture media.

Continuing Substratum Differences after Film Adsorption. Therefore, it becomes necessary to specify more carefully the actual similarities and dif-

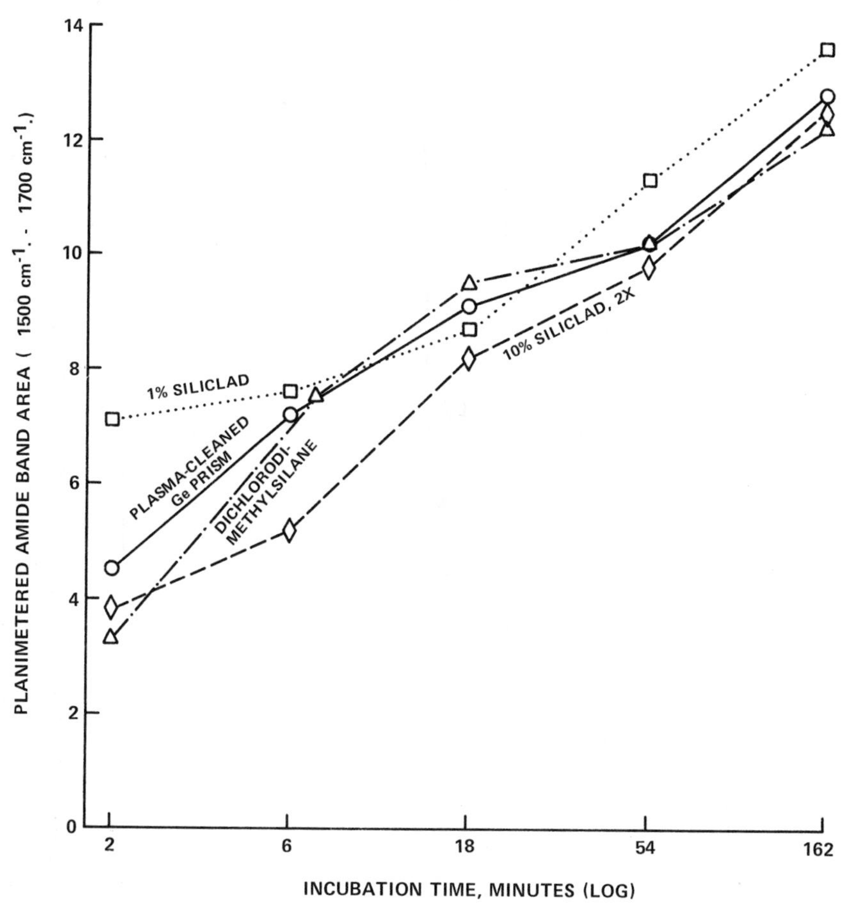

Figure 3.3 Amide band absorption as function of incubation time in medium containing 5% calf serum.

ferences at the surfaces of test substrata at the time approaching cells come into their "sphere of influence." Also one must attempt to more carefully discriminate the details of the varying cellular responses to these differing emplaced substrata. Returning to the parameter of critical surface tension as one indicator of possibly continuing substrata differences after "conditioning" film coverage, the changes of the critical surface tension measured for differing test materials versus incubation time at physiologic conditions in the culture media were examined. Inspection of Fig. 3.4 and comparison with the findings from the protein-adsorption experiments documented in Fig. 3.3 reveal that the highest energy substratum used (plasma-cleaned or, alternatively, glow-discharge-cleaned glass) undergoes dramatic decreases in

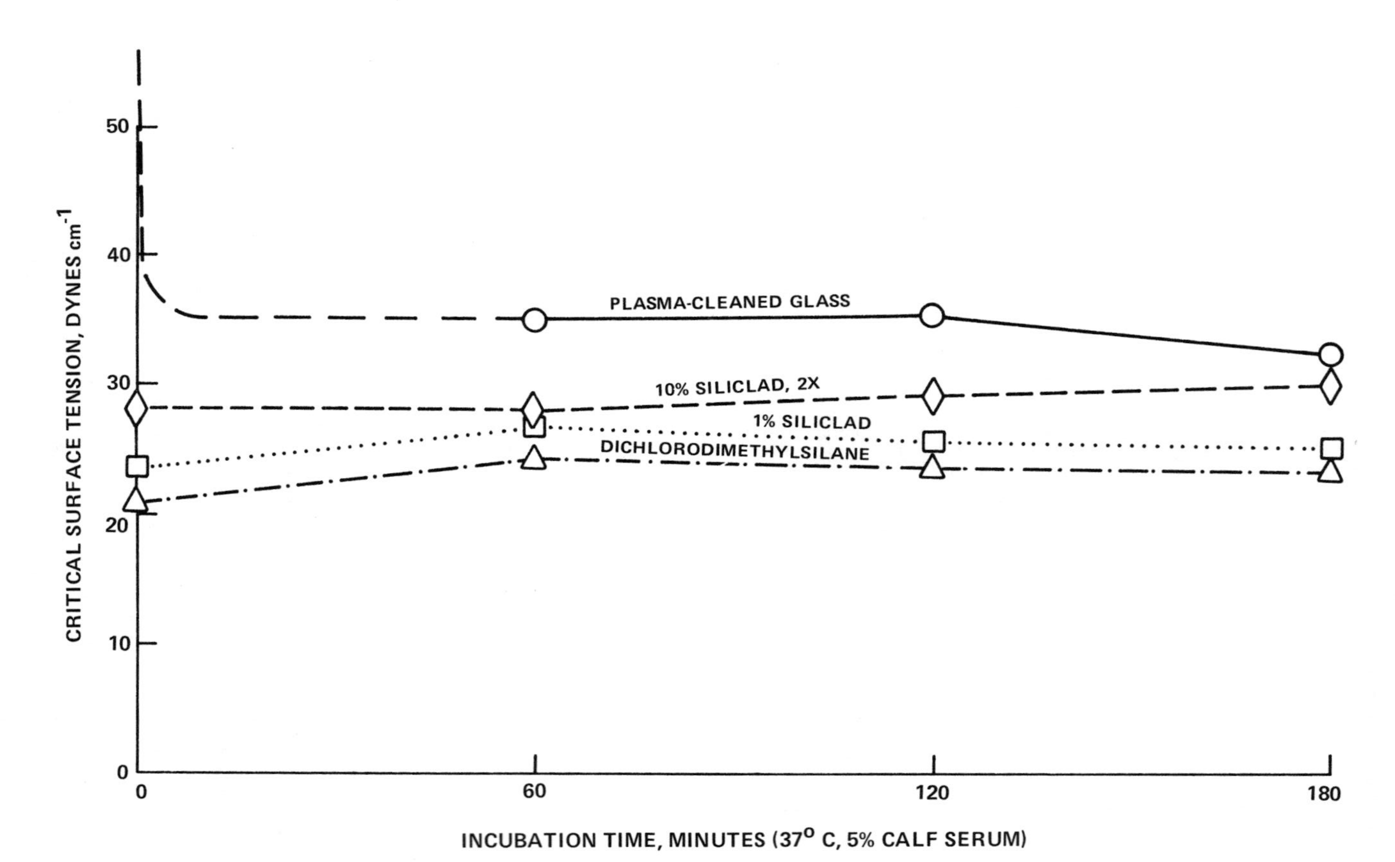

Figure 3.4 Critical surface tensions of test substrates before and after incubation in protein-containing cell culture medium.

critical surface tension within seconds of immersion into the incubation media. On the other hand, variously modified (with different "siliconizing" compounds) glass surfaces remain at their original critical surface tensions (20–30 dynes cm^{-1}, within experimental error) in spite of the independently demonstrated accumulation of about the same amount of proteinaceous material. One can then look to the cells and ask whether the number adsorbed to the various conditioning film-coated substrata of initially differing surface energies (but common surface texture) reflects the differences in the original samples. As shown in Fig. 3.5, the number of cells that are adsorbed at different incubation times ($\leq$4 hr), are only slightly different from one another and without clear relation to the surface properties of the substrata employed. This same finding had been made earlier with respect to the number density of blood platelets adsorbed to various substrata used in test flow cells conducting blood drawn from living animals [17]. Therefore, simple determination of the number density of adsorbed cells, as they randomly arrive at the substrata of interest, is not a sufficient discriminator of the different cell–substrata interactions actually occurring.

An improved discriminator, based on measures of the spread areas of the adsorbed cells, is demonstrated in Fig. 3.6. Different spread areas are found for the same cell types in the same tissue culture incubation medium. The cells exhibit 50% greater spread areas on the high-energy glass (plasma cleaned) than on the siliconized (i.e., methyl-group-dominated, low energy) specimens. The differentiation is even more striking when the analysis is carried to the level of assessing the number of filopodia, or irregular projections, on these same cells adsorbed to substrata varying in surface energy (Fig. 3.7). As for the "spread-cell" data in Fig. 3.6, evidence exists for a continuing influence, at an apparently considerable distance from the original substratum surface, of the initial surface chemistry (or surface energy) of differing materials on the events of cell adhesion.

3.2.3 Practical Consequences of Substrata Modifications

It only remains, then, to inquire of the practical consequences of these observed differences, as usefully described by simple "strength of adhesion" tests, for example. Data from the simplest "strength of adhesion" test imaginable are shown in Fig. 3.8. The test surfaces, with their gravitationally settled and supposedly adsorbed cells in place on the protein-coated substrata of differing initial quality, were dropped from a standard height onto the laboratory bench top. This was immediately followed by microscopic inspection for the number of cells remaining after such "impulse detachment." The results in Fig. 3.8 show, in an unambiguous way, that the

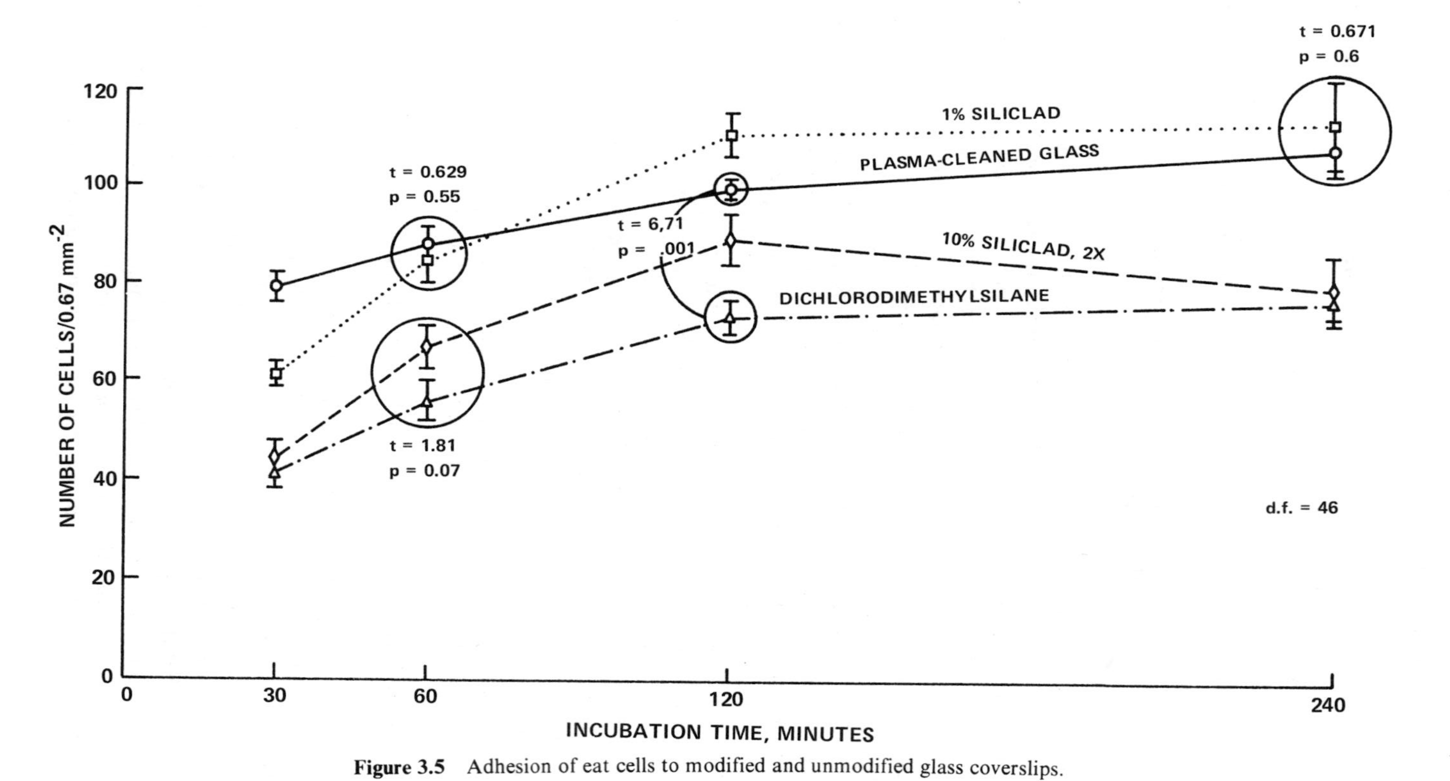

Figure 3.5 Adhesion of eat cells to modified and unmodified glass coverslips.

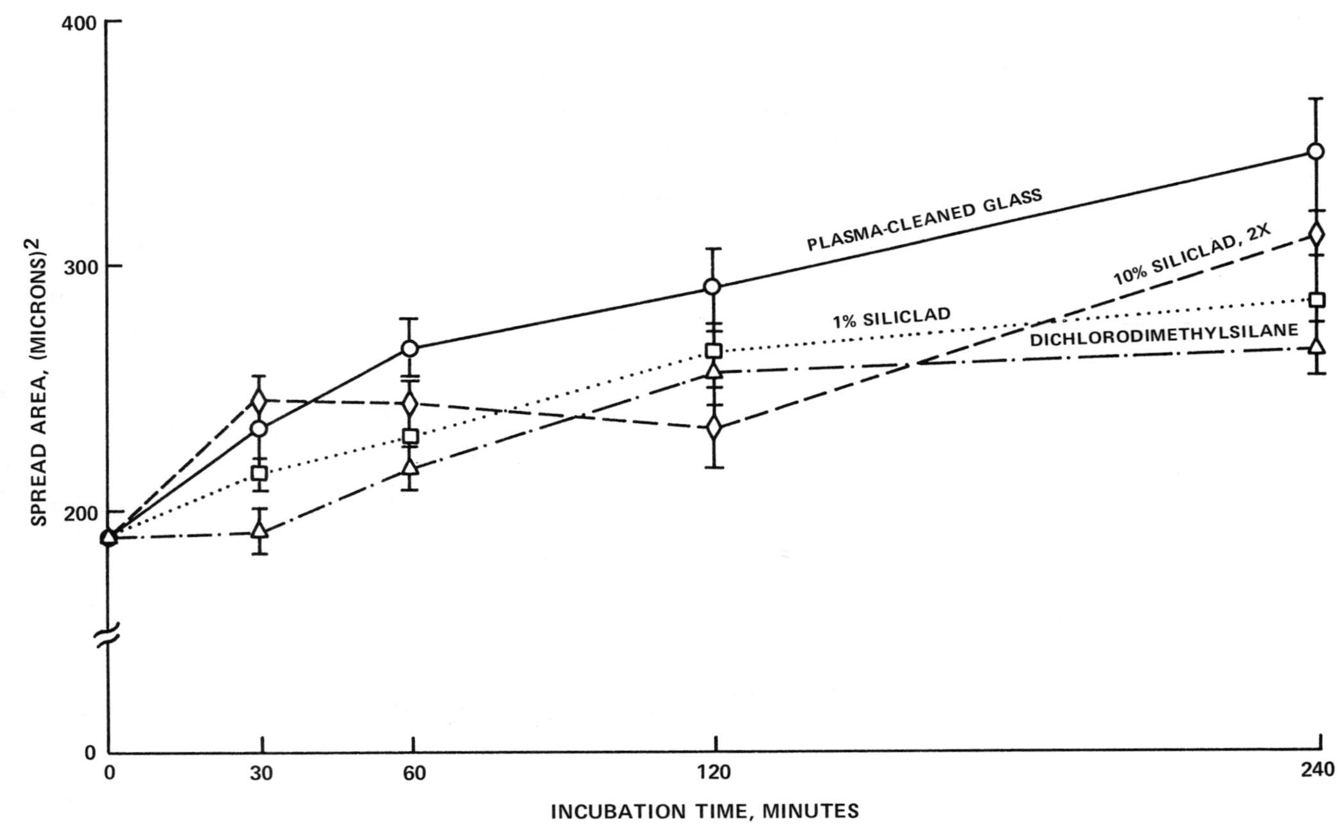

Figure 3.6 Spread areas of eat cells on the four types of substrates.

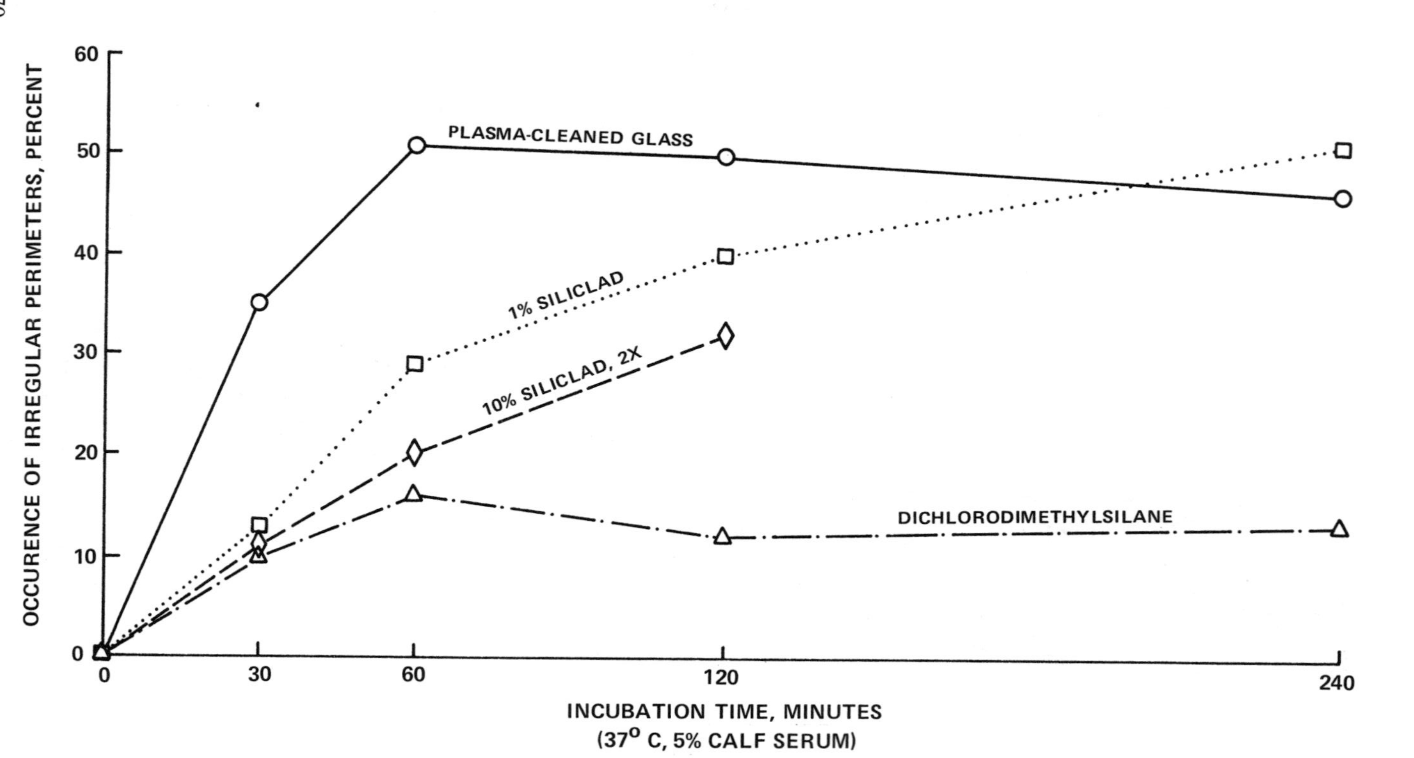

Figure 3.7 Percentage of eat cells having irregular perimeters on the four types of substrates, as function of incubation time.

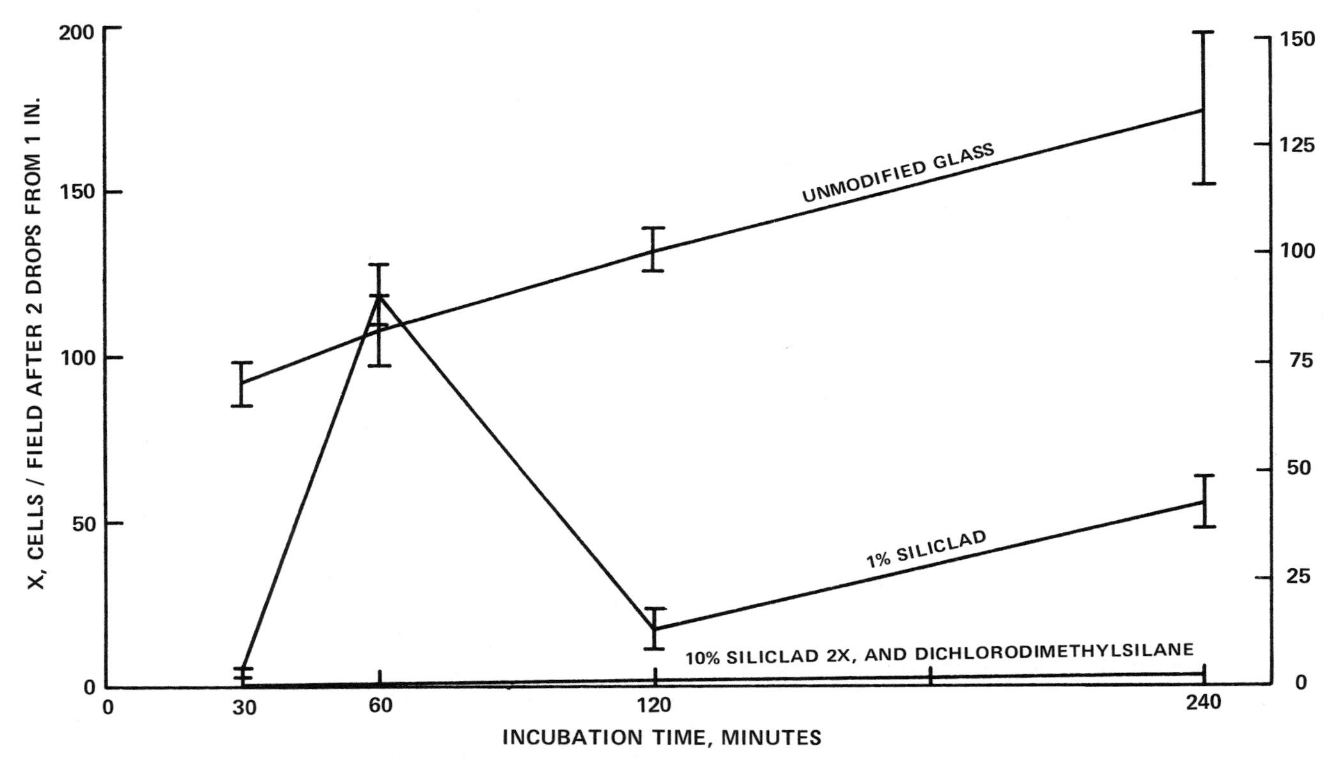

Figure 3.8 Impulse detachment of eat cells from test surfaces.

strength of cell adhesion as measured by the ability of the cells to remain in place after a very modest challenge (estimated to exceed no more than $\frac{1}{80}$th of a microdyne per cell) is highly dependent on the original substratum surface energy. Unmodified glass (alternatively, plasma-treated, or glow-discharge-treated glass) holds the cells originally arriving at its protein-coated surface most tenaciously; in fact, essentially quantitatively. Incomplete coatings on glass, produced with a 1% concentration of a siliconizing agent, allow significantly less early adhesive strength; but after 1 hr of incubation, an apparent "hold-fast" tendency almost equal to that of the unsiliconized glass develops. This increasing adhesive strength is attributed to "show through" of uncoated glass areas. Very well "methylated" and minimally bioadhesive substrata are produced by siliconization treatments using fundamentally different starting reagents. These substrata support essentially no cell adhesion when the substratum–cell bond is challenged by simple impulses or, as demonstrated in companion experiments, by modest increases in shear stress.

3.2.4 Generality of Findings

These background data strongly endorse results achieved previously in a completely different context, the cardiovascular system, regarding relative platelet adhesion, platelet spreading and distortion, and thrombogenic (fouling) tendencies. Thus it was predicted that the basic events of biological adhesion (fouling) in the sea, including initial colonization by microfouling organisms, as in bacterial slime films, and ultimate attachment of macrofouling organisms, such as barnacles, mussels, tube worms, and algae, would exhibit the same general dependency on critical surface tension [18]. Relatively strong adhesion and biofouling growth, in fact, was found for both the very low and high surface free energy (or critical surface tension) substrata, but minimal bioadhesion was noted in the zone of intermediate surface energies designated by critical surface tensions of 20–30 dynes cm^{-1} [19]. The recent work of Dexter, based on this hypothesis, verifies the prediction and reproduces the findings with bacteria adhering to various substrata [20–22].

Yet many questions remain before microbial adhesion to substrata of various engineering and synthetic types can be defined, because the same range of unknowns must be addressed as encountered in cell-adhesion studies. Namely, improved control is required for surface textural differences and improved definition must be developed for the relative influences of potentially differing amounts or kinds of spontaneously adsorbed biopolymers on the substrata tested. These latter films might mask, amplify, or simply transduce the substrata properties to the level at which microbial colonization first succeeds.

3.3 SPONTANEOUS ATTACHMENTS OF BACTERIA TO SMOOTH SUBSTRATA

3.3.1 Experimental Design

In cooperative experiments with Kevin Marshall (University of New South Wales, Kensington, Australia), information was sought on the spontaneous adsorptive events preceding and then influencing the adhesion of marine bacterial species in artificial seawater culture. Here just a small subset of the experimental results obtained is presented to illustrate the general features of these processes. The substrata used were equally smooth plates of essentially pure germanium, either cleaned by detergent washing to remove all organic debris or deliberately coated with a few molecular layers of polydimethylsiloxane, covalently bound to the germanium surface. Contact-angle measurements with exceptionally pure diagnostic wetting liquids were used to define the relative surface free energies, by the empirical descriptor of critical surface tension, and also by the mathematically dissected fractions of the total apparent surface free energy, that is, the polar and dispersion force components as previously described [23]. Briefly, the results showed that "clean" substrata had about equal contributions of dispersion- and polar-force-dominated sites at their original surfaces, whereas the siliconized ("methylated") substrata of essentially identical surface texture had their polar interfacial sites masked totally to provide a dispersion-force-dominated apparent surface free energy near 22 dynes cm^{-1}, at the lower end of the "minimally bioadhesive" range described earlier.

The analyses of surface-energy data presented later in Tables 3.1–3.4 require some further introductory remarks on the meaning and significance of the terms used. First, the critical surface tension γ_c is used as an empirically determined descriptor of the substratum that indicates which liquids will spontaneously wet and spread over its surface. Liquids with liquid–vapor surface tensions $\gamma_{lv} < \gamma_c$ will wet and spread over the substratum, whereas liquids with $\gamma_{lv} > \gamma_c$ will exhibit nonzero contact angles on the same surface. The slope, s, for the data used to determine γ_c is considered to be an empirical measure of the strength of the interaction between the substratum and any liquids placed on it. The slope is usually

Table 3.1 Typical contact-angle data acquired on replicate films of biologic materials, illustrating range and variability of measurements.

	ACTUAL CONTACT ANGLES (DEGREES) MEASURED FOR EACH LIQUID										
	WATER	GLYCEROL	FORMAMIDE	THIODIGLYCOL	METHYLENE IODIDE	SYM-TETRABROMO-ETHANE	1-BROMO-NAPHTHALENE	0-DIBROMO-BENZENE	α – METHYL NAPHTHALENE	DICYCLOHEXYL	n-HEXADECANE
WETTING LIQUIDS AND THEIR SURFACE TENSIONS (DYNES cm^{-1}) AT 20°C	72.8	63.4	58.2	54.0	50.8	47.5	44.6	42.0	38.7	33.0	27.7
AIR-DRIED FILMS OF INTACT TOBACCO MOSAIC VIRUS	62 66 62 61 49 48 46 48 PENETRATES QUICKLY	88 78 78 74 78 75 68 70 PENETRATES SLOWLY	57 58 60 56 45 48 49 48 PENETRATES QUICKLY	64 64 63 64 59 58 59 58 PENETRATES SLOWLY	58 59 58 59 55 57 57 55 55 56 55 55 STABLE	52 49 47 52 50 50 50 51 STABLE	44 43 42 43 39 38 44 43 STABLE	34 33 33 35 35 40 35 34 STABLE	32 28 28 28 26 30 30 31 STABLE	16 15 12 17 STABLE	0 0 0 0 SPREADS QUICKLY
AIR-DRIED FILMS OF RE-AGGREGATED BACTERIAL FLAGELLA	42 40 48 42 48 46 46 43 44 44 38 34 30 20 PENETRATES QUICKLY	70 72 74 72 66 70 70 68 60 64 64 64 62 64 63 64 PENETRATES SLOWLY	54 48 54 55 56 55 40 40 35 35 PENETRATES QUICKLY	52 50 50 52 48 48 50 50 50 48 51 50 47 48 48 45 PENETRATES SLOWLY	54 52 51 52 54 52 48 52 49 50 50 50 STABLE	46 44 46 44 44 44 STABLE	41 40 37 39 40 40 STABLE	40 39 38 38 36 34 STABLE	32 28 35 36 31 29 STABLE	18 18 17 14 STABLE	0 0 0 0 SPREADS QUICKLY

Table 3.2 Average contact angles (degrees) on uncontaminated surfaces of nutrient agars used for culturing microorganisms (completely hydrated).

WETTING LIQUIDS AND THEIR SURFACE TENSION (dynes cm^{-1}) AT 20°C		TRYPTICASE SOY AGAR	BLOOD AGAR	EMB AGAR	ENDO AGAR
WATER	72.8	9	17	8	4
GLYCEROL	63.4	18	15	10	–
THIODIGLYCOL	54.0	20	13	11	6
METHYLENE IODIDE	50.8	63	52	55	64
α-BROMO-NAPHTHALENE	44.6	59	50	44	57
1-METHYL-NAPHTHALENE	38.7	62	38	46	55
DICYCLOHEXYL	33.0	34	24	46	41
n-HEXADECANE	27.7	17	21	35	27
n-TETRADECANE	26.7	22	22	34	–
n-DODECANE	25.4	16	18	0	10
n-DECANE	23.9	0	0	0	0
CRITICAL SURFACE TENSIONS (dynes cm^{-1}) γ_C		24.1	23.3	22.0	23.5
SLOPE OF THE ZISMAN PLOT (cm $dyne^{-1}$)		-0.021	-0.014	-0.016	-0.022
DISPERSION FORCE COMPONENT OF TOTAL SURFACE ENERGY (dynes cm^{-1}) γ_D		29.0	26.1	23.8	26.3
POLAR COMPONENT OF TOTAL SURFACE ENERGY (dynes cm^{-1}) γ_P		2.4	7.0	8.6	2.3

given as a negative number, so that the more positive the slope, the stronger are the apparent interactions suggested between the substratum and any liquid applied to it.

The interaction of the test liquids (used to make contact angle measurements) with various substrata is then further resolved into terms called the "dispersion" and "polar" parts by dividing the measured liquid–vapor surface tensions of the test liquids into their dispersion γ_l^d and polar γ_l^p parts, and the calculated solid surface free energies into corresponding dispersion γ_s^d and polar γ_s^p parts. In many respects, γ_s^d and γ_s^p convey similar information as γ_c and s, respectively. The critical surface tension γ_c is numerically close to γ_l^d when the substratum–liquid interactions are primarily dominated by dispersion forces. Likewise, the more positive the

Table 3.3 Average contact angles (degrees) on confluent bacterial cultures in place on surfaces of nutrient agars (completely hydrated).

WETTING LIQUIDS AND THEIR SURFACE TENSIONS (DYNES cm^{-1}) AT 20°C		STREPTOCCUS FAECALIS ON TRYPTICASE SOY AGAR	STREPTOCCUS FAECALIS ON BLOOD AGAR	STAPHYLOCOCCUS ALBUS ON TRYPTICASE SOY AGAR	STAPHYLOCOCCUS ALBUS ON BLOOD AGAR	ESCHERICHIA COLI ON EMB AGAR	ESCHERICHIA COLI ON ENDO AGAR
WATER	72.8	13	11	0	12	13	14
GLYCEROL	63.4	14	20	15	21	19	15
THIODIGLYCOL	54.0	14	19	15	26	17	17
METHYLENE IODIDE	50.8	70	65	81	68	47	74
α-BROMONAPHTHALENE	44.6	71	56	78	69	43	70
1-METHYL NAPHTHALENE	38.7	70	49	78	67	35	49
DICYCLOHEXYL	33.0	46	29	67	58	16	33
n-HEXADECANE	27.7	0	26	44	29	0	28
n-TETRADECANE	26.7	0	19	43	32	0	20
n-DODECANE	25.4	0	11	30	15	0	10
n-DECANE	23.9	0	0	17	0	0	0
CRITICAL SURFACE TENSION (DYNES cm^{-1}) γ_c		27.1	24.2	21.0	23.7	28.8	24.4
SLOPE OF THE ZISMAN PLOT (cm $DYNE^{-1}$)		-0.040	-0.022	-0.034	-0.031	-0.015	-0.029
DISPERSION FORCE COMPONENT OF TOTAL SURFACE ENERGY (DYNES cm^{-1}) γ_D		29.6	25.4	20.7	23.8	33.1	24.0
POLAR COMPONENT OF TOTAL SURFACE ENERGY (DYNES cm^{-1}) γ_P		1.3	2.9	0.9	1.1	3.3	3.4

Table 3.4 Average contact angles (degrees) on air-dried confluent films of isolated bacterial cell walls, bacterial flagella, intact virus, and isolated virus components.

WETTING LIQUIDS AND THEIR SURFACE TENSIONS (DYNES CM^{-1}) AT 20°C		CELL WALLS OF BACILLUS MEGATERIUM DRIED AT BOTH 5°C AND 80°C, ON PLATINUM FROM WATER SUSPENSION	CELL WALLS OF MICROCOCCUS LYSODEIKTECUS DRIED AT BOTH 5°C AND 80°C, ON PLATINUM FROM WATER SUSPENSION	CELL WALLS OF ESCHERICHIA COLI AIR DRIED AT 5°C, ON PLATINUM, FROM WATER SUSPENSION	CELL WALLS OF ESCHERICHIA COLI AIR DRIED AT 80°C, ON PLATINUM FROM WATER SUSPENSION	RE-AGGREGATED BACTERIAL FLAGELLA AIR DRIED AT BOTH 5°C AND 80°C ON PLATINUM FROM WATER SUSPENSION	INTACT TOBACCO MOSAIC VIRUS AIR DRIED AT BOTH 5°C AND 80°C ON PLATINUM FROM WATER SUSPENSION	TOBACCO MOSAIC VIRUS PROTEIN AIR DRIED AT BOTH 5°C AND 80°C ON PLATINUM FROM WATER SOLUTION	TOBACCO MOSAIC VIRUS PROTEIN AIR DRIED AT BOTH 5°C AND 80°C ON PLATINUM FROM 10^{-5}M KOH SOLUTION
WATER	72.8	26	25	35	58	41	56	67	75
GLYCEROL	63.4	67	66	73	87	67	76	67	77
FORMAMIDE	58.2	25	43	55	60	48	53	67	67
THIODIGLYCOL	54.0	42	43	62	78	49	61	54	60
METHYLENE IODIDE	50.8	45	46	52	64	51	57	47	53
SYM-TETRABROMOETHANE	47.5	37	37	47	61	45	50	43	48
α-BROMONAPHTHALENE	44.6	30	30	41	55	39	42	36	39
O-DIBROMOBENZENE	42.0	26	28	37	54	37	35	30	33
1-METHYL NAPHTHALENE	38.7	17	19	30	50	32	29	25	30
DICYCLO HEXYL	33.0	7	0	0	35	17	15	12	11
n-HEXADECANE	27.7	0	0	0	24	0	0	0	0
n-DECANE	23.9	0	0	0	0	0	0	0	0
CRITICAL SURFACE TENSION (DYNES CM^{-1})		36.7	36.1	33.3	24.5	30.8	32.1	33.4	32.3
SLOPE OF THE ZISMAN PLOT (CM $DYNE^{-1}$)		-0.020	-0.020	-0.024	-0.024	-0.018	-0.023	-0.0019	-0.022

slope of the contact angle plot that leads to γ_c, the stronger is the polar bonding character at the substratum surface.

3.3.2 Test Substrata

The choice of germanium as the base material for these experiments was primarily because it could serve as a "light pipe" for multiply internally reflected IR waves and could be employed for the determination of the internal reflectance IR spectra of the first intimately adhering organic layers on the substrata surfaces. Further, in its "clean" state, it presented a distribution of surface-energy-influencing components similar to that expected for other intrinsically "high-energy" inorganic substrata, such as glass, ceramics, metals, and metal oxides [6,7] and, in its siliconized version, simulated the best of the polymeric or plastic materials that might be used in construction of biomedical, dental, or oceanographic devices where bacterial adhesion could be anticipated.

3.3.3 Results of Exposure to a Bacterial Culture

Figure 3.9 presents multiple attenuated internal reflection IR spectra of "clean" germanium substrata (called "prisms" because of their use as optical reflecting elements in the sample beam of an IR spectrophotometer), after exposure for as little as two hours and as long as two days to a bacterial culture in artificial sea water. Within 2 hr, an abundant proteinaceous film spontaneously adsorbed to the substratum surface (Fig. 3.9*a*). Vigorous rinsing of the prism with distilled water was not sufficient to remove the material completely, nor to cause extraction and/or detachment of anything but salts and loosely bound polysaccharide components of the film (Fig. 3.9*b*). Replicate specimens of clean germanium substrata exposed for two days to the same bacterial culture acquired significantly thicker interfacial deposits of essentially the same material (Fig. 3.9*c*).

Figure 3.10 provides scanning electron microphotographs of originally "clean" metallic (i.e., germanium) substrata, the same samples as characterized in Fig. 3.9, after 2-hr or 2-day exposure to marine bacteria in artificial seawater culture. These electron micrographs reveal that the scattered bacteria present in the attached film at 2 hr were completely embedded in a matrix of adsorbed or exuded glycoproteinaceous matter, covering the plates rather uniformly. The replicate specimens, incubated in the same media for 2 days, showed a considerably greater abundance of adsorbed bacteria, embedded still in a coherent biopolymeric film only moderately thicker than that on the 2-hr specimens.

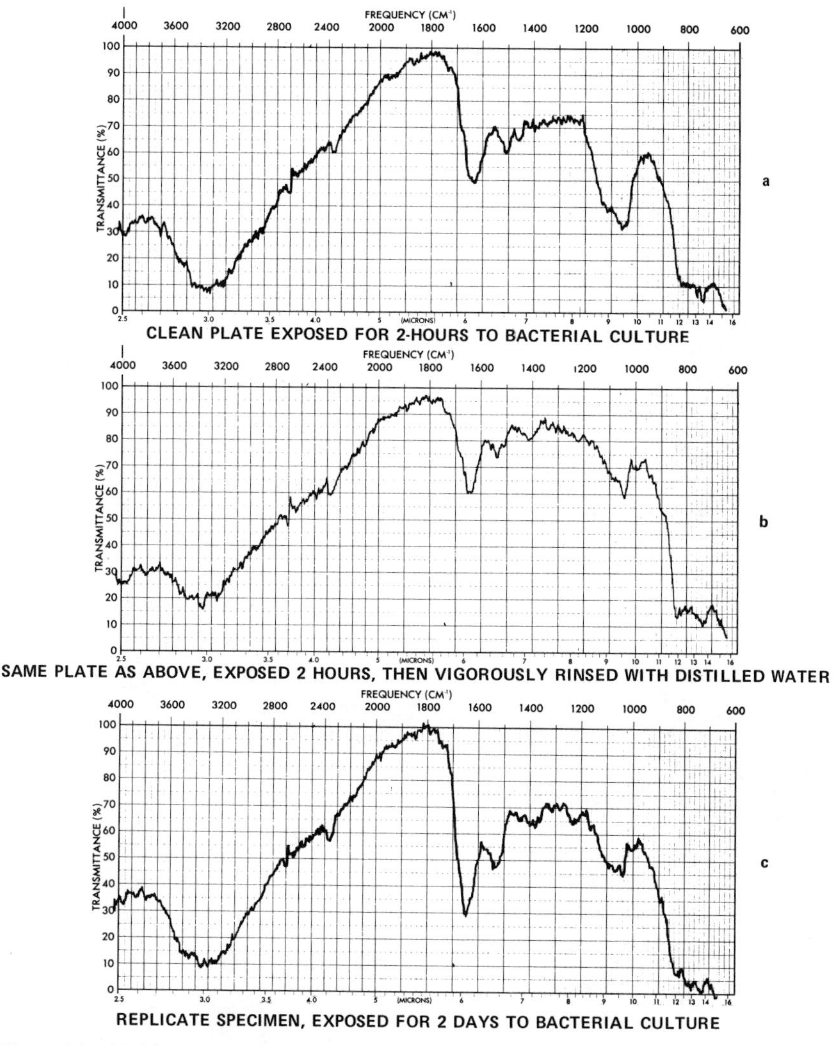

Figure 3.9 Multiple attenuated internal reflection IR spectra of clean germanium prisms after exposure for 2 hr and for 2 days to bacterial culture (luminous bacterium, M_{43}) in artificial seawater and glucose at 28°C. Source: K. C. Marshall, Australia.

Figure 3.11 provides multiple attenuated internal reflection IR spectra of well-siliconized (methylated), low surface energy substrata, exposed for 2 hr to the same cell culture at the same time as the "clean" substrata of germanium were exposed. Included are spectral traces for the same plate in its unexposed control condition (Fig. 3.11*a*), as withdrawn from the bac-

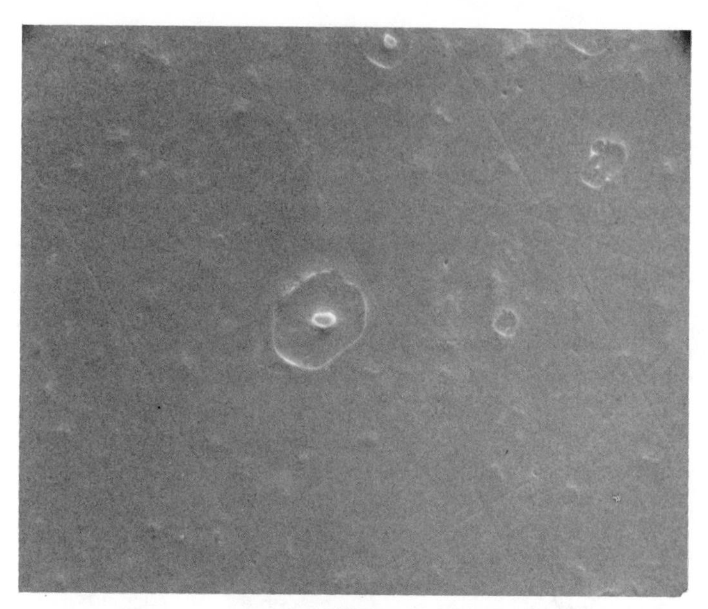

1200X
EXPOSED FOR 2 HOURS

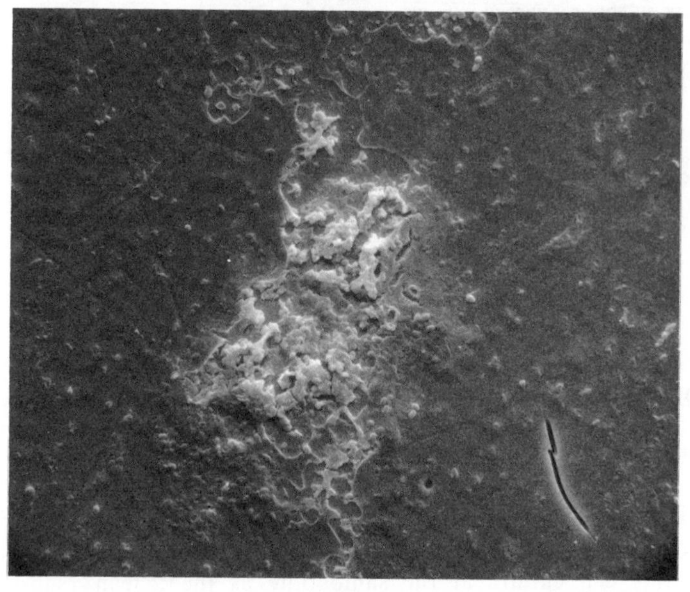

1200X
EXPOSED FOR 2 DAYS

Figure 3.10 Scanning electron micrographs of clean metallic (germanium) substrates after exposure for 2 hr and 2 days to luminous bacterium (M_{43}) in artificial seawater and glucose at 28°C. Source: K. C. Marshall, Australia.

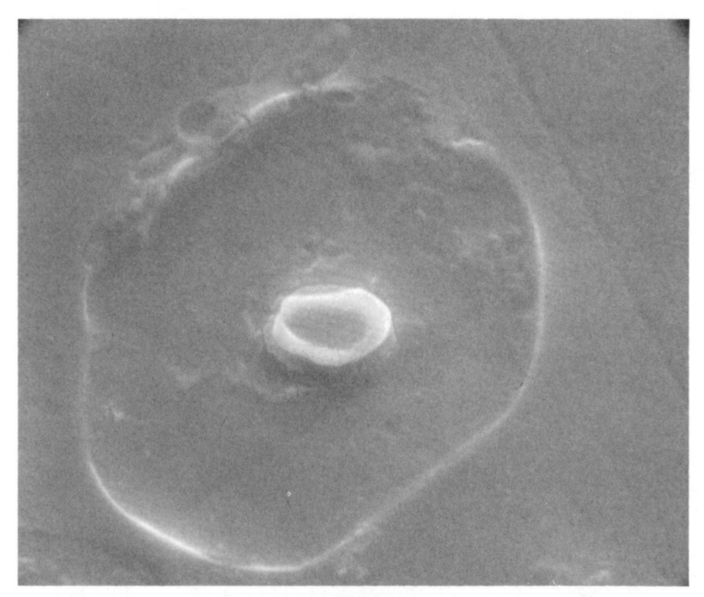

6000X
EXPOSED FOR 2 HOURS

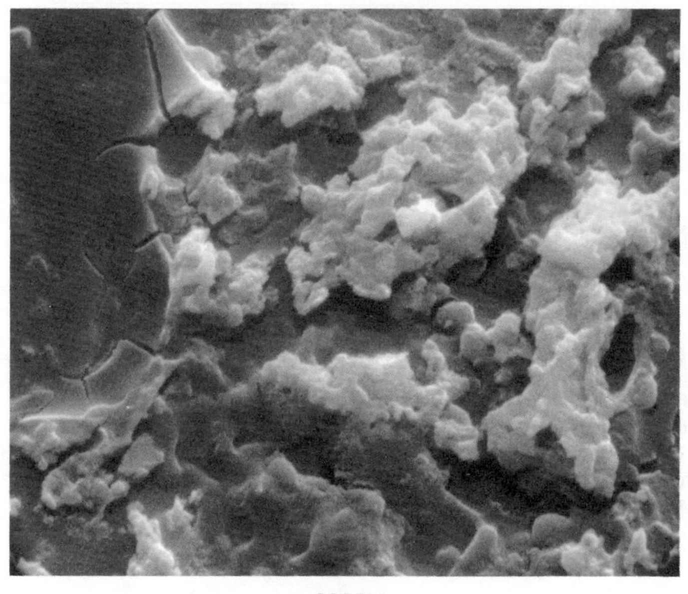

6000X
EXPOSED FOR 2 DAYS

(Continued)

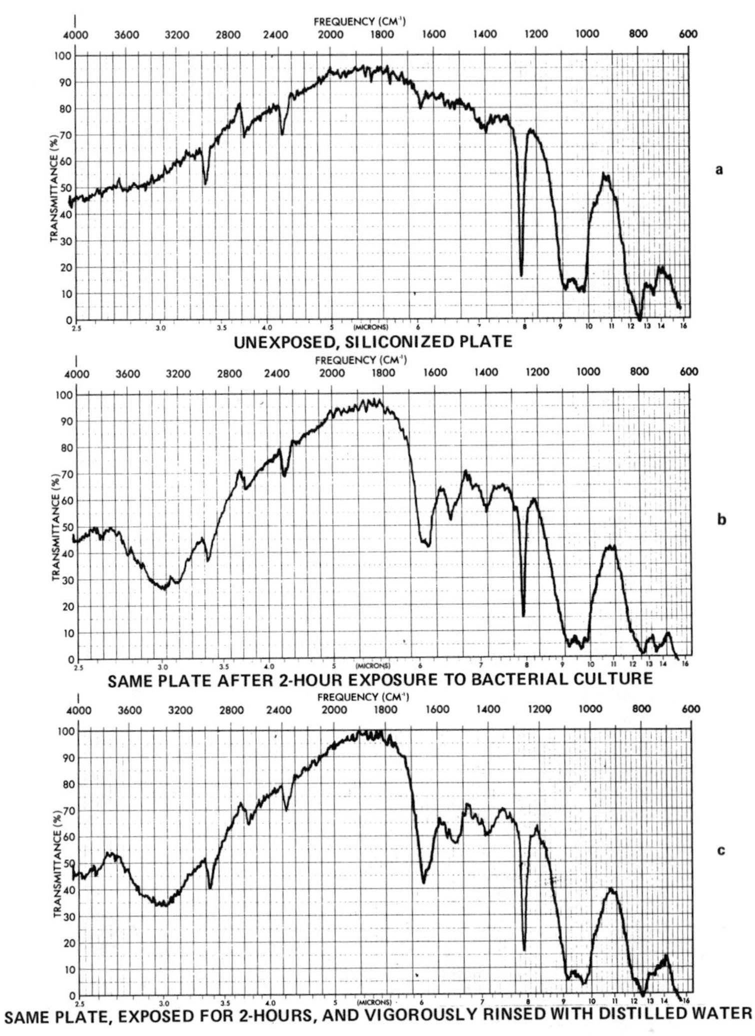

Figure 3.11 Multiple attenuated internal reflection IR spectra of unexposed, well-siliconized (methylated) germanium prism and same prism after 2 hr exposure to bacterial culture (luminous bacterium, M_{43}) in artificial seater and glucose at 28°C. Exposure of replicate specimen for 2 days did not lead to substantial increase in adsorbed matter. Source: K. C. Marshall, Australia.

terial culture system (Fig. 3.11*b*), and after vigorous rinsing with distilled water (Fig. 3.11*c*). Note in Fig. 3.11*a* the spectral absorption bands at 1260 cm^{-1}, 1090 cm^{-1}, and 1020 cm^{-1} typical of the polydimethylsiloxane coating covalentally bound to germanium–oxygen linkages at the substratum surface. Note also the persistence of these bands, in both position and intensity, for the test plate as withdrawn from 2 hr of incubation in the bacterial culture medium (Fig. 3.11*b*), and even after vigorous final rinsing with distilled water (Fig. 3.11*c*). Note further, comparing Figs. 3.11*b* and 3.11*c*, that the attached material on the substratum surface was of the same protein-dominated character found on the "clean" substrata of the same base material, as well as the excellent resistance to rinsing and/or detachment. As shown for the "clean" substrata in Fig. 3.9, the siliconized, low-energy substrata continued to accumulate modest additional amounts of film through 2 days of incubation in the same bacterial culture, at which point the incubation was terminated. Spectral traces obtained for specimens incubated for this longer period were not qualitatively different from that in Fig. 3.11*b*; quantitative changes were documented by slightly more intense IR absorption bands characteristic of the bound biological polymeric matrix.

Typical scanning electron micrographs of the bacterial and adsorbed macromolecular film components on the well-siliconized, low-surface-energy substrata after their exposure for both 2 hr and 2 days to artificial seawater culture of a marine bacterium are shown in Fig. 3.12. In contrast to the illustrations provided in Fig. 3.10 for the companion "clean" substrata, the adsorbed bacteria, although present in roughly the same abundance at 2 hr and 2 days on the siliconized substrata as on the "clean" substrata, are more clearly visualized. They appear to be residing on, rather than being embedded within, the uniformly dispersed glycoproteinaceous matrix. Given Dexter's results [20–22] that bacterial populations adsorbed to low energy surfaces of about the same critical surface tension have the poorest rate of success in surface colonization, these scanning electron micrographs confirm the influence of original substratum surface properties on the spreading, binding, and embedment of the bacteria within the surrounding "slime" film, rather than on the number density of originally adsorbed organisms.

3.3.4 Implications for Biological Fouling

It seems that original substrata surface properties, and particularly those properties experimentally described by contact-angle data and parameters of relative surface free energy, can and do influence the nature, rate, and

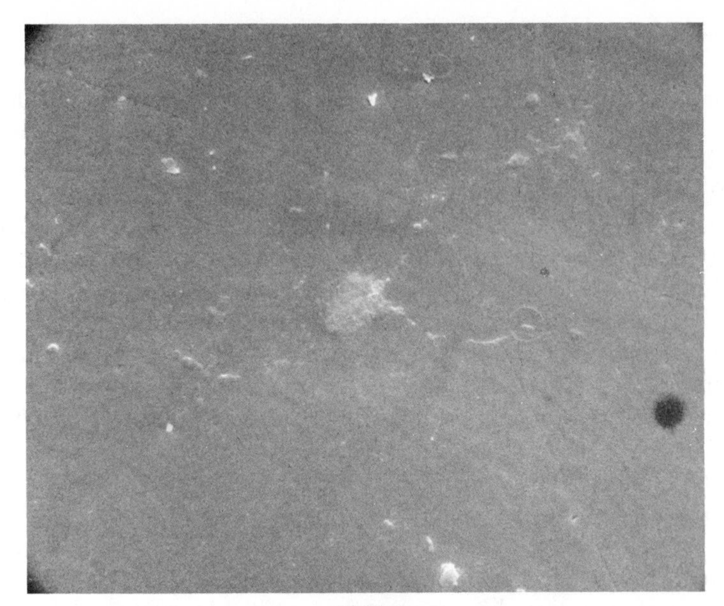

1200X
EXPOSED FOR 2 HOURS

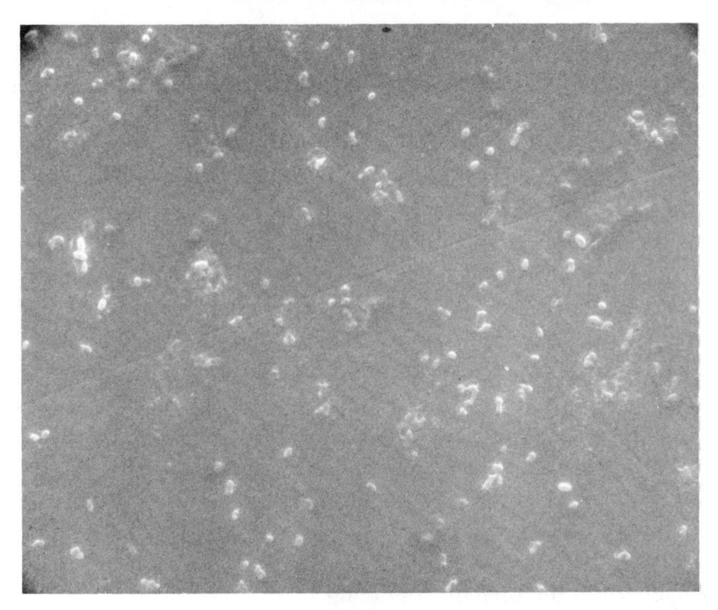

1200X
EXPOSED FOR 2 DAYS

Figure 3.12 Scanning electron micrographs of well-siliconized (methyl-group dominated) metallic substrates after exposure for 2 hr and for 2 days to luminous bacterium (M_{43}) in artificial seawater and glucose at 28°C. Source: K. C. Marshall, Australia.

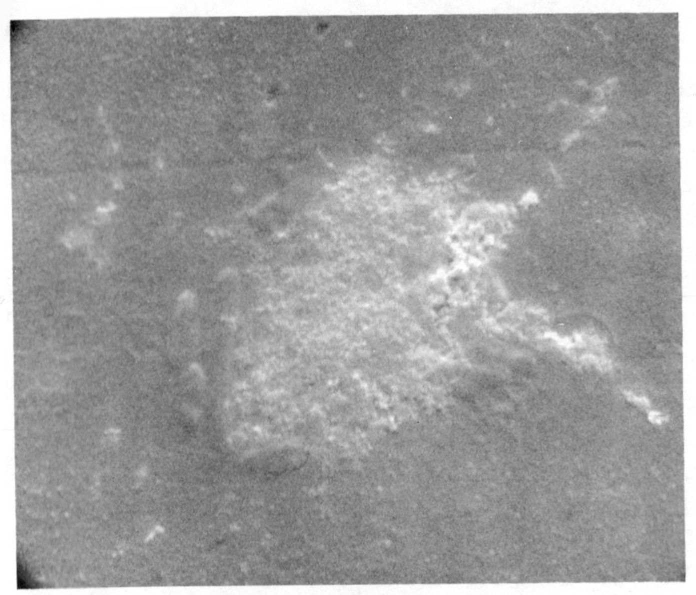

6000X
EXPOSED FOR 2 HOURS

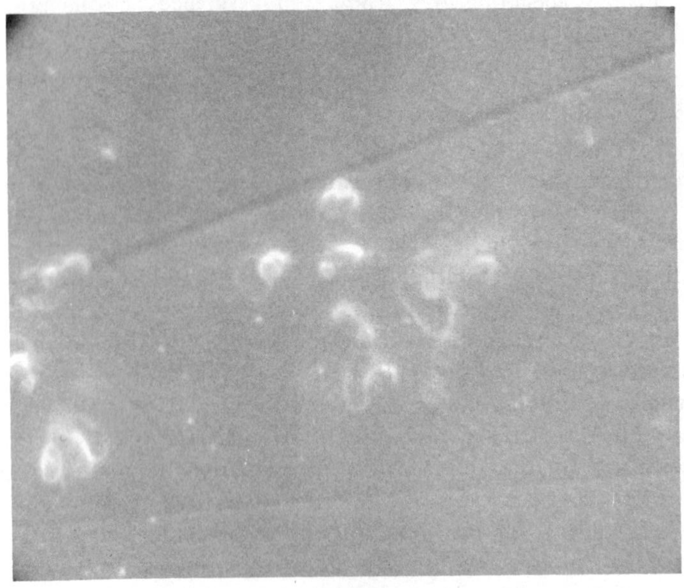

6000X
EXPOSED FOR 2 DAYS

(Continued)

colonization success of bacteria spontaneously adsorbed . logical circumstances. In most cases continued colonization strata in contact with favorable culture media leads to undesira effects, such as the degradation of heat-exchange capacity in ther converters proposed for deployment in the tropical oceans [24–. thriving layers of adsorbed microorganisms seem to favor the bui climax communities of other organisms and, ultimately, the gross fo with macroscopic bioadhesive clusters of barnacles, mussels, and mentous algae.

The strength of adhesion of these fouling layers and, in turn, of coincident mineralized deposits on the same surfaces must be controlled by the new surface properties of the bacterial–micronutrient matrices to which they have attached, since the original substrata surface properties are effectively masked by the adsorbed bacterial–slime films by that time. If one desires to decontaminate such surfaces or to create scrupulously clean substrata in ocean-dwelling heat exchangers, for example, one requires details of the new surface chemistry, surface texture, and relative wettabilities of the bacterial cell surfaces. These issues already have been addressed in the context of cleaning plaque accumulations from dental surfaces by using surface chemical displacing agents [27,28].

3.3.5 Additional Data Required

In other environments it is important to know the actual nature of the new surface properties displayed by adsorbed microorganisms, their cell walls, and their associated exudates. There are numerous cases in which residual microbial surface contamination has been related to the transmission of disease, such as in improper sanitization of food processing and distribution equipments, in hospital facilities, and in devices utilized by many individuals subject to cross contamination [29].

The remainder of this chapter, therefore, is dedicated to the description of applicable surface science techniques for inspection and reproducible characterization of the interfacial properties of microorganisms, their external coats, and/or remnants, in both living and dead preparations. Specific attention is given to the experimental requirements for valid measurements and interpretations of the data acquired by contact-angle techniques, recognizing the increasing interest in such direct measurements of cell and microorganism surface properties [30–32] and the difficulties expected when only limited data are available for intrinsically complex, water-swollen, metabolically active cells in culture.

3.4 SURFACE PROPERTIES OF ADHERENT MICROORGANISMS AND THEIR CELL WALLS

3.4.1 Specimen Preparation and Analytical Techniques

Schematic drawings (Fig. 3.13) show the typical methods by which monolayer cultures of living microorganisms are grown to confluence on nutrient gels and by which their relative surface qualities are measured by simple contact-angle observations using extremely pure diagnostic liquids. The method by which nondestructive analysis of the actual surface chemical constituents present at the outermost face of such microorganism "sheets" is accomplished, using an auxiliary "prism" of IR-conducting material in the internal reflection spectroscopic mode, is illustrated in Fig. 3.14. Specifications of the internal reflection materials applicable for this technique are provided in the figure legend. Note that, after removal of any biological sample from the prism surface, remaining deposits, cellular debris, or cellular exudates can be sensitively analyzed further. A schematic diagram suggesting how seven nondestructive, analytical methods are then applied, in turn, to such surface films is presented in Fig. 3.15. All these techniques have demonstrated sensitivity to matter deposited in even submicrogram amounts. These techniques range through determination of the thickness, refractive index, orientation of electrical dipoles, surface texture and morphology, and association with inorganic elements and specific crystalline compounds of the one film. All the instrumental methods noted in Fig. 3.15 can be applied directly, without need for manipulation, extraction, or even contact with the surface film. The final test applied is usually the measurement of the actual wettability (or operational surface free energy) of the deposited matter by the only method to contact the film in question, the placement of multiple small droplets of purified diagnostic liquids.

3.4.2 Demonstration of Applicability of Methods to Biological Surfaces

Surface-specific IR Techniques. Multiple attenuated internal reflection IR spectra revealing the surface composition of Bacto Endo agar (Difco) and of a confluent monolayer culture of *E. coli* in place on that same agar, as analyzed in various states of hydration, are given in Fig. 3.16. These spectra clearly indicate the ability of internal-reflection spectroscopic methods to sense and identify bacterial films, even in the presence of the intimately associated, subjacent fully hydrated nutrient gel that supports the microbial growth. Further, certain of the IR spectra in Fig. 3.16 reveal the nature of the components transferred to the analytical internal-reflection

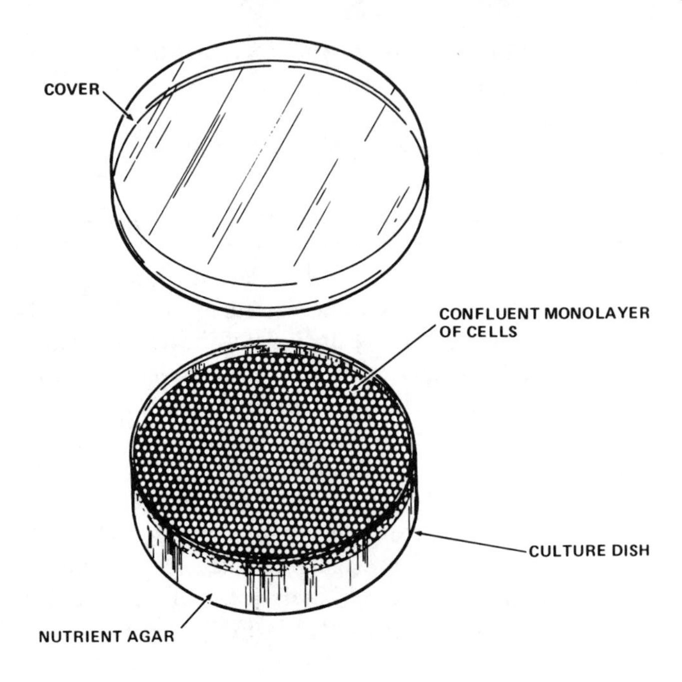

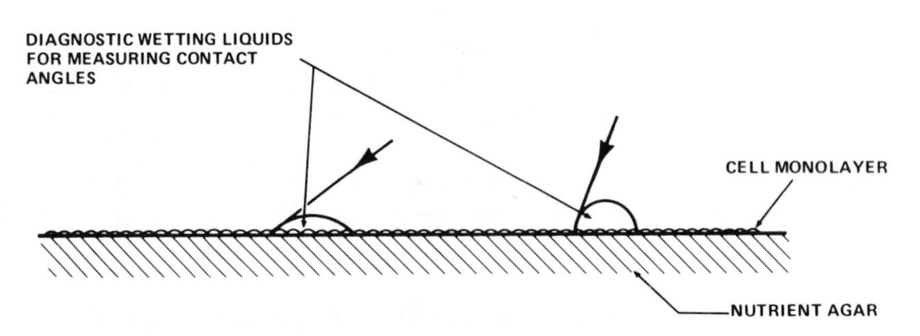

Figure 3.13 Schematic views of (top) monolayer cultures of living cells on nutrient agar and (bottom) varying contact angle profiles that reveal the relative adhesive characteristics of cell surface layers.

prisms during the contact period required for acquisition of the other spectra. Again, there is clear evidence of the presence and identity of the bacterial components in the interfacial layer, as easily differentiated from the agar-gel support structure. With this knowledge of the average chemistry and cellular integrity of the microbial monolayers, one can

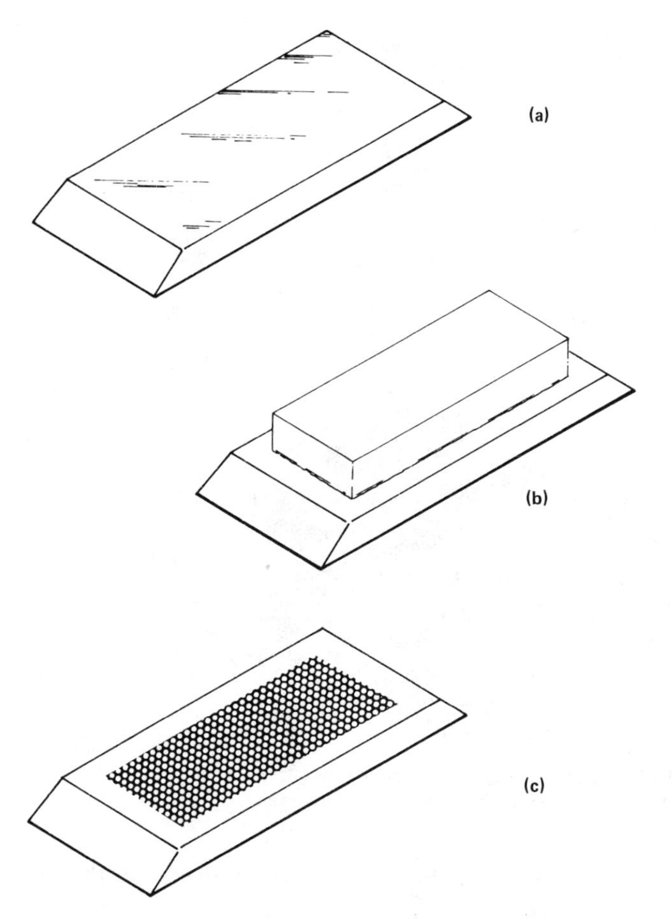

Figure 3.14 Illustration of placement of agar culture segment on face of internal reflection prism, for nondestructive analysis of molecular or cellular constituents present at contact surface: (a) a trapezoidal prism, slightly smaller than glass microscope slide and constructed of IR-transmitting salts (AgCl, AgBr, KRS-5, etc.) or IR-transmitting, reflective metals (germanium, silicon, etc.) is typically used; (b) agar segment placed in intimate contact with one or both faces of the trapezoidal prism.—a pressure-augmenting clamp can be used to increase degree of surface contact; (c) after removal of agar, any transferred residues (cultured cells, lipid deposits, etc.) can be further analyzed by techniques illustrated in Fig. 3.15. Alternatively, deposited films acquired by direct contact of prisms with cells or biological fluids can be sensitively analyzed.

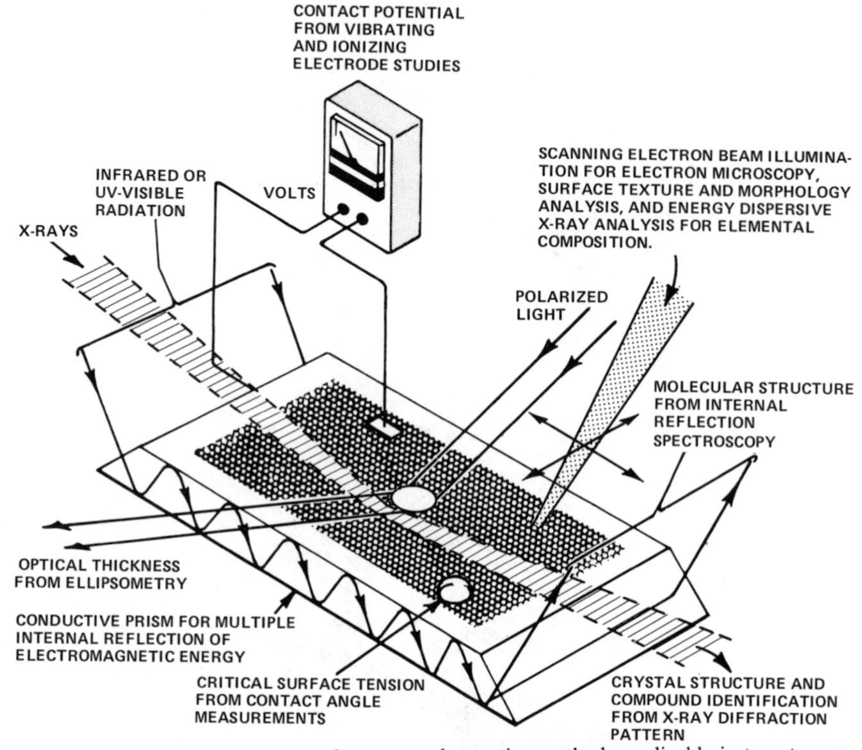

Figure 3.15 Schematic diagram of seven nondestructive methods applicable in turn to same surface deposit, with sensitivity at submicrogram level.

proceed with confidence to direct measurements of the resultant new wetting properties.

Contact-angle Measurements. It has been just over 10 years since the method of contact-angle measurement and critical-surface-tension determination was introduced to the complex substrata provided by living, *in situ* and freshly dissected biological tissues [33,34]. Since then the application of contact-angle methodology to assessing surface-energy parameters of microorganisms and other cells has become quite popular. Unfortunately, with only occasional exceptions, the contact-angle method has been applied in a very qualitative, uncritical manner, not reliably supporting the quantitative conclusions that the investigators have drawn from their wetting data. A major source of the difficulty with most investigations reported on biological surfaces has been the absolute dependence on the contact angles measured for water and water-miscible liquids. The supposition has been that these fluids would do the biological structures under analysis the least harm, erroneously dismissing from consideration the enormous variability intrinsic to such measurements because of the penetration, swelling, component extraction, and general instability of aqueous fluids placed on water-based substrata.

To demonstrate the expected data scatter that must be accounted for when applying contact-angle techniques to biological films, typical contact-angle data acquired on replicate films of biological materials are listed in Table 3.1 to illustrate the range of values obtained. Note that, in many cases, the variability ranges to as much as 30° for water on these films, but can be as little as 5° for water-immiscible fluids on the same surfaces. When these values are plotted in the typical Zisman format [35] (see Fig. 3.17), with every data point entered, it is found that critical surface tension intercepts of good reliability are obtained *only* if one ignores the spurious values given by water and formamide, both liquids exhibiting rapid penetration of the test film and both having high hydrogen-bond capacities as well as small molecular sizes. Thus in situations where one desires a judgment of the actual operational surface free energy of biological films or of surfaces of single cells (e.g., microorganisms occupying heat-exchange surfaces), with regard to their potential interactions with their neighbors or their resistance to organic cleaning fluids, contact-angle data obtained with water and other aqueous fluids can only be misleading. The extreme hydrophilic nature of such biological surfaces belies the fact that their average interactions are those of moderately high surface-energy biopolymers, typically giving critical surface-tension intercepts in the range 30–40 dynes cm^{-1}.

3.4.3 Results for Typical Microbial Surfaces

With these cautionary remarks in mind, one can turn in earnest to the subject of assessing the actual surface properties of confluent microbial films. The average contact angles measured on uncontaminated surfaces of the nutrient agars used for culturing the microorganisms are given in Table 3.2, along with the extrapolated critical surface tensions and slopes of the Zisman plots and the calculated dispersion force and polar force components of the total surface free energy as calculated by published methods [36,37]. The average contact angles measured on living, confluent *in situ* bacterial cultures on the surfaces of the same nutrient agars are given in Table 3.3. Again, the table provides critical surface-tension values extrapolated from the contact-angle values measured, the slopes of the empirical Zisman data plots, and the calculated dispersion and polar components of the total surface free energy. A selection of these data is

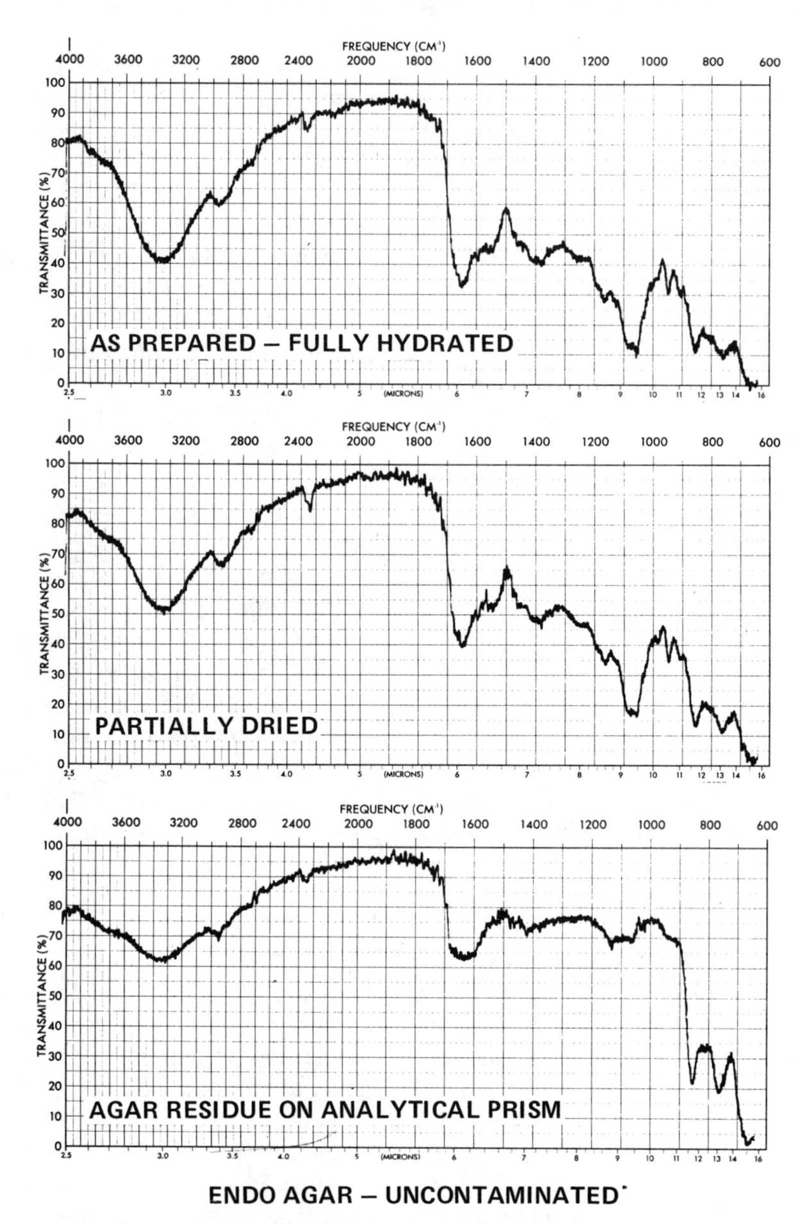

Figure 3.16 Multiple-attenuation internal reflection IR spectra revealing surface composition of endo agar and confluent monolayer culture of *Escherichia coli* grown on same nutrient agar.

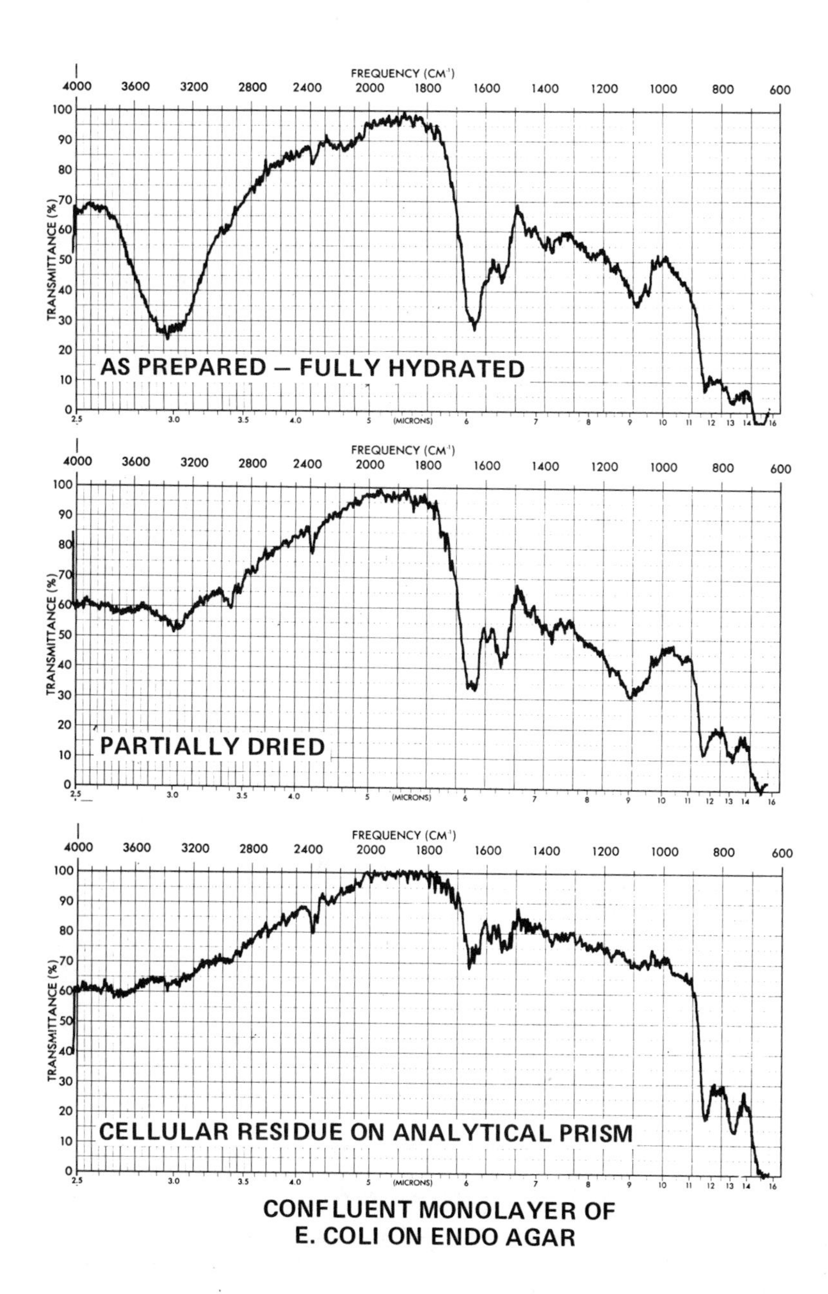

CONFLUENT MONOLAYER OF
E. COLI ON ENDO AGAR

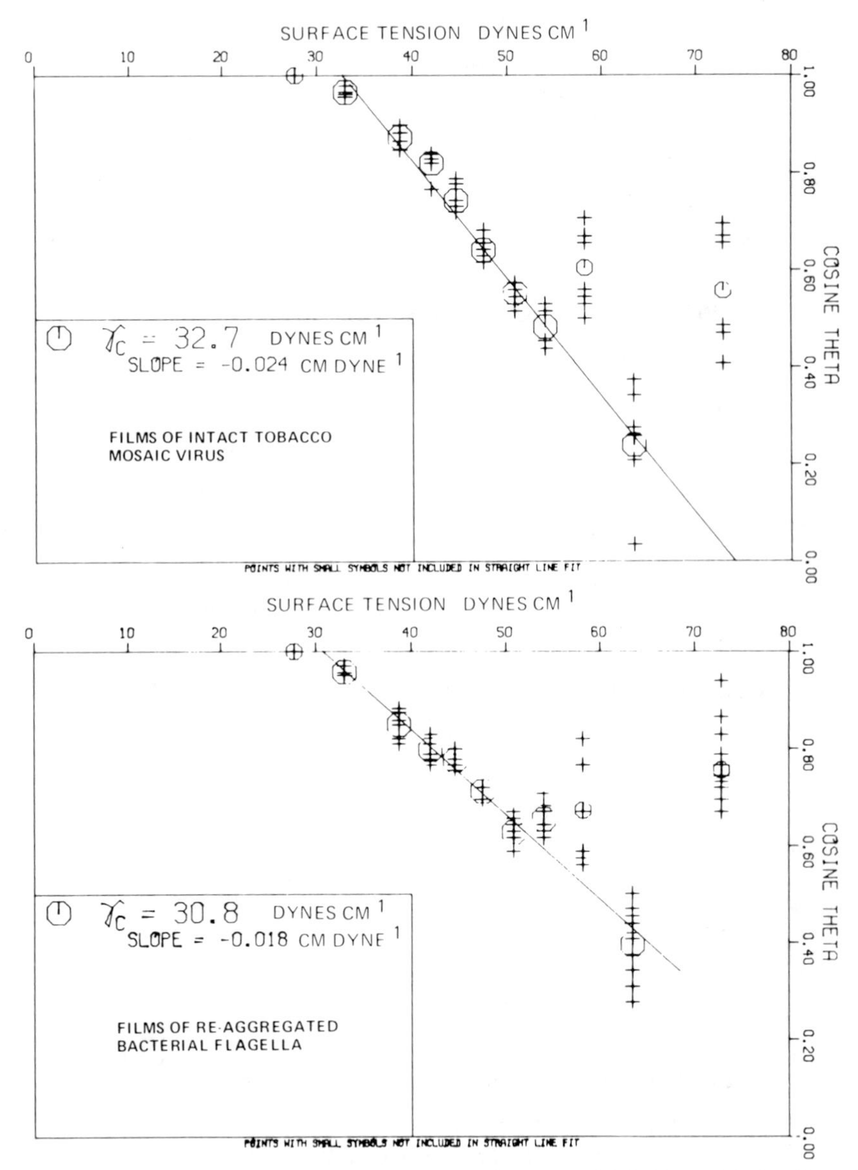

Figure 3.17 Zisman plots of typical contact-angle data (recorded in Table 3.1) for films of biologic materials. Note that every value measured is plotted, indicating range and variability of separate values for each wetting liquid.

presented in the plots collected in Fig. 3.18, which reveal the differential wettability of variously prepared surfaces of a single microorganism, *E. coli*, in comparison with the surface properties of an uncontaminated nutrient agar. It can be noted that the bacterial monolayer actually increased the contact angles obtained for most of the test liquids, indicating a lowering of the composite surface free energy while maintaining a critical surface-tension intercept in the zone 20–30 dynes cm^{-1} that typifies most thriving biological matter in its naturally hydrated form. The separated, isolated, and air-dried cell walls of these organisms, present in a thin confluent layer on a freshly cleaned platinum support plate, showed results more typical of those of the biopolymers of which most biomass is constructed (see Fig. 3.18*c*). The critical surface-tension value of 30–40 dynes cm^{-1} is, as often noted, typical for proteins, glycoproteins, proteoglycans, and like structures. On the other hand, heat denaturation of the isolated cell walls of *E. coli* does cause an easily perceived structural and surface chemical change. The contact-angle data plot (Fig. 3.18*d*) shows that, for the heat-denatured, air-dried, dehydrated, cell walls of *E. coli* in thin films on clean platinum support plates the critical surface-tension intercept is in the zone 20–30 dynes cm^{-1}, mimicking better than the undenatured cell walls the actual surface properties dominant in the living hydrated state.

Figure 3.19 provides multiple attenuated internal reflection IR spectra of these isolated cell walls of *E. coli*, as air dried from water suspension onto platinum foils at refrigerator temperature and heat denatured at 80°C. Note the apparent shift in the amide I and amide II absorption bands (1500–1700 cm^{-1}), which is associated with the heat denaturation and was first signaled by the dramatic change in contact-angle behavior on these same surfaces. It is speculated that heat denaturation of these bacterial cell walls caused an enrichment in the outermost layer of the air-dried films of the aliphatic hydrocarbon moieties of the structure, presenting an interface of closely packed methyl groups for the most part and blinding the external environment to the higher-energy protein backbone and glycoprotein sidechain components.

With the exception of the *E. coli* cell walls isolated in water suspension, there was no susceptibility of other cell-surface components to heat denaturation demonstrated in our experimental program. For example, Table 3.4 lists the average contact angles on air-dried, confluent films of isolated bacterial cell walls of *B. megaterium* and *M. lysodekticus*, in addition to those of *E. coli*, as well as reaggregated bacterial flagella, intact tobacco mosaic virus (TMV), and isolated tobacco virus proteins in water and alkaline solutions, as formed at refrigerator temperatures near 5°C and at increased, potentially heat-denaturating, temperatures near 80°C. With the sole excep-

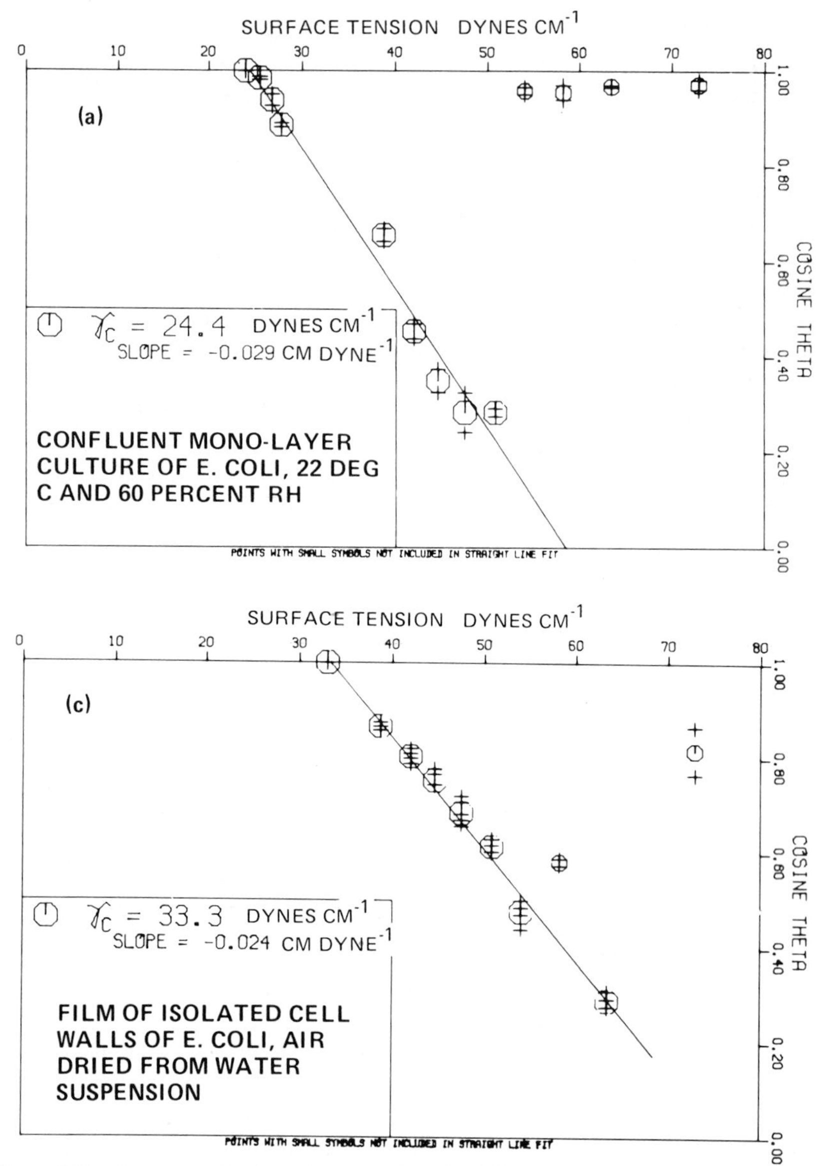

Figure 3.18 Contact-angle data plots revealing the differential wettability of variously prepared surfaces of *Escherichia coli*, compared with surface of uncontaminated nutrient agar.

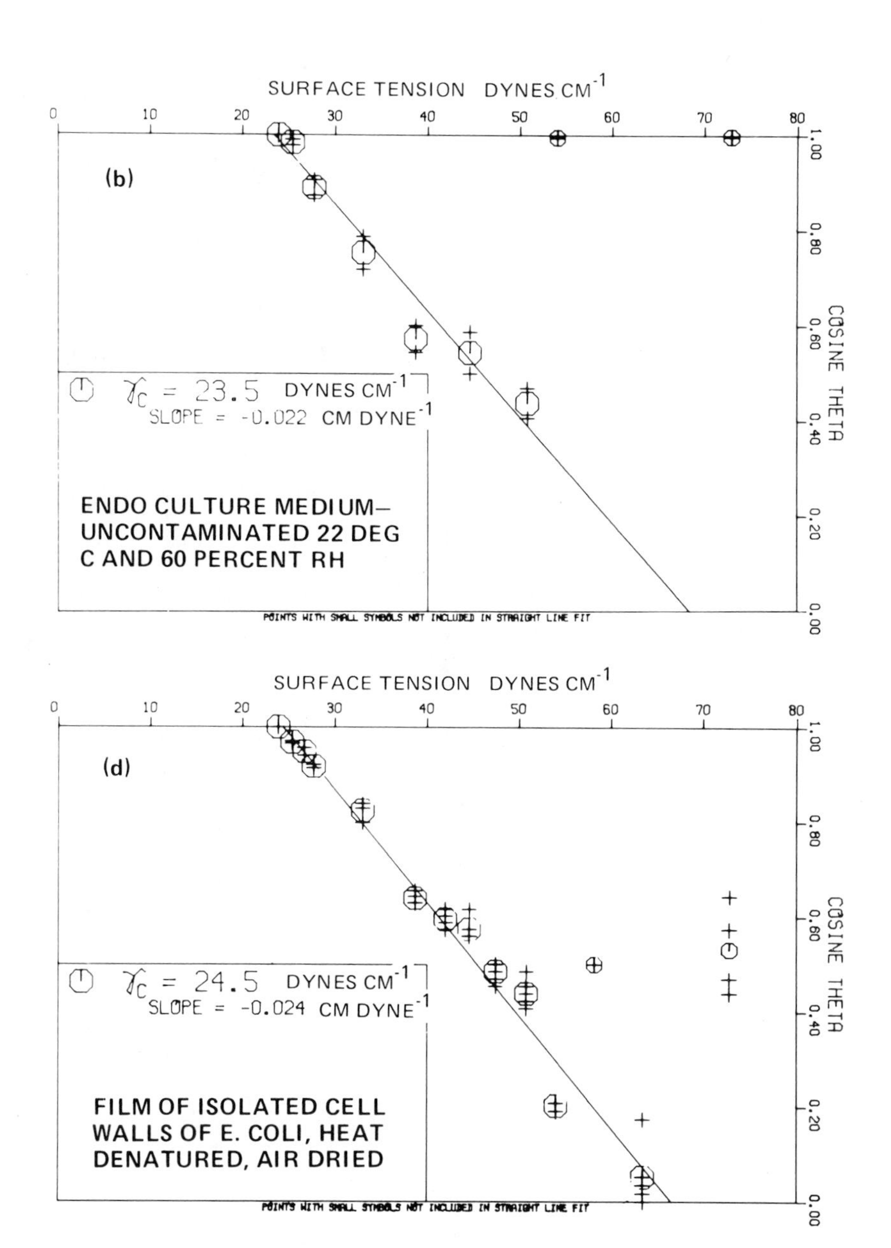
SURFACE TENSION DYNES CM-1
(b)
γc = 23.5 DYNES CM-1
SLOPE = -0.022 CM DYNE-1
ENDO CULTURE MEDIUM–UNCONTAMINATED 22 DEG C AND 60 PERCENT RH
POINTS WITH SMALL SYMBOLS NOT INCLUDED IN STRAIGHT LINE FIT
COSINE THETA
SURFACE TENSION DYNES CM-1
(d)
γc = 24.5 DYNES CM-1
SLOPE = -0.024 CM DYNE-1
FILM OF ISOLATED CELL WALLS OF E. COLI, HEAT DENATURED, AIR DRIED
POINTS WITH SMALL SYMBOLS NOT INCLUDED IN STRAIGHT LINE FIT
COSINE THETA

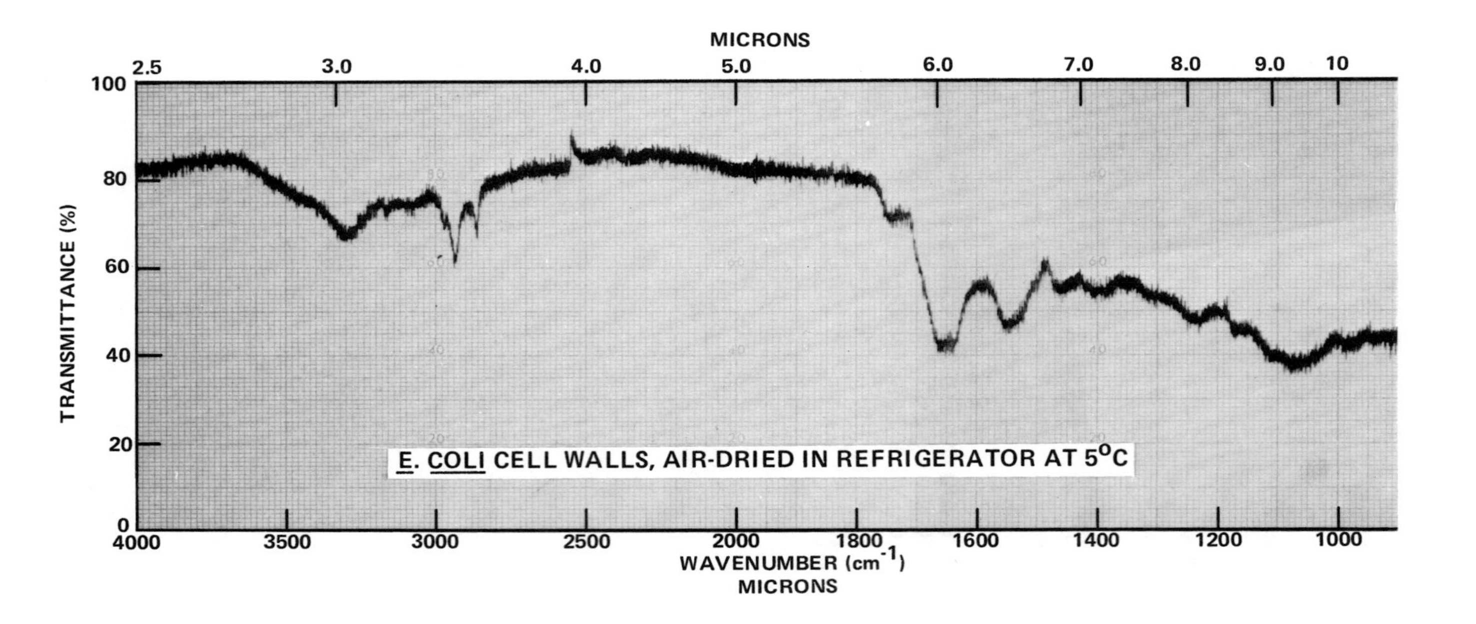
MICRONS
2.5
3.0
4.0
5.0
6.0
7.0
8.0
9.0
10
100
80
60
40
20
0
TRANSMITTANCE (%)
E. COLI CELL WALLS, AIR-DRIED IN REFRIGERATOR AT 5°C
4000
3500
3000
2500
2000
1800
1600
1400
1200
1000
WAVENUMBER (cm-1)
MICRONS

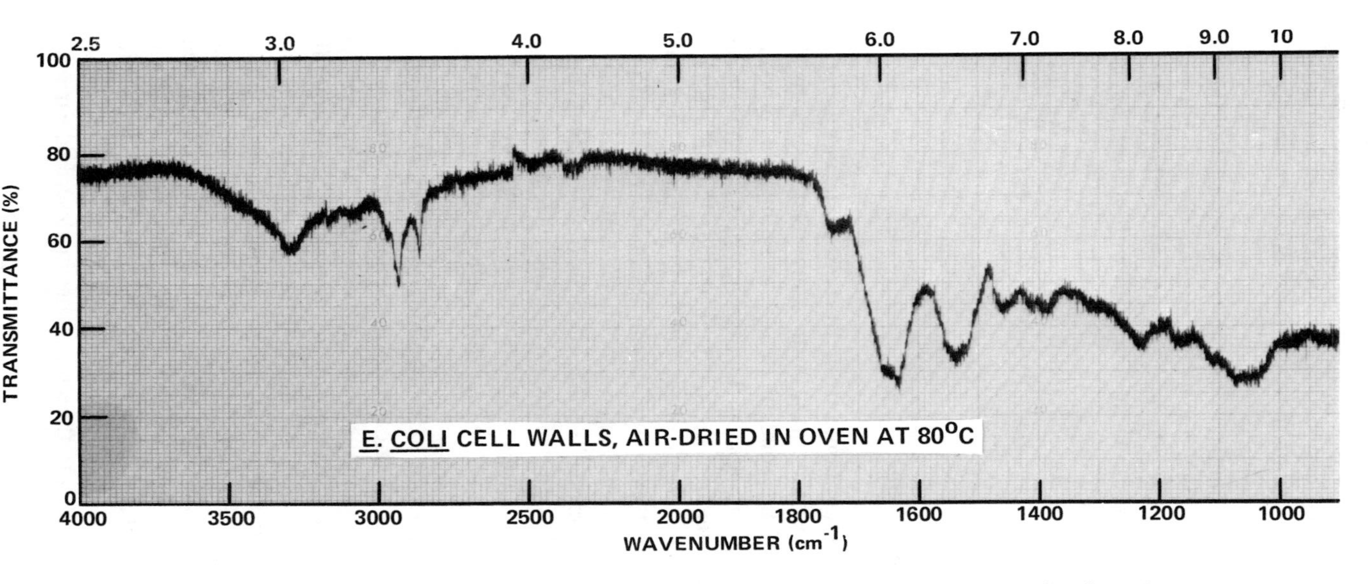

Figure 3.19 Multiple attenuated internal reflection IR spectra of isolated cell walls of *Escherichia coli*, air-dried from water suspension, onto platinum foils at 5°C and 80°C. Note that apparent heat denaturation indicated by amide I and amide II band shifts to 1630 cm^{-1} and 1530 cm^{-1}.

tion of the denatured *E. coli* cell walls, all these biological surface components exhibit critical surface tensions at 30–40 dynes/cm in air-dried films.

Kevin Marshall (private communication) has pointed out that the differences observed among the contact angles obtained on *E. coli* cell walls, before and after heat denaturation, almost certainly reflect the novelty of cell-wall structure for this organism when compared with viruses, *Bacillus*, and *Micrococcus* strains. As confirmed in the IR spectra of Fig. 3.19, there is strong evidence for lipid from the outer lipopolysaccharide layer of *E. coli*, a Gram-negative organism. The lipopolysaccharide layer is not present in the two Gram-positive organisms studied.

3.5 SUMMARY AND CONCLUSIONS

Recognizing that microbial contamination or deliberate culturing of microorganisms on solid substrata must bear some relationship to the original surface properties of the solid phase, we examined the probable influence of surface-energy parameters in the absence of significant variations of surface charge and texture. It is demonstrated that at least one empirical parameter of surface free energy, the "critical surface tension" determined from simple contact-angle measurements with a variety of pure liquids, provides a useful "label" for apparently bioadhesive versus abhesive solid substrata. Materials with critical surface tensions of 20–30 dynes cm^{-1} exhibit minimal biological adhesion, whereas materials with critical surface tensions above this zone (including most common engineering polymers) or below this zone (primarily fluorocarbons) support bioadhesion to greater degrees.

Examining the actual initial events at solid–biological media interfaces, it is found that all attachments of formed cellular elements are preceded by spontaneous adsorption of macromolecular "conditioning" films on the solids. These films are largely composed of the same, amphipathic polymers, adsorbing from aqueous media at about the same rates but in different configurations on substrata of differing original surface qualities. Initial rates of cellular or bacterial adhesion to such adsorbed films are also comparable, but the tenacity of adhesion varies with the degree of apparent "denaturation" of the first attached macromolecules. The critical surface-tension parameter provides a useful descriptor of the bioadhesive properties of these "conditioned" substrata, revealing continuing differences in practical adhesive events, according to the same ordering described earlier.

Specific examples of microorganism adhesion to smooth test substrata placed into marine bacterial cultures support the generalization of these findings to bacterial adhesion in natural circumstances and imply that nontoxic approaches to minimizing biological fouling (often initiated by bacterial slime layers) may be based on control of substrata surface properties.

In the absence of clearly demonstrated fouling-resistant surfaces, however, it is still necessary to consider methods for the scrupulous cleaning of bacterial layers and their exudates from numerous contaminated substrata. A prerequisite to designing effective cleaning systems is development of a more complete understanding of the new surface properties imparted to these substrata by the successful colonizing organisms. A series of surface chemical–physical analytical methods is described, as applicable to the same confluent monolayer culture of living bacteria in sequence, to determine the actual chemical components at the interfaces and their relative surface energies and stages of organization. Illustrations are provided of the utility of these methods for biological surfaces of a variety of bacteria, bacterial cell walls, bacterial flagella, intact viruses, and virus proteins, providing new data acquired by surface-specific IR spectroscopy and by contact-angle measurements.

Particular attention is given to the experimental requirements for making reliable determinations of surface-energy parameters from contact-angle measurements on living, hydrated, complex biological materials, demonstrating the inadequacy of using water or water-miscible fluids alone. Although water droplets will quickly penetrate, swell, or adsorb surface-active components when placed on biological substrata, exhibiting low contact angles or complete spreading, this complicated event cannot be taken to indicate an operationally high surface free energy for those substrata in contact with other environmental agents. In fact, contact angles measured with water-immiscible fluids of a variety of types on biological surfaces in equilibrium with their surroundings reveal the apparent surface free energy to be that of the biologically stable surface state, labeled by a critical surface tension of 20–30 dynes cm^{-1}, rather than the greater than the value of 70 dynes cm^{-1} normally attributed to free aqueous surfaces. Desiccated biological surfaces, as exemplified by isolated cell-wall preparations, by extracted proteins, polysaccharides, and their complex conjugates, typically display critical surface tensions of about 30–37 dynes cm^{-1}. The practical result of these findings is that chemical cleaning agents must, to be efficacious on bacterially contaminated solid substrata, have surface-active ingredients capable of depressing the systems' surface tensions to at least

20–30 dynes cm^{-1}. Otherwise, it would be difficult or impractical to remove all the adherent proteinaceous and bacterial residues with simple solvents or even by brushing with abrasives. Displacement of biological or organic matter coating solid surfaces is best accomplished by specific surface-chemical action, aided by simple solution or emulsification processes.

A marginally stable emulsion-type cleaner would be especially effective in meeting these requirements. The dispersed hydrocarbon component of the emulsion would readily penetrate the organic residues, and the water portion would serve to flush away the contaminants. Residual emulsion and contaminants could then be rinsed away with fresh water.

3.6 ACKNOWLEDGMENT

Dr. James Wilkins kindly gave permission to use the data cited herein, as gathered during his predoctoral experiments in the author's laboratory. Dr. Kevin Marshall granted permission for inclusion of examples of data acquired jointly with the author in a long-sustained investigation into the initial events of marine bacterial adhesion. Dr. Rex Neihof of the U.S. Naval Research Laboratory, Washington, D. C., donated the isolated bacterial cell walls, and personnel of the Fort Detrick biological laboratories provided the pure virus and virus-protein preparations. Dr. J. Berger of Roswell Park Memorial Institute, Buffalo, N. Y. kindly supplied the purified bacterial flagella. Three of the author's former coworkers, S. Pek, S. Perlmutter, and P. Schnizler, made important analytical contributions during the course of these studies, and the author's primary collaborators in biofouling investigations, Drs. V. A. DePalma and D. Goupil, offered constructive critical comments. B. J. Spahr prepared the final manuscript.

3.7 REFERENCES

1. R. E. Baier, "Surface Properties Influencing Biological Adhesion," in R. S. Manly, Ed., *Adhesion in Biological Systems*, Academic, New York, 1970, p. 15.
2. R. E. Baier and V. A. DePalma, "The Relation of the Internal Surface of Grafts to Thrombosis," in W. A. Dale, Ed., *Management of Arterial Occlusive Disease*, Year Book Medical Publishers, Chicago, 1971, p. 147.
3. R. E. Baier, "Applied Chemistry at Protein Interfaces," in R. E. Baier, Ed., *Advances in Chemistry*, Vol. 145, American Chemical Society, Washington, D. C., 1975, p. 1.
4. F. Grinnell, M. Milam, and P. A. Srere, *Arch. Biochem. Biophys.*, **153,** 193 (1972).
5. E. G. Shafrin, "Critical Surface Tension of Polymers," in J. Brandrup and E. H. Immergut, Eds., *Polymer Handbook*, Interscience, New York, 1966, pp. 111–113.

6. M. K. Bernett and W. A. Zisman, *J. Colloid Interface Sci.*, **28,** 243 (1968).
7. M. K. Bernett and W. A. Zisman, *J. Colloid Interface Sci.*, **29,** 413 (1969).
8. R. E. Baier, *Bull. N. Y. Acad. Med.*, **48,** 257 (1972).
9. J. W. Boretos, W. S. Pierce, R. E. Baier, A. F. Leroy, and H. J. Donachy, *J. Biomed. Mat. Res.*, **9,** 327, (1975).
10. R. E. Baier, C. K. Akers, S. Perlmutter, H. Dardik, I. Dardik, and M. Wodka, *Transact. Am. Soc. Artif. Int. Organs*, **22,** 514 (1976).
11. R. Baier, L. Weiss, J. Harlos, manuscript in preparation.
12. A. C. Taylor, "Adhesion of Cells to Surfaces," in R. S. Manly, Ed., *Adhesion in Biological Systems*, Academic, New York, 1970, p. 51.
13. V. A. DePalma, R. E. Baier, J. W. Ford, V. L. Gott, and A. Furuse, *J. Biomed. Mat. Res. Symp.* **3,** 37 (1972).
14. J. F. Wilkins, "Adhesion of Ehrlich Ascites Tumor Cells To Surface Modified Glass," dissertation submitted to the Faculty of Biophysics, Roswell Park Graduate Division, State University of New York at Buffalo, 1974.
15. J. F. Wilkins, private communication.
16. R. E. Baier, *Swed. Dent. J.*, **1,** 261 (1977).
17. L. I. Friedman, H. Liem, E. F. Grabowski, E. F. Leonard, and C. W. McCord, *Transact. Am. Soc. Artif. Int. Organs*, **16,** 63 (1970).
18. R. E. Baier, "Influence of the Initial Surface Condition of Materials on Bioadhesion," in R. F. Acker, B. F. Brown, J. R. DePalma, and W. P. Iverson, Eds., *Proceedings of Third International Congress on Marine Corrosion and Fouling*, Northwestern University Press, Evanston, Ill., 1973 p. 633.
19. D. W. Goupil, V. A. DePalma, and R. E. Baier, "Prospects for Nontoxic Fouling-Resistant Paints," in *Proceedings of Ninth Annual Conference on Marine Technology Society*, Washington, D.C., 1973, p. 445.
20. S. C. Dexter, J. D. Sullivan, Jr., J. Williams III, and S. W. Watson, *Appl. Microbiol.*, **30,** 298 (1975).
21. S. C. Dexter, "Control of Microbial Slime Films in OTEC Heat Exchangers," in G. L. Dugger, Ed., *Proceedings of Third Workshop on Ocean Thermal Energy Conversion*, Johns Hopkins University, Laurel, Md., 1975, p. 137.
22. S. C. Dexter, "Influence of Substratum Critical Surface Tension on Bacterial Adhesion—*in situ* Studies," Abstracts, 52nd Colloid and Surface Science Symposium, American Chemical Society, Washington, D. C., 1978; *J. Colloid Interface Sci.* (in press).
23. V. A. De Palma and R. E. Baier, "Microfouling of Metallic and Coated Metallic Flow Surfaces in Model Heat Exchange Cells," in R. Gray, Ed., *Proceedings of OTEC Biofouling and Corrosion Symposium*, Department of Energy, Washington, D. C., 1978.
24. E. C. Haderlie, "The Nature of Primary Organic Films in the Marine Environment and their Significance for Ocean Thermal Energy Conversion (OTEC) Heat Exchange Surfaces," Naval Postgraduate School Technical Report No. NPS-68Hc77021, Monterey, Calif., 1977.
25. A. F. Conn, M. S. Rice, and D. Hagel, "Utra-Clean Heat Exchangers—a Critical OTEC Requirement," in G. E. Ioup, Ed., *Proceedings of Fourth Annual Conference on Ocean Thermal Energy Conversion*, University of New Orleans, La., 1977, p. VII–11.
26. J. G. Fetkovich, G. N. Grannemann, L. M. Mahalingham, D. L. Meier, and F. C.

Munchmeyer, "Studies of Biofouling in Ocean Thermal Energy Conversion Plants," in G. E. Ioup, Ed., *Procceedings of Fourth Annual Conference on Ocean Thermal Energy Conversion*, University of New Orleans, La., 1977, pp. VII–15.

27. R. E. Baier, "Occurrence, Nature, and Extent of Cohesive and Adhesive Forces in Dental Integuments," in A. Lasslo and R. P. Quintana, Eds., *Surface Chemistry and Dental Integuments*, Thomas, Springfield, Ill., 1973, p. 337.
28. R. P. Quintana, A. Lasslo, J. W. Clark, and R. E. Baier, "Surface-Chemical Approach to the Prevention of Pathological Accumulations of Dental Deposits," in A. Lasslo and R. P. Quintana, Eds., *Surface Chemistry and Dental Integuments*, Thomas, Springfield, Ill., 1973, p. 392.
29. W. R. Sanborn, *Am. J. Public Health*, **53,** 1278, 1963.
30. C. J. van Oss, C. F. Gillman, and A. W. Neumann, *Phagocytic Engulfment and Cell Adhesiveness as Cellular Surface Phenomena*, Marcel Dekker, New York, 1975.
31. M. W. Cowden, B. J. Kinzig, and S. L. Schwartz, "Contact Angle Measurements on Drug Treated Mouse Peritoneal Macrophage Monolayers Using Different Spreading Liquids," Abstracts, 51st Colloid and Surface Science Symposium, American Chemical Society, Washington, D. C., 1977, p. 98.
32. G. Colacicco, C. F. Gillman, and A. K. Ray, "Cultured Mammalian Cells: Relation of Contact Angle to Pathologic, Pharmacologic and Experimental Conditions," Abstracts, 52nd Colloid and Surface Science Symposium, American Chemical Society, Washington, D. C., 1978.
33. R. E. Baier, E. G. Shafrin, and W. A. Zisman, *Science*, **162,** 1360 (1968).
34. R. E. Baier, R. C. Dutton, and V. L. Gott, "Surface Chemical Features of Blood Vessel Walls and of Synthetic Materials Exhibiting Thromboresistance," in M. Blank, Ed., *Surface Chemistry of Biological Systems*, Plenum, New York, 1970, p. 235.
35. W. A. Zisman, "Relation of the Equilibrium Contact Angle to Liquid and Solid Constitution," in F. W. Fowkes, Ed., *Advances in Chemistry*, Vol. 43, American Chemical Society, Washington, D. C., 1964, p. 1.
36. D. H. Kaelble, *J. Adhesion*, **2,** 66, (1970).
37. E. Nyilas, W. A. Morton, R. D. Cumming, D. M. Lederman, T.-H. Chiu, and R. E. Baier, *J. Biomed. Mat. Res. Symp.*, **8,** 51 (1977).

CHAPTER 4

Microbial Surface Components Involved in Adsorption of Microorganisms onto Surfaces

WILLIAM A. CORPE

Department of Biological Sciences, Columbia University, New York, New York

CONTENTS

4.1 INTRODUCTION

The isolation, chemical composition, and function of microbial surface structures have been studied in some detail over the past 25 years. We are beginning to understand how membranes, cell walls, flagella, cilia, and other surface structures function in the ecology of microorganisms. A relatively small amount of the accumulated information, however, has been related directly to the capacity of microorganisms to attach to surfaces.

Strong adhesion of a microbial cell to a solid surface of any sort requires intimate contact, but both cell surface and substratum usually carry a net negative charge, and close approximation of the two surfaces is not favored because of the electrical repulsion barrier [1]. Balanced against repulsion are attractive forces such as ionic bonding, van der Waals forces, and hydrophobic and hydrogen bonding, which function at short range [2]. Although individually weak, they become very effective when quantitatively increased; theoretically, at least, any mechanism proposed for the firm attachment of a microbial cell to a solid surface must take into account the establishment of intimate contact at the interface.

Surface structures of microbial cells that conceivably could participate in the attachment process include *cell walls* or *envelopes*, *capsules*, and *extracellular slime*. The latter have physical characteristics, such as high viscosity, gelling properties, and capacity to establish intimate contact with solid surfaces, making a "glue" concept of adhesion easy to accept. Many organisms, however, do not produce slime or capsules or, for that matter, other more structured "adhesive organelles" such as tactile *fimbriae* (pili) or *holdfasts*, yet they are able to attach to solid surfaces. The adhesive properties of cell walls of periphytic bacteria, fungi, and algae or the protozoan pellicle must be reexamined for adhesive qualities.

The contribution of the substratum to the interaction with cells is likely to be considerable in the establishment of firm adhesion. Gibbons [3] pointed out that the high degree of selectivity of different mammalian mucosal surfaces for bacteria may account for the host tropisms commonly observed. The same also may be true for viruses. A high degree of selectivity is also imposed by certain plant surfaces for strains of viruses and bacteria.

Fletcher [4] has shown that the rate of attachment may be determined by the affinity between bacterium and substratum. Negative surface charge density and wettability of the substratum both affect bacterial attachment

[5,6]. Adsorption of materials from the surrounding medium to the surface may affect the wettability and physical properties of the substratum, and the rate of response of organisms with a propensity for attachment is likely to be quite variable.

Since temperature, pH, E_h, specific ions, ionic strength, availability of nutrients and surface active materials are known to affect growth and viability of organisms, environmental conditions certainly must influence the rate of adsorption and probably the strength of adhesion as well. The secretion of adhesive material by the cell and the conformation of macromolecular surface materials are clearly under both physiological and environmental control.

Much of the literature relating to attached microorganisms is descriptive of populations of periphytes observed under natural or simulated natural conditions. Critical experiments to define the process of attachment and the nature of the adhesion in biochemical and molecular terms is restricted to a few systems. The attachment of viruses and procaryotic and eucaryotic microbial cells is discussed in this chapter, and an attempt is made to establish some understanding of the specific chemical and physical character of microbial surfaces that serve as the basis for adsorption to specific substrata.

4.2 ADSORPTION OF BACTERIOPHAGE AND OTHER VIRUSES

As viruses are nonmotile, they encounter host cells by random collision. They display great specificity for particular host cells and, before penetration, become irreversibly adsorbed to cell surfaces. Adsorption of viruses apparently is independent of temperature, except insofar as Brownian movement is likely to affect collisions [7].

4.2.1 Bacteriophage

Specific and irreversible adsorption implies establishment of strong bonding between virus and host cell surfaces. Most information about virus–host interaction is from work with the *T*-coliphage. The T-even phages adsorb by long tail fibers and possibly by tail-plate pins to specific receptor sites in specific regions on the surface of *E. coli* strains [8,9]. Lindberg [10] has reviewed the literature on bacteriophage receptors. He concluded that almost every structure exposed or originating on the bacterial cell surface can act as phage receptor. These include the fractions of the outer membrane of Gram-negative bacteria, the lipopolysaccharide, and the lipoprotein components. In Gram-positive bacteria, peptidoglycan is the dominant component and frequently is associated with receptor activity. Other polymers associated or linked to peptidoglycan, such as teichoic acids, teichuronic acids, protein in *Bacillus* and *Streptococcus* and *C*-carbohydrate in *Streptococcus*, have been implicated as part of specific receptor sites. Capsules, slime layers, and flagella of both Gram-positive and Gram-negative bacteria have been shown to adsorb phage. The basis for believing that phage receptors are associated with surface components is that isolated fractions can inactivate virus irreversibly and inhibit infection [11].

Lindberg [10] has suggested three mechanisms for firm attachment of bacteriophage. First, the large tadpole-shaped DNA phages become attached, probably reversibly, by protein tail fibers that recognize attachment sites. The tail spikes and base-plate organelles interact with the receptor on the host membrane to accomplish irreversible attachment. The nature of the interaction is not known. *Escherichia coli* lysogenic for the lambdoid phage PA_2 produces a major outer membrane protein absent in nonlysogenic cells. This protein has a reduced ability to bind phage. On the other hand, phage is inactivated by cell envelopes and envelope fractions containing the wild-type protein [12]. Some chemical and genetic characterization of the protein has been achieved. Second, phages with short, noncontractible tails, such as E^{15} and F^{22}, attach to cell-wall lipopolysaccharide of the host cell. An endoglycosidase associated with the phage partially degrades the receptor. Robbins and Uchida (13) have reported that lysogenization of *Salmonella anatrium* by phage E^{15} induces an altered specificity in a surface antigenic determinant. Residues containing 6-*O*-acetyl α-D-galactosyl are replaced by unacetylated β-D-galactosyl residues and the organism is no longer susceptible to phage E^{15} [14]. Third, isometric RNA and filamentous DNA phages attach to receptors on sex pili with the aid of the so-called *A*-protein [15,16]. The phage is uncoated, and the *A*-protein enters the cell with the nucleic acid. The mechanism by which this is accomplished is obscure. The *F*-pilus receptor sites of *E. coli* RNA phages were looked on by Brinton [17] as hollow tubes through which infecting RNA could pass. Bradley [18], on the other hand, believes that the phage-bearing pilus is shortened or retracted until the phage particles reach the cell wall. Retraction of *F*-pili of *E. coli* has been noted by Novotny and Fives-Taylor [19].

4.2.2 Other Viruses

Animal viruses may not have a specialized surface component capable of combining with host-cell surface receptors [20]. Many viruses are able to agglutinate red blood cells, which indicates they have two or more combining groups per particle. Brick-shaped vaccinia particles adsorb on both long and short axes, indicating that combining sites are distributed over the coat. Adsorption was abolished by pretreatment of substrate cells with neuraminidase, so the common receptor of these viruses would seem to be the *N*-acetyl neuraminic-acid side chain of cellular mucoprotein. Fawcett [21] has suggested that only a limited area of the host-cell surface is capable of adsorbing virus. The receptor substances are believed to be glycolipoproteins. The basis for such claims was the finding that infectivity was abolished by glycolipoprotein fractions isolated from host cells [22]. Purified receptor protein in some cases could release viral RNA from intact particles.

The receptor site for polio virus is a lipoprotein in the cell membrane of several cell types located in the brain, spinal cord and intestine of humans and primates [23], but there are differences in receptor affinities for various virus strains. The lipoprotein was removed from cultured host cells by trypsin treatment, but it was slowly regenerated [24]. During the process of adsorption of virus, subunits of the capsid may be changed or reoriented, since the capsid became sensitive to proteolytic enzyme and lost the ability to react with antiserum. In spite of these changes in the protein coat, it remained protective of the RNA against destruction by RNAase.

Laver and Vallentine [25] showed that the protein coat of the influenza virus is a glycoprotein with hemagglutinin activity. Initial attachment of avian tumor virus (ATV) to chicken-embryo fibroblast cells is through a trypsin-soluble cell surface component. The solubilized attachment-site material interacted directly with ATV *in vitro* and reduced its infectivity [26]. Both virus and host-cell attachment site are thought to be glycoproteins.

Lipoprotein-enveloped viruses, such as the myxoviruses, are derived in part from the host cells. Virus attachment occurs with the alignment of the virus and host-cell lipoprotein layers followed by fusion of membranes. The enveloped form of herpes virus is more infectious than the nonenveloped form, perhaps because of more rapid attachment and penetration [7].

4.3 ATTACHMENT OF BACTERIA

Most of the bacteria that become irreversibly attached to solid surfaces do so with no apparent involvement of special organelles discernable with the light microscope or the scanning electron microscope (SEM). Both techniques are indispensable, however, for determining the orientation of the cells to substratum, the possible location of the attachment site on the bacterium surface, the other useful information about the substratum–bacterium interface. These instruments are of greatest value for studying populations of attached organisms in natural environments. The transmission electron microscope (TEM) and associated techniques (ultrathin sectioning and staining) have provided and, undoubtedly in the future, will provide the greatest amount of information about the structural changes at the cell–substratum interface.

Many other methods, both direct and indirect, are available for measuring the qualitative and quantitative nature of the interaction and, together with microscopy, lead to localization of the cell surface components specifically concerned with adhesion.

Greatest advancement in this regard has been accomplished with the coliphage–host-cell interaction, as previously described. Bacterium–substratum systems are much more complex in view of the variety of surface components that may participate in an interaction. Our rather imperfect knowledge of the chemistry and biosynthesis of the components further adds to the difficulty.

In this chapter the roles of (1) surface polymers, (2) fimbriae or pili, and (3) holdfasts or other special organelles involved in attachment are discussed. For the sake of convenience, they are considered as separate categories, although this arrangement of presentation is somewhat artificial. The nature of the literature relating to the attachment of bacteria and the discussion of attachment by eucaryotic organisms that follow make this approach appropriate.

4.3.1 Role of Bacterial Cell Wall and Unstructured Exudates

Nonspecific Attachment. The irreversible attachment of bacteria to many different substrata submerged in natural waters was explored by ZoBell [27] and, subsequently, by various workers cited in a recent review by Fletcher [28]. Attachment occurs in a time-dependent process, probably involving biosynthesis and secretion of polymeric adhesives that provide for firm, irreversible attachment [29] to a conditioned surface [30,31]. Cells attached to electron microscope grids [32,33] showed fine, extracellular polymeric fibrils extending from the cell surface to that of the grid (Fig. 4.1). When cells were sheared from the grid surface, polymer "footprints" remained. The polymeric fibrils presumably served as a bridge from cell to substratum.

ZoBell [27] described microcolonies of attached bacteria, surrounded by exopolymers that stained lightly with the basic dye, crystal violet. All

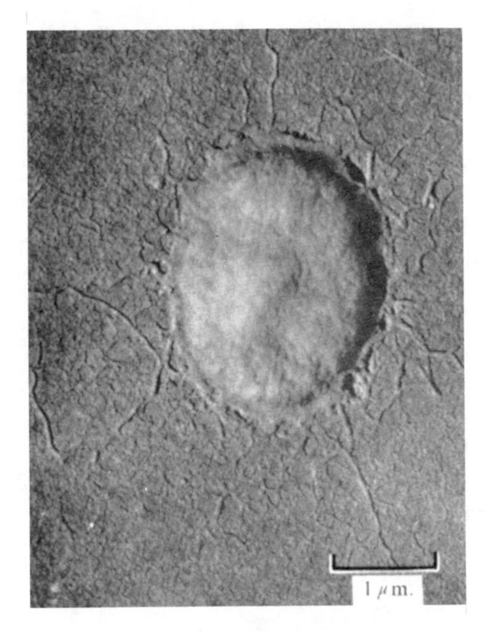

Figure 4.1 Extracellular polymeric fibrils on bacteria sorbed on grid immersed in artificial seawater + glucose (7 mg/liter) [32].

periphytic marine bacteria isolated [34,35] produce various quantities of extracellular polymer, which gave tests for polyanionic carbohydrate plus, in some cases, a substantial amount of protein [35]. Corpe [36] isolated a number of strains of *Pseudomonas atlantica* that produced large quantities of extracellular polymer that stained with crystal violet and alcian blue, a dye that forms insoluble complexes with polyanions [37]. The abundant slime was separated from cells, purified, and found to be a polyanionic carbohydrate [38].

Mutants of *P. atlantica* that had lost their ability to secrete the voluminous polysaccharide could still attach to glass slides, although the rate of attachment, as well as growth rate of the bacterium, was slower than that of the wild-type [39]. Mutants still produced a protein and carbohydrate-containing, alcian-blue-reactive substance (Matsuuchi and Corpe, unpublished observations), some of which was found free in the growth medium. Another fraction, washed from the cell surface with seawater, also appeared to be a glycoprotein. The materials facilitated adhesion of washed *Chlorella* cells to glass [40].

Fletcher and Floodgate [41], studying a marine *Pseudomonas* attached to cellulose and sectioned for electron microscopy, recognized two alcian-blue-stained surface layers in the bacterium. One was a compact layer on the outer surface of the cell wall and the other, a loose fibrous polysaccharide.

The former was believed to be responsible for initial cell adhesion, and the latter produced after attachment.

Marshall [42,43] has shown electron micrographs of thin sections of reembedded blocks of embedding material to which bacteria had become attached. The organisms showed end-on attachment through a radiating pattern of polymer fibrils. Oblique sections cut through the solid–liquid interface showed the arrangement of the secreted material (Figs. 4.2*a*,*b*) in relation to the substratum. The polymer seems to form a discoid type of holdfast structure, not dissimilar to those produced by sheathed bacteria [44], to be discussed later.

The voluminous acidic polysaccharide materials described by Corpe [38] and Fletcher and Floodgate [41] in different organisms may not be concerned with the primary attachment of the producing cells; but such polymers enhance aggregation of other bacteria under proper conditions of pH, polymer, and electrolyte concentration [45]. Friedman et al. [46,47] and Jones et al. [48] showed that extracellular polymeric fibrils produced by

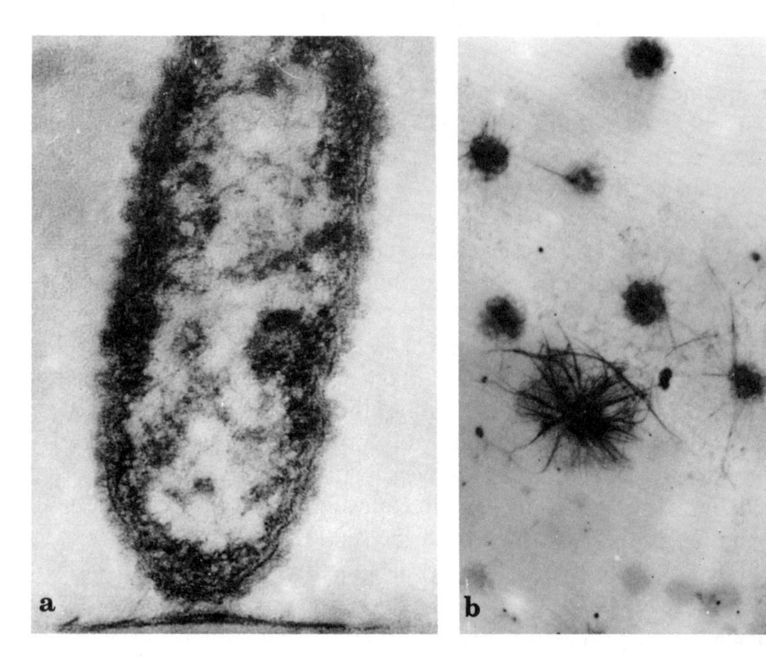

Figure 4.2 Electron micrographs of (a) thin section of reembedded araldite block showing attachment of marine bacterium to surface (×180,000) (courtesy of Dr. K. C. Marshall, Australia) and (b) thin section cut obliquely across original araldite–water interface revealing nature of polymer fibrils binding cell to solid surface (× 67,500) [43].

bacteria caused flocs to develop, presumably because of the formation of cell-to-cell bridges (Fig. 4.3). Glass slides coated with polymer from *P. atlantica* and suspended in the sea became fouled by microorganisms more rapidly than control slides [38].

Specific Attachment. The specific attachment and colonization of specific sites in the human mouth by various bacterial species is well documented [3,49,50]. The colonization of the gastrointestinal tract of animals by a bewildering array of microbial species also has been explored [51,52]. The contents, as well as epithelial surfaces, are colonized by both indigenous and nonindigenous organisms, existing in complex communities occupying ecological niches distributed throughout the system. The mechanisms of attachment, except for a few, have not been studied.

Gibbons [3] has reasoned that, because of possible loss through mutation, more than one surface component may be required for persistent attachment and colonization; but clear-cut examples are few. *Streptococcus mutans* attaches to teeth or experimentally to hydroxyapatite that has selectively adsorbed mucinous glycoprotein from saliva [53]. The organisms synthesize an extracellular glucan from sucrose. The glucose polymer is bound to the bacterial cell surface by a glucan-binding protein that causes the cells to aggregate [54,55]. Cell-bound glucosyltransferase activity and glucan synthesis at the cell surface is required for adhesion of *S. mutans* [56]. Other bacteria do not bind glucosyltransferase sufficiently [57] to be strongly adsorptive. Gibbons [3] suggests that *S. mutans* cells may have more than three recognition molecules involved in adhesion and colonization.

The phenomenon of interbacterial aggregation [58] may be an important means by which some bacteria, with little ability to attach and colonize a smooth surface, do attach when a missing component is supplied by another species. Whittenberger et al. [59] have shown that glucosyltransferase produced by *S. salivarius* greatly facilitates the adhesion of *Veillonella* to smooth surfaces. The precise mechanisms by which this is accomplished have not been explored.

Strains of *S. mitis*, *S. sanguis*, and *S. salivarius* attach to human buccal epithelial cells. The streptococci are known to bind selectively blood-group-reactive glycoproteins. Glycoproteins of host secretions are also known to competitively inhibit bacterial adsorption because they mimic the surfaces that the secretions bathe [3]. Receptors for these organisms on epithelial cell surfaces can be masked by concanavalin A and by antibodies to specific blood-group substances.

Although specific properties of bacterial capsules and envelopes are believed to mediate their attachment to animal surfaces in some cases [51],

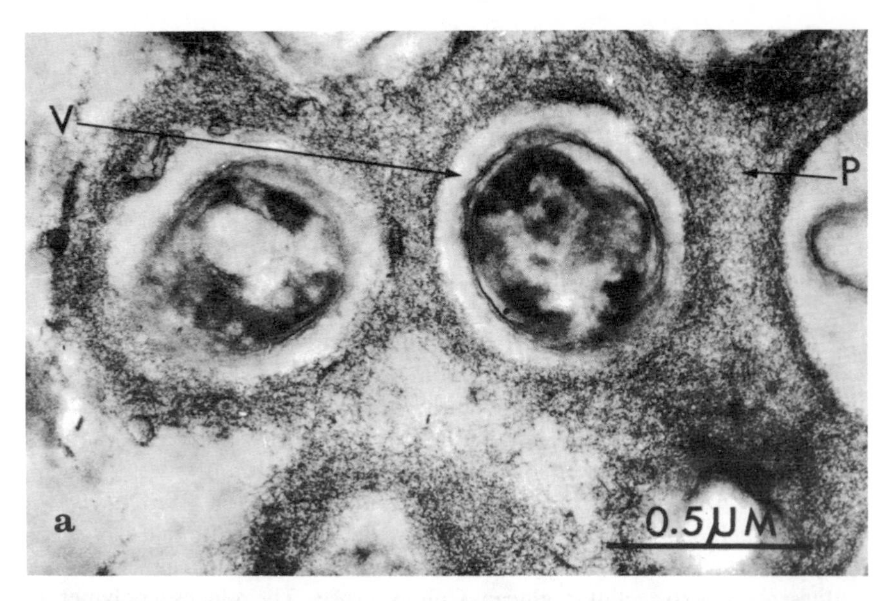

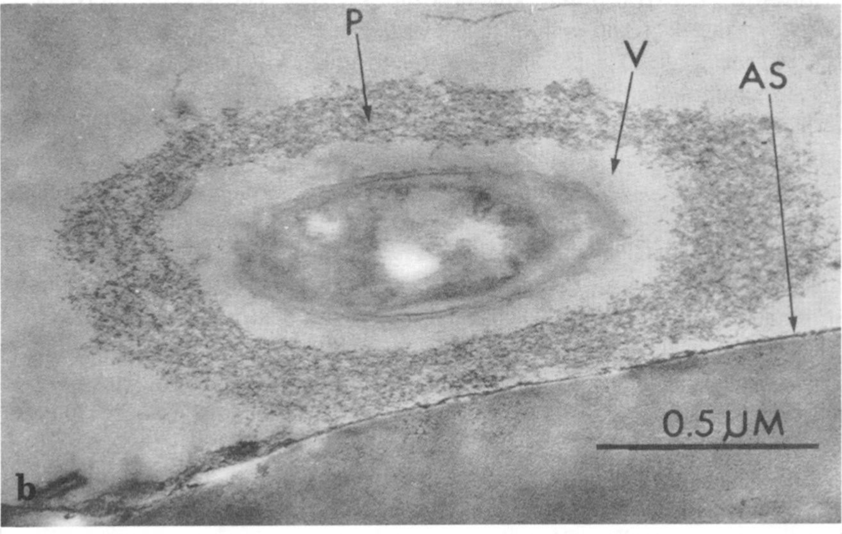

Figure 4.3 Thin sections of 9-day slime layer from microfouled surface: (a) cells surrounded by void (V) area and interlacing slime (P) connecting each cell to give form to slime layer; (b) slime (P) lying between cells and solid surface (AS); void area (V) between cell and slime layer was characteristic of 9-day samples [48].

relatively little experimental data are available to support that view. Savage [60] reported that an indigenous *Lactobacillus* species attached specifically to keratinized, squamous cells through a surface mucopolysaccharide. The orientation of the bacteria on the gastrointestinal epithelium suggests a specific site may be attached through a polarly located organelle (Fig. 4.4), but none was observed [61]. Savage and Blumershine [62] showed a number of SEM photomicrographs of rod-fusiform-shaped bacteria inhabiting the colonic mucosa of mice. Crisscrossing filaments that seem to connect the cell surface and epithelial substratum seem to be involved, but Savage [52] cautions that such conclusions are not warranted, since artifacts have not been ruled out. Fitzgerald et al. [63] have shown that *Treponema pallidum* cells readily attach to cultured mammalian cells, but no mechanism of attachment was suggested.

Nelson et al. [64] studied adhesion and colonization by *Vibrio cholera* of rabbit ileal loops. The bacteria became aligned horizontally with the epithelial surface, but a few were attached in an end-on manner with flagella extending into the lumen. Even though transmission electron micrographs of a thin section showed the outer membrane of the *Vibrio* in intimate contact with the surface of the microvilli, one cannot claim that special holdfast structures, fimbriae, or other structurally identifiable attachment organelles are involved. The junction of procaryotic and eucaryotic cells may show the presence of mucopolysaccharide fibrils in thin sections [65], but it is not clear whether the fibrils are of host or bacterial origin. Lankford [66] has shown that *V. cholera* can agglutinate erythrocytes, but whether this property is related to that of epithelial attachment is unknown.

Kirby [67] and Ball [68] have described numerous examples of bacteria adhering to protozoa. Some of the bacteria attach by means of special structures, which are described later. Bloodgood and Fitzharris [69], however, showed that one group of bacteria attaching to gut flagellate protozoa have no organelles specialized for that purpose. Beams and Kessel [70] have described rod-shaped bacteria that attach in shallow surface folds of the ciliate *Cyclidium*. Both the bacterium and the surface of the *Cyclidium* seem to possess an outer coating of a "sticky substance" that on contact, holds the bacterium to the protozoan. In view of the enormous number of organisms on animal surfaces, a great deal of experimentation and careful observation must be done before making generalizations about mechanisms of attachment.

The firm attachment of microorganisms to plant surfaces is well established [71]. Royale [72] has shown that the study of structure and development of the epiphytic microflora by SEM can be valuable if rigorous standards of specimen preparation are maintained. Whereas SEM photo-

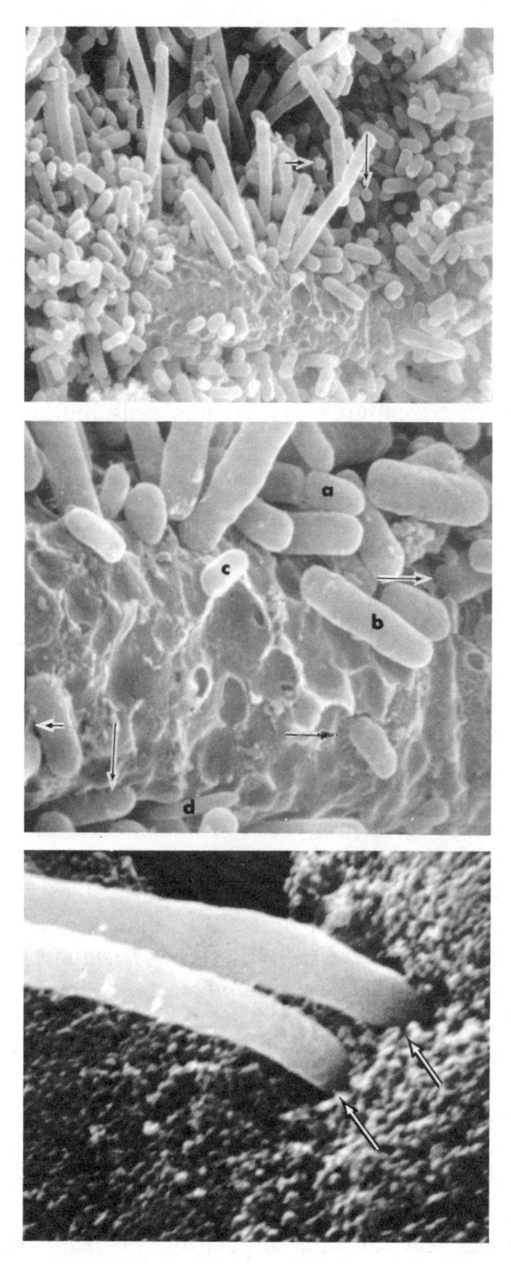

Figure 4.4 (Top) bacteria adhering to keratin layer of rat stomach epithelium by their sides and ends—both rods and cocci (arrows) are shown (×2225); (center) four morphological types of rod-shaped bacteria attached to keratinized surface (a–d)—arrows show filamentous material connecting bacteria to surface (×7625); (bottom) end-on attachment of two rod-shaped bacteria to rat ileal epithelial-cell surface.—arrows indicate site of attachment (×10,000).

micrographs of plant surfaces show the presence of colonizing microorganisms, it is generally difficult or impossible to determine much about the mode of attachment of individual cells.

Both pathogenic and nonpathogenic strains of bacteria associated with plants tend to produce extracellular polysaccharide slime that may serve to bridge the cells to substratum. Various species of *Agrobacterium*, *Corynebacterium*, *Erwinia*, *Pseudomonas*, and *Xanthomonas* are associated with plants and produce extracellular polysaccharides [73]. There is good reason to believe that the exopolymer may function in a specific interaction, rather than as a nonspecific adhesive.

Recent studies by Lippincott and Lippincott [74] suggest an interaction between *Agrobacterium tumefaciens* and a specific surface exposed at a plant wound site. A complementary surface charge or configuration between bacterium and host sites probably is involved. Bacterial cell envelopes and the envelop fraction extractable by detergent effectively inhibited initiation of infection. Cells of *E. coli* and *R. leguminosarum* did not interfere with *A. tumefaciens* infection.

The ability of *Rhizobium* species to establish a symbiotic relationship with specific leguminous plant groups has been known for some time. Bohlool and Schmidt [75] attempted to correlate lectin* binding to *Rhizobium* cells with ability of the bacteria to establish a symbiotic relationship. Soybean lectin combined specifically with all but three of 25 strains of soybean isolates. It did not bind to 23 *Rhizobium* cultures from other sources. Bohlool and Schmidt [76] had found that fluorescent-labeled soybean lectin accumulated preferentially on ends of *R. japonicum* cells. Tsien and Schmidt [77], using electron microscopy, recognized an extracellular polar body attached to the cells. The cells displayed end-on attachment. (Figs. 4.5*a,b*.) The extracellular polysaccharides of *R. japonicum* reacted with soybean lectin in gel-diffusion tests. The adsorption of infective strains of *R. trifolii* to clover roothairs was much greater than that of noninfective strains of other species. Attachment occurred in a polar, end-on fashion; and host specificity in *Rhizobium*–clover interaction involved a preferential adsorption of infective cells through a 2-deoxyglucose-containing receptor site, since adsorption was inhibited by that sugar [78,79].

The production of microbial extracellular slime may not be the only bridge of contact between plant surface and epiphyte. Singh and Schroth [80] have shown that fibrillar structures originating from the plant cell wall in the intercellular spaces of leaves of red kidney bean, *Phaseolus vulgaris*,

* *Lectins* (from Latin *legere*, to choose) are plant proteins or glycoproteins that have carbohydrate-binding specificity.

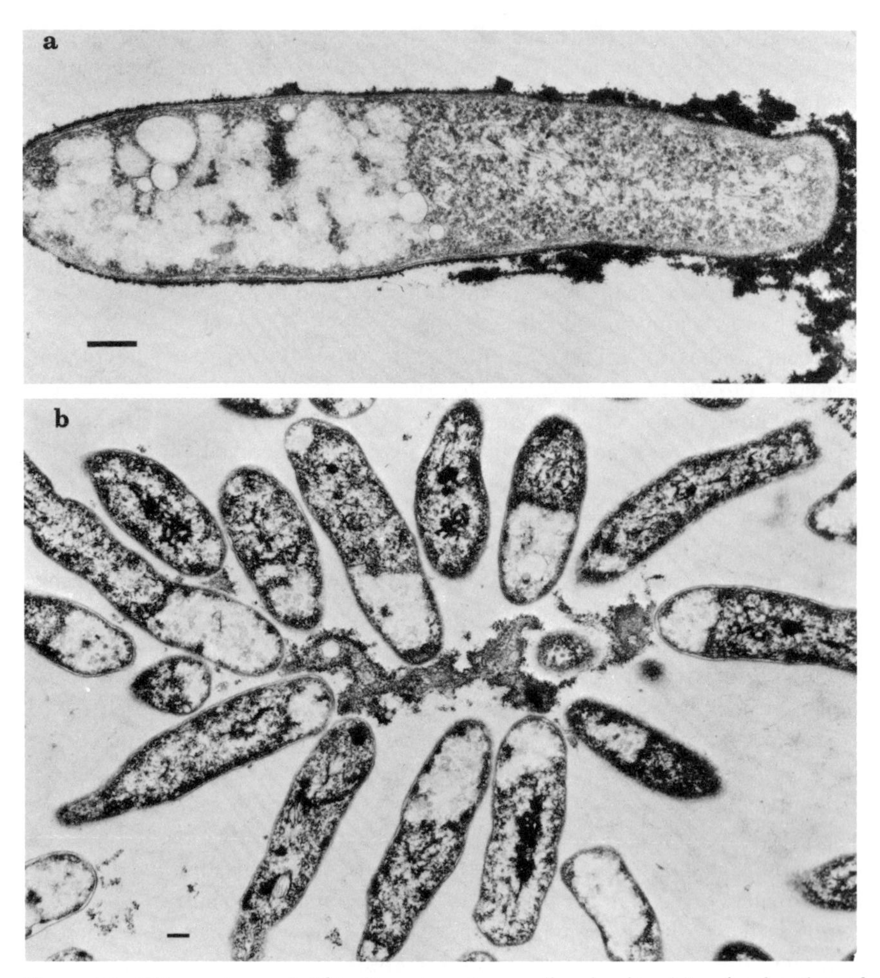

Figure 4.5 Thin sections of *Rhizobium japonicum* cells, showing (a) polar location of extracellular polysaccharide, stained with ruthenium red (bar = 0.2 μm) and (b) polar orientation of cells in aggregate with polymer-bearing ends toward center of aggregate (bar = 0.2 μm) [77].

engulfed a saprophytic bacterium, *Pseudomonas putida*, after its initial attachment to the host wall. Pathogens, on the other hand, did not adhere to the wall but divided extensively, filling intercellular spaces and causing a deterioration in the internal structures of the plant. The immobilization reaction directed against *P. putida* may be a plant lectin–bacterial lipopolysaccharide interaction. Only *P. putida* was agglutinated by lectin from *Phaseolus vulgaris*.

4.3.2 Role of Fimbriae (Pili) and Flagella in Attachment

Ottow [81] has discussed bacterial fimbriae or pili with regard to types, properties, and functions. The term *fimbriae* he reserves for nonflagellar appendages other than those clearly involved in transfer of viral or bacterial nucleic acids. The latter he designates with the term *pili*. The fimbriae tend to increase "active" surfaces of the cell, which according to Ottow [81] enhance respiration and nutrient uptake, but they also may facilitate attachment by overcoming the initial electrostatic repulsion barrier that exists between the cell and the substratum.

The fimbriae, like the flagella, probably originate in the cytoplasmic membrane and extend through the cell wall as hair-like filaments. Brinton [17] indicated that "tactile pili" are composed of protein and contain quantities of amino acids with nonpolar side chains, which presumably increase their hydrophobicity. The type 1 fimbriae [82] enabled enteric bacteria to stick to fungi, plant, and animal cells, including epithelial and red blood cells; but they may not be essential for colonization of the intestine or establishment of infections. The agglutination of erythrocytes by type 1 fimbriated cells of various enteric bacteria was blocked by D-mannose and methyl-α-D-mannoside [83]. The α configuration at the C-1 (first carbon) position in D-mannose was most important for effective blocking of the agglutination.

Many fimbriated bacteria tend to form a pellicle in broth at the liquid–air interface. *Pseudomonas*, *Agrobacterium*, and *Rhizobium* species form "stars" in liquid medium and, according to Ahrens et al. [84], polarly fimbriated cells first aggregate and then contract their polar fimbriae into star-shaped clusters. The contraction of fimbriae may have some significance as a mechanism of firm adhesion [85], and as a mechanism for phage infection of *P. aeruginosa* [18]. According to Mayer [85], the fimbriae disintegrate and form a number of parallel fibers, presumably increasing the strength of adhesion. Whether such a process could occur between a properly conditioned substratum and fimbriated cells is not known at this time. Many aquatic bacteria have been shown to possess fimbriae [39]. Recently, MacRae et al. [86] have shown the presence of possible fimbriae in 19 of 23 strains of gliding bacteria. Some similarity exists between such fimbriae and the extracellular, fibrous material secreted by the gliding bacterium, *Flexibacter polymorphus* [87], which is discussed in Section 4.3.3 under the subheading "Gliding bacteria." The blue–green bacterium, *Anabaena* sp., was shown by Leak [88] to be surrounded by a mucilagenous sheath. When negatively stained or examined in thin sections, it was found to be composed of fine, radiating fibrils or glycocalyx.

Swanson [89] suggested that pili or fimbriae enhance attachment of gonococci to epithelial cells, human sperm, and erythrocytes [90] but play a minor role in the interaction of gonococci with polymorphonuclear leucocytes. Novotny et al. [91] have suggested that, although examples of the involvement of pili in attachment of gonococci to tissue-culture cells could be demonstrated, there were numerous exceptions. The non pilus, gonococcus surface component (leukocyte association factor) seems to be more important for that interaction. The factor is believed to be at least in part a protein, presumably distinct from pilus protein [89].

The fibrillar surface coating of *S. pyogenes* and certain other *Streptococcus* species [92] may contain a trypsin sensitive, type-specific *M* antigen that may become attached to the pharyngeal epithelium [93]. Since *M*-negative mutants attached poorly and antibodies to *M*-protein-inhibited attachment, Ellen and Gibbons [94] suggested that *M* protein and the fibrils were one and the same. Beachey and Ofek [95], however, showed that proteolytic digestion of the streptococci removed stereotypically active *M* protein but had no visible effect on the fimbriae or their capacity to adhere to mucosa. The fimbriae contained both *M* protein and lipoteichoic acid; only the latter component is thought to function in adhesion [95].

The value of flagella to motile, periphytic bacteria may be to permit chemotactic responses to nutrients concentrated at solid surfaces and to help them remain in close proximity with the surface until firm attachment by some other means can be accomplished, at which time flagellar motion and Brownian motion cease [29,32]. Scheffers et al. [96] suggested that adhesion of bacteria to particles may be enhanced by the presence of flagella. The interaction could be maintained continually between flagella and solid substrates with the flagella actually participating in bridging the gap from cell to substratum. The swarmers of *Caulobacter* sp. are visualized as being able to settle with the flagellar end in close proximity to the substratum. As settlement begins, flagella are lost [97] and secretion of an adhesive substance and development of a stalk begins [98].

Monoflagellated *Bdellovibrio bacteriovorus* attacks host cells by striking them with great impact [99]. Then, with an "arm-in-socket" movement on the host cell, it creates an irreversible contact. Although no holdfast structure is observed [100], strong bonding rather than mechanical invagination is suspected. Ultimately entrance of the host cell is accomplished through a pore created by enzymatic digestion.

4.3.3 Organisms Reported to Have Morphologically Distinct Holdfasts

Sheathed Bacteria. The attachment of microorganisms to surfaces by means of holdfast structures has been described, but little is known about their chemical nature. Swarm cells of *Sphaerotilus natans* become attached by a special base on which a filament of cells develop [44]. The filament is held together by a sheath. The sheath of this organism has been studied by Romano and Peloquin [101], who found it to be composed of a protein–polysaccharide–lipid complex that was chemically and anatomically distinct from the cell wall and slime layer. The latter is composed of a polysaccharide [102]. Both the sheath and discoid base became encrusted with iron and gave a strong prussian-blue reaction. Dias et al. [103] found that both sheath formation and cell adhesion required calcium. It is possible that the composition of sheath and holdfast are the same. In a comprehensive review of the *Sphaerotilus–Leptothrix* group, Dondero [104] acknowledged the observation of holdfast structures by numerous workers but points out that cells of *Sphaerotilus* can adhere without sheaths or discernable holdfasts. He believes that holdfasts are probably formed after adhesion is established by some other means. The other sheathed bacterial genera (*Leptothrix*, *Phragmidiothrix*, *Crenothrix*, and *Clonothrix*) become attached to surfaces, but no more than superficial descriptions of the holdfast structures are available [105].

Corpe [36] suggested that iron- or manganese-oxide-depositing forms become embedded in mineral organic encrustations and form an effective union with solid substratum. The same is true of bacteria that secrete copious quantities of extracellular slime, as mentioned in a previous section. The voluminous film is impressive, but this may be misleading if a primary mechanism of adhesion is sought.

Stalked and Budding Bacteria. Species of stalked and budding bacteria have been identified as the numerically important part of the periphytic bacteria in seawater [35]. They are also found in fresh waters [106,107].

Swarm cells of *Caulobacter* sp. secrete a small mass of adhesive material at the flagellar end of the cell. As the organism matures and loses the flagellum, a stalk emerges as a structure continuous with the cell wall and membrane of the Gram-negative organism. The holdfast remains at the distal end of the elongating stalk. In laboratory culture, *Caulobacter* cells may adhere to each other in rosettes of perhaps 100 cells, joined through a single mass of holdfast material, 0.3–0.5 in diameter [106] (Fig. 4.6).

Species of the genus *Asticcaulis* secrete a small mass of adhesive material at a subpolar position, not at the end of their eccentric stalks. Poindexter [106] concluded that in *Asticcaulis* the flagellum and stalk develop at one site, with holdfast material secreted at another. She found the composition of holdfast material was neither a periodate-sensitive carbohydrate nor a trypsin-sensitive protein [106]. Its exact chemical nature is unknown.

Marine caulobacters, freshly isolated from surfaces submerged in seawater, showed a fairly prominent holdfast when attached cells were

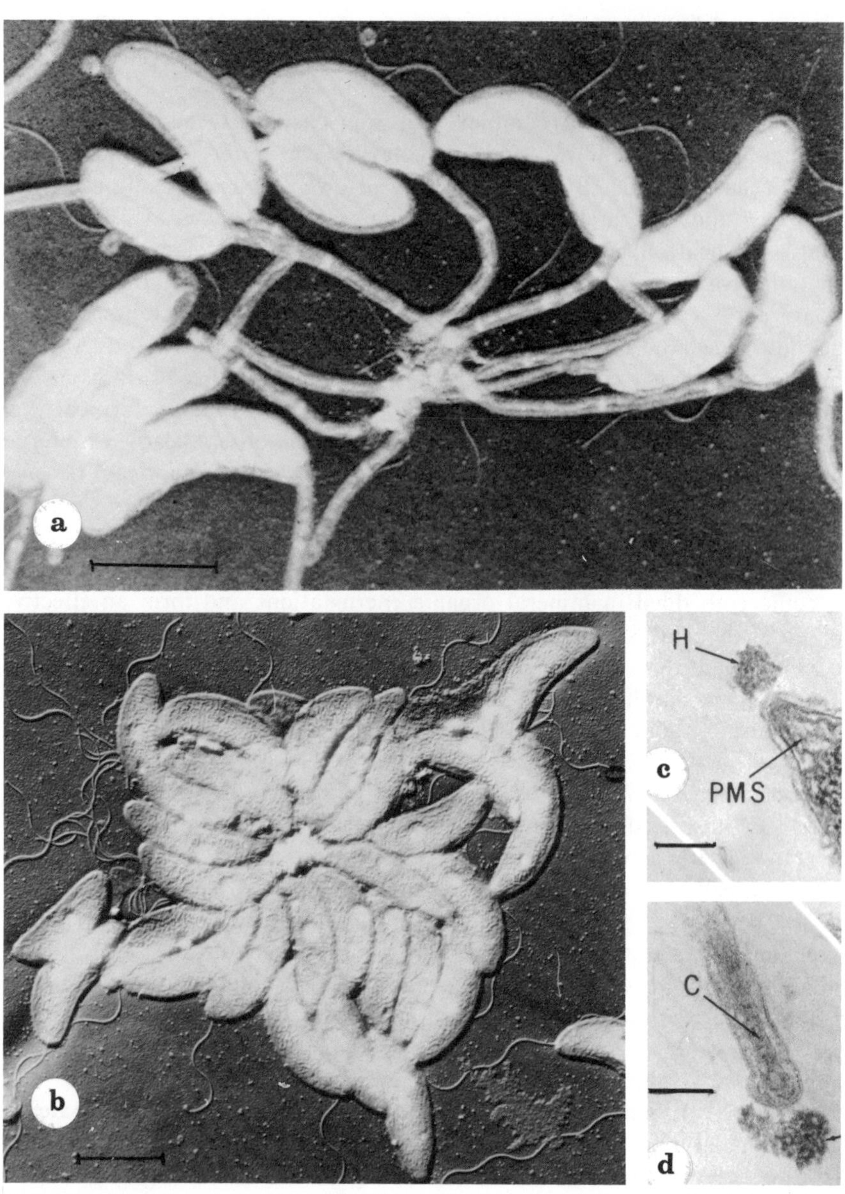

Figure 4.6 Electron micrographs of holdfast material in *Caulobacter* spp. [106]: (a) stalked cells; (b) swarm cells in rosettes, with cells adhering to common mass of holdfast material by tips of stalks or ends of cells (bar = 1 μm); (c) a median section of vibroid strain through developing stalk; (d) section of distal end of stalk showing granular holdfast material (H) section of distal end of stalk showing granular holdfast material (H) (PMS = membrane structure at the base of the stalk, C = core of stalk) (bar = 0.1 μm).

stained with 1% crystal violet after fixation in 2% acetic acid. The holdfast could also be stained with alcian blue and was tentatively considered to be acidic polysaccharide [35]. After some months in culture, holdfasts seemed less prominent, but attachment of the organism to glass was unaffected (Corpe, unpublished observations). Tyler and Marshall [108] reported that, whereas *Hyphomicrobium* cells were strongly adsorbed, holdfast structures were not observed with the light microscope. According to Marshall and Cruickshank [42], attachment of *H. vulgare* and perhaps other aquatic bacteria to solid surfaces is mediated by extracellular fibrous or amorphous materials, polysaccharide in nature.

Corpe [35] reported that caulobacters became most abundant on glass slides that had been immersed in the sea for a long enough period to be well populated with the film-forming pseudomonads. Whether the latter provided a special surface or specific nutrients has not been determined, but according to Poindexter [106], *Caulobacter* sp. attach more readily to other microbial cells than to other surfaces in nature. Stove and Stanier [109] found that adhesion of caulobacter cells into rosettes is not a strain specific phenomenon. Furthermore, it is conceivable that extracellular polymers secreted by pseudomonads greatly facilitate rate of adhesion of caulobacters to various surfaces.

Hyphomicrobium sp. have been recognized as numerically important periphytes that attach to glass and metal surfaces immersed in the sea. Like the marine *Caulobacter* species described above, they seem to require a surface prepared by pioneer pseudomonads [34,35], which, in turn, seem to require a chemically conditioned surface [6]. Marked improvement in the wetting properties of surfaces may be facilitated by sorption of monolayers of macromolecules from natural seawater [30].

Hirsch and Conti [110,111] observed the formation of rosettes by methanol-utilizing strains of *H. vulgare*. They suggested that rosette formation was due to excretion of a "holdfast" by young swarmer cells from one pole (Fig. 4.7). Marshall and Cruickshank [42], however, found no obvious physical contact between cells in a rosette when observed with the light microscope, but observation of serial thin sections of a rosette by electron microscopy suggested that cells were connected by the extracellular fibrous material mentioned previously. Firm adhesion of cells into rosettes or to conditioned surfaces seems to occur by hydrophobic bonding followed by reinforcement of the attachment site with excreted adhesive materials.

Species of *Gallionella*, *Metallogenium*, *Kusnezovia*, *Planctomyces*, *Nevskia*, *Seliberia*, *Blastobacter*, and *Pasteuria* are all described as having a propensity for attachment to surfaces [105], but the nature of the organelles of attachment has not been described [112]. *Gallionella ferrugenia* produces a stalk composed of ferric hydroxide strands that attaches to one side of the

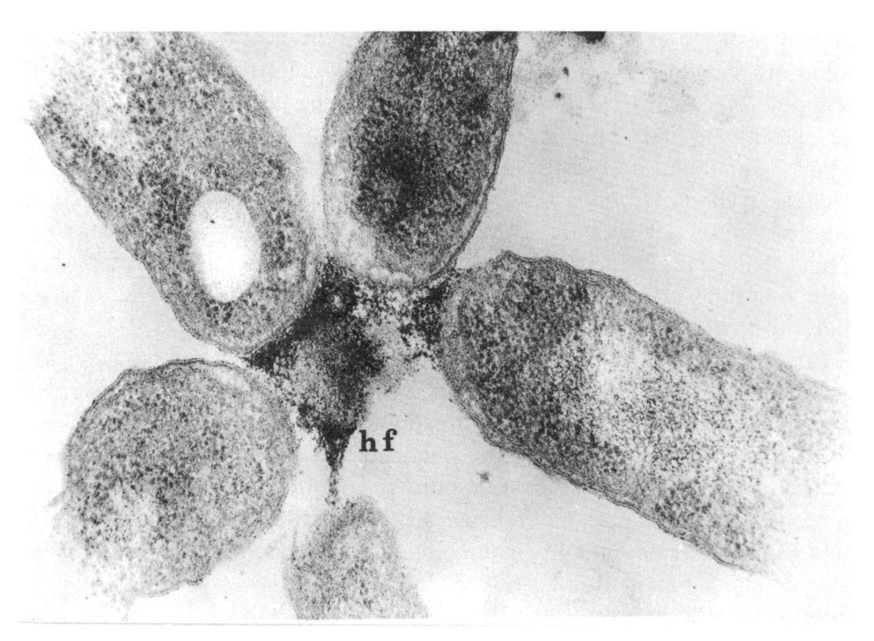

Figure 4.7 Holdfast material in *Hyphomicrobium* strain as seen in section through rosette of cells. Cells attached to each other by means of holdfast material (hf) (×60,125) [111].

bean-shaped cell [113]. Enrichment cultures prepared according to the procedure outlined by Wolfe [113] showed *Gallionella* colonies attached to the sides of the culture tubes. Cells are not attached firmly either to the secreted iron hydroxide or to the test-tube walls.

Gliding Bacteria. There are reported examples of adhesion by gliding bacteria to solid surfaces by inconspicuous holdfasts. Most of the organisms secrete quantities of slime, which presumably aid their movement but might also be involved in adhesion. Brock and Conti [114] described holdfasts in cultures of *Leucothrix mucor*. Rosettes of cells were joined through an unstructured holdfast, which could be observed when stained with congo red according to the method recommended by Pringsheim [115]. Ridgway et al. [87] described "goblet"-shaped particles on the surface of the marine gliding bacterium, *Flexibacter polymorphus*. The particles were associated with the outer lipopolysaccharide component of the cell envelope. Examination of negatively stained cells showed long fibrils generally oriented at right angles to the walls, which were believed to be secreted through a duct located in the stem of the goblet-shaped particles. These "macromolecular pores" (Fig. 4.8) apparently mediate extrusion of the fibrils, possibly as a means of

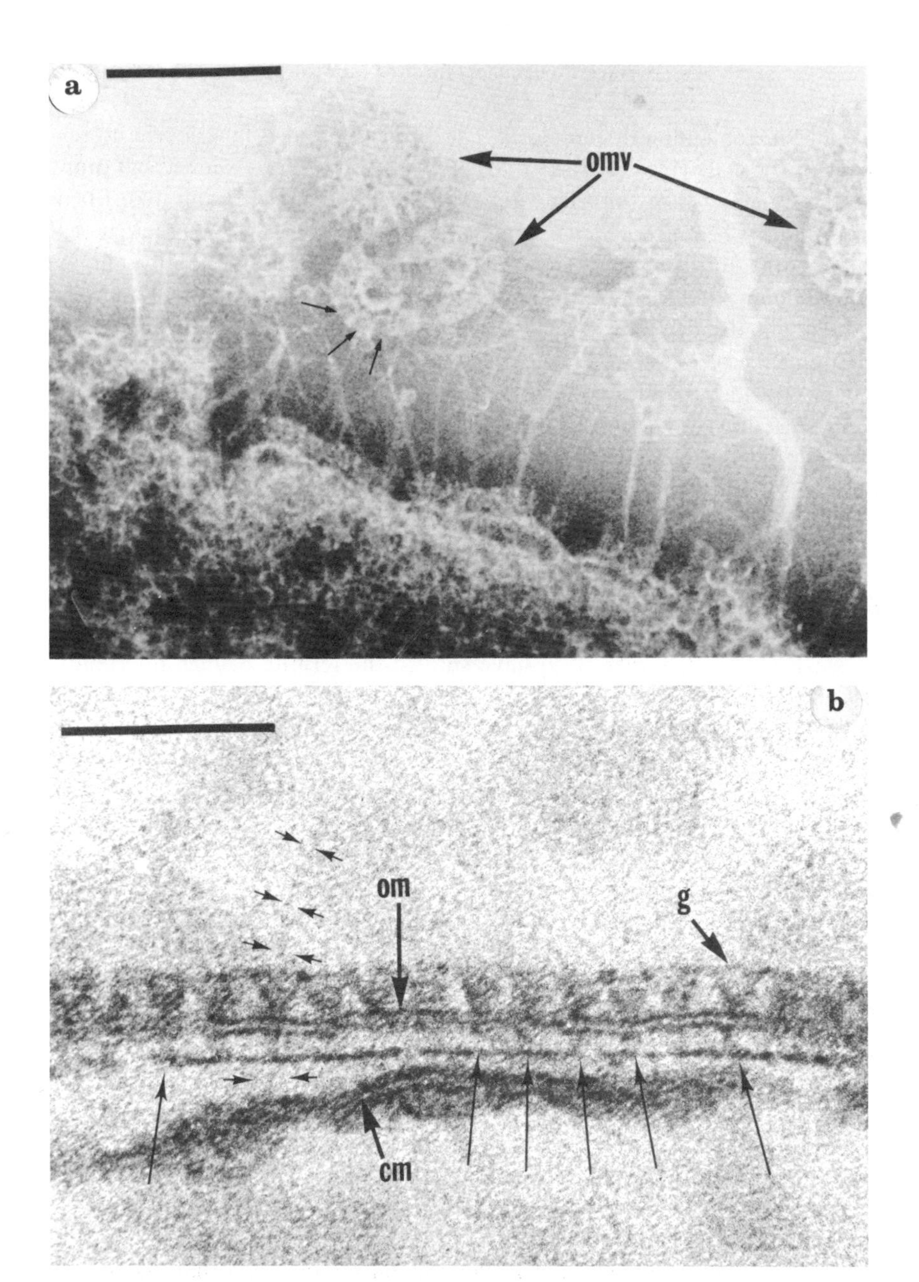

Figure 4.8 Electronmicrographs showing: (a) negatively stained autolyzed cells of *Flexibacter polymorphus* showing outer membrane vesicles (OMV) coated with goblet shaped structures, some bearing a goblet filament (unlabeled arrows) about 25 Å in diameter (bar = 200 nm) [87]; (b) section of envelope of autolyzed cell showing "dense bodies" (long unlabeled arrows) below each goblet arranged along outer membrane, apparently making contact with dense intermediate layer. In one case where a goblet filament appears to be present (short unlabeled arrows), a further extension to cytoplasmic membrane is evident (bar = 100 nm) [87].

propulsion for gliding locomotion. The fibrils may also function in attachment of the cells [116] (Fig. 4.9). MacRae et al. [86] have shown that many gliding bacteria have fimbriae, some of which look remarkably like fibrils produced by *F. polymorphus*. MacRae and McCurdy [117] suggested that the gliding motility of *Chondromyces crocatus* was facilitated by extracellular slime and polar fimbriae. They suggested that slime and fimbriae might also be related to cell attachment to solid surfaces.

Filaments of crescent-shaped cells of *Simonsiella spp.* glide when in contact with a substratum. The broad side of the crescent-shaped cell, referred to as the *ventral side* [118], makes intimate contact with substratum through fibrillar components. Presumably the fibrillar material is related to gliding motility and most actively motile strains are highly "capsulated."

Other Bacteria. Wood-eating insects exhibit a high incidence of procaryote–procaryote and procaryote–eucaryote associations. The organisms often are intimately associated to the extent they have evolved a specific attachment organelle.

Breznak and Pankratz [119] have studied the termite paunch microbiota in two termite species. Paunch epithelium was colonized densely by bacteria, many of which possessed holdfast elements that secured them to tissue and to other bacterial cells. Of particular interest are the abundant fibrils, observed in thin sections, that bridge small flattened rod-shaped bacteria to long filaments of spore-bearing bacterial cells (Figs. 4.10). Terminal fibrils are produced by some forms that show end-on attachment (Fig. 4.11). The association of spirochetes and rod-shaped bacteria with the surface of flagellate protozoa is of particular interest here because of the specialized structures they use to maintain permanent attachment.

Spirochetes associated with *Barbulanympha* from *Cryptocercus* and *Pyrsonympha* from *Reticulitermas flavipes* have a narrow nose-like appendage (60–70 nm wide and 0.4 nm long) that makes contact directly with the host-cell plasma membranes (Fig. 4.12). The other type of spirochete attachment specialization has been observed on *Pyrsonympha* from *Reticulitermes tribialis* [120]. The end of the spirochete in contact with the protozoan membrane is flattened and contains a thick layer of dense material *Pyrsonympha* from other *Reticulitermes* species have a different mode of attachment [121,122].

The flagellate protozoa *Urinympha* and *Barbulanympha* that occupy the hind gut of the wood-eating roach *Cryptocercus* have surfaces that are occupied almost entirely by bacteria; all but the flagellated regions are covered by a regular array of rod-shaped bacteria 3–4 μm in length and 0.5 μm in width (Fig. 4.13). The attachment sites on the host membrane are complex in structure. Every bacterium on the surface is in contact with a

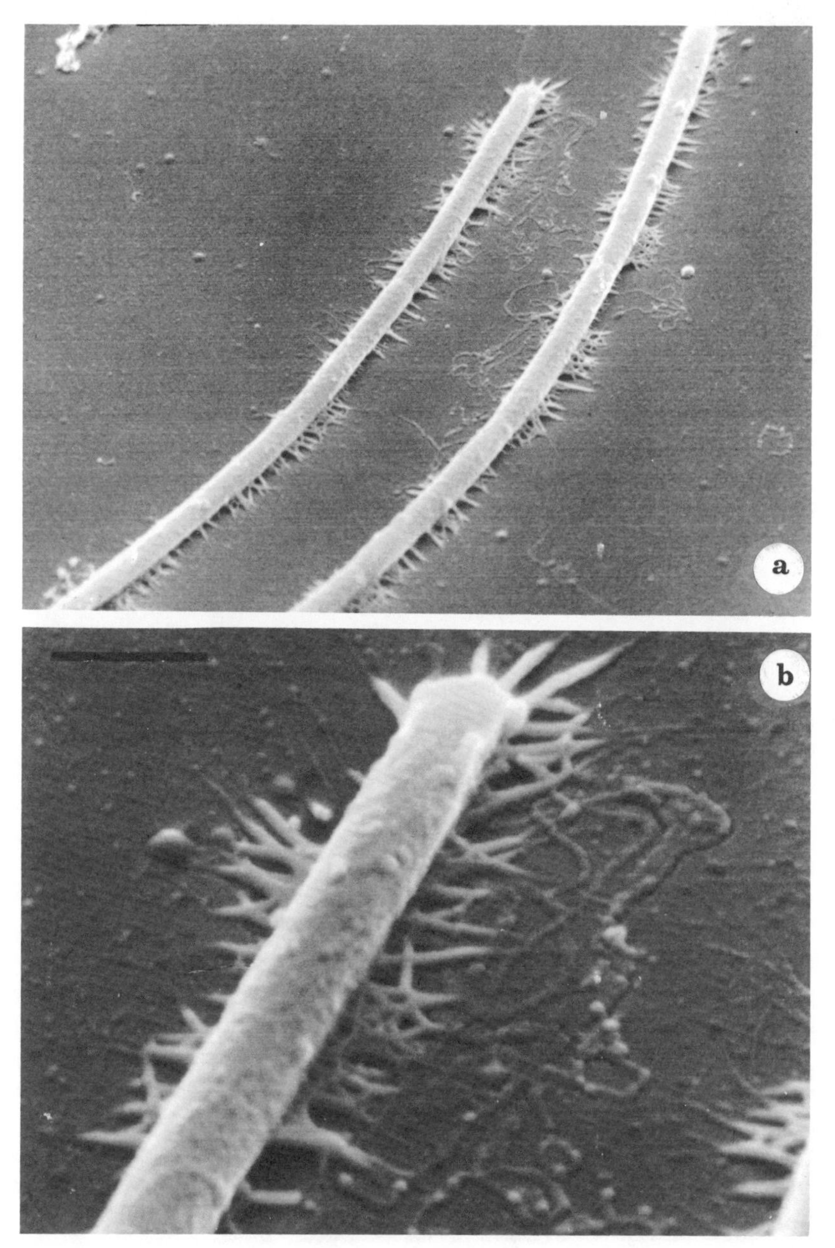

Figure 4.9 Scanning electron micrographs showing trichomes of *Flexibacter polymorphus* rapidly fixed with glutaraldehyde while gliding on surface of glass microscope slide. Bundles of axial goblet fibers appear to mediate attachment of cells to solid surface [(a) bar = 3 μm); (b) bar = 1 μm)]. Courtesy of H. F. Ridgway).

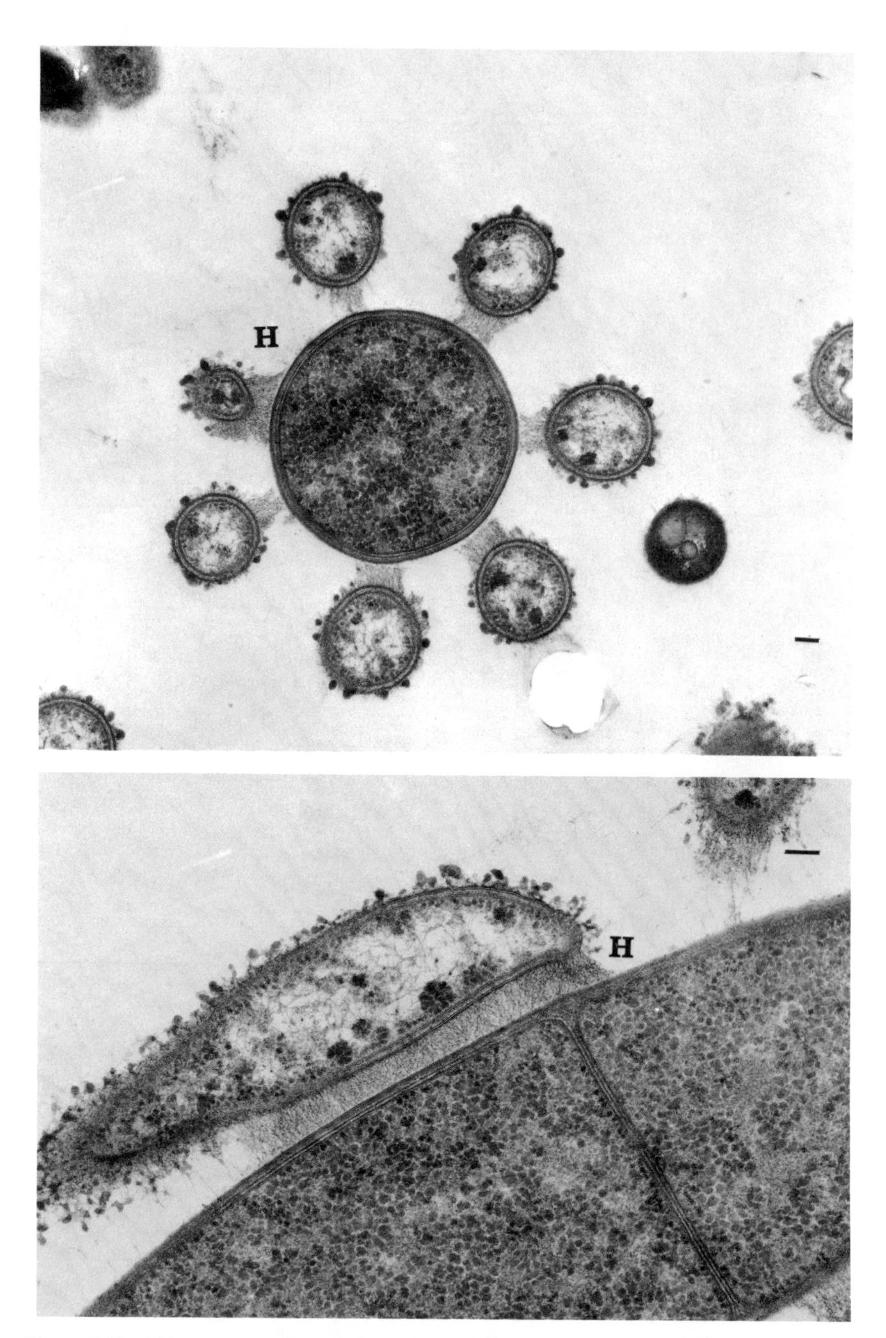

Figure 4.10 Thin sections through bacteria associated with paunch epithelium of termite, showing bacteria of one species attached to larger cells of another by means of fibrous holdfast material (H) produced by cells with smaller diameter. Sections were stained with ruthenium red (bar = 0.1 μm).

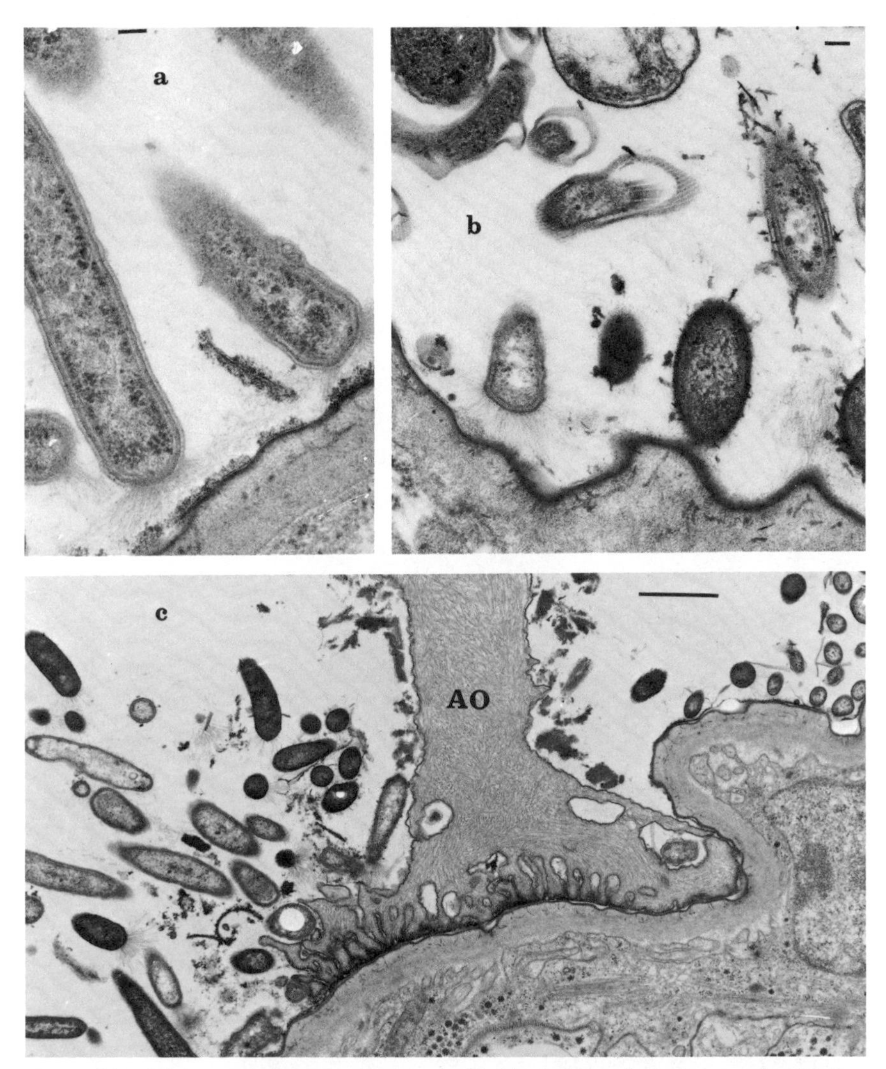

Figure 4.11 Electron micrographs showing: (a, b) thin sections of termite paunch epithelium showing bacterial morphotypes attached by polar fibrils (bar = 0.1 μm); (c) thin sections of protozoa (*Pyrsonympha vertens*) attachment organelle (AO) associated with termite paunch epithelium. Polarly attached bacteria adhere to both protozoa and host epithelium. Sections stained with ruthenium red (bar = 1 μm).

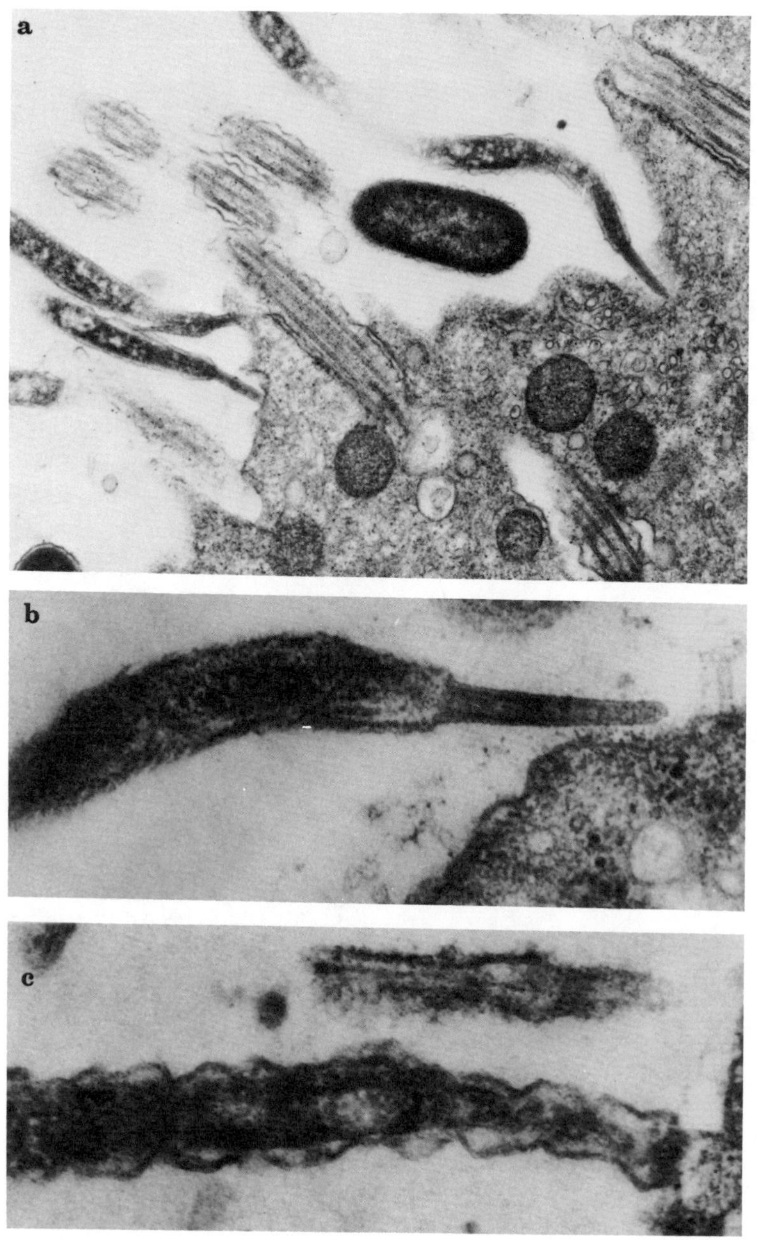

Figure 4.12 Electron micrographs showing: (a) attachment of spirochetes to surface membrane of flagellate protozoan *Barbulanympha* from wood-eating roach *Cryptocercus* (×23,750); (b) attachment of spirochete to cell surface membrane of protozoan *Pyrsonympha* from *Reticulitermes flavipes* (×70,000); (c) attachment of spirochete to cell surface membrane of *Pyrsonympha* from *Reticulitermes tibialis* (×70,000).

membrane specialization consisting of parallel electron-dense plates that seem to bracket the bacterium (Fig. 4.14). The brackets are observed only where the bacterium contacts the protozoan surface. The bacteria that occur on the surface are also found in the cytoplasm, where they exist as symbionts [69], and attached to membranes of cytoplasmic vacuoles. No destruction of bacteria or host tissue was observed.

It is interesting to note that the attachment of *Vibrio cholerae* to the surface of intestinal microvilli involves no special organelle nor any special accommodation by host cells. Fragmentation and a general destruction of the surface occurs [64].

4.4 ATTACHMENT OF FUNGI

Filamentous fungi are found widely distributed in soil, water, and on surfaces of almost every description. Fungus hyphae may spread across plant or animal surfaces or organically enriched mineral substrata and resist removal by rinsing. Jones and La Campion-Alsamard [123] have shown that fungi can attach to and colonize polyurethane panels submerged in the sea and degrade that polymer. Colonization by a large number of species of fungi of fresh wood exposed to rapidly moving cooling tower waters was observed by Eaton and Jones [124].

The attachment and growth of saprophytic fungi in rooted green leaves of *Antirrhinum majus* was studied by Collins [125]. The leaves supported populations of *Sporobolomyces roseus* and *Cladosporium cladosporioides*. No mechanism by which fungi became attached was suggested in *any* of these examples.

The fact that the fungi resist removal by rinsing suggests either that the fungus cells are protected by irregularities or surface structures characteristic of the substratum or that attachment is facilitated by some property of the fungus, or both.

The actual penetration by hyphae into "open" substrata by rhizoids may serve to anchor organisms to surfaces. The mechanism by which decay fungi penetrate material like wood has been reviewed by Wilcox [126]. Bore holes are produced by enzymatic digestion, which occurs in advance of the hyphal tip. Decomposition of cellulose and lignin in localized regions has been observed. Constriction of the hyphal strands allowed penetration of small bore holes into the host cells by various decay fungi.

Many of the filamentous fungi and yeasts secrete extracellular polysaccharides [127] with adhesive qualities. Crandall and Brock [128] showed that sexual agglutination in *Hansenula wingii* was achieved by the complementary interaction of two glycoproteins present on the cell surface of the

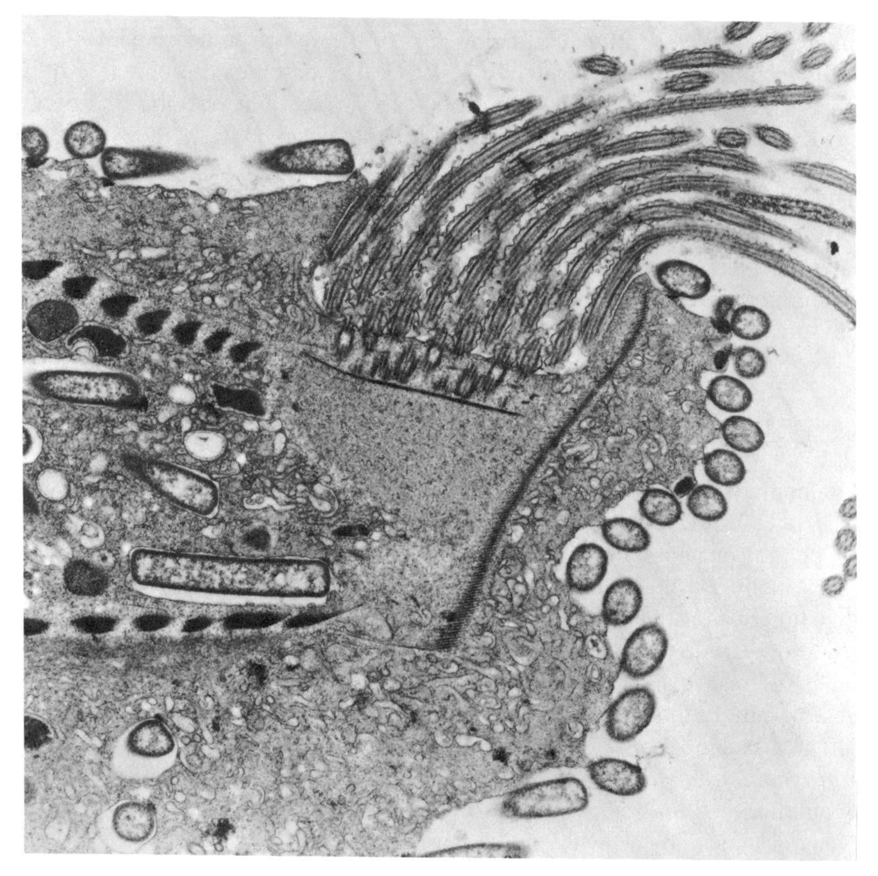

Figure 4.13 Anterior end of protozoan *Urinympha* from wood-eating roach *Cryptocercus* showing group of flagella and bacteria associated with cell surface and found within cytoplasm (×10,375).

respective mating types. The spores of many aquatic species are provided with appendages or gelatinous sheaths that, presumably, facilitate the attachment to substratum [129], an event that may be quite important for the initiation of the life cycle of the organism.

Zoospores of *Pseudoperonospora humuli* settle with the area of the flagella insertion facing the leaf surface. A germ tube that arises from the flagellar end penetrates directly between the stomatal guard cells, anchoring the fungus to the host. At high magnification the encysted *P. humuli* zoospore was anchored by a "network of strands" to the leaf surface. It is not certain whether these strands originate as secretory products of the spore or of the infested host [130].

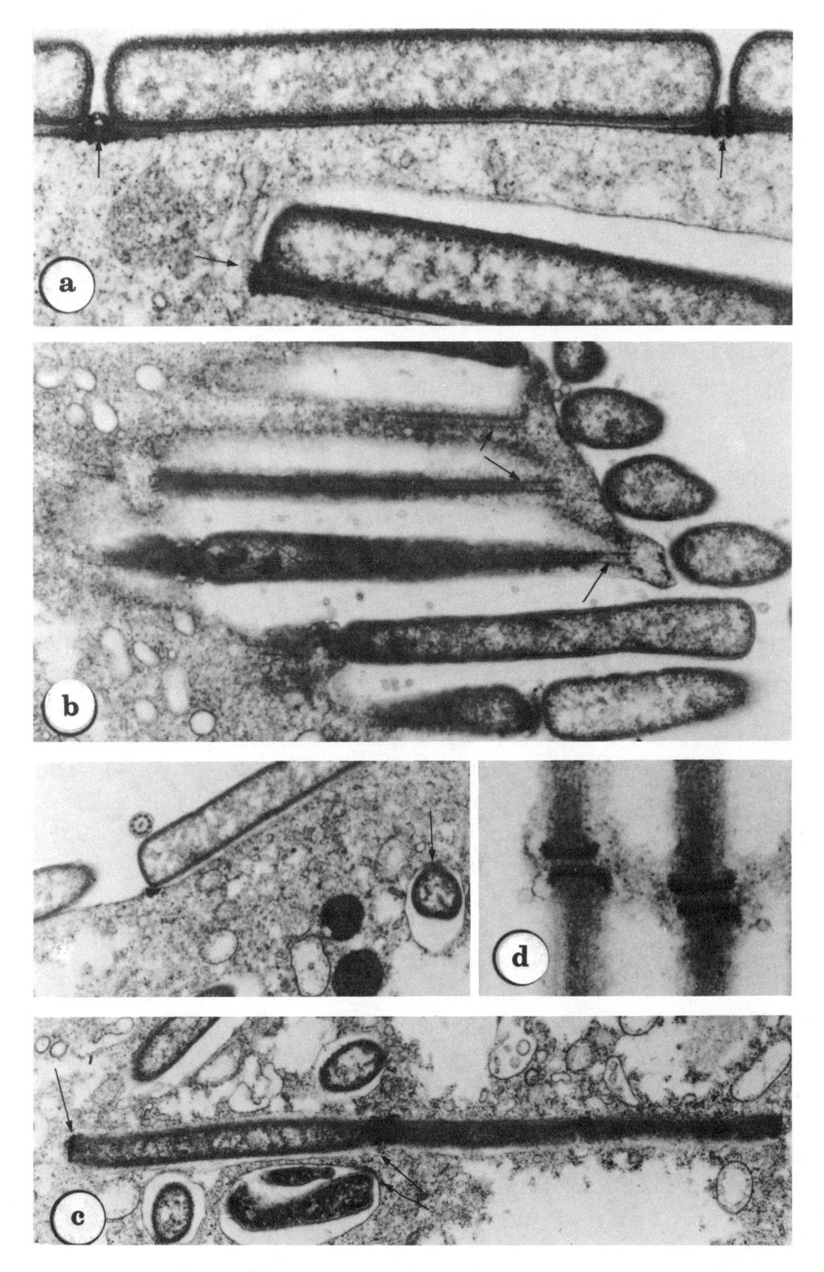

Figure 4.14 (a) Rod-shaped bacteria on surface and within cytoplasm of *Urinympha* from *Cryptocercus* are attached by means of densely staining specializations of protozoan membranes (arrows) (×235,000); (b) associations of rod-shaped bacteria with surface of *Barbulanympha* from *Cryptocercus*—section cut tangential to surface showing "railroad-track" array between bacterial surface and protozoan cell membrane (arrows) (×18,875); (c) rod-shaped bacteria within cytoplasm of *Urinympha* from *Cryptocercus* showing their association with densely staining attachment structures. (arrows) (×15,000); (d) high-magnification micrograph of structures anchoring rod-shaped bacteria to surface of *Urinympha* from *Cryptocercus* (×33,500).

Zoospores of *Phytophthora palmivora* tend to become attached to any substratum onto which they settle [131]. Adhesion occurs in the early stages of encystment through the partially discharged contents of peripheral vesicles. These were observed in thin electron-microscopic sections of encysting zoospores on plastic. It seems that the material contained in the peripheral vesicles serves in both cell adhesion and in construction of cyst walls [131].

Biflagellated zoospores of thraustochytrid fungi produce a holdfast "rhizoid" system that arises from an organelle similar to that forming slime tracks in *Labrynthula* [132]. The ectoplasmic nets of the thraustochytrids are quite different from chytrid fungus rhizoids, which are delimited by walls and may contain certain cytoplasmic organelles. Ectoplasmic nets consist of anastomosing membrane arrays containing no cytoplasmic organelles. Motile spindle cells of *Labyrinthula* are embedded in ectoplasmic nets with many cells sharing nets and restricting movements to the slime tracks developed on the solid substrata [132].

Penetration of insect integuments by pathogenic fungi is a matter of some interest in view of the possible use of fungi in control of insect pests. Zacharuck [133] has shown that the deuteromycete *Metarrhizium* forms an elaborate cushion of appressoria to attach the fungus to the host. Penetration of an insect larva is then effected by a fine-penetration peg formed by one of the appressoria. On the other hand, the insect pathogen *Eutomophthora apiculata* does not form an appressorium for attachment [134]; rather, a rapidly growing germ tube physically displaces the epicuticle. The spore may attach to the substratum before development of the germ tube by a tactile projection that bridges the two surfaces.

4.5 ATTACHMENT OF ALGAE

Algae are major primary producers of organic matter and occupy a position of enormous importance in the aquatic food web leading to the production of fish [135]. They occur in all types of soil and water in every latitude and range in size from microscopic to the enormous kelp. Whereas most of the important reference texts dealing with algae do not discuss the physiology and development of alga attachment or the organelles concerned, it is clear that the propensity for attachment is recognized as an ecologically important property.

Algae referred to as *haptophytes* become attached to substrata by means of basal cells, disks, or haptera and do not penetrate the substratum to any extent [136]. Their substrata are large stones or rock surfaces. In contrast, the term *rhizophyte* refers to rooting algae that produce a holdfast structure of varying size and complexity that is able to penetrate mud, sand, and loose gravel and in some cases, effectively enclose the loose substratum, providing an anchor. Whereas the structures have a superficial resemblance to plant roots, they do not have their adsorptive and conducting function.

Among the sessile diatoms, several attachment systems of varying morphology have been described by Chamberlain [137]. These involve acidic polysaccharide mucilaginous materials that are distinguished mainly by the site from which they are secreted. The arrangements include forms with an adhesive pad, a mucilaginous capsule, a flattened tubular sheath surrounding colonies and mucilaginous branching stalks. Motile diatoms attach only by raphe material that is deposited in tracts on the substratum, facilitating movement. The presence of sulfate or uronic acid or both in diatom polysaccharides has been confirmed by several groups, as discussed by Chamberlain [137].

Other algal groups also possess adhesive surfaces [138,139]. The settlement and germination of zoospores from various algal groups (Chlorophyceae, Phaeophyceae, and Rhodophyceae) and the subsequent development of attachment systems have been explored by Fletcher [140]. Mucilaginous material was secreted from the apical end of a basal filament, which made good contact with substrata and functioned as an adhesive (Fig. 4.15).

Adhesion to surfaces by carpospores and tetraspores of marine Rhodophyceae involves secretion of mucilaginous substances, which are believed to be produced in the golgi apparatus: (1) a fibrillar polysaccharide and (2) a polysaccharide–protein component. A refractile, hyaline pad that is formed established intimate contact between the spore and solid substratum [137].

4.6 ATTACHMENT OF PROTOZOA

Colonization by protozoa on natural or artificial substrata [141] is a matter of considerable interest to ecologists [142]. Protozoa that employ both temporary adhesion and "permanent" attachment to solid substrata are observed in such studies, but the mechanism of attachment or association by the protozoa has not been extensively explored.

The freshwater suctorian, *Tokophaya infusionum*, has an adhesive disk that keeps the organism attached in a suitable location throughout its life [143]. The organelle is developed by the ciliated, spinning, motile "embryo" that settles with the anterior end toward the substratum. An adhesive disk is secreted as a dense meshwork of fine fibrils only 30 sec after settlement, and a stalk is formed at the same time. Both are secreted from dense bodies located near the cell periphery and are believed to be of common origin.

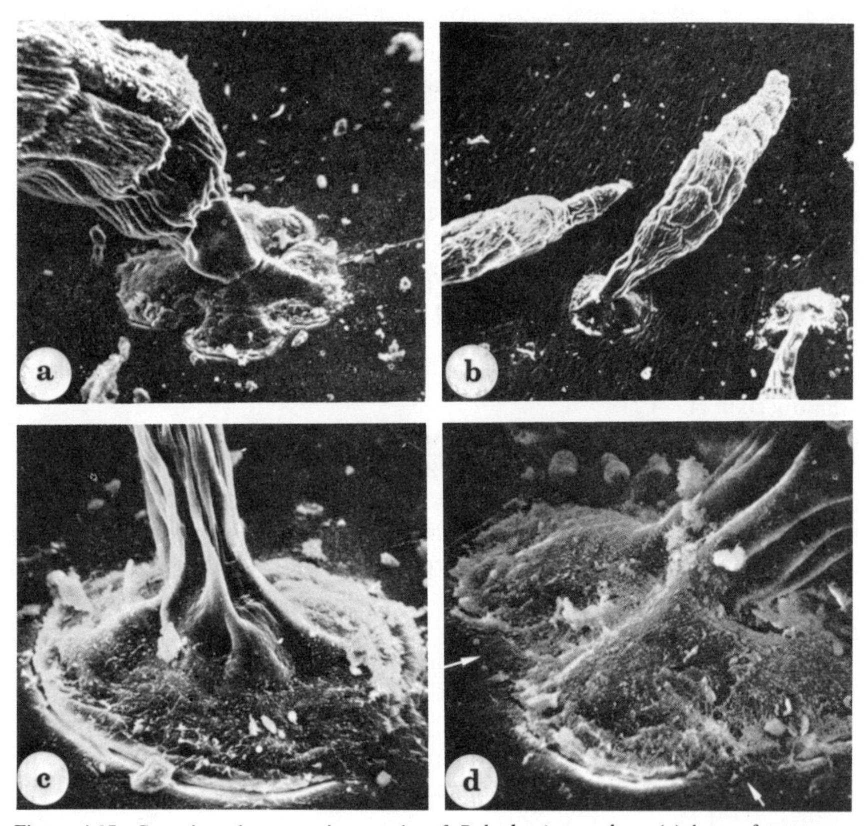

Figure 4.15 Scanning electron micrographs of *Polyphonia ureolata:* (a) base of tetraspore germling showing lobed structure of attachment disk (× 2950); (b) 7-day-old tetraspore germling grown on petri dish floor (×1225); (c) basal disk (×6625); (d) basal disk. Note extension of disk material onto substrate (arrowed) (×5875).

The stalk is connected to the body of the cell with only short bridges of fibrous material. The nature of the adhesion of the disk to substrata has not been clarified. Hascall [144] has shown, on the basis of histochemical staining reactions, that the stalk is probably a sulfated protein–polysaccharide complex.

The attachment of the gut flagellate *Giardia* has been studied by Holberton [145]. *Giardia* trophozoites are attached by a suction disk that lies beneath the cell, the rim of which presses into and interdigitates with the intestinal microvilli of the host. The cell is held in position by suction pressure, which may arise from the viscous stress of a fluid flow led beneath the cell by flagellar activity. Suction pressure is estimated to reach 10^2 dynes

cm^{-2}. Attachment and detachment from epithelial cells of intestinal microvilli can be controlled by the organism.

The ciliate *Epididinium ecaudatum*, found in the sheep rumen, is able to attach and colonize fragments of plant materials undergoing digestion. The protozoa were found in large numbers beneath the epidermis on the cut surface of plant material or where physical damage was obvious [146]. The protozoa apparently have no attachment organelles.

The three examples described in this section may represent types of attachment in the protozoa. They are: (1) actual secretion of a specific adhesive by an immature form that makes intimate contact with the surface and gives rise to a holdfast that persists in the adult animal, (2) the adhesion by some suction mechanism with force generated by motion producing thrust and/or cell deformation with surface contact made through a coating of surface slime, and (3) the capacity to seek out a protected niche and maintain position therein, mainly through constant motion.

4.7 CONCLUSIONS

In seeking surface components of microbial cells that function in attachment processes, one must consider those that will form adhesive bonds with an opposing site on a solid surface. The closeness with which a cell can approach an opposing surface is, according to Weiss [147], a most important consideration. Electrostatic, covalent, and hydrogen bonding are chemical interactions that act over distances of 2–3 Å and van der Waals physical bonds are effective over distances of 3–4 Å. However, in view of the overlapping of electrical double layers that surround charged surfaces bathed in electrolyte solutions, an electrical repulsion force develops that keeps surfaces apart. Some mechanism must then be available to the cell so this barrier can be breached [29].

In the foregoing sections, studies have been cited that describe various microbial surface components that probably can bridge the electrostatic repulsion barrier. In some instances, this is done by the cell exerting the mechanical force due to cell motility. Movement resulting in thrusting and flattening the cell periphery against the substratum [96,100] may provide contact adequate to affect attachment.

Biosynthesis of specific polymeric materials, secreted when the cell encounters a suitable surface, may be required for firm adhesion. On the other hand, cell movement and perhaps orientation of surface layers or minor conformational changes might be sufficient to accomplish adhesion. In any event, energy-requiring processes are probably involved. The formation of major quantities of polymer that are observed in some organisms would require biosynthesis and/or secretion [140,145]. In the case of bacteria, holdfast formation or production of copious quantities of slime may occur as the weight or bulk of the attached cell mass or filament length increases. It may serve to protect the cells, prevent diffusion away or rapid leaching of nutralites, and trap and concentrate nutrients from the bulk water, for use by the slimy cell mass.

The existence of fibrillar surface layers in the bacteria are common [46,88,95,148]. These materials give staining reactions characteristic of polyanionic carbohydrates that may be coupled to proteins [39,41]. The structural complexities and functions of Gram-negative bacterial cell envelopes have been reviewed [149]. Fibrillar materials seem to be associated with the gliding bacteria [116,118] and, presumably, function in motility but also may serve as an adhesive. Bacteria associated with certain invertebrates [119] also display well-developed fibrillar surface structures, which clearly seem to function in adhesive processes. It is interesting that the fibrils seem to occur only when in contact with the substratum. They do not have the structural appearance of the tactile fimbriae.

The "fibrillar net" produced by eucaryotic *Labyrinthula* and related organisms seems to function in motility [131] but also may serve to retain enzymes and nutritive substances within the mass of communicating cells. Such a multifunctional slime may have greatest expression among the cellular slime molds. Cells of *Dictyostelium discoideum*, for example, associate into streamlike assemblages of intercommunicating cells, which may contact each other through sites that are distributed on cell surfaces but may extend into the intercellular space. Regulation of receptor sites for cyclic AMP and cell adhesion occur during development by a common genetic system [150].

The chemistry and structure of the substratum in relation to microbial attachment has not been evaluated thoroughly. The attachment of bacteria to glass or other biologically inert surfaces is more complex than it might seem at first glance. Surface charge, wettability, nutritive conditioning, and other properties are affected by adsorption of soluble materials from surrounding waters. The specific nature of the interaction of the bacterial surface with a conditioned substratum is poorly understood.

Glycoproteins of virus coats may interact with specific carbohydrates on the host-cell surface and be bound irreversibly [13,14]. Surface polysaccharides of *Rhizobium* species attach to root hairs of specific leguminous plant hosts determined by specific lectins [151] or carbohydrate binding proteins exposed most intensely at the tips of root hairs [79]. The *Rhizobium* surface polysaccharides are located, at least in some cases, at a pole of the cell [78]. Cell-bound glycosyl transferases of certain oral streptococci can bind specifically to carbohydrates on the host-cell surfaces [3]. Such an enzyme mechanism may be responsible for clumping of bacteria as well, as the controlling factors are the presence of surface-bound enzyme and the complementary saccharide side chain.

The substratum then can supply specific sites for attachment but in other circumstances, the substratum becomes *structurally adapted* to the attached organism as well. Spiral-shaped bacteria in the gut of the termite have evolved a rather specialized structure by which they attach to flagellated protozoa, coinhabitants of the gut [69]. The protozoa membrane accomodated the elongated holdfast without injury. The membrane also accommodated rod-shaped cells. One presumes that the organisms have evolved these specialized adaptations in perfection of the complex symbiotic relationship. In contrast, the attachment of distinctly pathogenic organisms induces no structural accomodations by the host cells, but host-cell destruction is clearly apparent [60].

4.8 ACKNOWLEDGMENT

I would like to express my appreciation to Dr. Donald Ritchie and Dr. Philip Ammirato for their constructive criticism of this chapter and their valuable suggestions.

Studies in the author's laboratory were supported by the National Science Foundation.

4.9 REFERENCES

1. E. J. W. Verwey and J. Th. G. Overbeek, *Theory of the Stability of Lyophobic Colloids*, Elsevier, New York, 1948.
2. L. Weiss, *The Cell Periphery, Metastasis and Other Contact Phenomena*, Wiley, New York, 1967.
3. R. J. Gibbons, "Adherence of bacteria to host tissue," in D. Schlessinger, Ed., *Microbiology—1977*, American Society for Microbiology, Washington, D. C., 1977, pp. 395–406.
4. M. Fletcher, *J. Gen. Microbiol.*, **94,** 400 (1976).
5. S. C. Dexter, J. D. Sullivan, J. Williams, III, and S. W. Watson, *Appl. Microbiol.*, **30,** 298 (1975).
6. M. Fletcher and G. I. Loeb, "The Influence of Substratum Surface Properties on the Attachment of a Marine Bacterium," in M. Kerker, Ed., *Colloid and Interface Surface*, Vol. 3, Academic, New York, 1976, pp. 459–469.
7. F. Fenner, B. R. McAuslan, C. A. Mimms, J. Sambrook, and D. O. White, *The Biology of Animal Viruses*, 2nd ed., Academic, New York, 1974.
8. M. E. Bayer, *J. Virol.*, **2,** 346 (1968).

9. L. D. Simon and F. F. Anderson, *Virology*, **32,** 279 (1967).
10. A. A. Lindberg, *Annu. Rev. Microbiol.*, **27,** 205 (1973).
11. W. Weidel, *Ann. Rev. Microbiol.*, **12,** 27 (1958).
12. P. J. Bassford, D. A. Diedrich, C. L. Schnaitman, and P. Reeves, *J. Bacteriol.*, **131,** 608 (1977).
13. P. W. Robbins and T. Uchida, *J. Biol. Chem.*, **240,** 375 (1965).
14. R. Losick and P. W. Robbins, *J. Molec. Biol.*, **30,** 445 (1967).
15. J. W. Roberts and J. E. A. Steitz, *Proc. Nat. Acad. Sci.* (*USA*), **58,** 1416 (1967).
16. D. A. Marvin and B. Hohn, *Bacteriol. Rev.*, **33,** 172 (1969).
17. C. C. Brinton, *Transact. N. Y. Acad. Sci.*, **27,** 1003 (1965).
18. D. E. Bradley, *J. Gen. Microbiol.*, **72,** 303 (1972).
19. C. P. Novotny and P. Fives-Taylor, *J. Bacteriol.*, **117,** 1306 (1974).
20. A. Newton, "Adsorption and penetration of viruses," in H. A. Charles and B. C. J. G. Knight, Eds., *20th Symposium, Society of General Microbiology*, Cambridge U. P., Cambridge, U. K., 1970.
21. D. W. Fawcett, *J. Histochem Cytochem.*, **13,** 75 (1965).
22. L. Philipson, S. Bengtson, S. Brishamm, L. Svennerholm, and O. Zetterquist, *Virology*, **22,** 580 (1964).
23. J. J. Holland, *Bacteriol. Rev.*, **28,** 3 (1964).
24. J. J. Holland and L. C. McLaren, *J. Expl. Med.*, **114,** 161 (1961).
25. W. G. Laver and R. C. Vallentine, *Virology*, **38,** 105 (1969).
26. C. F. Moldow, P. Volberding, M. McGrath, and J. J. Lee, *J. Gen. Virol.*, **37,** 385 (1977).
27. C. E. ZoBell, *J. Bacteriol.*, **46,** 39 (1943).
28. M. Fletcher, "Attachment of Marine Bacteria to Surfaces," in D. Schlessinger, Ed., *Microbiology—1977*, American Society for Microbiology, Washington, D. C., 1977, pp. 407–410.
29. K. C. Marshall, "Mechanism of Adhesion of Marine Bacteria to Surfaces," in R. F. Acker, B. F. Brown, J. R. DePalma, and W. P. Iverson, Eds., *Proceedings of Third International Congress on Marine Corrosion Fouling*, Northwestern U. P., Evanston, Ill., 1973, pp. 625–632.
30. R. E. Baier, "Influence of the Initial Surface Condition of Materials on Bioadhesion," in R. F. Acker, B. F. Brown, J. R. DePalma, and W. P. Iverson, Eds., *Proceedings of Third International Congress on Marine Corrosion and Fouling*, Northwestern U. P., Evanston, Ill., 1973, pp. 633–639.
31. R. Neihof and G. Loeb, "The Interaction of Dissolved Organic Matter in Sea Water with Metallic Surface," in V. Romanovsky, Ed., *Proceedings of Fourth International Congress on Marine Corrosion and Fouling*, Antibes, France, 1977, pp. 359–361.
32. K. C. Marshall, R. Stout, and R. Mitchell, *J. Gen. Microbiol.*, **68,** 337 (1971).
33. K. C. Marshall, R. Stout, and R. Mitchell, *Can. J. Microbiol.*, **17,** 1413 (1971).
34. W. A. Corpe, "Microfouling: The Role of Primary Film Forming Marine Bacteria," in in R. F. Acker, B. F. Brown, J. R. DePalma, and W. P. Iverson, Eds., *Third International Congress on Marine Corrosion and Fouling*, Northwestern U. P., Evanston, Ill., 1973, pp. 598–609.
35. W. A. Corpe, "Periphytic Marine Bacteria on the Formation of Microbial Film on Solid

Surfaces," in R. Colwell and R. Morita, Eds., *Effect of the Ocean Environment on Microbial Activities*, University Park Press, Baltimore, 1974, pp. 397–417.

36. W. A. Corpe, "Attachment of Marine Bacteria to Solid Surfaces," in R. S. Manley, Ed., *Adhesions in Biological Systems*, Academic, New York, 1970, pp. 73–87.
37. J. E. Scott, G. Quintarelli, and M. C. Dellovo, *Histochemie*, **4,** 73 (1964).
38. W. A. Corpe, *Dev. Ind. Microbiol.*, **11,** 402 (1970).
39. W. A. Corpe, L. Matsuuchi, and B. Armbruster, "Secretion of Adhesive Polymers and Attachment of Marine Bacteria to Surfaces," in J. M. Sharpley and A. M. Kaplan, Eds., *Proceedings of Third International Biodegradation Symposium*, Applied Science Publishers, London, 1975, pp. 433–442.
40. T. R. Tosteson and W. A. Corpe, *Can. J. Microbiol.*, **21,** 1025 (1975).
41. M. Fletcher and G. D. Floodgate, *J. Gen. Microbiol.*, **74,** 325 (1973).
42. K. C. Marshall and R. H. Cruickshank, *Arch. Mikrobiol.*, **91,** 29 (1973).
43. K. C. Marshall, *Interfaces in Microbial Ecology*, Harvard U. P., Cambridge, Mass., 1976.
44. E. G. Pringsheim, *Phil. Transact. Roy. Soc.* (*Lond.*), **233,** 453 (1949).
45. P. L. Busch and W. Stumm, *Environ. Sci. Technol.*, **2,** 49 (1968).
46. B. A. Friedman, P. R. Dugan, R. M. Pfister, and C. C. Remsen, *J. Bacteriol.*, **96,** 2144 (1968).
47. B. A. Friedman, P. R. Dugan, R. M. Pfister, and C. C. Remsen, *J. Bacteriol.*, **98,** 1328 (1969).
48. H. C. Jones, I. L. Roth, and W. M. Sanders, III, *J. Bacteriol.*, **99,** 316 (1969).
49. R. J. Gibbons and J. Van Houte, *Infect. Immun.*, **3,** 567 (1971).
50. R. J. Gibbons and J. Van Houte, *Annu. Rev. Microbiol.*, **29,** 19 (1975).
51. D. C. Savage, *Soc. Gen. Microbiol.*, **22,** 25 (1972).
52. D. Savage, "Electron Microscopy of Bacteria Adherent Epithelia in the Intestinal Canal," in D. Schlessinger, Ed., *Microbiology—1977*, American Society for Microbiology, Washington, D. C., 1977, pp. 422–426.
53. T. Ericson and I. Magnusson, *Caries Res.*, **10,** 8 (1976).
54. R. J. Gibbons and R. J. Fitzgerald, *J. Bacteriol.*, **98,** 341 (1969).
55. J. Kelstrup and T. D. Funder-Nielsen, *J. Gen. Microbiol.*, **81,** 485 (1974).
56. G. R. Germaine and C. F. Schachtele, *Infect. Immun.*, **13,** 365 (1976).
57. H. D. Slade, "Cell Surface Antigenic Polymers of *Streptococcus mutans* and Their Role in Adherence of the Microorganism *in vitro*," in D. Schlessinger, Ed., *Microbiology—1977*, American Society for Microbiology, Washington, D. C., 1977, pp. 411–416.
58. R. J. Gibbons and M. Nygaard, *Arch. Oral Biol.*, **15,** 1397 (1970).
59. C. L. Whittenberger, A. J. Beaman, L. N. Lee, R. M. McCabe, and J. A. Donkersloot, "Possible Role of *Streptococcus salivarius* Glucosyltransferase in Adherence of *Veillonella* to Smooth Surfaces," in D. Schlessinger, Ed., *Microbiology—1977*, American Society for Microbiology, Washington, D. C., 1977, pp. 417–421.
60. D. C. Savage, "Electron Microscopy of Bacteria Adherent to Epithelia in the Murine Intestinal Canal," in D. Schlessinger, Ed., *Microbiology—1977*, American Society for Microbiology, Washington, D. C., 1975, pp. 120–123.

61. C. P. Davis, *Appl. Environ. Microbiol.*, **31,** 304 (1976).
62. D. C. Savage and R. V. H. Blumershine, *Infect. Immun.*, **10,** 240 (1974).
63. T. J. Fitzgerald, P. Cleveland, R. C. Johnson, J. N. Miller, and J. A. Sykes, *J. Bacteriol.*, **130,** 1333 (1977).
64. E. T. Nelson, J. D. Clements, and R. A. Finkelstein, *Infect. Immun.*, **14,** 527 (1976).
65. R. C. Wagner and R. J. Barrnett, *J. Ultrastruct. Res.*, **48,** 404 (1974).
66. C. E. Lankford, *Ann. N. Y. Acad. Sci.*, **88,** 1203 (1960).
67. H. Kirby, "Organisms Living on and in Protozoa" in E. N. Calkins and F. N. Summers, Eds., *Protozoa in Biological Research*, Columbia U. P., New York, 1941, pp. 1009–1113.
68. G. H. Ball, "Organisms Living on and in Protozoa," in T. T. Chen, Ed., *Research in Protozoology*, Vol. 3, Pergamon, Oxford, 1969, pp. 565–718.
69. R. A. Bloodgood and T. P. Fitzharris, *Cytobios*, **17,** 103 (1976).
70. H. W. Beams and R. G. Kessel, *Z. Zellforsch.*, **139,** 303 (1973).
71. F. W. Beech and R. R. Davenport, "A Survey of Methods for the Quantitative Examination of the Yeast Flora of Apple and Grape Leaves," in T. F. Preece and C. H. Dickinson, Eds., *Ecology of Leaf Surface Microorganisms*, Academic, London, 1971, pp. 139–157.
72. D. J. Royale, "Scanning Electron Microscopy of Plant Surface Microorganisms," in C. H. Dickinson and T. F. Preece, Eds., *Microbiology of Aerial Plant Surfaces*, Academic, New York, 1976.
73. M. L. Schuster and D. P. Coyne, *Annu. Rev. Phytopathol.*, **12,** 199 (1974).
74. J. A. Lippincott and B. B. Lippincott, *Annu. Rev. Microbiol.*, **29,** 377 (1975).
75. B. B. Bohlool and E. R. Schmidt, *Science*, **185,** 269 (1974).
76. B. B. Bohlool and E. L. Schmidt, *J. Bacteriol.*, **125,** 1188 (1976).
77. H. C. Tsien and E. L. Schmidt, *Can. J. Microbiol.*, **23,** 1274 (1977).
78. F. B. Dazzo, C. A. Napoli, and D. H. Hubbell, *Appl. Environ. Microbiol.*, **32,** 168 (1976).
79. F. B. Dazzo and W. J. Brill, *Appl. Environ. Microbiol.*, **33,** 132 (1977).
80. V. O. Singh and M. N. Schroth, *Science*, **197,** 759 (1977).
81. J. C. G. Ottow, *Annu. Rev. Microbiol.*, **29,** 79 (1975).
82. J. P. Duguid, *Arch. Immun. Ther. Exp.*, **16,** 173 (1968).
83. D. C. Old, *J. Gen. Microbiol.*, **71,** 149 (1972).
84. R. Ahrens, G. Moll, and G. Rheinheimer, *Arch. Microbiol.*, **63,** 321 (1968).
85. F. Mayer, *Arch. Mikrobiol.*, **76,** 166 (1971).
86. T. H. MacRae, W. J. Dobson, and H. D. McCurdy, *Can. J. Microbiol.*, **23,** 1096 (1977).
87. H. F. Ridgway, R. M. Wagner, W. T. Dawsey, and R. A. Lewin, *Can. J. Microbiol.*, **21,** 1733 (1975).
88. L. V. Leak, *J. Ultrastruct. Res.*, **21,** 61 (1968).
89. J. Swanson, "Adherence of Gonococci," in D. Schlessinger, Ed., *Microbiology—1977*, American Society for Microbiology, Washington, D. C., 1977, pp. 422–426.
90. J. Swanson, M. D. Stephen, J. Kraus, and E. C. Gotschlich, *J. Exp. Med.*, **134,** 886 (1971).
91. C. P. Novotny, J. A. Short, and P. D. Walker, *J. Med. Microbiol.*, **8,** 413 (1975).
92. R. J. Gibbons, J. Van Houte, and W. F. Liljemark, *J. Dent. Res.*, **51,** 424 (1972).

93. J. Swanson, K. C. Hsu, and E. C. Gotschlich, *J. Exp. Med.*, **130,** 1063 (1969).
94. R. P. Ellen and R. J. Gibbons, *Infect. Immun.*, **5,** 826 (1972).
95. E. H. Beachey and I. Ofek, *J. Exp. Med.*, **143,** 759 (1976).
96. W. A. Scheffers, W. E. DeBoer and A. M. Looyard, *Society for Applied Bacteriology Summer Conference, 1976*, at Lancaster, U. K. (1976) (abstr.).
97. L. Shapiro and J. V. Maizel Jr., *J. Bacteriol.*, **113,** 478 (1973).
98. R. Whittenbury and C. S. Dow, *Bacteriol. Rev.*, **41,** 754 (1977).
99. H. Stolp and M. P. Starr, *Antonie Van Leeuwenhoek J. Microbiol. Serol.*, **29,** 217 (1963).
100. M. P. Starr and N. L. Baigent, *J. Bacteriol.*, **91,** 2006 (1966).
101. A. H. Romano and J. P. Peloquin, *J. Bacteriol.*, **86,** 252 (1964).
102. E. Gaudy and R. S. Wolfe, *Appl. Microbiol.*, **9,** 580 (1961).
103. F. F. Dias, N. C. Dondero, and M. S. Finstein, *Appl. Microbiol.*, **16,** 1191 (1968).
104. N. C. Dondero, *Annu. Rev. Microbiol.*, **29,** 407 (1975).
105. R. E. Buchanan and N. E. Gibbons, Eds., *Bergey's Manual of Determinative Bacteriology*, 8th Ed., Williams and Wilkins, Baltimore, 1974.
106. J. S. Poindexter, *Bacteriol. Rev.*, **28,** 231 (1964).
107. J. T. Staley, *Appl. Microbiol.*, **22,** 496 (1971).
108. P. A. Tyler and K. C. Marshall, *Arch. Mikrobiol.*, **56,** 344 (1967).
109. J. L. Stove and R. Y. Stanier, *Nature*, **196,** 1189 (1962).
110. P. Hirsch and S. F. Conti, *Arch. Mikrobiol.*, **48,** 339 (1964).
111. S. F. Conti and P. Hirsch, *J. Bacteriol.*, **89,** 503 (1965).
112. P. Hirsch, *Annu. Rev. Microbiol.*, **28,** 391 (1974).
113. R. S. Wolfe, *J. Am. Water Works Assoc.*, **50,** 1241 (1958).
114. T. D. Brock and S. F. Conti, *Arch. Mikrobiol.*, **66,** 79 (1969).
115. E. G. Pringsheim, *Bacteriol. Rev.*, **21,** 69 (1957).
116. H. F. Ridgway, *Can. J. Microbiol.*, **23,** 1201 (1977).
117. T. H. MacRae and H. D. McCurdy, *Can. J. Microbiol.*, **21,** 1815 (1975).
118. J. Pangborn, D. A. Kuhn, and J. R. Woods, *Arch. Microbiol.*, **113,** 197 (1976).
119. J. Breznak and H. S. Pankratz, *Appl. Environ. Microbiol.*, **33,** 406 (1977).
120. R. A. Bloodgood, K. R. Miller, T. P. Fitzharris and J. McIntosh, *J. Morphol.*, **143,** 77 (1974).
121. H. E. Smith and H. J. Arnott, *Transact. Am. Microsc. Soc.*, **93,** 180 (1974).
122. H. E. Smith, H. E. Bultse, and S. J. Stamler, *BioSystems*, **7,** 374 (1975).
123. E. B. G. Jones and T. Le Campion-Alsumard, *Int. Biodetn. Bull.*, **6,** 119 (1970).
124. R. A. Eaton and E. B. G. Jones, *Mat. Organismen*, **6,** 2 (1971).
125. M. A. Collins, "Colonization of Leaves by Phylloplane Saprophytes and Their Interactions in This Environment," in C. H. Dickinson and T. F. Preece, Eds., *Microbiology of Aerial Plant Surfaces*, Academic, New York, 1976, pp. 401–418.
126. W. W. Wilcox, *Bot. Rev.*, **36,** 1 (1970).
127. S. M. Martin and G. A. Adams, *Can. J. Microbiol.*, **2,** 715 (1956).
128. M. A. Crandall and T. D. Brock, *Bacteriol. Rev.*, **32,** 139 (1968).

129. C. E. ZoBell, "Substratum as an Environmental Factor for Aquatic Bacteria, Fungi and Blue-Green Algae," in O. Kinne, Ed., *Marine Ecology*, Vol. 1, *Environmental Factors*, Wiley, New York, 1972, pp. 1252–1270.
130. D. J. Royale, "Scanning Electron Microscopy of Plant Surface Microorganisms," in C. H. Dickinson and T. F. Preece, Eds., *Microbiology of Aerial Plant Surfaces*, Academic, New York, 1976, pp. 569–606.
131. V. O. Sing and S. Bartnicki-Garcia, *J. Cell Sci.*, **18,** 123 (1975).
132. F. O. Perkins, *Arch. Mikrobiol.*, **84,** 95 (1972).
133. R. Y. Zacharuk, *J. Inver. Path.*, **15,** 81 (1970).
134. J. T. Lambiase and W. G. Yendol, *Can. J. Microbiol.*, **23,** 452 (1977).
135. F. E. Round, *Br. Phycol. Bull.*, **2,** 456 (1965).
136. C. Den Hartog, in O. Kinne, Ed., *Marine Ecology*, Vol. 1, *Environmental Factors*, Pt. 3, Wiley-Interscience, New York, 1972, pp. 1277–1322.
137. A. H. L. Chamberlain, "Algal Settlement and Secretion of Adhesive Materials," in J. M. Sharpley and A. M. Kaplan, Eds., *Proceedings of Third International Biodegradation Symposium*, Applied Science Publishers, London, 1975, pp. 417–432.
138. L. V. Evans and A. O. Christie, *Ann. Bot.*, **34,** 451 (1970).
139. J. R. J. Baker and L. V. Evans, *Protoplasma*, **77,** 181 (1973).
140. R. L. Fletcher, "Post-germination Attachment Mechanisms in Marine Fouling Algae," in J. M. Sharpley and A. M. Kaplan, Eds., *Proceedings of Third International Biodegradation Symposium*, Applied Science Publishers, London, 1975, pp. 443–464.
141. A. Sládečková, *Bot. Rev.*, **28,** 286 (1962).
142. J. Cairns, Jr., Ed., *The Structure and Function of Fresh-Water Microbial Communities*, Research Division monograph No. 3, Virginia Polytechnic Institue and State University, Blacksburg, Va., 1971.
143. G. Hascall and M. A. Rudzinska, *J. Protozool.*, **17,** 311 (1970).
144. G. K. Hascall, *J. Protozool.*, **20,** 701 (1973).
145. D. V. Holberton, *J. Exp. Biol.*, **60,** 207 (1974).
146. T. Bauchop and R. T. J. Clarke, *Appl. Environ. Microbiol.*, **32,** 417 (1976).
147. L. Weiss, "A Biophysical Consideration of Cell Contact Phenomena," in R. S. Manley, Ed., *Adhesion in Biological Systems*, Academic, New York, pp. 1–8.
148. A. Cassone and E. Garaci, *Can. J. Microbiol.*, **23,** 684 (1977).
149. J. W. Costerton, J. M. Ingram, and K. J. Cheng, *Bacteriol. Rev.*, **38,** 87 (1974).
150. G. Gerisch, H. Beug, D. Malchow, H. Schwartz, and A. V. Stein, "Receptors for Intercellular Signals in Aggregating Cells of the Slime Mold, *Dictyostelium discoideum*," in E. Y. C. Lee and E. E. Smith, *Biology and Chemistry of Eucaryotic Cell Surfaces*, Miami Winter Symposium 1, Academic, New York, pp. 49–66.
151. H. Lis and N. Sharon, *Annu. Rev. Biochem.*, **42,** 541 (1973).

CHAPTER 5

Normal Flora of Animal Intestinal Surfaces

ADRIAN LEE

School of Microbiology, The University of New South Wales, Kensington, N. S. W. 2033, Australia

CONTENTS

5.1 INTRODUCTION

The microorganisms of the gastrointestinal tract have been studied since the early days of bacteriology. The conviction of Elie Metchnikoff that longevity of Bulgarian peasants was linked with the colonization of the gut with lactobacilli following ingestion of yogurt provided a stimulus for this early work. It has since been established that stable populations of intestinal bacteria are essential for the healthy existence of man, animals, and birds in normal environments. These bacteria alter the morphology of intestinal tissue, contribute growth factors, and protect the host from invasion by intestinal pathogens [1].

Until recently most interest in the gut microorganisms was focused on the bacteria of the lumen, represented by the feces excreted at the end of the intestinal tract. However, these populations are not representative of the inhabitants of the many ecological niches that exist over the entire length of this complex ecosystem [2].

A major paper by Savage et al. [3] demonstrated that the surfaces of the intestine of rodents were colonized by populations of bacteria different from those organisms found in the lumen content. These mucosa-associated bacteria were considered to be members of the autochthonous flora, that is, organisms that have evolved with the host and are found in any normal rodent [4,5]. Similar organisms are found in the intestines of all animals studied. The purpose of this chapter is to review current knowledge on the mucosa-associated flora. A mucosa-associated organism is defined as any organism that is seen in significant numbers in specimens of intestinal tissue following vigorous washing. Colonization of the mucosa of the rodent intestine by microorganisms is described, in the first section of this chapter, as an example of a typical gut ecosystem. Then the mechanisms of association with intestinal surfaces are discussed. Not only do different organisms colonize the different surfaces of the gut, but mechanisms of association vary. The concept of autochthony, mentioned earlier in this paragraph is useful when considering reasons for these differences. A major hypothesis put forward in this chapter is that, because the mucosa-associated flora have evolved with their animal host, the mechanism of association can be explained by considering the optimum conditions necessary for survival at the site of association. The oral cavity has not been included in this discussion as it is covered in Chapter 7.

5.2 ASSOCIATION OF MICROBIAL POPULATIONS WITH GASTROINTESTINAL-TRACT SURFACES OF RODENTS

The bacteria associated with the intestinal mucosa of mice and rats are very similar, but particular areas have been studied more in one animal than in the other (e.g., the stomach has been most studied in mice, whereas the ileum has been investigated more thoroughly in rats). In both species it is clear that the epithelial tissues of the different regions of the gastrointestinal tract are colonized by different microbial types of organisms. These regions are described in turn.

5.2.1 The Stomach

Unlike that in man, the surface of the rodent stomach is divided into two portions, the nonsecreting and the secreting areas. Keratinized stratified squamous epithelium lines the mucosa of the nonsecreting area [3]. This epithelium is colonized heavily by Gram-positive bacteria (Fig. 5.1), the majority of which were identified as lactobacilli [6]. The need for attachment to the keratin layer probably is related to the fact that the stomach can be empty for prolonged periods if the animal does not have continuous access to food. Thus only the organisms remaining in the region are those that can attach to epithelial surfaces [2]. A local factor influencing this mechanism of attachment is the fact that the superficial cells of the stomach lining are sloughed off continually, taking the attached lactobacilli with them. A release mechanism may be necessary, as newly exposed cells need to be recolonized by bacteria from the lumen.

The attachment of the lactobacilli is perpendicular to the surface (Fig. 5.1). This is a common form of association in the intestinal tract. There are several possible advantages of perpendicular attachment [2,7]. Protrusion of the main body of the cell into the lumen would certainly facilitate utilization of stomach contents. This mode of adhesion also permits more bacteria to colonize a given unit of surface area. An environmental influence that presumably favors colonization by lactobacilli is the stomach acidity, since this genus is one of the best able to survive low pH conditions.

The secretory mucosa of the rodent is composed of columnar epithelial cells and often is covered with a layer of another aciduric organism, the yeast *Torulopsis pintolopesii* [8]. These yeast cells are found only on the

Figure 5.1 Microorganisms associated with stratified squamous epithelium of stomachs of adult CD-1 mice. Note end-on attachment of bacteria to epithelium (SEM, ×3800). Micrograph courtesy of D. Savage; reproduced from *Infect. Immun.* **10,** 242 (1974) by copyright permission of the Rockefeller University Press, New York.

secretory section of the stomach, unless the more highly adapted lactobacilli of the nonsecretory epithelium are removed by penicillin treatment [9].

5.2.2 The Small Intestine

The flow of material through the small bowel is very rapid. Van Liere et al. [10] observed that a suspension of charcoal, administered by stomach tube to unanesthetized dogs, was propelled through more than half the length of the small intestine in 15 min. This rapid emptying of the intestine was given as the reason for the low number of bacteria found by Dixon [11] in the small bowel. Most of these studies were of lumen content. Thus there are two factors that would act as selective pressures on organisms adapting to tissue association in the small intestine: continual emptying, as in the stomach, and peristaltic movement.

For microorganisms to survive in this environment of sporadic fluid flow, they would need to either firmly attach to the intestinal surface or colonize an area protected from the mainstream. Both types of organism are found in the rat small intestine, although only in the lower regions, mainly in the ileum. Two morphologically distinct groups of bacteria are found deep in the crypts of the ileal mucosa, a short curved rod and a plump spiral organism with a characteristic outer surface (Fig. 5.2). Even though this location is protected from the main flow of the small-bowel content, there is presumably considerable outpouring of fluid from the crypt. To prevent being flushed out, the small curved bacteria apparently attach to the surface. The motility of the spiral organism must allow the organism to resist the outflow of mucus. An ability to survive in mucus is the main mechanism of association with surfaces other than adhesion; this is discussed in greater detail later.

The most studied organism–surface association in the intestinal tract of any animal is that of a filamentous organism on the surfaces of the rodent ileum (Figs. 5.3*a*,*b*) [12]. Evolution of host and organism together has resulted in the most sophisticated example of specialized adhesion in the gut. Before the ingenious nature of this mechanism of association can be appreciated, it is necessary to consider the conditions to which the organism has had to adapt. The rapid flow has been mentioned previously. The organism overcomes this by a firm attachment to epithelial cells. However, these tissue cells are turning over at a rapid rate. Cells migrate from the base of the crypt to the tips of the villi at a rate of 1.2 cell positions per hour. Thus crypt migration time in the rat is 20 hr, and it takes the cell a further 20–30 hr to move up the villus [13]. It appears that, in the rat, this organism has evolved a complex life cycle to capitalize on this continual turnover of the host-cell population. This life cycle has been deduced by observation of numerous electron micrographs of rat ileum. The organism has yet to be cultured [12, 14–16].

A small body called a *holdfast* firmly attaches to cells at the bottom of the intervillar space (Fig. 5.4*a*). As the host cell migrates up the villus, the microorganism multiplies forming a filamentous stage of the life cycle, but remaining firmly attached to the epithelial cell and well suited to obtaining nutrient from the lumen content into which it protrudes (Figs. 5.3*b*, 5.5*a*). This process of division is interesting in itself, since classical binary fission is impossible because the end cell is fixed to the tissue. As the epithelial cell reaches the end of its migration, new holdfasts develop within the filament (Figs. 5.4*b*, 5.5*b*). The holdfasts are released and thus are ready to attach to a younger epithelial cell at the start of its progression up the villus (Fig. 5.4*c*). This is a remarkable process of self-conservation. For survival outside the host, and propagation in other animals, an occasional spore is formed in the filaments. Filaments of all stages are seen attached to cells on the floor of the intervillus space before the cells migrate up the villus. Therefore,

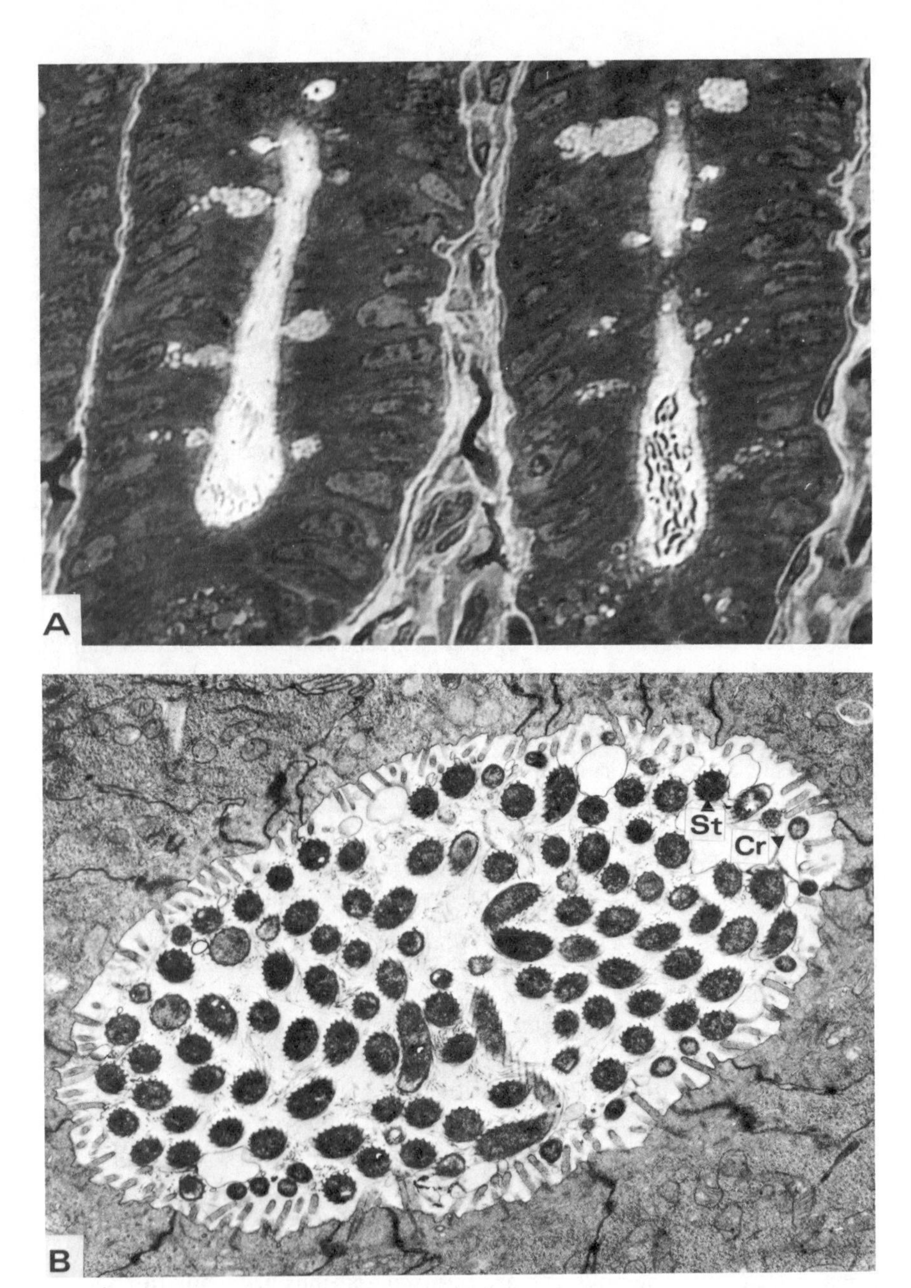

Figure 5.2 Mucosa-associated flora of rat ileum: (A) Section through two crypts showing localization of bacteria (×1050); (B) Section through base of crypt showing cross section of two morphological types of bacteria: (i) spiral-shaped organisms with characteristic surface structure (St) mainly seen in lumen; (ii) small crescent-shaped organisms only seen closely associated with tissue surface (Cr) (×7600). Micrographs courtesy of M. Phillips, University of New South Wales.

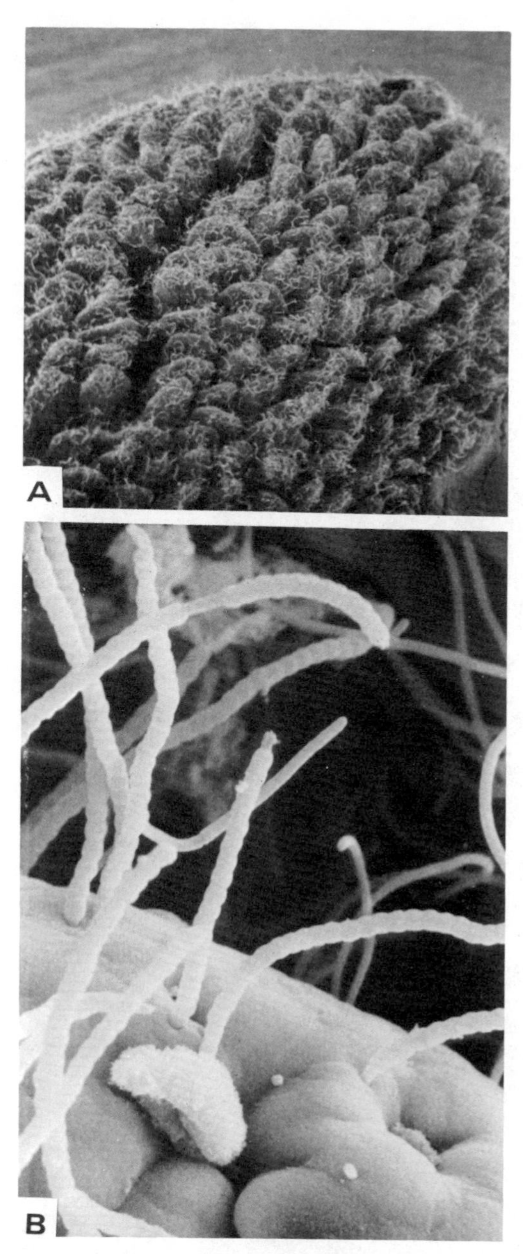

Figure 5.3 Association of filamentous microorganism with surface of rat ileum: (A) picture at low magnification showing heavy colonization of epithelial surfaces (SEM, ×68) (micrograph courtesy of C. Garland, University of New South Wales); (B) higher magnification showing filaments inserted into epithelial surface (SEM, ×19,000) [micrograph courtesy of D. G. Chase; reproduced from *J. Bacteriol.* **127,** 574 (1976) by copyright permission of the American Society for Microbiology].

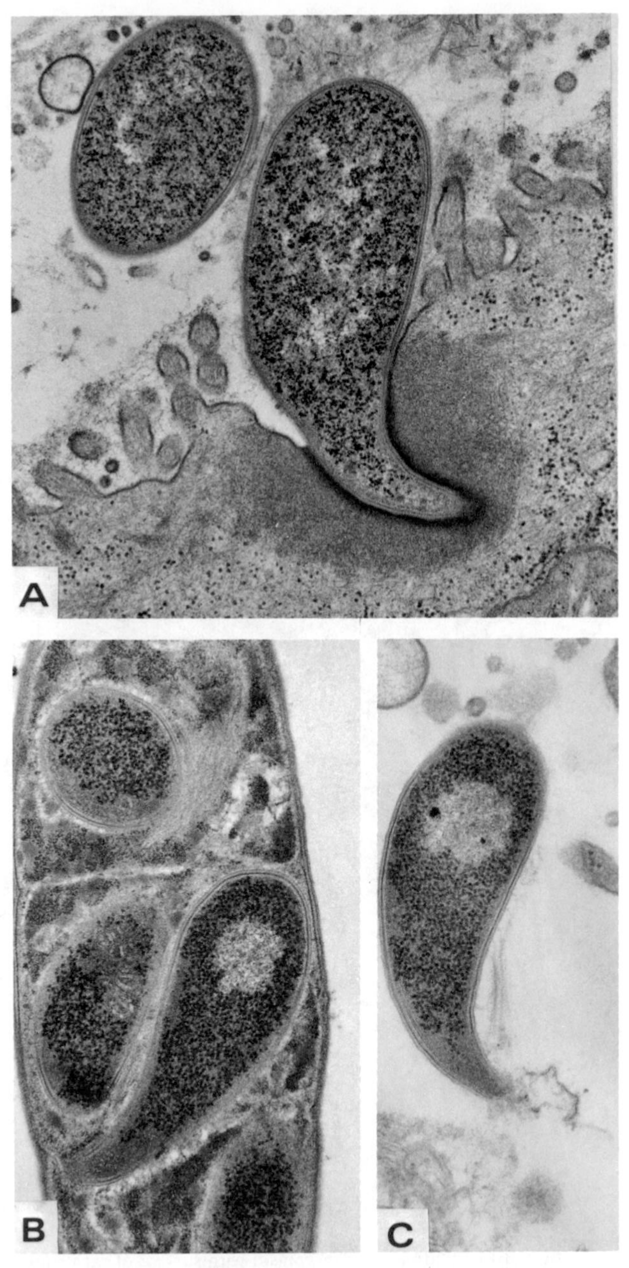

Figure 5.4 Stages in life cycle of filamentous microorganism associated with rat ileal surface: (A) early "holdfast" stage (×34,750); (B) development of "holdfasts" within cells during filamentous stage of life cycle (×34,750); (C) individual "holdfast" segment free in intervillar space (×34,750). Micrographs courtesy of D. G. Chase; reproduced from *J. Bacteriol.* **127,** 576 (1976), by copyright permission of the American Society for Microbiology.

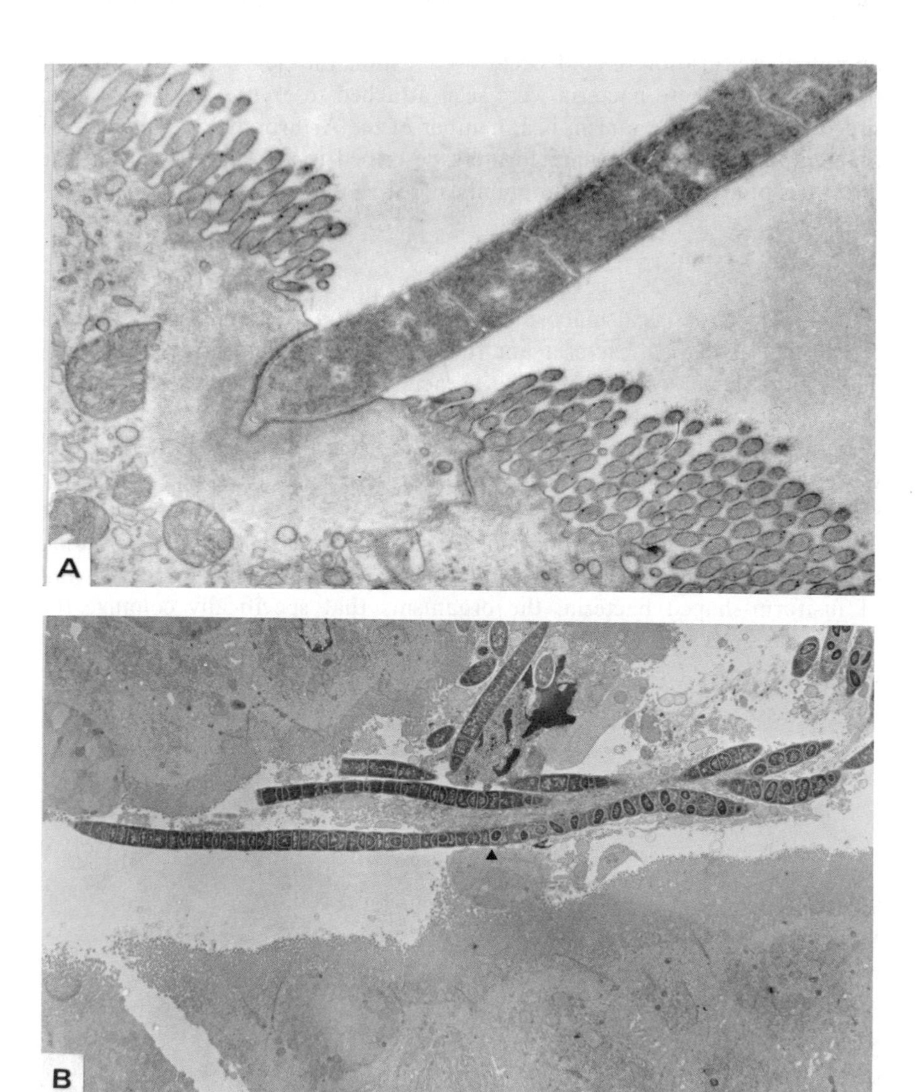

Figure 5.5 Stages in life cycle of filamentous microorganism associated with rat ileal surface: (A) early state in growth of filament—note clearly defined attachment site (×15,375) [micrograph courtesy of J. Hampton; redproduced from *Lab. Invest.* **14,** 1468 (1965) by copyright permission of American Gastroenterological Association]; (B) long chain of reproductive segments showing various stages in holdfast formation ranging from tertiary segments on left to dividing C-shaped inclusions on right, with abrupt transition in stage of segment maturirty at arrow, right are advanced C-shaped inclusions to the right, and stages preceding spherical inclusion to the left (×86,250) [micrograph courtesy of D. G. Chase; reproduced from *J. Bacteriol.* **127,** 574 (1976) by copyright permission of the American Society of Microbiology].

microbial maturation should occur well within the 20–30 hr villus transit time. None of these bacteria were seen attached to crypt cells. It has been suggested that this organism is a member of the Arthromitaceae, a group of filamentous, segmented spore formers described many years ago from the intestines of arthropods and amphibians [15].

5.2.3 The Cecum

The crypts of the cecal mucosa of mice and rats are colonized by characteristic spiral-shaped bacteria not found elsewhere in the intestinal tract. These organisms are often closely associated with the epithelial cells, but there is no apparent attachment to the surface (Fig. 5.6).

Fluid flow in the lumen of the cecum is considerably less than in the small bowel [17], and thus attachment would not necessarily be a requirement for surface association. Certain fusiform-shaped bacteria do form layers on the cecal-mucosa surface; these organisms simply colonize the mucus coating covering the epithelium. As the lumen of the cecum of adult rodents is full of fusiform-shaped bacteria, the organisms that specifically colonize the mucus are seen best in 10-day-old animals before the full population has developed [18].

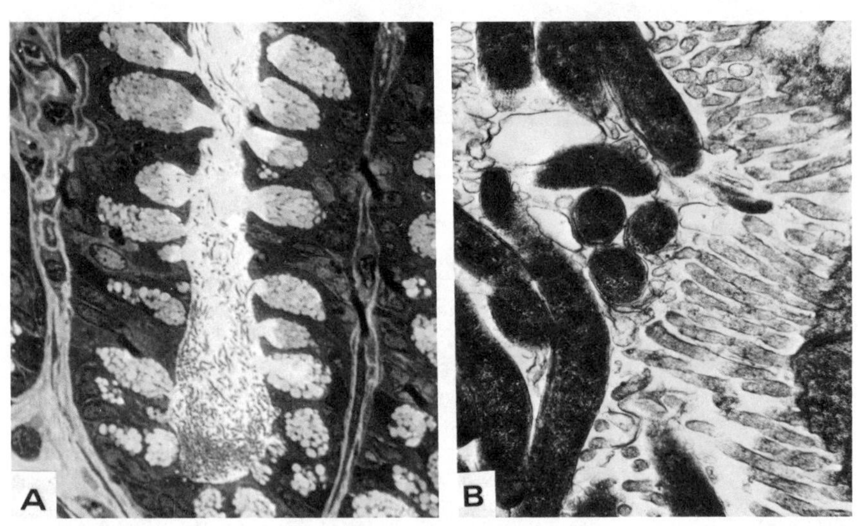

Figure 5.6 Mucosa-associated flora of rat cecum: (A) section through crypt showing large numbers of spiral bacteria especially at base of crypt (× 650); (B) Spirochetes lying close to epithelial cells in cecal crypt (×15,000). Micrographs courtesy of M. Phillips, University of New South Wales.

5.2.4 The Colon and Rectum

Despite the close proximity of the cecum and colon, the mucosa-associated flora at these two sites in the rodent is very different. Presumably, this is due to different environmental conditions on the two surfaces. The spiral organisms seen in the mucus of the crypts have a different morphology to the cecal spirals; and, unlike the cecum, there are spiral-shaped bacteria attached firmly to the base of the colonic crypts [19] (Fig. 5.7). This need for attachment could result from a greater flow of mucus in the colon. Lubrication of the fecal pellet in the colon requires large volumes of mucus.

Thus it is seen that the lining of the gastrointestinal tract of rodents has a wide variety of different populations in different sites, each possibly playing an important role in the well-being of the host. The distribution of these bacteria is so constant that I can identify which of six areas of the intestinal tract a specimen comes from simply by recognition of bacteria in crude scrapings of washed mucosa examined by phase-contrast microscopy. Before having a closer look at the mechanisms of the association of these populations with tissue surfaces, I intend to establish that mucosa-associated microorganisms are a constant feature of most animals.

5.3 MUCOSA-ASSOCIATED BACTERIA IN OTHER ANIMALS

Mucosa-associated bacteria have been found on the intestinal surfaces of many animal species, such as bacteria on the epithelial surface of the sheep rumen [20], lactobacilli on the pig-stomach surface [21], and chicken crop [22,23], and spirochetes on the guinea-pig and dog cecal mucosa [24,25]. Chickens also have a filamentous organism similar to that in the rat ileum, associated with the crop epithelium [22]. Gram-positive cocci were found even on the gut mucosa of a worm, *Aspiculoris tetraptera*, inside a mouse intestine [26].

The least convincing association between a gut surface and flora has been demonstrated in man. This probably relates to the difficulty in obtaining fresh normal specimens for investigation. Nelson and Mata [27] observed Gram-positive cocci in the mucosa of the human jejunum and saw Gram-positive rods and cocci in six colonic samples studied. Plaut et al. [28] previously had seen organisms in jejunal mucosa of man but had concluded that these bacteria did not constitute a population distinct from that residing in the intestinal fluids.

Further reports have described spiral-shaped bacteria attached to the human stomach epithelium or deep in the foveae [29]. Takeuchi and his

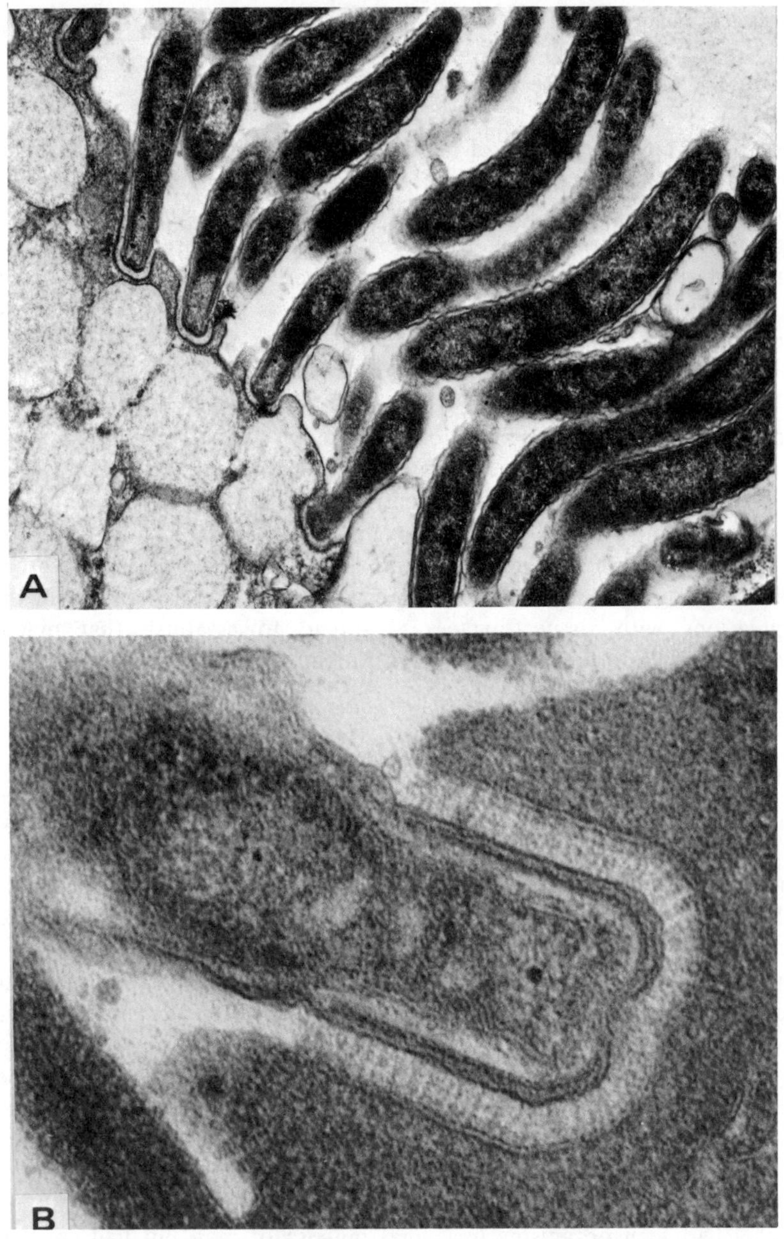

Figure 5.7 Microorganisms associated with surface of rat colon: (A) spiral organisms attached to surface of goblet cell at base of colonic crypt (×21,000) (micrograph courtesy of M. Phillips, University of New South Wales); (B) attachment site of spiral organism associated with surface of rat colon (×245,000) [Micrograph courtesy R. C. Wagner; reproduced from *J. Ultrastruct. Res.*, **48,** 411 (1974) by copyright permission of Academic Press, New York.

collaborators estimated that 3% of normal human subjects have spirochetes attached to the colonic surface [30,31]. Of a group of rhesus monkeys examined, 28% also had spirochetes on the colonic surface. These organisms had a very intimate association with the tissue surface.

The increasing study of animal-gut-surface-associated bacteria is motivated partly by the assumption that similar organisms occur in man. Possibly, the major priority in gastrointestinal microbiology today is for a more extensive investigation of human intestinal surfaces.

5.4 MECHANISMS OF MICROBIAL ASSOCIATION WITH INTESTINAL SURFACES

There are very few papers that specifically consider the mechanisms of association of microorganisms described in the preceding section. The following discussion attempts to collate isolated comments and observations to summarise our present knowledge in this area. There are two main modes of microbial association with intestinal surfaces: (1) firm attachment to the surface and (2) adaption to survival in the surface environment, usually in mucus.

5.4.1 Attachment of Microorganisms to Intestinal Surfaces

The surfaces of bacteria and most solid materials are negatively charged; for these surfaces to come close together, there must be a mechanism that overcomes the repulsion between like charges [32]. The same is true of intestinal organisms and the epithelial-cell surface. In most cases of firm attachment to surfaces in other systems, extracellular polymers are produced that bridge this gap and allow anchoring of a bacterium to the surface [33]. Often this is a one-way process; for instance, pseudomonads produce a compact polysaccharide layer that facilitates adhesion to almost any surface [34]. In a similar manner, streptococci produce dextrans that allow adhesion to a tooth surface (see Chapter 7). As is seen in the text that follows, there is no doubt that polysaccharides are involved in adhesion of intestinal bacteria, but the phenomenon is even more interesting as the epithelial-cell surfaces are lined with polysaccharides and other materials that could also contribute to attachment. Before considering these specific mechanisms, it is useful to review the structure of the epithelial surface.

Nature of Epithelial Surfaces within the Gastrointestinal Tract. Apart from the stratified squamous epithelial cells in the rodent stomach, cells in the intestinal tract of other animals and birds are mainly either columnar cells, with a striated border, or goblet cells. The outer surface of these columnar cells is composed of thousands of fine microvilli (Fig. 5.8). Coating the microvilli is a conspicuous layer composed of fine filaments radiating from the outer dense leaflet of the plasma membrane [35]. This "fuzzy coat," also called the *mucus perimatrix* or the *glycocalyx*, is rich in polysaccharide and glycoprotein [35–38]. The goblet cells produce secretions that are rich in acid mucopolysaccharide that also contribute to the surface layer [39,40]. Clearly, this coating of the cell surfaces must play a role in attachment. Two general observations on cell attachment emphasize that it is variations in these cell surfaces that lead to variations in attachment.

Specificity of Attachment.

Interspecies Specificity of Attachment. A remarkable specificity is found in the adhesion of certain organisms to the same types of cells, that is, rat stomach cells and avian crop cells. Thus avian lactobacilli only attach to avian epithelial cells, whereas rat lactobacilli attach only to rat cells [41,42]. Presumably, this is due to different molecular groupings providing specific attachment sites on the cell surface in a manner analogous to virus–host cell specificity.

Specific Localization within Intestinal Tract. As has been described in detail in the preceding paragraphs, certain bacteria only attach in one small area of the gut. Thus holdfasts of the filamentous organism in the mouse small bowel are found only in the lower duodenum or ileum and attach only to cells on the intervillous floor or the villus, not deep in the intestinal crypts. Similarly, spirochetes in the monkey, cat, or guinea-pig large bowel attach only to the outer surface, not in the crypts.

Intestinal-tissue cells have different surface constituents at different positions in the gut; this is probably the basis for the specificity of localization at these different positions.

Etzler [43] presented data suggesting that (1) throughout most of the adult rat small intestine, changes occur in the carbohydrate portion of the microvillous surface of the epithelial cells as they differentiate and move up the villi, (2) these changes in cell surface are related to the region of the bowel in which the villi are located, and (3) the carbohydrate nature of the secretory material of the goblet cell varies in the different regions of the small intestine. These conclusions are based on the differential reactivities of the cell surfaces and secretory components with four different lectins (plant agglutinins) that had been labeled with fluorescein isothiocyanate. They were *Dolichos biflorus* lectin with specificity for terminal nonreducing α-*N*-acetyl-D-galactosamine residues [44], the wheat-germ agglutinin with a

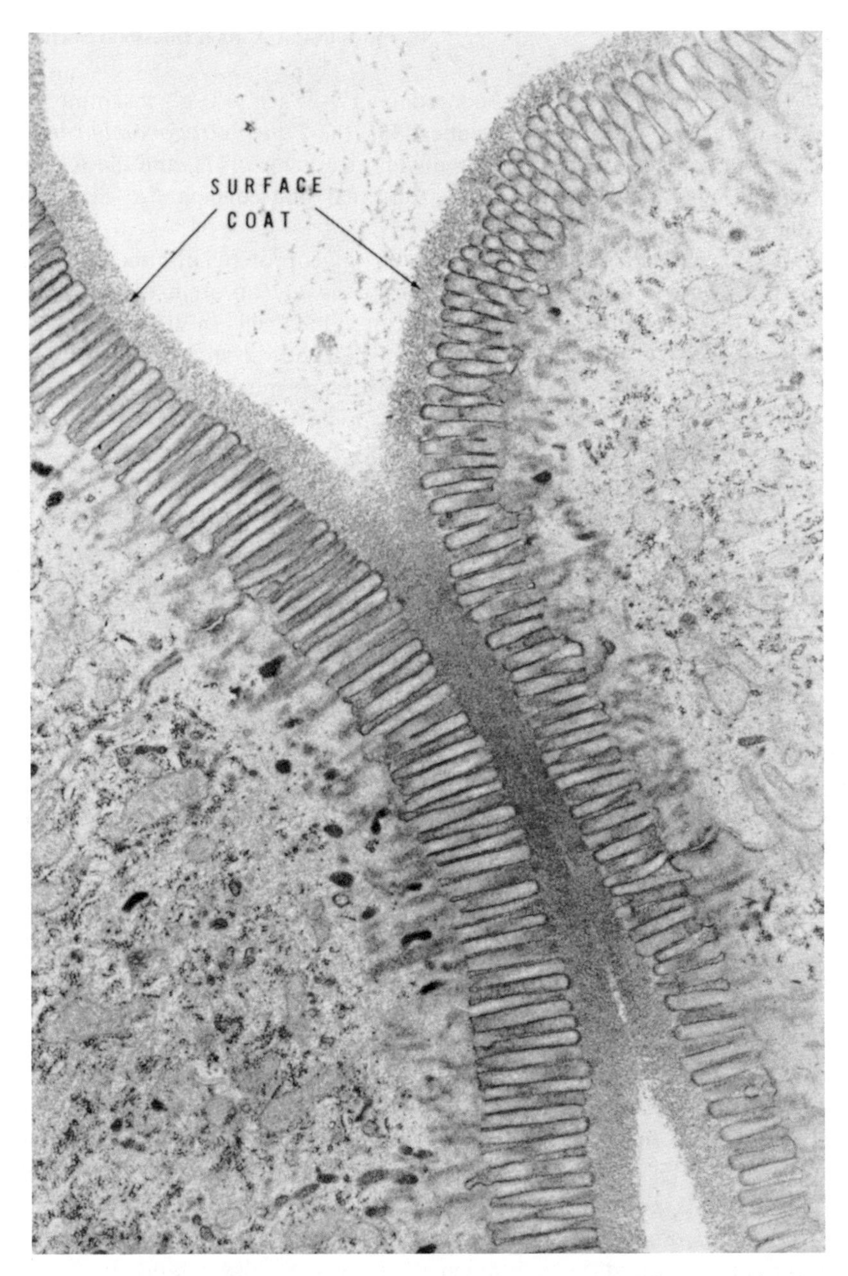

Figure 5.8 Striated borders of cat intestinal absorptive cells. Microvilli are topped with conspicuous surface coating about 0.4 μm thick. In regions where coats are tightly juxtaposed, filamentous meshwork appears to be compacted (×13,250). Micrograph courtesy of S. Ito; reproduced from *J. Cell Biol.* **27,** 477 (1965) by copyright permission of the Rockefeller University Press, New York.

reported specificity for terminal nonreducing β-*N*-acetyl-D-glucosamine [45] and *N*-acetyl-neuraminic acid residues [46], the *Lotus tetragonolobus* lectin with a specificity for terminal nonreducing α-L-fucose [47], and the *Ricinus communis* lectin with specificity for terminal nonreducing β-D-galactosyl residues [48].

Another study provided evidence that the glycoproteins of human and rat crypt cells were incomplete in contrast to the glycoprotein of the villous region [49]. Differences also were found in villous cells in different regions along the intestine; the upper area of the small bowel in the rat has a high alkaline-phosphatase activity, whereas levels are much lower on the ileal surface [50]. Other examples of variation in surface properties at different sites in the gut have been reported [51–54].

All these data support the idea of very specific reactions between the microorganisms and the tissue surfaces. The common presence of mucopolysaccharides and glycoproteins at these surfaces supports the mechanism of polymer bridging as described for aquatic bacteria [34,55,56], but with adhesion in the gastrointestinal tract we are dealing with specific interactions with the possibility of complementary molecular structures contributed by both host and microbial cell.

Evidence for Polymer Bridging. Electron micrographs of sections of intestinal surfaces (Fig. 5.9) clearly show threadlike structures between adsorbed bacteria and the epithelium similar to those seen between the hydrophobic end of a flexibacter cell and an araldite surface [56]. Note the layer of material outside the bacterial cell wall stained with ruthenium red and colloidal iron. Using an *in vitro* system of lactobacilli and crop cells, adhesion was inhibited by periodate [23] as with the flexibacter, but here the similarity ends. As mentioned earlier, the adhesion of the lactobacilli to the epithelium is specific.

A slight but consistent inhibition of adhesion after treatment of the bacterial cells with proteolytic enzymes, as well as with periodate, suggested the involvement of a glycoprotein. Protein alone was not considered to be the determinant. Adhesion was inhibited by the binding to the bacteria of Concanavalin A (Con A) [58], a lectin known to bind specifically to certain sugar residues [59]. Heating the lactobacilli at 60°C for 1 hr or treatment with detergents prevented adhesion to stratified squamous epithelial cells from the rat stomach. Formalin treatment had no effect. Addition of chondroitin sulfate A and gastric mucin to the system also inhibited adhesion [42]. These findings provide evidence for the contribution of acid mucopolysaccharides and/or glycoproteins, with exposed glucose, mannose, fructose, or arabinose moieties on the bacterial surface, to the specific attachment of lactobacilli to epithelial cells in the gut. The recent work of Suegara et al. [42], however, makes it difficult to explain the contribution of

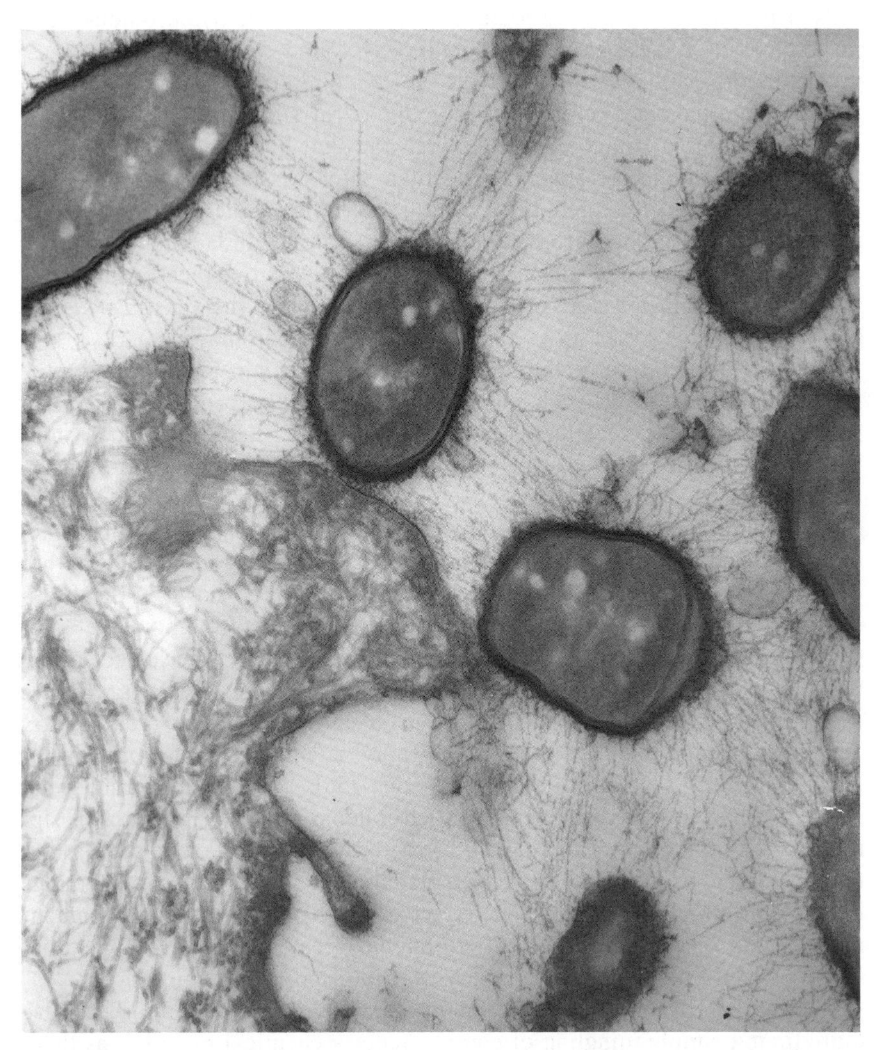

Figure 5.9 Lactobacilli attached to chicken crop epithelium fixed in presence of 0.15% ruthenium red. Three layers of bacterial cell wall are stained. Two of these correspond approximately to position of layers visualized by conventional methods of fixation and staining. Third and superficial layer of cell wall is believed to be of primary importance in bacterial adhesion. Numerous filaments radiate from this surface layer and extend to both neighboring bacilli and crop cell. Thin ruthenium red-positive layer is visible in surface of crop cell membrane (×29,000). Micrograph courtesy R. Fuller; reproduced from *J. Ultrastruct. Res.* **52,** 25 (1975) by copyright permission of Academic Press, New York.

the epithelial-cell surface to this specific adhesion. Keratinized epithelial cells of the rat stomach were pretreated with trypsin and hyaluronidase, and no change in adhesion of lactobacilli was found, whereas urea and 2-mercaptoethanol treatment resulted in a 50% inhibition of adhesion. These results suggested that keratinized cells of the rat stomach have no specific surface proteins or mucopolysaccharides and that hydrogen bonds mediate the adhesion of the lactobacilli. The filamentous extensions from the lactobacilli (Fig. 5.9) are thought to reinforce primary adhesion but were not considered essential because stationary-phase lactobacilli lacking such structures still adhered. Polymer fibrils are seen in many other examples of attachment within the gut, such as the attachment of the yeast *Torulopsis pintolopesii* to the secretory epithelium of the rodent stomach [60].

A recent observation made by us may provide further information on the mechanism of these cell–cell interactions. During the process of scraping washed mucosal epithelia, organisms are often dislodged from their permanent location in the tissues. It was noted that many bacteria found in the large bowel, especially spirochetes attached perpendicularly to the gut surface, spontaneously formed rosettes in the scrapings, whereas spiral bacteria showing no attachment in tissue did not form rosettes. Another example was found in the small bowel of the rat and is shown in the cross-section in Fig. 5.2. A fat spiral organism can be seen throughout the crypt (St), whereas a smaller crescent-shaped rod is seen only at the surface of the tissue (Cr). Only this latter organism formed rosettes in scrapings. Rosette formation has been described in *Flexibacter* sp. and *Hyphomicrobium* sp., and it was suggested that this process may resemble the phenomenon of micelle formation in surfactant molecules [56]. Rosette formation may result from an inward orientation of hydrophobic portions of cells. It is significant that the portion of the cell attracted to an interface always forms the inner region of the rosette. The explanation for rosettes in gut scrapings is uncertain and is difficult to reconcile with the specific attachment mechanisms discussed earlier. Hydrophobic areas of these gut bacteria, however, could facilitate and complement the specific adhesion process.

Attachment of many of the organisms described earlier must involve more than a simple mingling of complementary chemical moieties, as there is the definite formation of an attachment site in the epithelial cell surfaces that remains even when the microorganisms detach from the surface. The most clearly demonstrated attachment sites are found in the ileum and colon of rats.

Attachment Sites in Intestinal Epithelia

The Rat Ileum. The pointed end of the attached holdfasts in filamentous organisms on the rat ileum are inserted into an invagination in the epithelial-cell membrane. There is a nipple-like extension at the very end of the bacterial cell (Fig. 5.10). The microvilli are displaced at the attachment site. In some cases many filaments attach to one cell and the microvillous border is disrupted completely. However, there is *no* penetration of the bacterium through the host-cell membrane (Fig. 5.10*a*). At the bacterium–host -cell junction there is an alteration in the thickness and number of layers of the bacterial cell wall and in the density of the cytoplasm underlying the epithelial cell membrane. The outermost layer, represented by a dense plaque, finely textured and 30–50 nm thick, immediately underlies the area of contact between bacterial cell and epithelial-cell membrane. This area is surrounded by a fibrillar layer. Thin fibrillar structures may be part of the terminal web, a normal component of the cell surface. Boyd and Parsons [61] suggested that similar fibrils in the normal cell could be contractile elements, leading Erlandsen and Chase [62] to suggest that fibrils in the attachment site may contribute to adhesion by some form of contractile mechanism. Whatever the mechanism, the host cell certainly appears to contribute structurally to the attachment site. This is a unique situation in microbial–surface associations but is not surprising in light of the concept of the dual evolution of the bacterium with the host as proposed in Section 5.1. It has been suggested that the host cell could be contributing to the nutrition of the bacteria [14]. A natural question to ask is "What is the benefit of this association to the host?"

The Rat Colon. Wagner and Barrnett [19] described in detail the attachment between bacteria and colonic mucosal epithelial cells in the rat and considered the association to be permanent. Our studies have confirmed the details of attachment but showed the organism to be one of the spiral-shaped bacteria described earlier (Section 5.2.4) rather than a member of the Enterobacteriaceae. As many as six bacteria can be attached to a single cell with points of attachment regularly spaced 0.5–1 μm apart (Fig. 5.7*a*). The ends of the organism are inserted into invaginations in the epithelial cells (Fig. 5.7*b*). The bacterial cell wall appears twice as thick in the region of the insertion tip and modified to form a rigid plug accommodating a tight fit in the epithelial cells. Unlike the ileal filament-attachment site, there is no significant alteration in host-cell structure. The gap between the procaryotic and eucaryotic organisms is 300–400 Å, and appears striated due to the presence of thin filaments of 30–50 Å diameter (Fig. 5.7*b*). This suggests another example of interaction between components of surface layers of the two organisms (e.g., glycoprotein and mucopolysaccharide). Wagner and Barrnett [19] suggested that antibody may play a role in this mucosal association. The presence of filaments in the region of contact may result from microprecipitation between bacterial capsular material and anti-

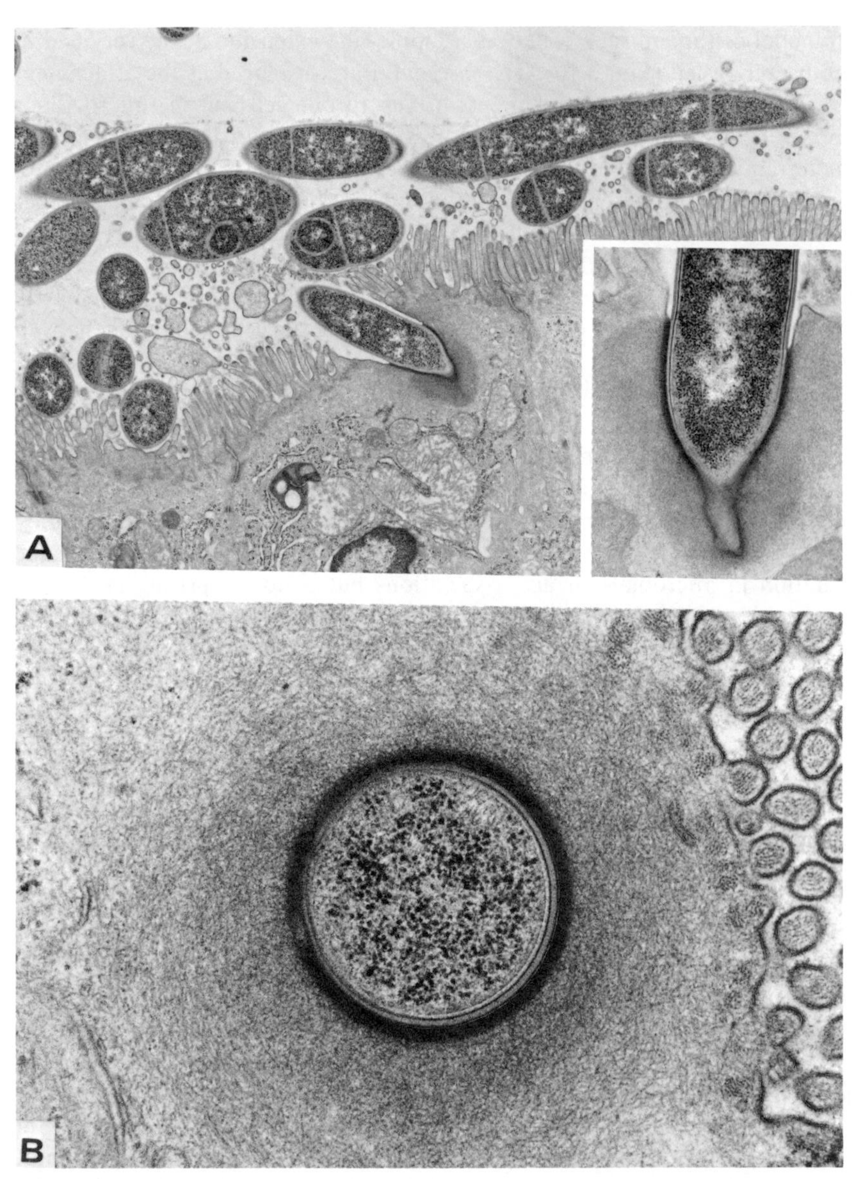

Figure 5.10 Attachment of filamentous microorganisms to rat ileal surface: (A) section passing obliquely through holdfast and segmented filaments in intervillar space—inset shows longitudinal section through attachment site (×5500, inset ×15,750); (B) plane of section passes transversely through holdfast—note arrangement and abundance of cytoplasmic filaments encircling attachment site (×32,000). Micrographs courtesy of S. L. Erlandsen; reproduced from *Am. J. Clin. Nutr.* **27,** 1277 (1974) by copyright permission of *The American Journal of Clinical Nutrition.*

body of the apical cell surface. Certainly, other workers have shown that immunoglobulins are present in the mucus perimatrix of intestinal epithelial cells [63–65].

Other Attachment Sites. There are other associations within the intestinal mucosa that have a more severe impact on epithelial-cell integrity. The colonization of large-bowel epithelium in humans, monkeys, and guinea pigs is so massive (Fig. 5.11) that the microvilli are almost completely replaced and most of the fuzzy coat of the glycocalyx is lost [66–70].

Unlike the attachment of spiral organisms in the rat colon (Section 5.2.4), the attachment sites in these animals often were so close together that a "plug-in-socket" appearance could not be seen (Fig. 5.11*b*). This attachment of spirochetes in primates and guinea pigs was seen only at intestinal surfaces facing the lumen.

All the microbial associations discussed so far concern bacteria, but there are some supposedly nonpathogenic protozoans normally attached to intestinal surfaces. As the surfaces of protozoa and bacteria are so different it is not surprising that different attachment structures are seen. Thus *Cryptosporidium parvum* in the mouse and guinea-pig ileum [66,71] attaches via fusion of the membranes of both cells. The microvilli are either removed or pushed aside, and the attachment site appears as complex foldings of membranes.

5.4.2 Adaption to Survival on Surface Environments

There are various organisms that colonize the intestinal surfaces without apparently forming a permanent attachment to epithelial cells. It is proposed that an ability of bacteria to survive in the local environment closest to the epithelial surface (especially in mucus) is the major mechanism of association with gut surfaces other than that by attachment. Over time, adaption to this environment has given these organisms a selective advantage over the allotochthonous flora, the non-indigenous organisms that normally cannot colonize intestinal surfaces.

In the rat and mouse large intestine there are populations of fusiform-shaped bacteria that grow in densely packed sheets within the mucus layer lining the epithelial surfaces. When fed to germ-free animals, these bacteria specifically colonize this ecological site [2]. They probably utilize mucins as an energy source. Within the crypts of the large bowel, the ability to multiply in mucus is clearly not enough to ensure localization. The organism must move freely in mucus to withstand the continual outward flow from the base of the crypt to the lumen. Possibly, this may explain the successful colonization of the crypts by spiral bacteria. Even though there are many different types of bacteria in the ileal, cecal, and colonic crypts, they all

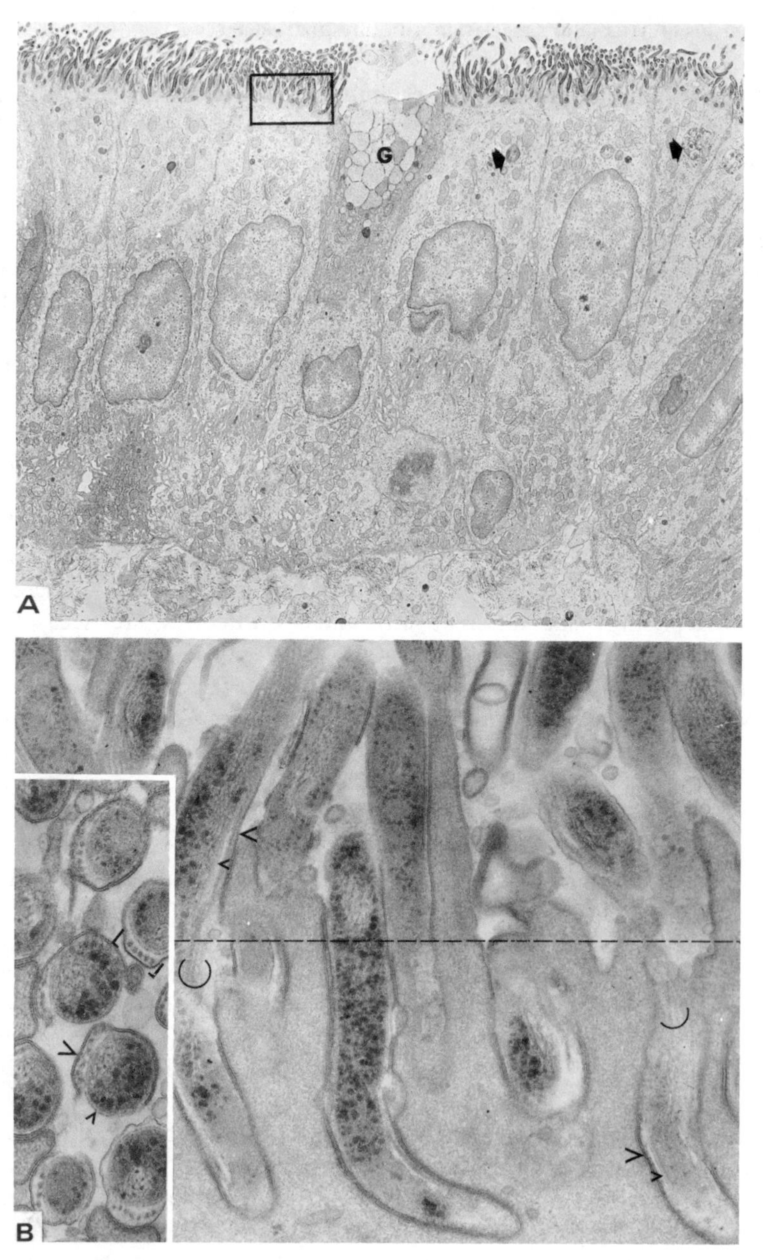

Figure 5.11 Spirochete-covered surface epithelium of colon of rhesus monkey: (A) multiple spirochetes localized at brush border, totally replacing microvilli at this low magnification—spirochetes characteristically oriented in long axis of epithelial cells; cytoplasmic organelles of colonized cells surprisingly well preserved; occasional autophagocytic vesicles (arrows) present at apical cytoplasm; and goblet cell (G) in process of secretion of mucus granules [square refers to (B)] (×2200); (B) high-magnification of square in (A)—microvilli

have a spiral shape. The peculiar motility of spirochetes has been studied in detail [72], and this form of motility should suit movement through the viscous mucus. Phase-contrast microscopic examination of mucosal scrapings reveals that mucus certainly influences the motion of the intestinal spirals. Organisms dislodged from the mucus in an aqueous environment have an aimless random motion. Once they come into contact with mucus, however, they move with remarkable speed, tracking up and down the mucus strands. The attachment of the bacteria to these strands could be by a type of temporary or Stefan adhesion [73], where the bacterial and mucus surfaces are difficult to separate, yet lateral movement of the organism across the mucus occurs with ease. Experiments in which the flow of mucus is being manipulated are in progress. Preliminary results show that depression of mucus production in the crypt environment will allow non-spiral-shaped bacteria to colonize.

5.5 CONSEQUENCES OF MUCOSAL ASSOCIATION OF BACTERIA IN GASTROINTESTINAL TRACT

The major part of this chapter has been concerned with the nature of mucosa-associated populations in the gut and the mechanism of these associations. However, it is also important to determine the possible consequences to the host of the presence of mucosa-associated bacteria. The obvious effects of normal gut microorganisms on the host has been the reason for so much investigation in this field. The contribution of the mucosa-associated bacteria to these effects, however, has yet to be defined.

5.5.1 Intestinal Morphology and Histology

The appearance of the villi in the small intestine of germ-free rats is different from that in the conventional animal. In germ-free animals, the villi are much longer and the overall surface area is 30% less than in the conventional rat [74]. It is possible that the presence of ileal filamentous and spiral organisms in the small bowel could modify this morphology, but it is deemed unlikely because monocontamination of germ-free animals with nonmucosal bacteria such as *E. coli* results in a change in intestinal mor-

---

mostly absent with occasional remnants remaining; terminal web obscure; axial fibrils of spirochete seen between cytoplasmic membrane (small arrow) and cell wall (large arrows); and fibrils (semicircular line) appear to be covered around cytoplasm (inset, cross section showing axial fibrils) (×40,000, inset ×46,500). Micrographs courtesy of A. Takeuchi; reproduced from *Curr. Top. Pathol.* **54,** 22, 23 (1971), by copyright permission of Springer-Verlag, New York.

phology. In the cecum, however, the mucosal integrity could be maintained by the associated bacteria. Removal of the fusiform bacterial layer and crypt spiral organisms with penicillin results in dramatic enlargement of the cecum and possible interference in the regulation of the water transport mechanism across the intestinal epithelium [75].

5.5.2 Enzymatic Activity

Material on the surfaces of the small bowel has significant enzymatic activity. This activity is much higher in germ-free animals, as demonstrated with alkaline phosphatase and Mg^{2+}-dependent ATPase in duodenal homogenates from germ-free and conventional mice [76,77]. The mucosa-associated microorganisms are unlikely to be responsible for this activity, as there are not many bacteria on the duodenal surface, and experiments have shown that activity was depressed in germ-free animals by colonization with *Bacteroides* sp. and *Lactobacillus* sp., which do not associate with the mucosa.

5.5.3 Host Resistance to Infection

Investigation of the importance of mucosal microorganisms in protection against invasion by intestinal pathogens has been hampered by an inability to culture many of the organisms. Thus the filamentous bacteria and spiral organisms of the rat ileum cannot be established experimentally in the intestines of germ-free animals to measure their effect. The covering of the intestinal surface by these bacteria is not sufficient to prevent passage of *Salmonella* sp. through the ileal mucosa, its normal portal of entry. Establishment in germ-free animals of organisms that only colonize the large bowel increases the resistance of rodents to *Salmonella* infection (G. W. Tannock, unpublished observations). A conclusion from this observation is that the bacteria associated with the rodent ileum almost certainly are not involved in host defence against *Salmonella*.

In the large bowel it is more likely that the mucosa-associated bacteria will contribute to resistance to infection because of the dense covering of these organisms on all surfaces. Several of the spiral-shaped organisms from the cecal and colonic mucosa recently have been cultured [78,79]. Now it should be possible to carry out experiments with gnotobiotic animals colonized with fusiforms and spiral-shaped bacteria to assess their resistance to *Shigella* sp. and other pathogens, compared to the germ-free animal.

5.5.4 Immune and Inflammatory Responses to Mucosa-associated Microorganisms

The normal response of tissue to close contact with foreign organisms is an initial inflammatory response followed by an immunological response, especially if tissue-cell integrity is disturbed. The organisms described in the preceding paragraphs stimulate no cellular response, even though they have very close association with the epithelial-cell surfaces. This could be because the extent of attachment is such that no foreign material passes into the lamina propria of intestinal tissue. Alternatively, the host could be unresponsive to antigens of the autochthonous microorganisms. This latter hypothesis was initially proposed by Dubos [4]. Certainly, some intestinal bacteria do not stimulate an immune response after parenteral administration into their natural hosts. The organisms used in these studies were *Bacteroides* sp., which are lumen inhabitants [80,81]. A parenteral response has been produced against some fusiform bacteria and spiral organisms, but these are only preliminary results and give no information as to local immune responses at the mucosal surface (Lee, unpublished data). The possibility of associated organisms producing antiinflammatory products also cannot be excluded. Absence of an inflammatory response to the massive infestation of spirochetes shown in Fig. 5.11 would appear to support this hypothesis. An important observation is that Paneth cells, at the base of small intestinal crypts, can phagocytose the spiral organism shown in Fig. 5.2. It is suggested that these cells play a role in the regulation of the intestinal microorganisms. This would appear a very inefficient defense mechanism, and so this observation is at present unexplained [82].

5.6 MODIFICATION OF MUCOSA-ASSOCIATED FLORA

As with any established ecosystem, external influences can alter the balance of microbial populations in the intestinal tract. Thus the mucosa-associated flora can be removed by treatment with antibiotics causing some of the changes mentioned previously, such as changes in resistance to infection and alterations in morphology of intestinal mucosa [75,83]. Two other isolated observations worth recording may provide clues as to the nature of some of the microbial associations already discussed. There are two reports that starvation can affect the bacteria on intestinal surfaces. Thus Fuller showed that lactobacilli disappeared from the chicken-crop epithelium if animals were starved [41]. Similarly, in mice deprived of food, water, and bedding, yeasts and lactobacilli were not found on the stomach lining; filamentous bacteria in the ileum were reduced; and numbers of fusiforms in mucus layers were reduced [84]. It is interesting that major changes in diet did not influence mucosal populations in either the mice or the birds. These observations could be evidence for a significant contribution of the host cells to the nutrition of mucosa-associated microorganisms.

The other observation concerns irradiation. When mice are subjected to high doses of irradiation, the integrity of the intestinal epithelium is lost and the filamentous bacteria in the ileum enter the lamina propria [85]. Irradiated animals are very susceptible to infection by *Pseudomonas* sp. These bacteria are thought to enter tissue via the gut. Changes in the tissue allowing this breakdown are unknown, but further study of the mucosa-associated microorganisms of irradiated animals might prove useful. Nitrogen mustard has an effect similar to that of X rays on villous epithelial cells. Hampton [86] noted that the filamentous organisms normally attached to ileal epithelium were commonly observed in the lumen of treated animals.

5.7 CONCLUSIONS

Despite the studies quoted in this chapter, investigation of the mucosa-associated microbial populations of the intestinal tract is still in its infancy. As Savage has commented, "the majority of the research into the role of intestinal bacteria is done on the fecal flora with little appreciation of the multitude of different ecological niches that comprise the intestinal tract" [2]. The importance of attachment of microorganisms to tissue surfaces has been established with pathogens in the oral cavity and the respiratory and gastrointestinal tracts (see Chapters 6 and 7). It is likely that the autochthonous flora of these sites are attaching via similar mechanisms. I believe that there are several key areas requiring more extensive study. These are:

1. Development of improved culture methods for the mucosa-associated flora.
2. Investigation of the specificity of microbial–host cell attachment.
3. Further investigation on the composition of intestinal mucus and the behavior of organisms in this substance.
4. Mechanisms of manipulating *in vivo* the adhesion of organisms to the intestinal surface.
5. Demonstration of a mucosa-associated flora in man.

6. Establishment of variously defined mucosa-associated microbial populations in germ-free animals.

If there are more than 300 distinct microbial species in the intestinal tract [87], it may be more rewarding to concentrate on those species that colonize the gut surfaces.

5.8 REFERENCES

1. H. A. Gordon and L. Pesti, *Bacteriol. Rev.*, **35,** 390 (1971).
2. D. C. Savage, *Annu. Rev. Microbiol.*, **31,** 107 (1977).
3. D. C. Savage, R. Dubos, and R. W. Schaedler, *J. Exp. Med.*, **127,** 67 (1968).
4. R. Dubos, R. W. Schaedler, R. Costello, and P. Hoet, *J. Exp. Med.*, **122,** 67 (1965).
5. D. C. Savage, *Am. J. Clin. Nutr.*, **25,** 1372 (1972).
6. D. C. Savage and Blumershine, *Infect. Immun.*, **10,** 240 (1974).
7. A. Takeuchi and J. A. Zeller, *J. Ultrastruct. Res.*, **40,** 313 (1972).
8. D. C. Savage and R. J. Dubos, *J. Bacteriol.*, **94,** 1811 (1967).
9. D. C. Savage, *J. Bacteriol.*, **98,** 1278 (1969).
10. E. J. Van Liere, D. W. Northup, and J. C. Stickney, *Am. J. Physiol.*, **141,** 462 (1944).
11. J. M. S. Dixon, *J. Pathol. Bacteriol.*, **79,** 131 (1960).
12. D. G. Chase and S. L. Erlandsen, *J. Bacteriol.*, **127,** 572 (1976).
13. M. Lipkin, *Physiol. Rev.*, **53,** 891 (1973).
14. C. P. Davis and D. C. Savage, *Infect. Immun.*, **10,** 948 (1974).
15. C. P. Davis and D. C. Savage, *Infect. Immun.*, **13,** 180 (1976).
16. J. C. Hampton and B. Rosario, *Lab. Invest.*, **14,** 1464 (1965).
17. T. D. Luckey, *Am. J. Clin. Nutr.*, **27,** 1266 (1974).
18. C. P. Davis, J. S. McAllister, and D. C. Savage, *Infect. Immun.*, **7,** 666 (1976).
19. R. C. Wagner and R. J. Barrnett, *J. Ultrastruct. Res.*, **48,** 404 (1974).
20. T. Bauchop, R. T. Clarke, and J. C. Newhook, *Appl. Microbiol.*, **30,** 668 (1975).
21. G. W. Tannock and J. M. Smith, *J. Comp. Pathol.*, **80,** 359 (1970).
22. R. Fuller and A. Turvey, *J. Appl. Bacteriol.*, **34,** 617 (1971).
23. R. Fuller and B. E. Brooker, *Am. J. Clin. Nutr.*, **27,** 1305 (1974).
24. C. P. Davis, D. Cleven, E. Balish, and C. E. Yale, *Appl. Environ. Microbiol.*, **34,** 194 (1977).
25. W. D. Leach, A. Lee, and R. P. Stubbs, *Infect. Immun.*, **7,** 961 (1973).
26. G. W. Tannock and D. C. Savage, *Infect. Immun.*, **9,** 475 (1974).
27. D. P. Nelson and L. J. Mata, *Gastroenterology*, **58,** 56 (1970).
28. A. G. Plaut, S. L. Gorbach, L. Nahas, L. Weinstein, G. Spanknebel, and R. Leviton, *Gastroenterology*, **53,** 868 (1967).
29. S. Ito, in *Gastric Secretion. Mechanisms and Control*, Pergamon, Oxford, 1967.

30. A. Takeuchi and J. A. Zeller, *Infect. Immun.*, **6,** 1008 (1972).
31. A. Takeuchi, H. R. Jervis, H. Nakayawa, and D. M. Robinson, *Am. J. Clin. Nutr.*, **27,** 1287 (1974).
32. K. C. Marshall, "Interfaces in Microbial Ecology", Harvard U. P., Cambridge, Mass., (1976).
33. H. C. Jones, I. L. Roth, and W. M. Sanders, *J. Bacteriol.*, **99,** 316 (1969).
34. M. Fletcher and G. D. Floodgate, *J. Gen. Microbiol.*, **74,** 325 (1973).
35. S. Ito, *J. Cell Biol.*, **27,** 475 (1965).
36. H. S. Bennett, *J. Histochem. Cytochem.*, **11,** 14 (1963).
37. H. S. Bennett, *Saibo Kagaku Shimpoziumu*, **14,** Suppl., 529 (1964).
38. J. G. Swift and T. M. Makherjee, *J. Cell Biol.*, **69,** 491 (1976).
39. J. B. Arbuckle, *J. Pathol.*, **104,** 93 (1971).
40. L. C. Hoskins and N. Zamcheck, *Gastroenterology*, **54,** 210 (1968).
41. R. Fuller, *J. Appl. Bacteriol.*, **36,** 131 (1973).
42. N. Suegara, M. Morotomi, T. Watanabe, Y. Kawai, and M. Mutai, *Infect. Immun.*, **12,** 173 (1975).
43. M. E. Etzler and M. O. Branstrator, *J. Cell Biol.*, **62,** 329 (1974).
44. M. E. Etzler and E. A. Kabat, *Biochemistry*, **9,** 899 (1970).
45. M. M. Burger and A. R. Goldberg, *Proc. Nat. Acad. Sci.* (*USA*), **57,** 359 (1967).
46. P. J. Greenaway and D. LeVine, *Nat. New Biol.*, **241,** 191 (1973).
47. W. T. J. Morgan and W. M. Watkins, *Br. J. Exp. Pathol.*, **34,** 94 (1953).
48. G. L. Nicolson and J. Blaustein, *Biochim. Biophys. Acta*, **266,** 543 (1972).
49. S. N. Rao, T. M. Mukherjee, and A. W. Williams, *Gut*, **13,** 33 (1972).
50. H. R. Jervis, *J. Histochem. Cytochem.*, **11,** 692 (1963).
51. T. M. Mukherjee and A. W. Williams, *J. Cell Biol.*, **34,** 447 (1967).
52. T. M. Mukherjee and C. A. Staehelin, *J. Cell Sci.*, **8,** 573 (1971).
53. M. K. Rifaat, O. A. Iseri, and L. S. Gottlieb, *Gastroenterology*, **48,** 593 (1965).
54. K. J. Isselbacher, *Ann. Int. Med.*, **81,** 681 (1975).
55. K. C. Marshall, R. Stout, and R. Mitchell, *J. Gen. Microbiol.*, **68,** 337 (1971).
56. K. C. Marshall and R. H. Cruickshank, *Arch. Mikrobiol.*, **91,** 29 (1973).
57. R. Fuller and B. E. Brooker, *J. Ultrastruct. Res.*, **52,** 21 (1975).
58. R. Fuller, *J. Gen. Microbiol.*, **87,** 245 (1975).
59. N. Sharon and N. Lis, *Science*, **177,** 949 (1972).
60. D. C. Savage, in H. M. Stiles, W. J. Loesche, and T. C. O'Brien, Eds. *Proceedings of Microbial Aspects of Dental Caries; Microbiol. Abstr.* **1,** (spec. suppl.), 33 (1976).
61. C. A. R. Boyd and D. S. Parsons, *J. Cell Biol.*, **41,** 646 (1969).
62. S. L. Erlandsen and D. G. Chase, *Am. J. Clin. Nutr.*, **27,** 1277 (1974).
63. W. D. Allen and P. Porter, *Immunology*, **24,** 365 (1973).
64. W. D. Allen and P. Porter, *Clin. Exp. Immunol.*, **21,** 407 (1975).
65. G. C. Schofield and A. M. Atkins, *J. Anat.*, **107,** 491 (1970).
66. A. Takeuchi, *Curr. Top. Pathol.*, **54,** 1 (1971).

67. M. V. Voino-Iasenetskii, E. S. Snigirevskaia, and V. L. Belianin, *Arkh. Patol.*, **37,** 34 (1975).
68. W. A. Harland and F. D. Lee, *Br. Med. J.*, **3,** 718 (1967).
69. D. L. Harris and J. J. Kinyon, *Am. J. Clin. Nutr.*, **27,** 1297 (1974).
70. F. D. Lee, A. Krasjewski, J. Gordon, J. G. R. Howie, D. McSeveny, and W. A. Harland, *Gut*, **12,** 126 (1971).
71. J. C. Hampton and B. Rosario, *J. Parasitol.*, **52,** 939 (1966).
72. H. C. Berg, *J. Theor. Biol.*, **56,** 269 (1976).
73. B. A. Humphrey, M. R. Dickson, and K. C. Marshall, *Arch. Microbiol.*, **120,** 231 (1979)
74. H. A. Gordon and E. Bruckner-Kardoss, *Am. J. Physiol.*, **201,** 175 (1961).
75. D. C. Savage and R. Dubos, *J. Exp. Med.*, **128,** 97 (1968).
76. D. P. Yolton, C. Stanley, and D. C. Savage, *Infect. Immun.*, **3,** 768 (1971).
77. D. P. Yolton and D. C. Savage, *Appl. Environ. Microbiol.*, **31,** 880 (1976).
78. A. Lee and M. Phillips, *Appl. Environ. Microbiol.*, **35,** 610 (1978).
79. C. P. Davis, D. Cleven, J. Brown, and E. Balish, *Internat. J. Syst. Bact.*, **36,** 498 (1976).
80. M. C. Foo and A. Lee, *Infect Immun.*, **6,** 525 (1972).
81. M. C. Foo and A. Lee, *Infect Immun.*, **9,** 1066 (1974).
82. S. L. Erlandsen and D. G. Chase, *J. Ultrastruct. Res.*, **41,** 319 (1972).
83. A. Lee and E. Gemmell, *Infect. Immun.*, **5,** 1 (1972).
84. G. W. Tannock and D. C. Savage, *Infect. Immun.*, **9,** 591 (1974).
85. J. C. Hampton, *Rad. Res.*, **28,** 37 (1966).
86. J. C. Hampton, *Rad. Res.*, **30,** 576 (1967).
87. W. E. C. Moore and L. V. Holdeman, *Appl. Microbiol.*, **27,** 961 (1974).

CHAPTER 6

Colonization by and Survival of Pathogenic Bacteria on Intestinal Mucosal Surfaces

DWAYNE C. SAVAGE
Department of Microbiology, University of Illinois, Urbana

CONTENTS

6.1 INTRODUCTION

In several mammalian species, epithelial surfaces in various areas of the gastrointestinal tract are habitats for bacterial and yeast members of the indigenous microbiota [1,2] (see also Chapter 5). Similarly, certain bacterial species able to cause intestinal diseases in man or other animals, that is, enteropathogenic bacteria, associate with intestinal epithelial surfaces in their hosts (Table 6.1). Some of these pathogenic species induce diarrheal disease by producing and excreting enterotoxins that cause epithelial cells to secrete body fluids into the intestinal lumen [29]. Others induce diarrhea by entering, multiplying in, and destroying epithelial cells. Still others penetrate through and around the epithelial cells into subepithelial spaces and then travel via lymphatics and blood to deep organs in the body, where they induce systemic disorders [30]. In some of these cases, the bacteria may colonize (i.e., multiply on) the intestinal surface.

In this chapter some recent evidence concerning some of these intestinal pathogens is discussed. The discussion is organized to amplify some speculative generalizations about the mechanisms by which the microorganisms associate with the epithelial surfaces. The generalizations can be summarized as follows:

1. To reach intestinal surfaces and then survive on them, bacterial pathogens must resist or bypass host defences, of which interference by the indigenous microbiota is an important component.
2. Some bacterial pathogens colonize intestinal surfaces while adhering to the glycocalyx of the microvillus border of the epithelial cells.
3. Some bacterial pathogens colonize intestinal surfaces by multiplying in mucin overlying the epithelial cells.
4. To colonize intestinal surfaces, pathogens must be able to fill niches available on those surfaces.

To facilitate discussion of these issues in reference to specific pathogens, some general information is first presented on the intestinal epithelial surface as a microbial habitat.

Table 6.1 Bacterial pathogens reported to associate with mammalian intestinal surfaces *in vivo* or *in vitro*[a]

Bacterium	Cells	Location	Reference
Vibrio cholerae	Human small bowel[b]	*in vivo*	3
	Baby mouse small intestinal	*in vivo*	4
	Rabbit ileal epithelial	*in vivo*	5
	Rabbit ileal	*in vitro*	6
	Rabbit intestinal "brush borders"	*in vitro*	7
Escherichia coli	Infant swine intestinal	*in vivo*	9,10,11,12, 13,14
	Baby swine intestinal[c]	*in vivo*	15
	Baby swine intestinal	*in vitro*	16,17
	Rabbit ileal	*in vivo*	18
	Rabbit small intestinal	*in vitro*	19
	Infant rabbit intestinal	*in vivo*	20
	Human fetal intestinal (fragments)	*in vitro*	21
	Guinea-pig intestinal "brush borders"	*in vitro*	21
	Rabbit intestinal "brush borders"	*in vitro*	21
	Calf intestinal	*in vivo*	22
Clostridium perfringens	Infant swine intestinal	*in vivo*	23
Vibrio parahaemolyticus	Human fetal intestinal[d]	*in vitro*	24
LW-613 Chlamydial strain	Infant calf intestinal	*in vivo*	25
Spirochetes and "*Campylobacter*-like" bacteria	Infant swine intestinal[e]	*in vivo*	26
Salmonella sp.	Mouse intestinal (gnotobiotic)	*in vivo*	27
Shigella sp.	Guinea-pig intestinal	*in vivo*	28

[a] The words "associated with" are used to leave open the issue as to whether the bacteria adhere to the animal cells or rather occupy the mucin overlying the cells (see text).

[b] Postulated.

[c] Ligated segments.

[d] Cell line.

[e] In lesions of swine dysentery.

6.2 INTESTINAL EPITHELIAL SURFACE

Indigenous microorganisms [1,2,31] and enteropathogenic bacteria (see Section 6.5) associate with intestinal epithelial surfaces in complex ways. Much of the complexity arises because the gut surface is complicated architecturally and physiologically.

In both the small and large intestine (the target areas of the pathogens is discussed later), the mucosal lining shares common structural features [32]. In both areas the epithelium of the mucosa is a single layer of columnar cells lying on a basement membrane overlying a lamina propria. The lamina propria consists of a reticulum network containing smooth-muscle cells and cells involved in host resistance to disease, such as lymphocytes, macrophages, and plasma cells. In the small intestine the mucosa is organized so that the epithelium covers finger- or leaf-shaped villi that protrude into the lumen (Fig. 6.1). Villi are not found in the large intestine, although the mucosa may fold in that area when the lumen is empty. In both areas the epithelium lines depressions or pits in the mucosa. These, the "crypts of Lieberkuhn," are located at the bases of the villi in the small bowel (Fig. 6.2) and are spaced periodically in the mucosa of the large bowel.

The epithelium consists of cells of a variety of types, including secreting–absorbing cells (enterocytes), goblet (mucus) cells, and Paneth cells. The Paneth cells are found only in the crypts in the small intestine [32]. The predominating cell type is the enterocyte. These cells and the goblet cells are produced through mitosis in the crypts of Lieberkuhn (Fig. 6.2) [32]. They then move on the basement membrane up the crypt wall and out onto the luminal surface of the intestine [33]. Mitosis normally cannot be detected in the cells after they pass out of the crypt onto the luminal surface proper (Fig. 6.2). The cells continue to glide slowly on the basement membrane until they reach areas where they shed into the lumen. These areas, called extrusion zones (Fig. 6.2), are located on the tips of the villi in the small bowel and between crypts on the mucosal surface in the large intestine. The process by which the epithelial cells are created by mitosis in the crypts and migrate to extrusion zones, where they are discarded, is a mechanism for the normal renewal of the epithelium. The process requires about 2 days in man and some animals [33].

The cytoplasmic membrane on the side of the epithelial cells exposed to the lumen of the bowel is ramified into rod-shaped structures called *microvilli* [32]. These structures are beyond the resolution of the ordinary light microscope and appear in histological preparations as a thin striated layer, the so-called brush borders, on the lumenal surface of the epithelium

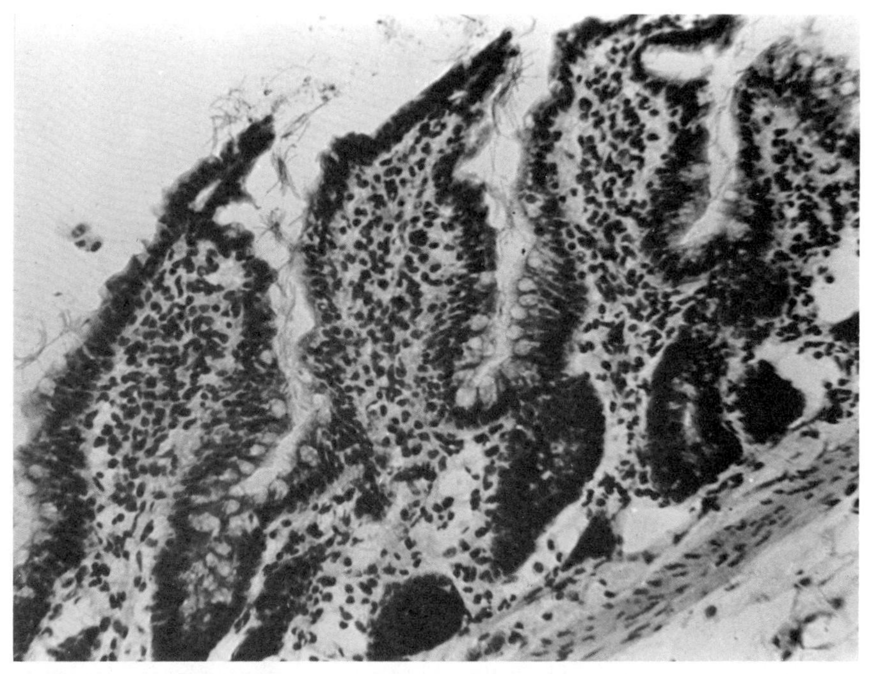

Figure 6.1 Villi in small intestine (ileum) of adult rat; hematoxylin and eosin (×515).

(Fig. 6.3). Microvilli are easily seen in profile in epithelial preparations examined in the transmission electron microscope (Fig. 6.4). Likewise, they can be seen in some preparations of the epithelial surface of gut mucosa examined in SEMs (Fig. 6.5). The membranes of the microvilli carry enzymes that function in digestion and absorption [34].

The transmission electron microscope (TEM) reveals that the membranes of the microvilli are covered by a thin coat (Fig. 6.4) called the *glycocalyx* [35]. This coat is composed of glycoproteins and glycolipids [36] and, depending on the preparation techniques used, appears as a granular or fuzzy layer in preparations examined in the electron microscope (Fig. 6.4). This layer separates the membranes of the microvilli from the lumen of the tract. The glycocalyx is covered by mucus in most areas, but especially in the crypts. The mucus is composed of glycoproteins of various types, some of which are sulfated [36] and is synthesized by the goblet cells [32]. It can be seen as a layer over the microvilli in preparations of intestinal mucosa frozen with contents intact before being prepared for SEM [37].

In the small bowel of the living animal, villi move in such a way that the mucus and lumenal content move downstream [32]. That action is enhanced

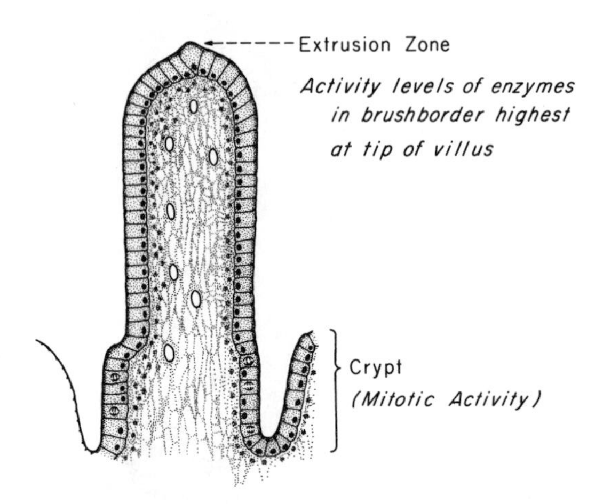

Figure 6.2 Representation of longitudinal section of villus and crypt of Leiberkuhn showing mitotic activity in crypt and extrusion zone. Enzymes in brush border are alkaline phosphatase, disaccharidases, peptidases, and so on.

greatly by peristalsis, in which the entire mucosa moves vigorously due to muscular contraction [38]. In the large bowel a similar but less vigorous peristalsis also distally propels the luminal content and mucus.

To colonize habitats in the bowel, microorganisms must have means of overcoming villous and peristaltic motion as well as epithelial turnover. They could overcome villous motion and peristalsis by producing substances that stop such motion, by adhering physically to and multiplying on membranes of the epithelial cells or by multiplying in the overlying mucins at a rate more rapid than that at which the motion carries them away. As noted, pathogenic bacteria of several types associate in some way with the epithelial membranes. Some of these organisms penetrate into or through the epithelial cells and need only survive on the surface long enough to make the penetration. Such organisms might multiply while associated with the membranes but need not do so because they multiply elsewhere in the body. Pathogens that do not penetrate into or through the epithelial cells by contrast must not only survive on the surface, but also must multiply in the vicinity and then move to fresh epithelium appearing during normal turnover. To survive on the intestinal surface, bacteria must resist normal host defense mechanisms. To multiply on the surface, they must have the capacity to obtain nutrients with which to generate energy and satisfy other needs. In other words, to colonize the habitat, a pathogen must be able to fill a niche in it [1].

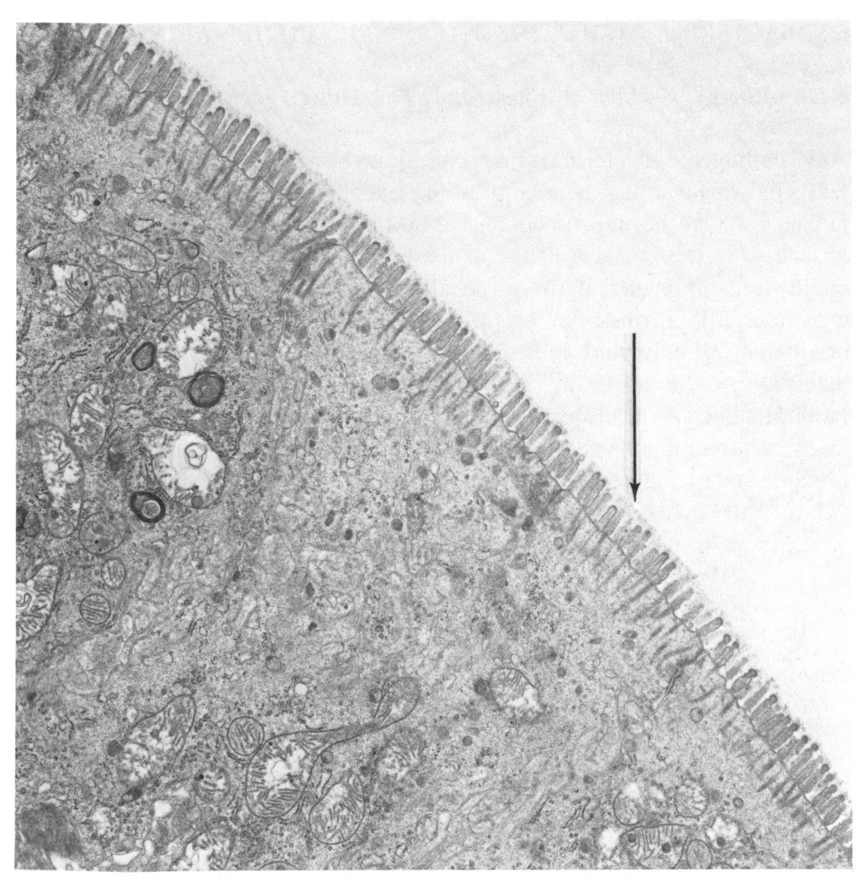

Figure 6.4 Epithelial cell of mouse colon showing microvilli; arrow indicates glycocalyx (TEM, ×9063. Courtesy of Ruth V. Blumershine.

The glycoproteins and glycolipids of the glycocalyx and glycoproteins of mucins may be important nutrient sources for pathogens of mucosal surfaces [1]. Any bacterial species producing enzymes to hydrolyze such macromolecules could obtain fermentable saccharides from the polysaccharide moieties, fatty acids, and other compounds from the lipids and amino acids and peptides from the proteins. Thus the compounds would be ready sources of energy, carbon, nitrogen, and sulfur. Glycosidases that hydrolyze some such compounds are produced by indigenous microbes of certain types [42] and also by some bacterial pathogens [43]. However, neither indigenous bacteria nor bacterial pathogens that associate with intestinal epithelial surfaces are known to degrade complex macromolecules

6.3 BACTERIAL NICHES ON INTESTINAL EPITHELIAL SURFACI

6.3.1 Energy, Carbon, Nitrogen, and Some Other Essentials

Any pathogenic bacterium that can occupy a habitat on an intest epithelial surface must be able to utilize nutrients available at the site. Si nutrients must be substances either not absorbed by the epithelium absorbed by the bacteria at the expense of the epithelial cells. The forn could be components of the glycocalyx, mucinous glycoproteins, digesti enzymes [39], intrinsic factor [40], and, perhaps, some compounds fro desquamating epithelial cells. The latter could be foodstuffs ingested an digested by the host and certain substances produced by the host t facilitate digestive transport, such as components of the bile [41]. Pathogen might also compete with the host epithelium for any end products of the metabolism of indigenous microorganisms in or near the habitat, such as volatile fatty acids, alcohols, and lactic acid [1].

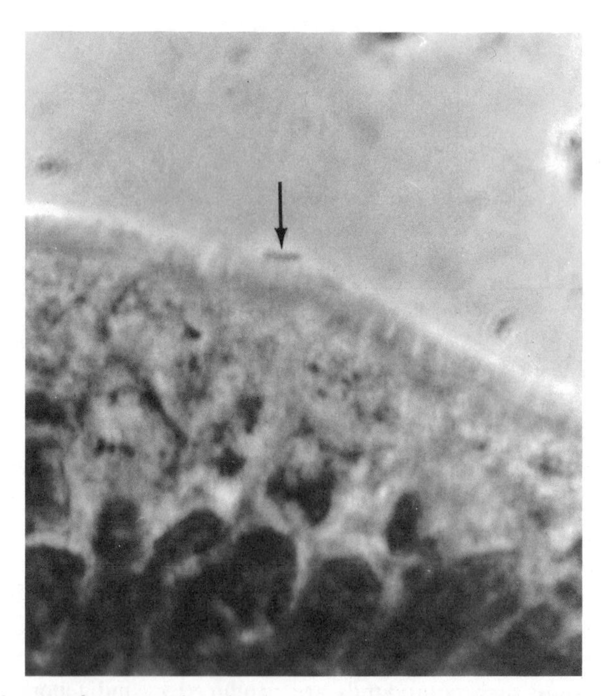

Figure 6.3 Mucosa of mouse colon showing microvilli (brush borders) on epithelial cells. Arrow indicates bacterial cell (*Salmonella typhimurium*) in contact with microvilli (tissue Gram stain, ×4500). Reprinted from Tannock et al. [27], with permission of the American Society for Microbiology.

while on such surfaces. Some microorganisms may obtain nitrogen as ammonia through ureolysis [1], obtaining the urea from host secretions into the gastrointestinal tract, although most undoubtedly gain much of their nutrition by competing with the epithelium for products of the host's own digestive system.

Much, if not all, of a pathogen's energy, carbon, nitrogen, sulfur, and mineral needs can be satisfied by mono- and disaccharides, amino acids and peptides, purines and pyrimidines, and other products of the host's digestive enzymes. The critical issue is that the pathogen can compete successfully for a particular substance with the epithelium. An interesting case in point is how bacterial pathogens may compete with host cells for iron by producing chelating compounds (enterochelins) with affinities for binding iron greater than those of the iron-binding proteins of mammalial tissues [44]. In like manner, bacteria may deplete the epithelium of other substances by

Figure 6.5 Epithelial surface of human large bowel showing microvilli (SEM, ×13,000). Courtesy of Ruth V. Blumershine.

establishing transport gradients [45] more efficient than those of the epithelial cells. Little direct information on such matters is available, unfortunately, from studies of such eucaryotic–procaryotic interactions.

6.3.2 Oxygen

Indigenous bacteria able to colonize epithelial habitats in the gastrointestinal tract share the common characteristic of generating energy for multiplication by utilizing metabolic systems not dependent on oxygen as a terminal electron acceptor; that is, they are capable of anaerobic growth [1]. Therefore, habitats on epithelial surfaces in the gut rarely may contain much oxygen.

Most bacterial pathogens of the gut are capable of anaerobic growth. For example, all members of the family Enterobacteriaceae (*Escherichia*, *Salmonella*, and *Shigella*) are facultative anaerobes. Likewise, *Vibrio cholerae* can grow anaerobically as well as aerobically [46]. Certainly such pathogens can, and probably do, multiply anaerobically in the bowel [47]. However, all such bacteria multiply *in vitro* at fastest rates when growing aerobically [48]. An important virulence mechanism for such pathogens could be an ability to travel to aerobic environments or attract O_2 to themselves.

Salmonellae and shigellae may travel to aerated environments by penetrating into tissues supplied with blood [30]. *Vibrio cholerae* and enteropathogenic strains of *E. coli* that do not penetrate the epithelium (see Section 6.5.2) may bring O_2 to themselves by inducing the intestinal epithelial cells to secrete copious quantities of fluids containing O_2. No experimental evidence exists to substantiate such speculations. Nevertheless, pathogens able to obtain oxygen from the host's hemoglobin would have a strong selective advantage over strictly anaerobic indigenous microbes in habitats in the gastrointestinal canal.

6.4 HOST DEFENSES ON INTESTINAL EPITHELIAL SURFACES

6.4.1 Nonspecific Resistance Mechanisms

In an immunologically naive host, that is, an animal with no prior experience with the antigens of a particular bacterial pathogen, the pathogen colonizing a habitat in the intestinal tract would have to contend only with nonspecific resistance mechanisms functioning in the tract (Table 6.2). It would have to resist or bypass the low pH in the stomach [49,50] and the peristalsis and villous sweep in the small bowel (see Section 6.2). In addition, it would have to resist the toxicity of conjugated or deconjugated bile acids [50]. Deconjugated bile acids are particularly toxic to bacteria *in vitro* and have been regarded classically as important in restricting bacterial growth in the intestinal canal [51]. Indigenous bacterial types that deconjugate bile acids [51] could thus contribute to resistance against bacterial pathogens. Some bacterial pathogens of the bowel, however, are relatively resistant to the toxic effects of bile acids [52]. Moreover, those that are sensitive may avoid exposure to the acids by colonizing the upper small bowel, where the concentrations of deconjugated bile acids may be low [53]. At that level in the bowel, in man at least, little deconjugation takes place and conjugated acids are either propelled distally in the lumen or are transported actively by epithelial cells. Thus the concentrations of deconjugated bile-acids, especially at the epithelial surfaces, may be below levels that are toxic for bacterial pathogens.

Mucin is also thought to be a nonspecific resistance mechanism in the bowel and elsewhere in the body. Generally, the mucin is thought to trap particles, such as bacteria, and then carry them away from vulnerable areas as it is swept out by cilia in the respiratory tract and by peristalsis in the stomach and intestines [54]. Nevertheless, some indigenous bacteria [1] and intestinal pathogens have developed ways to overcome that action. As has been discussed (see Section 6.3.1) and is amplified further (see Section 6.6), certain pathogens may colonize mucins and exploit them as sources of carbon, energy, and other needs.

Phagocytes, with access to the intestinal mucosal surface, may also be important in preventing access of bacterial pathogens to that surface [55]. Little evidence supports an hypothesis, however, that polymorphonuclear leukocytes and macrophages can be effective phagocytes in the lumen or on the epithelial surface of the intestinal tract [55]. Paneth cells, now known to be phagocytes [56], may function to keep the crypts of Leiberkuhn in the small bowel free of bacteria and other microorganisms. Nevertheless, bacterial pathogens that associate with the mucosal epithelium would not be influenced by Paneth cells unless they entered the crypts.

Indigenous microorganisms also may prevent, or at least limit, growth of bacterial pathogens in the gastrointestinal canal [1,30]. The mechanisms of such microbial interference undoubtedly are complex, involving activities of the indigenous inhabitants that affect the pathogens both directly and indirectly (Table 6.2). The processes do not function well, however, in hosts with gastrointestinal ecosystems perturbed by antimicrobial or other drugs [30,57] or by starvation or other stress [58]. For example, emotional stress may alter the indigenous biota in humans [59]. Such stress in travelers could

Table 6.2 Nonspecific resistance mechanisms functioning in the gastrointestinal tract[a]

Mechanism
Gastric pH
Bowel motility (peristalsis)
Bile acids
Mucin
Paneth cells (phagocytes)
Microbial interference[b]
Direct
Bacteriocins
Nutritional competition
Toxic metabolic end products[c]
Maintenance of low oxidation–reduction potentials
Indirect
Enhancement of host mechanisms[d]

[a] Adapted from Savage [1,30]; references cited in text.
[b] By indigenous microorganisms.
[c] Volatile fatty acids (especially at low oxidation–reduction potentials), H_2S, and other compounds toxic to bacteria.
[d] Indigenous microbes may stimulate peristalsis; deconjugate bile acids, rendering them more toxic to bacteria than conjugated ones; stimulate mucin secretion; and otherwise enhance nonspecific resistance mechanisms in the gastrointestinal canal.

influence their susceptibility to diarrheal diseases by perturbing their gastrointestinal ecosystems [60]. Thus pathogens may establish disease only in hosts with such perturbed ecosystems.

Of course, pathogens entering a host at quite high population levels might establish infection even in hosts with established indigenous ecosystems. At high population levels the pathogens may have a capacity to alter the ecosystem, perhaps by inducing diarrhea [61]. Chemically induced diarrhea is known to alter the mucosal ecosystem in dogs and rats [62]. Diarrheal fluids could act simply to flush out the indigenous microbiota. As postulated earlier, however, such fluids may contain oxygen (see Section 6.3.2). If so, the oxygen, possibly in addition to the flushing action, could act to perturb severely the strictly anaerobic components of the biota.

6.4.2 Immunospecific Mechanisms

Bacterial pathogens entering hosts that have had prior experience with antigens of the pathogen may encounter specific, as well as nonspecific, resistance mechanisms. A host's experience with antigens need not involve interaction with the pathogen itself. Macromolecules of similar or identical antigenic specificities are found in many quite dissimilar bacterial types [63] Pathogens may share antigens with presumably nonpathogenic indigenous bacterial types [64]. Thus enteropathogenic bacteria entering a host for the first time might encounter immunospecific mechanisms, even though the host had had no direct experience with that particular pathogen.

A comprehensive review of immunospecific mechanisms that may function on gastrointestinal epithelial surfaces is beyond the scope of this discussion (see Tomasi [65] for a review). However, the manner in which such mechanisms may influence bacteria associating with an intestinal surface is discussed.

The only antibodies that could influence pathogens associating with such surfaces would be those secreted onto the surfaces [65]. These so-called secretory antibodies probably function, at least in part, to prevent pathogenic bacteria from attaching to the epithelium [66]. They could function as well, however, by inhibiting mobility of motile pathogens [67]. Either way, they would be functioning to limit the access of the bacteria to the surface. In conjunction with interference exerted by the indigenous microbiota [68], such antibodies, especially those of the IgA molecular class [65], may constitute powerful defense mechanisms on intestinal epithelial surfaces.

The indigenous microorganisms may contribute more than just interference activities to the defense system. In mice with a conventional microbiota, the levels of IgA in the intestinal contents are high in comparison with the levels in such contents from germfree mice [69]. Thus, at least in mice, indigenous intestinal microorganisms may inhibit destruction of secretory IgA by host digestive enzymes.

Some bacterial pathogens of mucosal surfaces other than intestinal surfaces have evolved ways of destroying secretory antibodies. *Neisseria gonorrhoeae* [70] and *Streptococcus mutans* [71] produce proteolytic enzymes that hydrolyze secretory IgA molecules. These enzymes are specific for the antibody proteins, catalyzing hydrolysis of them at specific points in the peptide chains [70]. Such enzymes could be an effective means by which bacteria neutralize the specific immunological responses in mucosal secretions but have not been reported to be present in bacterial pathogens of intestinal mucosal surfaces.

6.5 BACTERIAL PATHOGENS THAT ASSOCIATE WITH INTESTINAL EPITHELIAL SURFACES

6.5.1 Indigenous Intestinal Bacteria: Association, Adhesion, and Colonization

As noted earlier (Section 6.2), some indigenous bacterial types associate with epithelial surfaces in the gastrointestinal canals of animals [1,2]. Some of these organisms effect the association by adhering to and altering the ultrastructure of the epithelial cells with which they associate. In mice, rats, and chickens [1,2], and also dogs under certain conditions [72], segmented, filamentous bacteria adhere end on to the membranes of the epithelial cells of the jejunal and ileal segments of the small bowel. The bacteria and the microvillous membranes of the epithelial cells interact to form an elaborate attachment site in which a segment of the bacterium inserts into a socket formed by the membranes [2,73]. Microvilli around the attachment site are distorted [2] and even fused together [73]. Microvilli disappear from the membranes of colonic epithelial cells, to which adhere spirochetes and spiral-shaped bacteria in rhesus monkeys [74] and unusual rod-shaped bacteria in rats [75].

In all these cases the bacteria appear to interact intimately with the epithelial-cell membranes. As assessed by transmission electron microscopy, the organisms are located quite closely to the surface of the outer layer of the membrane. Little material recognizable as glycocalyx can be seen between the outer-cell envelope of the bacteria and the epithelial-cell membrane. The influence of these bacteria on microvilli, therefore, may be related in some way to the intimacy of contact between the bacteria and the membrane.

Some indigenous organisms associate with intestinal epithelial cells without obviously altering the microvilli. For example, anaerobic fusiform- and spiral-shaped bacteria are found in layers on the epithelium of the colons in mice and rats [1,2]. These organisms do not alter in any obvious way the ultrastructure of the microvillous border, including its glycocalyx. They may remain in the habitat by adhering rather loosely to the glycocalyx (M. Gilchrist, personal communication). If they do adhere to the surface, then they do not interact with the microvillous membranes as intimately as do the microorganisms just discussed.

Whatever the mechanism by which these indigenous microorganisms associate with the intestinal epithelial surface, they undoubtedly multiply on and colonize the surfaces. Definitive proof of such multiplication is not yet available. Nevertheless, microscopic and other evidence that the organisms specifically adhere to and form thick layers of unique composition on those surfaces [1,2] and have evolved remarkable mechanisms by which to overcome peristalsis, villous motility, and epithelial migration strongly support the concept that they are colonizing their epithelial habitats [1].

The same is surely true of some intestinal bacterial pathogens that associate with epithelial surfaces, at least those types that normally do not penetrate into or through the epithelium. These latter types may colonize the surface without obviously altering the microvilli. By contrast, some pathogens that penetrate epithelial cell membranes do alter the ultrastructure of the microvilli [76]. As is seen later (Section 6.6), these interactions at the interface between bacterium and animal host are complex mechanistically.

6.5.2 Enteropathogenic Bacteria That May Colonize the Intestinal Surface

Vibrio cholerae. Cholera is a disease of the small bowel of man [77]. The chief symptoms of the disease, massive watery diarrhea and consequent dehydration, are induced by an enterotoxin produced by the bacteria [78]. The toxin, a protein [79], attaches to a specific receptor on the membranes of enterocytes. It induces in the cells high levels of adenylate cyclase, the enzyme that catalyzes the reaction in which cyclic adenosine monophosphate (cAMP) is produced [78]. As a consequence, cAMP is produced in the cells in greater than normal levels. Enterocytes having high levels of cAMP secrete more fluids than normal; hence the diarrhea [78].

To deliver the toxin to the enterocytes in the small intestine, the bacteria remain and proliferate in the jejunum and proximal ileum [77]. They do so primarily by associating intimately with the intestinal mucosa but rarely, if ever, penetrating into or through the intestinal epithelial cells (Table 6.1). Interestingly, little published evidence specifically supports the hypothesis that the microbe associates with the epithelium in humans. Available evidence indicates only that the bacteria do not penetrate into the epithelium or deeper tissues of the bowel [3].

By contrast, *V. cholerae* is well known to associate with the gut epithelium in infected animals (Table 6.1). In such experimental systems the bacteria are usually injected into ligated segments of the small intestine of rabbits [5] or instilled into the stomachs of infant rabbits [5] and mice [4]. In all such systems the bacteria associate intimately with the mucosal surface and, rarely, penetrate into or through the epithelium. Conventional histological techniques, fluorescence microscopy, and SEM and TEM have been used to demonstrate the association [4,5]. The histological methods show the bacteria in close relationship with the epithelium, possibly in the mucous gel on the surface of the enterocytes [4]. Scanning electron microscopy reveals the bacteria closely juxtaposed to the microvillous surface, with long, thin strands extending from the vibrios to the surface [5] or adhering to mucus strands and balls [4]. Transmission electron microscopy suggests that the bacterial cells adhere to the glycocalyx of the microvilli and do not alter in any obvious way the microvilli or their membranes or the glycocalyx [5].

The microscopic techniques neither indicate whether the vibrios multiply on the surface nor point unequivocally to the mechanisms by which the organism associates with the surface. The long strands seen in preparations examined by SEM may indicate that the bacterial cells adhere to the glycocalyx. However, they could also be artifacts of the process of preparation for microscopy. Such strands appear frequently in photographs of bacteria on surfaces as visualized by SEM [2] and may be dried remains of mucin or envelope material rather than true filaments. Similar artifacts can occur in preparations made for TEM. Thus microscopic evidence must be interpreted with extreme care in studies of the mechanisms by which microbes adhere to surfaces [2]. Fortunately, some approaches involving techniques other than microscopy have been used to study the mechanisms. These findings are discussed in Section 6.6.1.

Escherichia coli. This is a most versatile intestinal pathogen. Depending on the bacterial strain and the host, the organism is capable of inducing enteritis by a variety of pathogenic mechanisms (Table 6.3). As a pathogen, it can behave similarly to *V. cholerae* and not penetrate the epithelium or to *Shigella* or *Salmonella* species and pass into or through the epithelial cells. Wherever it ultimately multiplies in causing disease, that is, on or in the epithelial cells or submucosally, it must first associate in some way with the epithelial surface (Table 6.1).

Such association has been well documented by conventional histological procedures, fluorescence microscopy, and electron microscopy for the strains that produce enteritis in neonatal swine [9,10] and calves [22]. It has also been demonstrated for a strain that produces enteritis in rabbits [18]. The phenomenon is less well documented directly for man. Nevertheless, strains known to cause human disease probably associate with small-bowel epithelia in rabbits infected experimentally [20]. There is no doubt, therefore, that association with intestinal epithelia is an important factor in the pathogenesis of enteritis produced by *E. coli.* Interestingly, indigenous

Table 6.3 Modes of induction of enteritis by *Escherichia coli*[a]

Associate with mucosal epithelium; rarely or never penetrate into epithelium or submucosa; induce disease by multiplying on epithelium of small bowel and producing enterotoxin that induces hypersecretion by enterocytes; similar to *Vibrio cholerae*[b]
Associate with and penetrate through mucosal epithelium into subepithelial tissues; induce disease by multiplying in submucosal tissues of small bowel primarily; similar to *Salmonella typhimurium*
Associate with and penetrate into mucosal epithelia; rarely penetrate into subepithelial tissues; induce disease by multiplying within and killing epithelial cells of large bowel primarily; similar to *Shigella dysenteriae*

[a] Adapted from Table 2 in Savage [30].
[b] Some strains may attach to epithelium and induce diarrhea without producing enterotoxin [18].

strains of *E. coli* associate with the epithelial surface in the large bowels of infant mice during succession of the indigenous microbiota [80] and may do so in climax communities in adults as well [81].

As with *V. cholerae*, the microscopic techniques neither prove that the *E. coli* multiplies on the surface nor point unequivocally to the mechanisms by which it associates with the epithelial cells. As revealed by electron microscopy, strains that cause enteritis in swine [10,13] associate closely with the microvilli, but do not alter them ultrastructurally, and may adhere to the glycocalyx via fibrillar material on their surfaces. As with *V. cholerae*, these concepts have been tested by techniques other than microscopic ones. For certain strains of *E. coli*, some of these latter approaches have provided evidence that the bacteria adhere to the epithelial surface via macromolecular structures on their surfaces. These findings are discussed in Section 6.6.2.

6.5.3 Enteropathogenic Bacteria That May or May Not Colonize the Intestinal Surface

Microscopic techniques, involving both conventional light, fluorescence, and electron microscopy [27] demonstrate that *Salmonella typhimurium* associates with the epithelium of the small intestine in monoinfected ex-germ-free mice (Fig. 6.3). Likewise, conventional light and fluorescence microscopy has been used to demonstrate *Clostridium perfringens* associating with the small bowel epithelium in swine infected experimentally [23], *Vibrio parahaemolyticus* associating with human fetal intestinal tissue in cell culture [24], and *Shigella flexneri* associating with the intestinal epithelium in guinea pigs infected experimentally [28]. Some other bacterial types, less well characterized as pathogens than *S. typhimurium*, *C. perfringens*, and *V. parahaemolyticus*, such as certain chlamydia in calves [25] and the bacteria that cause swine dysentery [26], have also been reported to associate intimately with the intestinal cells in animals infected experimentally.

These bacterial types may all penetrate into or through the epithelial cells or destroy them, while inducing disease in their natural hosts. Some of them (e.g., *S. typhimurium*) are known to alter the ultrastructure of the microvilli during such penetration, at least under experimental conditions [76]. Still, little evidence is available on the mechanisms by which these important pathogens associate with and penetrate the intestinal epithelium, and even less is known about whether they multiply on the surface of the epithelial cells. As noted earlier (Section 6.2), studies have been made, and are still underway, on the mechanisms mediating the associations of *V. cholerae* and enteropathogenic *E. coli* with intestinal epithelia. Findings from such efforts may serve as models for study of the mechanisms mediating epithelial association of these other enteropathogens.

6.6 MECHANISMS BY WHICH BACTERIAL PATHOGENS COLONIZE INTESTINAL EPITHELIAL SURFACE

6.6.1 *Vibrio cholerae*

This species is well adapted for colonizing the small intestine. It can survive passage through the stomach, although the mechanisms by which it resists gastric acidity are not clear. Food tends to raise the pH in the stomach [50] and thus may protect bacteria in it from the toxicity of gastric acid. Likewise, persons with achlorhydria may be susceptible to cholera because the pH in their stomach is higher than normal due to lack of HCl [50]. Once in the small bowel, the microorganisms find the environment hospitable. They are resistant to the antibacterial toxicity of conjugated bile acids [82], grow well at the moderately alkaline pH levels [82] found in the small bowel content [50], and can overcome peristalsis by associating with the mucosal surface (Table 6.1).

No matter how well adapted *V. cholerae* is to colonizing the small bowel, however, it may do so only in hosts in which the indigenous microbiota either is not established or has been perturbed in some way [83]. Cholera vibrios are cleared readily from the unmanipulated bowels of normal adult mice with intact indigenous microbiotas [57]. They can multiply in the tracts of germ-free animals [84], in those of certain conventional animals treated with antibacterial drugs [83] or substances that slow bowel motility [85], and in segments of intestines tied off at each end in the bowels of conventional animals [5]. In those models, nevertheless, the indigenous biota is either missing (i.e., in germ-free animals) or undoubtedly perturbed in some way [1]. The vibrios can also proliferate in the tracts of some neonatal animals before the succession of the indigenous biota is complete [4,5]. Therefore, the intact indigenous flora undoubtedly interferes with *V. cholerae* in its efforts to colonize its habitat in the small bowel, at least in some experimental animals.

The specific mechanisms by which the interference is exerted are not known in detail. As noted (Table 6.2), the indigenous biota enhances many of the various mechanisms by which animals resist enteropathogens and also contributes some influences of its own. Thus in a host with a fully functioning indigenous biota, the vibrios may not survive because many interference mechanisms contribute to preventing them from developing a population size sufficient to become self-sustaining. Of course, the organism does enter human hosts in which it can cause cholera. In such persons the indigenous gastrointestinal biota may quite possibly be perturbed by nutrition or other factors [1]. Only further research can resolve this issue.

As has been noted, (Table 6.1) when the organism establishes a successful population in the small bowel of a susceptible host, it associates with the surface of the villous epithelium. It may effect that association by either colonizing the mucous gel overlying the glycocalyx, by adhering to and colonizing the microvillous membrane, or by colonizing the lumen of the tract and migrating to the surface. In the first case the bacteria would have to multiply more rapidly than villous and peristaltic motion carried away the mucus. In the second case they would have to multiply on and spread over the surface to which they adhere to avoid being carried off by normal turnover of the epithelium. In the third case mechanisms must exist by which the bacteria stop peristalsis and are attracted or forced to enter the mucous gel overlying the surface and then pass through the gel to the glycocalyx. In all cases the pathogen would have to utilize nutrients available in the areas. Taken together, findings from electron-microscopic studies do not distinguish among these three possibilities. The vibrios have been seen juxtaposed to the brush borders in contact with the glycocalyx [5] and also in mucus strands and clumps [4]. The results obtained in such studies depend on the animal model used and support only the hypothesis that the vibrios associate in some way with the surface.

Similary, findings from studies where attempts have been made to explore the biochemistry and physiology of the association do not distinguish the possibilities (Table 6.4). The organisms adhere *in vitro* to brush-

Table 6.4 Chronology of some hypotheses and findings concerning the mechanisms by which *Vibrio cholerae* associates with the small-bowel epithelium

Year	Finding	Reference
1916	Adhesion of vibrios to the intestinal wall may be an important feature in human cholera	Stoerk (as cited by Freter et al. [3]
1960	Hypothesis: cholera vibrios attach to and proliferate on the surface of the intestinal epithelium	77
1961	Hypothesis: cholera vibrios adhere to the small bowel epithelium when causing disease in man	3
1965	Cholera vibrios associate with the intestinal epithelium without penetrating it in experimentally infected guinea pigs	85
1969, 1970	Antivibrio antibody in rabbit intestinal segments reduced the number of vibrios adsorbed to mucosal surface, but not the number of the organisms in the intestinal lumen	86,87
1972	Viable vibrios adhere better than nonviable ones to the mucosal surface in segments of rabbit ileum	88
1975	Nonmotile mutant vibrios are less effective than motile parent strains in killing suckling mice and associating with their intestinal surfaces	89
1976	Large masses of vibrios establish in the mucus in the intervillous spaces and crypts in segments of rabbit ileum	90
1976	Nonmotile mutant vibrios are less virulent than parental motile strains when administered intraperitoneally to adult mice	91
1976	Vibrios adhere to brush-border membranes isolated from rabbit intestinal epithelial cells; adhesion required divalent cations and was temperature dependent	92
1976	Nonmotile vibrio mutants do not adhere to rabbit intestinal brush-border membranes, although motile strains revertant from the nonmotile ones do so; however, adhesion did not depend on motility as such and was inhibited by L-fucose	92

Table 6.4 (Continued)

Year	Finding	Reference
1976	Nonmotile vibrio mutants adhere *in vitro* less well than do motile parent strains to slices of rabbit ileum; the nonmotile strains failed to adhere because they had lost substances on their surfaces that mediate adherence of bacteria to surfaces rather than motility	6
1976	Vibrios can be seen by SEM in patches and layers on the villous epithelium in ileal segments in adult rabbits and in the small bowels of baby rabbits; the bacteria were located predominantly on the tips of the microvilli	5
1977	Antibody to a crude preparation of flagella from vibrios protected suckling mice against oral infection	67
1977	In suckling mice motile vibrios populate the upper small bowel to higher levels and penetrate the intervillous spaces and crypts of Lieberkuhn more effectively than do nonmotile ones	93
1977	Motile vibrios are attracted to intestinal mucosal surfaces by chemotaxis	94
1977	Motile vibrios in suckling mice can be seen by SEM in material interpreted as mucin on the small-bowel epithelium	4

border (microvillous) membranes from rabbit ileal cells [7] or to slices of rabbit ileum [6]. In the brush-border model the bacteria adhered to the membranes in highest numbers when divalent cations (Ca^{2+}) were present and in lowest numbers when L-fucose, various glycosides of L-fucose, and, to a lesser extent, D-mannose, were present [92]. In the ileal-slice model, such compounds did not influence the interaction in any obvious way. Again, the findings seem to have varied with the experimental model used and support only the hypothesis that the vibrios can associate in some way with the epithelial surface.

Much experimental evidence, largely from one model system [93], suggests that *V. cholerae* must be motile to associate with the bowel surface. By contrast, evidence gained from experiments with the models involving brush-border membranes [7] and ileal slices *in vitro* [6] would tend to discount motility as important in the association. Yet again, the findings seem to depend much on the experimental system.

Other more recent findings, however, support the hypothesis that motility is important to the organism in becoming established on the epithelial surface. These findings indicate that the vibrios may be attracted into the mucous gel and to the glycocalyx by chemotaxis [94]. Thus the intestinal wall, not unlike some surfaces in aquatic environments [95,96] and the root hairs of leguminous plants [97], may transmit chemical signals that attract *V. cholerae* and perhaps other bacteria.

Even if *V. cholerae* is attracted chemotactically to enter the mucous layer overlying the intestinal surface, it still need not colonize (i.e., multiply on) that surface. Thus none of the evidence discussed distinguishes between the three possible modes by which the organism could colonize the small bowel. The question of whether the organism colonizes the bowel surface in some way must be left to speculation for the present. Nevertheless, that particular speculation seems to be a safe one. If the organism did not colonize the surface, then it would presumably be prevented by peristalsis and villous motility from remaining and multiplying in the lumen of the upper regions of the small bowel. Unless the organism has mechanisms for stopping bowel motility, which is highly unlikely considering the nature of the diarrhea in the disease, it must multiply in association with the bowel epithelium.

Moreover, *V. cholerae* has attributes that could equip it to occupy habitats on the bowel surface. It produces several enzymes that could allow it to obtain nutrients from macromolecules in the mucin and glycocalyx. It is known, for example, to produce enzymes that catalyze hydrolysis of certain mucinous glycoproteins [43], a neuraminidase that hydrolyzes sialic-acid residues from glycoproteins [98] such as occur in the glycocalyx [99], and a lecithinase that hydrolyzes choline moieties from lipoproteins [100] also occurring in the glycocalyx [35]. The hydrolysis products of all such reactions might be utilizable as sources of carbon and energy, and even of nitrogen and sulfur, by a fermentative bacterium such as *V. cholerae* [100]. Thus the cholera vibrio is well equipped to colonize a habitat on the bowel surface.

If it does colonize such a habitat, then, it could do so while adhering to receptors containing fucose [92] in the glycocalyx or by multiplying in the mucin at a rate more rapid than that at which the mucin is carried away. If it does not colonize the surface but rather multiplies in the lumen, to associate with the surface it certainly must adhere to some structure on it. If so, it may adhere either to the mucus [6] or to the glycocalyx [92] or, as is most likely, to both structures. Studies of ultrastructural aspects and physiology and biochemistry of adhesion of the organism to the surface may give mixed results, depending on the experimental model used, because the organism adheres by more than one mechanism [101].

The intestinal epithelial surface is a dynamic environment for bacteria. As discussed earlier (see Section 6.5.1), indigenous microorganisms of several types that occupy habitats on gastrointestinal epithelia have developed several mechanisms for adhering to the surfaces in their habitats. Successful bacterial pathogens of the intestinal canal (e.g., *V. cholerae*) may utilize several mechanisms as well for remaining in contact with the surface. Further investigation of these issues is needed.

6.6.2 *Escherichia coli*

As has been noted, *E. coli* is a successful and versatile enteropathogen of man and other animals, especially neonates (Table 6.3). As with *V. cholerae*, however, little evidence reveals how the organism transits the stomach. In adults, the circumstances that prevail to allow the organism to pass through the gastric lumen may be the same as for *V. cholerae*. In infants malnutrition may cause gastric pH to rise to levels high enough to allow the organism to survive and pass into the small intestine [50]. As are virulent strains of *V. cholerae*, some strains of *E. coli* thrive in the upper small bowel. In general, the organisms are resistant to bile [52], grow at moderately alkaline pH [102], and can overcome peristalsis by associating with the mucosal epithelium (Table 6.1).

Similarly to the cholera vibrio, *E. coli* enteropathogens are undoubtedly susceptible to microbial interference [30] in the tract. Such interference, especially from indigenous bacteria that are metabolically strict anaerobes, is believed to control the levels of populations of indigenous *E. coli* during succession in babies [103] and adults [30]. Enteropathogenic strains of *E. coli* tend to cause disease in infants [104] who may have incomplete successions [1] and in adults traveling in foreign countries [60] who may have biotas perturbed by emotional and other stress [59]. Thus such strains may be inhibited in colonizing the small bowel by indigenous microbes. In this case, however, little direct evidence supports that hypothesis.

Some strains of *E. coli*, nevertheless, do cause a cholera-like disease in man and a variety of other animals and do so in close association with the epithelial surface of the small intestine (Table 6.1). The evidence is stronger for *E. coli* than for the cholera vibrios, however, that the bacteria effect the association by adhering to the surface, probably to the glycocalyx. Moreover, for neonatal swine (Table 6.5) and calves and lambs (Table 6.6), the macromolecules that mediate adhesion for many enteropathogenic strains have been characterized partially. In such animals, the bacteria of

Table 6.5 Chronology of some findings concerning the mechanisms by which enteropathogenic *Escherichia coli* (strains) associate with the small-bowel epithelium in neonatal swine

Year	Finding	Reference
1963	Bacteria proliferate in the small intestine	105
1968	Anti-O antibodies confer no resistance	106
1971	Most diarrheagenic serotypes produce K88 antigen	107
1970, 1971	Bacteria associate with villi in small intestine	9,108
1971	K88 Plasmid is essential for bacteria to multiply in the upper intestine[a]	109
1971	Enteropathogenic, but not nonenteropathogenic, strains adhere to small-intestinal epithelium	13
1972	Anti-K88 antibody probably protective	110
1972	$K88^+$ Bacteria and cell-free K88 antigen adhere *in vitro* to mucosa from small intestine; K88-negative avirulent mutants do not; anti-K88 antibodies inhibit adhesion; $K88^+$ bacteria adhere contiguous to the brush-border from base to top of the villi *in vivo*[b]	11
1973	Antibody to purified K88 antigen is protective	111
1974	K88 Bacteria from swine adhere *in vitro* to brushborders of epithelial cells isolated from small bowel of piglets but not calves; anti-K88 antibodies inhibit adhesion; K88 strains enteropathogenic in swine do not adhere to the piglet cells; ovine and bovine bacterial strains do not adhere to piglet or calf cells; human strains do not adhere to piglet cells[b]	17
1974	K88 Antigen agglutinates guinea-pig red blood cells	112
1975	Hemagglutination of guinea pig red blood cells by K88 antigen is inhibited by mucinous glycoproteins; terminal β-D-galactosyl in heteroside side chain of glycoprotein may combine specifically with K88 antigen	113
1975	$K88^+$ Strain proliferated in anterior small bowel; $K88^-$ strain did not; both strains cause severe diarrhea[b]; both adhere to epithelium in posterior small bowel; surface appendages of two strains differ ultrastructurally	10
1975	Two pig phenotypes, one in which K88-bearing bacteria adhere in anterior small bowel and one in which such bacteria do not adhere; the former are more susceptible to diarrhea than the latter	14,114
1975	In mice, $K88^+$, Ent^- bacteria[b] inhibit $K88^+$ Ent^-- bacteria from inducing fluid accumulation in small bowel[b]	115
1976	Anti-K88 antibodies (and possibly some anti-O antibodies) in colostrum and milk protected piglets	116

Table 6.5 (Continued)

Year	Finding	Reference
1976	Some K88$^-$ enteropathogenic[b] strains adhere to ileal epithelium	16
1976	K88$^+$, Ent$^-$ Bacteria[b] inhibit K88$^+$ Ent$^-$ bacteria from inducing diarrhea	117
1977	Some strains of ETEC[c] from swine, but not from humans, have K99 antigen. Swine and calf K99-bearing ETEC[c] adhere to ileal epithelium and cause diarrhea in swine	118
1977	Unless they are richly piliated, K88$^-$ ETEC[c] do not adhere *in vitro* to epithelial cells isolated from the small bowels of piglets; pili may mediate adhesion of such strains to small-bowel epithelium *in vivo*	119
1977	Piliated, K88$^-$ porcine ETEC[b,c] colonize ileum; some such strains have K99 antigen; *d*-capsular, piliated and capsular, nonpiliated mutants of these strains do not colonize ileum	

[a] The genetic determinant for K88 antigen is transmitted from bacterium to bacterium by a plasmid [120]. The antigen is a protein in the form of fine filaments covering the surface of the bacteria cell [121].
[b] In K88$^+$, bacteria bear K88 antigen; in Ent$^+$, bacteria produce diarrheagenic enterotoxin; in K88$^-$, bacteria do not bear K88 antigen; in Ent$^-$, bacteria do not produce enterotoxin.
[c] Enterotoxigenic *E. coli*.

such strains frequently adhere, via so-called K-antigens on their surfaces, to specific receptors probably located in the glycocalyx [11,122,124]. In swine, at least, some strains may adhere to the surface via some surface structure other than K-antigens, perhaps pili [15]. Nevertheless, K-antigens are clearly important determinants of the virulence of many strains of *E. coli* enteropathogenic for neonatal animals of certain types.

Strains of *E. coli* enteropathogenic for man also may associate in some way with the small bowel epithelium. In this case, the bacteria may adhere to the surface via "pilus-like" structures on their surfaces (Table 6.7). However, no direct evidence supports this hypothesis. In suckling infants the organisms may adhere via a mechanism that can be inhibited by mannose derivatives (Table 6.7).

The concept that enteropathogenic strains of *E. coli* adhere to the bowel epithelium in inducing cholera-like disease in man and animals recently has received much experimental attention. In comparison, little attention has been paid to the mechanisms by which the organism obtains nutrients, if it does so, while adhering to the glycocalyx. Even less attention has been paid to an hypothesis that some strains may not adhere at all to the glycocalyx, but rather colonize the mucin overlying the glycocalyx. In fact, little evidence supports an hypothesis that any of the strains actually colonize the epithelium when they adhere to it. As could be the case with *V. cholerae*, *E. coli* could multiply in the lumen and simply migrate to epithelium because it is motile and attracted by chemotaxis [95].

Unlike *V. cholerae*, *E. coli* is not known as an active producer of neuraminidases [98] or lecithinases [100]. It has been reported to produce mucinases [43] that could facilitate its utilization of moieties of those glycoproteins as nutrients. Moreover, it is a nutritionally versatile organism, able

Table 6.6 Chronology of some findings concerning the mechanisms by which enteropathogenic *Escherichia coli* strains associate with the small-bowel epithelium in calves and lambs

Year	Finding	Reference
1972	Diarrheagenic strains share common K, referred to as "Kco", but not common O antigens	122
1974	Swine strains with K88ac or K88ab antigens adhere *in vitro* to brushborders of epithelial cells from small bowels of piglets; calf strains do not adhere to either piglet or calf cells *in vitro*	17
1975	Fifty-eight of 67 ETEC[a] had O and K antigens not belonging to recognized groups	123
1975	"*Kco*" Antigens established as K99	124
1976	Twenty-eight of 35 isolates of ETEC[a] had K99 antigen; none of 10 isolates of non-ETEC has the antigen	125
1976	Seventy of 74 ETEC isolates have K99 antigen; none of 10 non-ETEC[a] had the antigen	126
1976	Seventy-six to 95% of ETEC[a] have K99 antigen; none to 14% of non-ETEC have the antigen	127
1976	K99 Antigen hemagglutinates guinea-pig red blood cells; the hemagglutination is not inhibited by mannose	128
1977	K99 Antigen is proteinaceous, pilus, or pilus like and does not hemagglutinate guinea-pig red blood cells	129

[a] Enterotoxigenic *E. coli*.

Table 6.7 Chronology of some findings concerning the mechanisms by which enteropathogenic *Escherichia coli* associate with the intestinal epithelium in man

Year	Finding	Reference
1971	Diarrheagenic strains attain large populations in the small intestine in natural disease	130
1975	Human strains adhere *in vitro* to intestinal epithelial cells from the small bowels of rabbits; bovine colostrum-containing anti-*E. coli* antibody inhibited the adhesion	19
1975	The ETEC, but not non-ETEC[a], strains adhere *in vitro* to the mucosa of human fetal small intestine; adhesion not due to common fimbria	21
1975	An ETEC strain with certain piluslike surface structures, but not a derivative of the strain lacking the surface structures, reaches a high population level in the small bowel and induces diarrhea in infant rabbits; antiserum to the pilus-like structure protects the rabbits; although its genetic determinant is on a plasmid, the structure is not K88 antigen	20
1977	*Escherichia coli* 0111/B4, which causes neonatal enteritis in humans, attaches to the intestinal epithelium of preweanling rats by a mechanism that may be reversed by mannose derivatives	131

[a] Enterotoxigenic *E. coli.*

to oxidize a wide variety of compounds as carbon and energy sources [100]. Thus it may have the capacity to utilize the macromolecules abundant on the intestinal surface and occupy a habitat on that surface. Still, as noted for *V. cholerae*, little direct evidence supports such speculations.

6.7 SUMMARY AND CONCLUSIONS

The intestinal surface is a complex and dynamic environment for bacteria. Some bacterial pathogens have evolved mechanisms by which they can occupy and perhaps colonize that environment. Some of these organisms may adhere via macromolecules on their surfaces to specific receptors in the glycocalyx surrounding membranes of the microvillous border of intestinal epithelial cells. Some of them may enter and occupy the mucous layer overlying the glycocalyx. Whatever the mechanism by which they remain in the space near the epithelial cells, they must multiply there or in the lumen. Bacteria that adhere to the glycocalyx or occupy the mucin overlying it may multiply in those habitats, utilizing macromolecules in the glycocalyx and mucin as nutrients. Bacteria that occupy the mucin must multiply at a rate more rapid than intestinal motion carries the mucin away. Bacteria that multiply in the lumen or on the epithelial surface may be attracted to that surface by chemotaxis. Whether they multiply in the mucin while adhering to the glycocalyx or in the intestinal lumen, the organisms may also compete with the host for nutrients either from the host's own digestive processes or as produced by indigenous microbes. The pathogens can be prevented from colonizing any of these habitats by interference mechanisms exerted by an intact indigenous microbiota and also by immunospecific mechanisms, possible acting in concert with the interference mechanisms.

6.8 REFERENCES

1. D. C. Savage, *Annu. Rev. Microbiol.*, **31,** 107 (1977).
2. D. C. Savage, "Electron Microscopy of Bacteria Adherent to Epithelia in the Murine Intestinal Canal," in D. Schlessinger, Ed., *Microbiology-1977*, American Society for Microbiology, Washington, D. C., 1977, p. 422.
3. R. Freter, H. L. Smith, and F. J. Sweeny, *J. Infect. Dis.*, **109,** 35 (1961).
4. M. N. Guentzel, L. Field, G. T. Cole, and L. J. Berry, "The Localization of *Vibrio Cholerae* in the Ileum of Infant Mice," in *Scanning Electron Microscopy, 1977*, Vol. II, IIT Research Institute, Chicago, 1977, p. 275.
5. E. T. Nelson, J. D. Clements, and R. A. Finkelstein, *Infect. Immun.*, **14,** 527 (1976).
6. R. Freter and G. W. Jones, *Infect. Immun.* **14,** 246 (1976).
7. G. W. Jones, G. D. Abrams, and R. Freter, *Infect. Immun.*, **14,** 232 (1976).
8. E. H. LaBrec, H. Sprinz, H. Schneider, and S. B. Formal, "Localization of Vibrios in Experimental Cholera: A Fluorescent-antibody Study in Guinea Pigs," in *Proceedings of the Cholera Research Symposium*, publication No. 1328, U.S. Public Health Service, Washington, D. C., 1965, p. 272.
9. J. B. R. Arbuckle, *J. Med. Microbiol.*, **3,** 333 (1970).
10. A Hohmann and M. R. Wilson, *Infect. Immun.* **12,** 866 (1975).
11. G. W. Jones and J. M. Rutter, *Infect. Immun.*, **6,** 918 (1972).
12. T. E. Staley, E. W. Jones, and L. D. Corley, *Am. J. Pathol.*, **56,** 371 (1969).
13. H. U. Bertschinger, H. W. Moon, and S. C. Whipp, *Infect. Immun.*, **5,** 595 (1972).
14. R. Sellwood, R. A. Gibbons, and J. M. Rutter, *J. Med. Microbiol.*, **8,** 405 (1975).
15. R. E. Isaacson, B. Nagy, and H. W. Moon, *J. Infect. Dis.*, **135,** 531 (1977).
16. B. Nagy, H. W. Moon, and R. E. Isaacson, *Infect. Immun.*, **16,** 344 (1977).
17. M. R. Wilson and A. W. Hohmann, *Infect. Immun.*, **10,** 776 (1974).
18. J. R. Cantey and R. K. Blake, *J. Infect. Dis.*, **135,** 454 (1977).

19. G. Demierre, D. Rivier, H. Hilpert, H. Gerber, and R. Zinkernagel, *Pathol. Microbiol.*, **42,** 137 (1975).
20. D. C. Evans, R. P. Silver, D. J. Evans, Jr., D. G. Chase, and S. L. Gorbach, *Infect. Immun.*, **12,** 656 (1975).
21. A. S. McNeish, P. Turner, J. Fleming, and N. Evans, *Lancet* (November 15, 1975), 946 (1975).
22. H. W. Smith, *Ann. N. Y. Acad. Sci.*, **176,** 110 (1971).
23. J. B. R. Arbuckle, *J. Pathol.*, **106,** 65 (1972).
24. M. M. Carruthers, *J. Infect. Dis.*, **136,** 588, (1977).
25. A. M. Doughri, K. P. Altera, J. Storz, and A. K. Eugster, *Exp. Molec. Pathol.*, **18,** 10 (1973).
26. G. A. Kennedy and A. C. Strafuss, "Scanning Electron Microscopy of the Lesions of Swine Dysentery," in *Scanning Electron Microscopy, 1977*, Vol. II, IIT Research Institute, Chicago, 1977, p. 283.
27. G. W. Tannock, R. V. H. Blumershine, and D. C. Savage, *Infect. Immun.*, **11,** 365 (1975).
28. E. H. LaBrec, H. Schneider, T. J. Magnani, and S. B. Formal, *J. Bacteriol.*, **88,** 1503 (1964).
29. J. P. Craig, A. S. Benenson, M. C. Hardegee, N. F. Pierce, and S. H. Richardson, *J. Infect. Dis.*, **133** (March suppl.), S1 (1967).
30. D. C. Savage, "Survival on Mucosal Epithelia, Epithelial Penetration and Growth in Tissues of Pathogenic Bacteria," in H. Smith and J. H. Pearce, *Microbial Pathogenicity in Man and Animals*, Cambridge U. P., Cambridge, U. K., 1972, p. 25.
31. D. C. Savage, *Am. J. Clin. Nutr.*, **25,** 1372 (1975).
32. W. Bloom and D. W. Fawcett, *A Textbook of Histology*, 9th ed., Saunders, Philadelphia, 1968, p. 560.
33. J. S. Trier, and T. H. Browning, *New Engl. J. Med.*, **283,** 1245 (1970).
34. N. R. Stevenson, F. Ferrigni, K. Parnicky, S. Day, and J. S. Fierstein, *Biochim. Biophys. Acta*, **406,** 131 (1975).
35. S. Ito, *Proc. Fed. Am. Soc. Exp. Biol.*, **28,** 12 (1969).
36. S. Roseman, "Complex Carbohydrates and Intercellular Adhesion," in E. Y. C. Lee and E. E. Smith, Eds., *Biology and Chemistry of Eucaryotic Cell Surfaces*, Academic, New York, 1974, p. 317.
37. C. P. Davis, *Appl. Environ. Microbiol.*, **31,** 304 (1976).
38. J. J. Misiewicz, "Motility of the Gastrointestinal Trace," in J. M. Dietschy, Ed., *Disorders of the Gastrointestinal Tract, The Science and Practice of Clinical Medicine*, Vol. 1, Grune and Stratton, New York, 1976, p. 27.
39. B. S. Reddy, J. R. Pleasants, and B. S. Wostmann, *J. Nutr.*, **97,** 327 (1969).
40. S. Tabaqchali, H. Schjonsby, and D. Gompertz, "Role of Microbial Alterations in the Pathogenesis of Intestinal Disorders," in A. Balows, R. M. Dehaan, V. R. Dowel, Jr., and L. B. Guze, Eds., *Anaerobic Bacteria: Role in Disease*, Thomas, Springfield, Ill., 1974, p. 99.
41. A. C. Guyton, *Textbook of Medical Physiology*, 4th ed., Saunders, Philadelphia, 1971, p. 862.

42. A. A. Salyers, S. E. H. West, J. R. Vercellotti, and T. D. Wilkins, *Appl. Environ. Microbiol.*, **34,** 529 (1977).
43. C. Ross, *J. Pathol. Bacteriol.*, **77,** 642 (1959).
44. J. J. Bullen, H. J. Rogers, and E. Griffiths, "Bacterial Iron Metabolism in Infection and Immunity," in J. B. Neilands, Ed., *Microbial Iron Metabolism—A Comprehensive Treatise*, Academic, New York, 1974, p. 518.
45. A. Koch, *Adv. Microbial. Physiol.*, **6,** 147 (1971).
46. R. Y. Stanier, E. A. Adelberg, and J. Ingraham, *The Microbial World*, 4th ed., Prentice-Hall, Englewood Cliffs, N. J., 1976, p. 612.
47. G. H. Bornside, W. E. Donovan, and M. B. Myers, *Proc. Soc. Exp. Biol. Med.*, **151,** 437 (1976).
48. T. Bauchop and S. Elsden, *J. Gen. Microbiol.*, **23,** 457 (1960).
49. M. A. Franklin and S. C. Skoryna, *Can. Med. Assoc. J.*, **105,** 380 (1971).
50. B. S. Drasar and M. J. Hill, *Human Intestinal Flora*, Academic, London, 1974, pp. 10, 44, 189.
51. H. J. Binder, B. Filburn, and M. Floch, *Am. J. Clin. Nutr.*, **28,** 119 (1975).
52. G. S. Wilson and A. A. Miles, *Topley and Wilson's Principles of Bacteriology and Immunity*, Vol. 1, 5th ed., Williams and Wilkins, Baltimore, 1964, p. 813.
53. A. Mallory, F. Kern, Jr., J. Smith, and D. Savage, *Gastroenterology*, **64,** 26 (1973).
54. H. W. Florey, *J. Pathol. Bacteriol.*, **37,** 283 (1933).
55. H. Smith, *Bacteriol. Rev.*, **41,** 475 (1977).
56. S. L. Erlandson and D. B. Chase, *J. Ultrastruct. Res.*, **41,** 296 (1972).
57. R. Freter, *J. Exp. Med.*, **104,** 411 (1956).
58. G. W. Tannock and D. C. Savage, *Infect. Immun.*, **9,** 591 (1974).
59. L. V. Holdeman, I. J. Good, and W. E. C. Moore, *Appl. Environ. Microbiol.*, **31,** 359 (1976).
60. E. G. Shore, A. G. Dean, K. J. Holik, and B. R. Davis, *J. Infect. Dis.*, **129,** 577 (1974).
61. S. L. Gorbach, J. G. Banwell, B. D. Chatterjee, B. Jacobs, and R. B. Sack, *J. Clin. Invest.*, **50,** 881 (1971).
62. W. D. Leach, A. Lee, and R. P. Stubbs, *Infect. Immun.*, **7,** 961 (1973).
63. R. L. Myerowitz, R. E. Gordon, and J. B. Robbins, *Infect. Immun.*, **8,** 896 (1973).
64. R. L. Myerowitz, and C. W. Norden, *Infect. Immun.*, **17,** 83 (1977).
65. T. B. Tomasi, "The Concept of Local Immunity and the Secretory System," in *The Secretory Immunologic System*, U.S. Dept. of Health, Education and Welfare, Public Health Service, National Institutes of Health, Washington, D. C., 1971, p. 3.
66. E. S. Fubara, and R. Freter, *J. Immunol.*, **111,** 395 (1973).
67. E. R. Eubanks, M. N. Guentzel, and L. J. Berry, *Infect. Immun.*, **15,** 533 (1977).
68. S. Shedlovsky and R. Freter, *J. Infect. Dis.*, **129,** 296 (1974).
69. E. S. Fubara and R. Freter, *Infect. Immun.*, **6,** 965 (1972).
70. A. G. Plaut, J. V. Gilbert, M. S. Artenstein, and J. D. Capra, *Science*, 190, 1103 (1975).
71. R. J. Genco, A. G. Plaut, and R. C. Moellering, *J. Infect. Dis.*, **131,** S17 (1975).
72. C. P. Davis, D. Cleven, E. Balish, and C. E. Yale, *Appl Environ. Microbiol.*, **34,** 194 (1977).

73. J. E. Snellen and D. C. Savage, *J. Bacteriol.*, **134,** 1099 (1978).
74. A. Takeuchi and J. A. Zeller, *Infect. Immun.*, **6,** 1008 (1972).
75. R. C. Wagner and R. J. Barrnett, *J. Ultrastruct. Res.*, **48,** 404 (1974).
76. A. Takeuchi, *Am. J. Pathol.*, **50,** 109 (1967).
77. C. E. Lankford, *Ann. N. Y. Acad. Sci.*, **88,** 1203 (1960).
78. J. P. Craig, and N. F. Pierce, *J. Infect. Dis.*, **133** (March suppl.) S3 (1976).
79. J. J. Lospalluto and R. Finkelstein, *Biochim. Biophys. Acta*, **257,** 158 (1972).
80. D. C. Savage, R. Dubos, and R. W. Schaedler, *J. Exp. Med.*, **127,** 67 (1968).
81. C. P. Davis, J. S. McAllister, and D. C. Savage, *Infect. Immun.*, **7,** 666 (1973).
82. P. B. Fernandes and H. L. Smith, *J. Gen. Microbiol.*, **98,** 77 (1977).
83. R. Freter, "Host Defense Mechanisms in the Intestinal Tract," *Rec. Adv. Microbiol.*, **10,** 333 (1970).
84. C. E. Miller, K. H. Wong, J. C. Feeley, and M. E. Forlines, *Infect. Immun.*, **6,** 739 (1972).
85. E. H. LaBrec, H. Sprinz, H. Schneider, and S. B. Formal, "Localization of Vibrios in Experimental Cholera: A Fluorescent-Antibody Study in Guinea Pigs," in *Proceedings of the Cholera Research Symposium*, USPHS Publication No. 1328, Washington, D. C., 1965, p. 272.
86. R. Freter, *Tex. Rep. Biol. Med.*, **27** (suppl. 1), 299 (1969).
87. R. Freter, *Infect. Immun.*, **2,** 556 (1970).
88. R. Freter, *Infect. Immun.*, **6,** 134 (1972).
89. M. N. Guentzel and L. J. Berry, *Infect. Immun.*, **11,** 890 (1975).
90. G. D. Schrank and W. F. Verwey, *Infect. Immun.*, **13,** 195 (1976).
91. E. R. Eubanks, M. N. Guentzel, and L. J. Berry, *Infect. Immun.*, **13,** 457 (1976).
92. G. W. Jones and R. Freter, *Infect. Immun.*, **14,** 240 (1976).
93. M. N. Guentzel, L. H. Field, E. R. Eubanks, and L. J. Berry, *Infect. Immun.*, **15,** 539 (1977).
94. B. Allweiss, J. Dostal, K. E. Carey, T. F. Edwards, and R. Freter, *Nature* (*Lond.*), **266,** 448 (1977).
95. E. P. Greenberg and E. Canale-Parola, *J. Bacteriol.*, **130,** 485 (1977).
96. I. Chet and R. Mitchell, *Annu. Rev. Microbiol.*, **330,** 221 (1976).
97. W. W. Currier and G. A. Strobel, *Science*, **196,** 434 (1977).
98. A. Rosenberg and C.-L. Schengrund, "Sialidases", in A. Rosenberg and C.-L. Schengrund, Eds., *Biological Roles of Sialic Acid*, Plenum, New York, London, 1976, p. 295.
99. S. Roseman, "Complex Carbohydrates and Intercellular Adhesion," in E. Y. C. Lee and E. E. Smith, Eds., *Biology and Chemistry of Eucaryotic Cell Surfaces*, Academic, New York, 1974, p. 317.
100. J. M. Shewan and M. Veron, "Genus I. *Vibrio Pacini* 1854, 411," in R. E. Buchanan and N. E. Gibbons, Eds., *Bergey's Manual of Determinative Bacteriology*, 8th ed., Williams and Wilkins, Baltimore, 1974, p. 340.
101. D. C. Savage, "Indigenous Microorganisms Associating with Mucosal Epithelia in the Gastrointestinal Ecosystem," in D. Schlessinger, Ed., *Microbiology—1975*, American Society for Microbiology, Washington, D. C., 1975, p. 120.

102. T. D. Brock, *Biology of Microorganisms*, 2nd ed., Prentice-Hall, Englewood Cliffs, N. J., 1974, p. 309.
103. A. Lee and E. Gemmel, *Infect. Immun.*, **5,** 1 (1972).
104. B. Tennant, Ed., *Ann. N. Y. Acad. Sci.*, **176,** 1 (1971).
105. H. W. Smith and J. E. T. Jones, *J. Pathol. Bacteriol.*, **86,** 387 (1963).
106. J. B. R. Arbuckle, *Br. Veter. J.*, **124,** 274 (1968).
107. W. J. Sojka, *Veter. Bull.*, **41,** 509 (1971).
108. J. B. R. Arbuckle, *J. Pathol.*, **104,** 93 (1970).
109. H. W. Smith and M. A. Linggood, *J. Med. Microbiol.*, **4,** 467 (1971).
110. H. W. Smith, *J. Med. Microbiol.*, **5,** 345 (1972).
111. G. W. Jones and J. M. Rutter, *Nature* (*Lond.*), **242,** 531 (1973).
112. G. W. Jones and J. M. Rutter, *J. Gen. Microbiol.*, **84,** 135 (1974).
113. R. A. Gibbons, G. W. Jones, and R. Sellwood, *J. Gen. Microbiol.*, **86,** 228 (1975).
114. J. M. Rutter, M. R. Burrows, R. Sellwood, and R. A. Gibbons, *Nature* (*Lond.*) **257,** 135 (1975).
115. J. N. Davidson and D. C. Hirsh, *Infect. Immun.*, **12,** 134 (1975).
116. J. M. Rutter, G. W. Jones, G. T. H. Brown, M. R. Burrows, and P. D. Luther, *Infect. Immun.*, **13,** 667 (1976).
117. J. N. Davidson and D. C. Hirsh, *Infect. Immun.*, **13,** 1773 (1976).
118. H. W. Moon, B. Nagy, R. E. Isaacson, and I. Ørskov, *Infect. Immun.*, **15,** 614 (1977).
119. B. Nagy, H. W. Moon, and R. E. Isaacson, *Infect. Immun.*, **16,** 344 (1977).
120. I. Ørskov, and F. Ørskov, *J. Bacteriol.*, **91,** 69 (1966).
121. S. Stirm, F. Ørskov, I. Ørskov, and A. Birch-Anderson, *J. Bacteriol.*, **93,** 740 (1967).
122. H. W. Smith and M. A. Linggood, *J. Med. Microbiol.*, **5,** 243 (1971).
123. L. L. Myers, *Infect. Immun.*, **11,** 493 (1975).
124. I. Ørskov, F. Ørskov, H. W. Smith, and W. J. Sojka, *Acta Pathol. Microbiol. Scand.*, *Sect. B.*, **83,** 31 (1975).
125. L. L. Myers and P. A. M. Guinée, *Infect. Immun.*, **13,** 1117 (1976).
126. P. A. M. Guinée, W. H. Jansen, and C. M. Agterberg, *Infect. Immun.*, **13,** 1369 (1976).
127. H. W. Moon, S. C. Whipp, and S. M. Skartvedt, *Am. J. Veter. Res.*, **37,** 1025 (1976).
128. M. R. Burrows, R. Sellwood, and R. A. Gibbons, *J. Gen. Microbiol.*, **96,** 269 (1976).
129. R. E. Isaacson, *Infect. Immun.*, **15,** 272 (1977).
130. S. L. Gorbach, J. G. Banwell, B. D. Chatterjee, B. Jacobs, and R. B. Sack, *J. Clin. Invest.*, **50,** 881 (1971).
131. M. Hirschberger, D. Mirelman, and M. M. Thaler, *Pediatr. Res.*, **11,** 500 (1977).

CHAPTER 7

Retention of Bacteria on Oral Surfaces

HUBERT N. NEWMAN
Institute of Dental Surgery, Eastman Dental Hospital (British Postgraduate Medical Federation), University of London

CONTENTS

7.1 INTRODUCTION

To understand the relationship of oral bacteria to disease, an intensive international effort is underway to explain the mechanisms whereby organisms colonize and are retained on oral surfaces. Oral bacteria, like those of the rest of the alimentary tract, consist of floating and attached populations. Microorganisms colonize the mouth initially via the external environment, mainly inhaled air and ingested food (for review, see Burnett et al. [1]). Although there are variations in the microorganisms found at different sites even in the same mouth [2], the spectrum in each oral locus shows a degree of specificity typical of microbial habitats in general.

Various oral surfaces are colonized: teeth, external gingivae (gums), tongue (especially dorsum), and nonkeratinized and keratinized mucosa. The dental portion (plaque) has been most studied on account of its involvement in disease. With the advance of caries and chronic periodontitis, other surfaces become colonized, notably dentine, cementum, and possibly the gingival epithelium lining the tooth surface in the crevice region.

In dental health one may speak of a dynamic equilibrium between the forces of retention and those of removal. Increasing bacterial accumulation is generally associated with a shift toward disease [3,4]. The principal mechanisms considered as favoring retention of organisms are selective adhesion and stagnation. The latter condition may be associated with soft diet texture, inadequate oral hygiene, reduced salivary flow, dental appliances and poorly contoured restorations, and anatomical factors. The mechanisms counteracting accumulation include swallowing, frictional removal by diet, tongue and oral hygiene implements, aggregation by salivary factors or binding to desquamated epithelial cells and swallowing, the inflammatory or immune response, antagonistic activity of other organisms, and mimicking of bacterial receptors for oral surfaces by salivary factors.

Of particular importance is the accumulation of organisms at stagnant sites on teeth and in relation to the gums or gingivae, for it is in relation to these locations that disease is initiated. Both caries and chronic inflammatory periodontal disease are relatively uncommon in dentate animals, including man, that consume diets of natural texture [5–13]. This may be due to the prolonged and vigorous chewing required for such diets, which appears to minimize the accumulation of organisms at otherwise stagnant sites [14] (see Figs. 7.1 and 7.2). It may be noted in passing that these changes in diet texture associated with human civilization or domestication of animals appear to have affected general health [15–22]. At the same time it is evident that there are no benefits to oral health from the consumption of a single item of fibrous food daily [23].

Stagnant sites of importance in the mouth include the pits and fissures of occlusal tooth surfaces and the region around contiguous approximal surfaces. At a microscopic level organisms occur in natural pits in the enamel surface [24,25], in epithelial-cell surface pits, and in similar hollows in the dorsum of the tongue. Clinical studies in which oral hygiene has been performed effectively by the patient at home have shown that prevention of plaque accumulation at stagnant sites is associated with a low incidence of caries [4,26,27] and severity of chronic inflammatory periodontal disease [3,28,29].

Gibbons and van Houte [30] have shown that the experimentally observed affinity of some strains for given oral surfaces correlates with the proportions in which these strains occur naturally. They consider this phenomenon of selective adsorption to be a "powerful ecological determinant" in loci such as the mouth, where surfaces available for colonization are "subjected to a fluid flow." Colonization by a given strain is also influenced by the available numbers of that strain [30]. Selective adhesion may be of primary importance on oral epithelial and dental self-cleansing surfaces or in the border regions between those aspects and undisturbed sites. As stated earlier, however, stagnation is a primary factor in the retention of organisms at sites prone to disease. Selective adhesion may be a factor determining the attachment ability of organisms not capable of adhering directly to the host surface but able to attach to a primary selectively adsorbed microbial layer [30]. Selective adhesion may also confer benefits on the host, since many oral strains have been shown to be capable of inhibiting the growth of known pathogens [31–35].

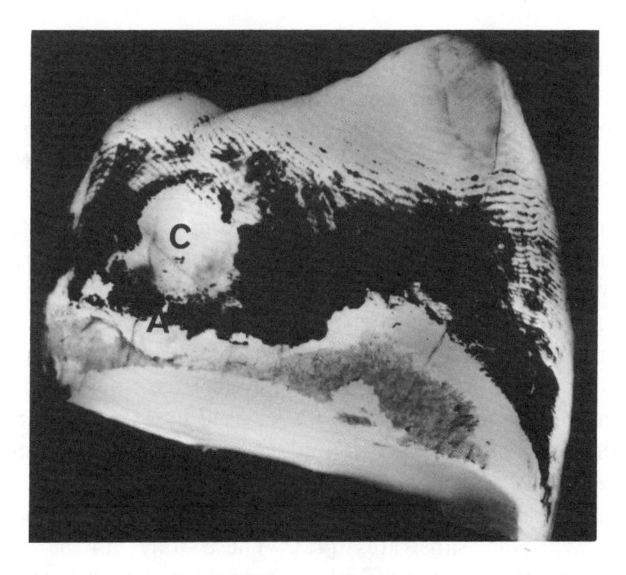

Figure 7.1 Approximal surface, child's premolar. Much of surface is covered by plaque (stained black) except for round contact area (C). Border apical to this is located at site of onset of both approximal caries and chronic gingivitis.

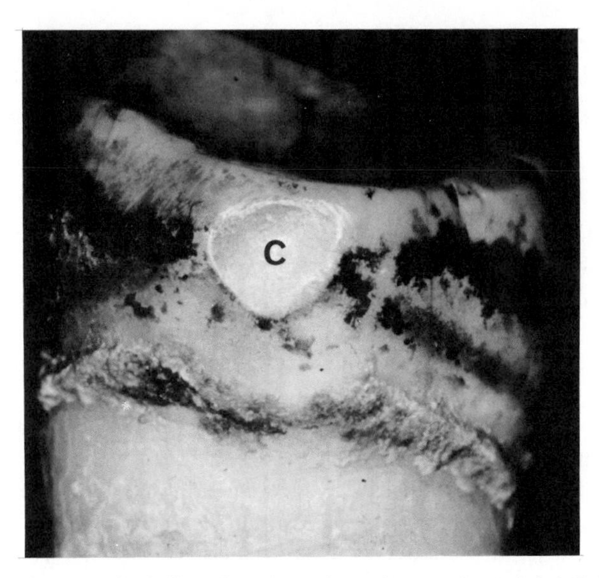

Figure 7.2 Approximal surface, Eskimo molar. Vigorous natural dental function produces occlusal and approximal wear, reducing extent and bulk of plaque. Note reduction in amount of plaque in relation to facetted contact area. Courtesy of the Society for Applied Bacteriology; reproduced from F. A. Skinner and J. G. Carr, Eds., *The Normal Microbial Flora of Man*, Academic Press, New York, 1974.

It is possible that organisms with the capacity to adhere to the different oral sites derive ecological advantages thereby. Savage [36] suggested that large numbers of animal-gut commensals found in the mucin layer on various epithelial surfaces of rats and mice must be highly adapted to that substance and also able to use it as a carbon and energy source. He supported this observation with the finding that germ-free mice excreted more intact intestinal mucin than did conventional animals. Similarly, mucin collection in grass leaf sheaths serves as a growth medium for many bacteria and fungi adept at nitrogen fixation [37], and rhizosphere organisms occur frequently in mucilage [38,39]. Salivary glycoproteins are rapidly degraded in the mouth by bacterial (and other) enzymes [40]. The carbohydrate components are thought to be released by extracellular bacterial glycoside hydrolases such as neuraminidase [41]. It has also been shown that both *Streptococcus sanguis* and *Streptococcus mutans* are capable of obtaining required amino acids from salivary proteins in the absence of more readily available substrates [42]. These may all be considered instances of the phenomenon of affinity of organic for inorganic matter subserving the nutrient requirements of microorganisms [43]. By the same means, organisms may be brought into contact with essential inorganic elements. It may be observed that organisms on mineral surfaces generally obtain such mineral nutrients by acid dissolution of the host mineral [44].

7.2 BACTERIAL ATTACHMENT

Study of various interfaces between bacteria and host surfaces has revealed a number of common factors that may derive from a shared and distant origin. It is very probable that microorganisms were the first form of life to evolve. Calcified forms analogous to dental calculus have been found in geological formations of the Precambrian era [45,46]. Certainly they have evolved in conjunction with metazoans [47] and may significantly modify the metabolism of the host on which they live. It has become evident recently that a microbial population exerts its effects not only at a distance, but often through close contact with its host surface. Many and varied factors, for instance, exposed and polished surfaces, rough water, and gut peristalsis almost certainly counteract attachment mechanisms [48]. In the mouth, as stressed previously, the sites most abundantly colonized are those least subject to friction and salivary flow. Many organisms are swallowed. The significance of frictional removal of organisms is shown by a comparison of human- and rat-tongue populations. The latter appear to contain significantly lower counts of organisms than the former [49] possibly due to the more vigorous masticatory function of the rat.

Cells in general show a random behavior in their ability to adhere to solid surfaces, determined partly by the surface properties of the different solids [50,51]. Organic matter has a marked physicochemical affinity for inorganic surfaces, to which it will bind [43]. As soon as a clean surface is placed in an aqueous environment, it is coated with a monolayer of organic polymeric material [51,52]. In this way many living organisms may be brought into contact with nutrient matter that would otherwise be dispersed. Such a phenomenon may be of importance in the aqueous environment of the mouth.

Bacteria at a surface initially may show flagellar or brownian movement, but this is lost after a few minutes, and the bacteria become attached to the surface along their length [53]. Unlike mammalian cells, attachment of bacteria in aqueous media appears to depend on cell-wall integrity, so that killed organisms are unlikely to attach. In natural waters solid surfaces tend to be colonized by living rather than dead bacteria [53]. The most apically located organisms in the dentogingival region on children's teeth are usually ghost cells [54]. The question arises as to whether they adhered before or after lysis. As the outer organisms in the plaque immediately coronal to this border also include many ghost cells [54], it seems likely that attachment of the apical cells to the tooth may have occurred while the organisms were vital and that cell death may be attributed to the host response in this region.

Recent reports [51,52] emphasize that adsorption of macromolecules at solid–liquid interfaces is related to the prevailing critical surface tension, which is the highest surface tension a liquid can have and still spread on a surface. Mitchell [52] pointed out that the deposition of polymers or the proximity of bacteria releasing such polymers on a surface in an aqueous medium usually lowers the critical surface tension at the interface. This prompts one to ask how the organisms reach the host surface in the first instance. Chemotaxis appears responsible for attraction of organisms to surfaces coated with adsorbed nutrients [43,52]. Many oral forms, unlike the marine bacteria described by Mitchell, are nonmotile. Motile organisms are abundant in periodontal pockets where disease is established, but these are stagnant sites, so it is doubtful that motility facilitates initial attachment in this region. In the marine situation a reversible adsorption phase has been demonstrated following initial attachment [55,56]. Bacteria that are held close to marine or oral surfaces by van der Waals forces for some hours are usually firmly adsorbed [30,52,55,56].

The theoretical basis of cell adhesion and attachment has been reviewed by Weiss [57]. In the absence of direct evidence as to the mechanisms in natural oral bacterial films, one must assume that short-range (2–3-Å) chemical bonds and that longer-range (3–4-Å) London–van der Waals forces operate [57]. Chemical bonds include electrostatic, covalent, and hydrogen bonds. The van der Waals forces are dependent on fluctuations in charge that lead to attraction by polarising other molecules [57]. What is evident from these general considerations is that oral bacteria must come very close to a host surface before adhesion can take place. Extracellular polysaccharide (EPS), which is considered later in more detail, may in some instances serve to attach organisms to surfaces, overcoming the natural repulsion barrier due to the presence of negative electrostatic charges on both host and bacterial surfaces. Extracellular polysaccharide may in this manner enable the approximation of cell and surface to a point at which forces such as London–van der Waals become effective [58]. Olsson et al. [58] have confirmed the importance of electrostatic forces in adhesion by showing that ions increasing repulsive forces reduce adhesion, whereas ions decreasing these forces increase the number of attached bacteria. Variations in the ability of different organisms to adhere suggested to them that, once electrostatic forces of repulsion had been overcome, species or type-specific properties of the individual cell surface determined the degree of adhesion. Negative electrostatic forces present on opposing host and bacterial surfaces may be reduced by lowering pH as well as by increasing ionic strength due to divalent cations [59].

As yet there is no clear evidence to explain the order of colonization of oral surfaces by organisms. Nor is it clear why initial colonizers are almost invariably cocci [60–67]. Apart from various attachment mechanisms and nutrient and growth inhibitory factors, it may be that cells of coccal shape are more likely to adhere because a diminishing radius of curvature of a cell or cell process reduces the potential-energy barrier to contact, although at the same time reducing the attraction potential promoting contact. Alternatively, bacteria may attach to oral surfaces by hydrophobic mechanisms [40]. If so, it is likely that specific sites on the bacterial cell surface are involved: rod-like organisms, generally the first to attach in aquatic media, show perpendicular orientation to interfaces [55], probably becoming attached by the hydrophobic portions of their surface [68]. Palisading in dental plaque and other microbial communities may provide another example of this phenomenon. Similarly, rosette formation where organisms attach to a central filament in a radial manner [48] "may result from an inward orientation of hydrophobic portions of cells (toward the filament), giving an aggregate with an hydrophilic outer surface" [68]. There is evidence that the hydrophobic pole of rod-shaped organisms is attracted to any nonaqueous phase [69]. Also, sorption of bacteria in an artificial seawater medium was found to be more extensive on hydrophobic siliconized than on normal (germanium prism) surfaces [68].

Weiss [57] suggests that reduction in radius of curvature facilitates very close approach of cells, permitting passage of the total potential energy-of-interaction "barrier." The closest contacts between organisms in human dental plaque are usually between bacteria of box-like morphology. This cell-wall deformation may be observed in specimens prepared by freeze etching and, presumably, not distorted by the processing techniques for routine transmission electron microscopy [70]. The distortion to be observed in coccal or coccobacillary cells is surprising in view of the well-known rigidity of bacterial cell walls. The phenomenon may be observed in both Gram-positive and Gram-negative forms. Weiss [57] concluded from several of his studies that some cells may be deformed more easily following treatment with neuraminidase, proteolytic enzymes, and calcium chelation. Weiss also noted that specific areas of the cell periphery are involved in contact phenomena, an observation substantiated by a number of recent studies, particularly of oral streptococci, showing that there are specific carbohydrate receptors on the outer cell surface mediating adhesion to epithelium and tooth [71,72]. McCabe et al. [71] have shown that at least some oral streptococci possess a mechanism involving extracellular bacterial cell-associated receptors for carbohydrate capable of binding specifically bacterial extracellular carbohydrates, carbohydrates present on host cells or carbohydrates adherent to host surfaces. They suggest that the ligands that these receptors bind may be oligosaccharides in glycoproteins or glycolipids, or in polysaccharides. One receptor they have identified tentatively was a heat-labile protein. The receptor is not dextransucrase (the enzyme synthesizing one form of EPS), although in *S. mutans* it may bind dextran (glucan). *Streptococcus sanguis* and *S. salivarius* possess receptors with specificities for various carbohydrate ligands present in human saliva and on human epithelial-cell surfaces [71]. On the basis of their findings, McCabe and Smith [71] suggest that *S. mutans* has evolved a mechanism utilizing dietary peculiarities of the host, that is, sucrose consumption, whereas *S. sanguis* and *S. salivarius* utilize host carbohydrate elements.

In the case of *S. mutans*, cell–surface adhesion may depend on the binding of cells by dextran receptors to dextrans previously synthesized by the cell or adjacent bacteria. This may be facilitated, as mentioned previously, by initial momentary adsorption of the cell to the host surface or by the primary adsorption of dextrans or other polysaccharides on the surface [72]. McCabe et al. [72] suggested "that in both intercellular and cell to surface adhesion, dextransucrase serves primarily to synthesise dextran while binding of dextran occurs at a unique cell surface site." Dextranase-producing organisms such as *S. miteor* may block sucrose-dependent adhesion between *S. mutans* and *S. sanguis* or between streptococci and tooth surface by competing for dextran (glucan) binding sites on bacterial cell surfaces [73]. However, Bleiweis [74] observed that, by selecting for cariogenic bacteria in this way, dextranase producers may actively enhance the pathogenicity of plaque.

Similarly, the serotype antigen is also located at specific sites on the cell surface [75]. This type-specific carbohydrate is not the binding site for dextran, since mutant cells of *S. mutans* attach although they do not produce dextran [75]. Nonetheless, the dextran-binding site may be located close to the type-specific antigen site, since antibody to the latter may block the former by steric hindrance. Iacono et al. [75] suggest that this type of inhibition of adhesion may be related to similar phenomena manifested by *S. mutans in vivo*. They also note that the serotype antigen location differs from the site on the cell wall of *S. mutans* to which lysozyme binds, producing aggregation. They suggested that lysozyme, too, could interfere with adhesion by steric hindrance of glucan binding by *S. mutans*.

Regarding the immune response of the host and the adhesion of organisms to oral surfaces, Gibbons and van Houte [29] have reviewed available information concerning the possible role of antibodies in preventing such adhesion. They suggested that antibodies in secretions may react with bacteria on the surface of mucosal epithelial cells, resulting in an impaired ability of organisms to reattach when they become dislodged. Prior exposure to the relevant surface antigen might impair attachment on reintroduction into the mouth of an organism with that surface antigen. They pointed out that many salivary organisms are coated with IgA and that parotid salivary IgA can inhibit the attachment of oral streptococci to epithelial cells [76]. Their study indicated that host antibody titers and antigenic composition of oral organisms may change with time, which may help to explain variations in the microbial population of a given surface.

Initial attachment or retention of bacteria is no guarantee of continued survival, nor does it indicate the significance of the presence of such organisms, especially in a mass as complex as dental plaque. Organisms that attach initially, particularly in the crevice region between gingiva and tooth, may be killed by host antimicrobial activity. Stagnation, excess fermentable carbohydrate, and host factors may severely limit growth of surviving organisms. Oxygen may cause growth inhibition or enlargement of cells, notably in *S. sanguis*, an early plaque colonizer and an abundant organism in plaque [77]. Numerous factors of microbial origin may inhibit growth of other forms. These range from simple chemical metabolites, most commonly acids or hydrogen peroxide [78,79] to bacteriocins and bacteriophages. Synergistic effects between different plaque isolates have also been observed [80–84].

7.3 ATTACHMENT TO EPITHELIUM

Organisms may be separated from epithelial surfaces by layers of mucin. They may also come into direct contact with the local epithelial cells or lie in hollows, such as hair follicles or the villi and crypts of the alimentary tract [85,86]. Organisms on gut epithelial surfaces, other than those of the oral cavity, are often "inserted" into depressions in the epithelial-cell surface [87,88]. On oral epithelial surfaces the attachment is often by fine fibrils present only on the cell wall aspect proximal to the epithelial cell (Fig. 7.3). Coating bacterial cells on epithelium are predominantly coccal and often contain granules, presumably polysaccharide, in relation to the same aspect of the cell wall [88,89]. A clear or nonstaining zone may be present between cell wall and surrounding interbacterial matrix, even when stained with specific reagents for glycoprotein or polysaccharide. This clear zone is a common finding in plaque and various other habitats, such as pond slime [90], the bovine rumen [91], marine flora [92], and chicken-crop epithelium [93]. It is unlikely that this zone is void of bacterial products. It is more probable that it contains capsular material normally lost during

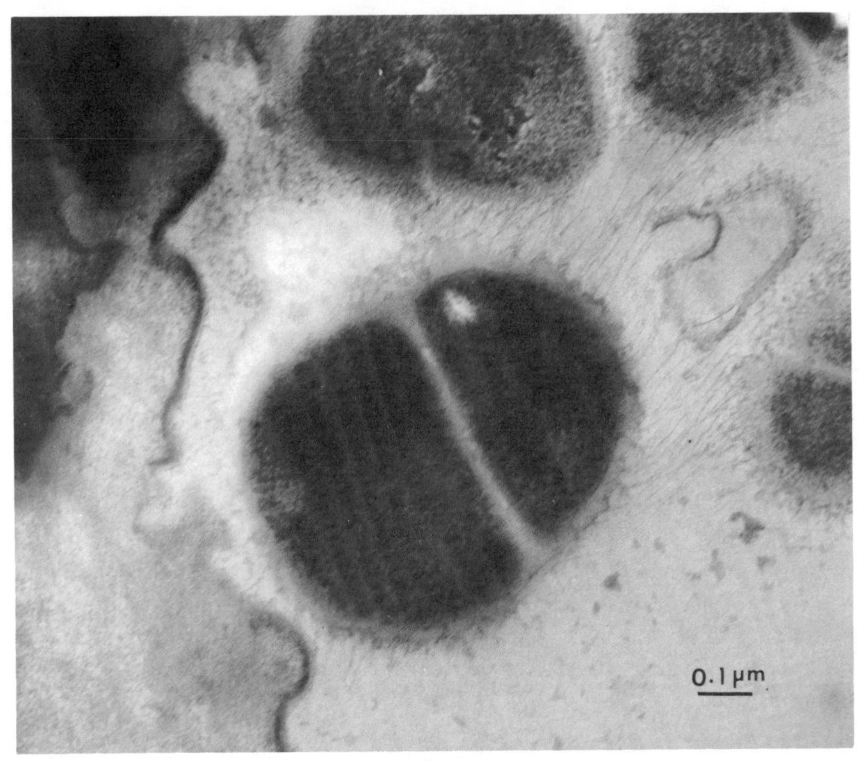

Figure 7.3 Bacterial attachment by fine fibrils to tongue epithelium and other organisms (×67,500). Courtesy of Col. J. M. Brady, U. S. Army Institute of Dental Research, Washington D. C.

dehydration prior to embedding for routine electron microscopy. Such polysaccharide matter may be preserved by freeze etching or by reaction with capsule-specific antibody [94].

A "fuzzy" polysaccharide or polysaccharide–protein fibrillar coat mediates the attachment of bacteria, mainly cocci, to the apparently uncoated dorsal papillary surface of rat tongue [95,96] and of healing human tooth-extraction wound epithelium [86]. Wagner and Barrnett [88] suggested that the presence of filaments at the site of contact between the bacterium and the mucosal cell may result from a microprecipitation reaction elicited by antibody with bacterial antigen at the host-cell surface. This may explain the absence of filaments on the surface of the bacterial cell distal to its point of contact with the mucosal cell. A fibrillar "fuzzy" coat may have a common ecological function in mucous membrane-colonizing bacteria, serving to mediate attachment to such surfaces. Recent evidence suggests that some such fibrils may be protein rather than polysaccharide, being removed partly by trypsin [97]. Veillonellae, neisseriae, and streptococci are among the oral organisms capable of binding to oral epithelium in this way [29,98]. It has also been proposed that mucosal attachment of organisms may be due to a mutual and highly specific interaction of carbohydrate polymer of bacterial and host origin [88,99,100].

It has been suggested that oral epithelial cells may be a vehicle for bacterial deposition on the surfaces of teeth [29,101,102], since such cells are often colonized by bacteria (Fig. 7.4). Epithelial cells are present in 24-hour plaque, but rarely after 2–3 days [103]. It would seem unlikely, therefore, that microbial attachment by this means is common. Indeed, attachment of organisms to epithelial cells that subsequently desquamate may provide one means of cleaning the vulnerable gingival crevice region, and of controlling the bacterial population of the mouth.

Bacteria are more abundant on the dorsal surface of the tongue than elsewhere on the oral mucosa [30]. Except in cases of general debilitation and certain systemic diseases, commonly upper respiratory tract infections and xerostomia (dry mouth), the various mucosal surfaces do not accumulate bacteria in large amounts. Bacteria are usually limited to the outer surface of an intact epithelium but may penetrate the deeper tissues, particu-

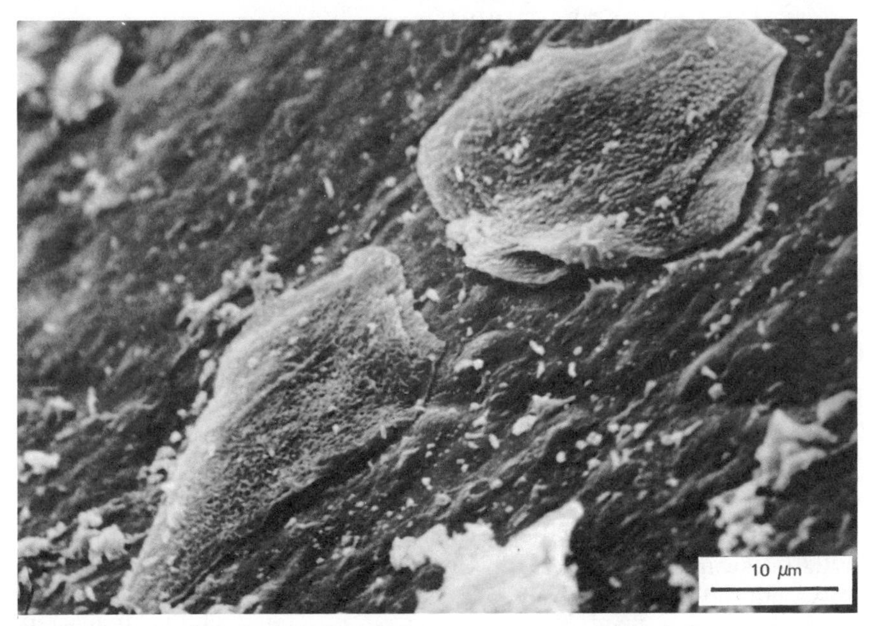

Figure 7.4 Scanning electron micrograph of desquamated epithelial cells on enamel near occlusal limit of approximal plaque on adult's premolar (×1700).

larly in areas of ulceration [104,106], although it is not known how organisms enter these tissues [105]. The microbial population is controlled doubtless by salivary flow and antibacterial activity and epithelial-cell desquamation [30]. There do not appear to be differences between the microbial populations of oral epithelial cells in normal mouths and those with a pathological condition [101]. There seems to be a distinct pattern of selective adhesion in the distribution of organisms found on oral epithelium. *Streptococcus salivarius* and *S. sanguis* adhere well; *S. mutans* slightly or not at all. *Actinomyces naeslundii*, *Fusobacterium*, and *Neisseria* strains adhere; *S. faecalis*, *A. viscosus*, *Rothia*, *Nocardia*, and *E. coli* do not [106]. The proportions of potential colonizing organisms available appear to be as important as their specific adhesive capacities, since the number of bacteria adsorbed per buccal epithelial cell (or per unit area of enamel) is related directly to the prevailing oral bacterial-cell concentration [30,107].

7.4 ATTACHMENT TO TEETH

7.4.1 Primary or Preeruptive Cuticle

Bacteria may attach directly to enamel hydroxyapatite [60, 108–110], possibly by means of EPS [110]. However, the enamel surface is rarely exposed directly to bacteria [111], except immediately following abrasion, and this is unlikely in the most significant stagnant dental sites, namely, the occlusal fissures and at the dentogingival junction. Baier [112] observed that "since surface contaminants in biological environments are difficult to remove, it should be accepted generally that hard dental surfaces–no matter how thoroughly cleaned–are *never pure*" (his italics).

The organic integuments of human enamel have been classified as follows: primary enamel cuticle, acquired pellicle, food debris, dental plaque, and calculus [113]. Neither the source nor chemical composition of the primary or preeruptively formed cuticle have yet been defined adequately. It is thought to be composed of the remains of the germinative layer of tooth enamel, the enamel organ [25,114–118]. It varies considerably in thickness, (Figs. 7.5–7.7), is continuous with the organic matrix of enamel [114,115] and may contain glycoproteins [119]. The significance of this layer in bacterial retention is that it remains on the tooth surface after lining cells of the junctional epithelium lose their attachment to it, so that initial attachment of bacteria to tooth is not generally to hydroxyapatite (HA), but to this preeruptively formed cuticle (Figs. 7.5–7.7) or to superimposed acquired pellicle of salivary origin. Enamel organ breakdown in occlusal fissures seems to follow a similar pattern, with subsequent "inva-

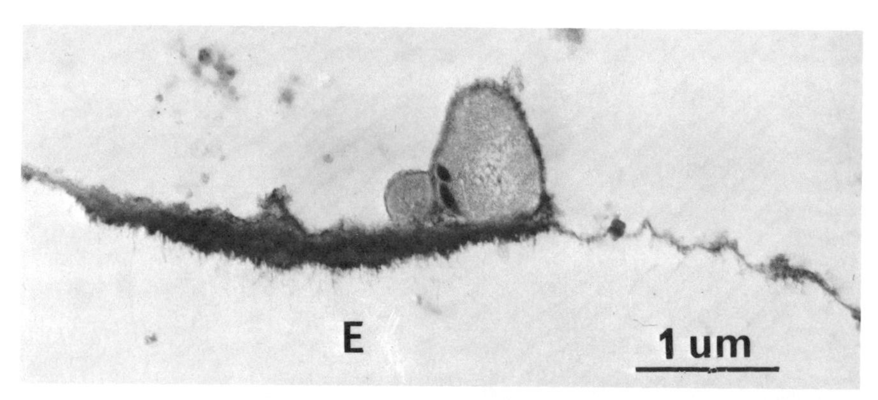

Figure 7.5 Organisms on cuticle of varying thickness, contrasted with ruthenium red. Note positive reaction of cuticle and external surface of bacteria (×19,125) (E = enamel space).

sion" of the fissure by bacteria. As on smooth surfaces, the organisms colonizing this site initially are mainly Gram-positive cocci with some rods [120].

7.4.2 Acquired Pellicle

The acquired pellicle consists mainly of salivary glycoprotein [65,113, 121–123] and contains some lipid [124,125]. Little is known of the mechanism of adhesion of the pellicle to the primary enamel cuticle, and bacteria do not appear to be essential for pellicle formation [126]. The adhesion of bacteria to teeth by means of the acquired pellicle (Fig. 7.8) is currently subject to intensive physicochemical research [127,128]. There are considerable variations in the results so far obtained because of variability in the behavior of the different strains and because of the difficulties inherent in attempting to correlate *in vitro* findings with the natural situation.

The acquired pellicle on enamel hydroxyapatite is composed of selectively adsorbed salivary components [114,126, 129–133], mainly specific parotid [134], and sublingual [135] glycoproteins. Minor salivary glands also seem to be an important source of pellicle material [128]. Serum components may be present [136] as well as immune factors [76,137,138]. Pellicle also contains bacterial cell-wall material [129,139]. Formation of acquired pellicle may be due to precipitation of protein from which terminal components, such as sialic acid, have been removed by bacterial or salivary enzymes [130,140–142]. Carbohydrate moieties released in this way may be used as nutrients by oral bacteria [143–145]. Some organisms can utilize the protein component, as both *S. mutans* and *S. sanguis* are able to obtain

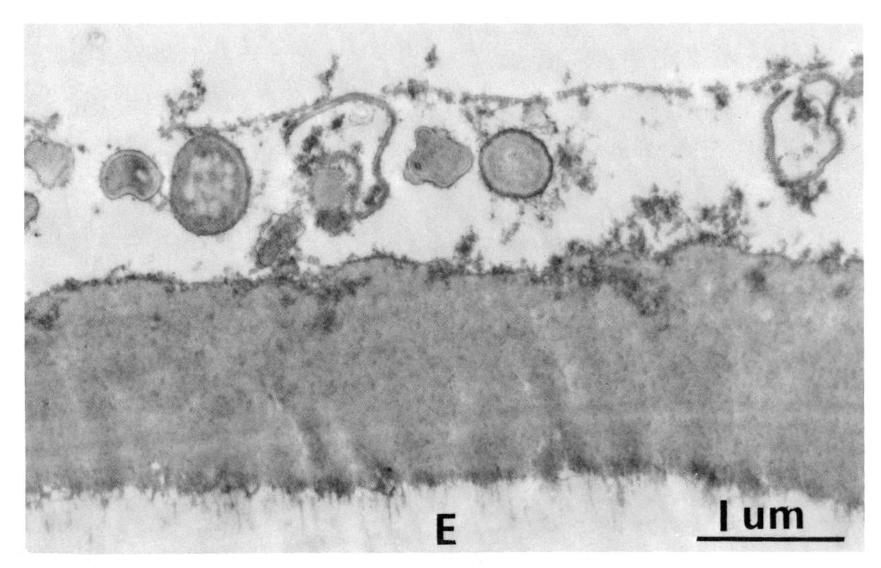

Figure 7.6 Organisms in relation to thick cuticle. Contrasted with ruthenium red. Separation of cells may have occurred during polymerization of embedding resin. Remnant of plasma membrane, probably from desquamated crevicular epithelial cell, is apposed to surface of bacterial monolayer. Organisms show varied response to environment. Two cells have lysed; one has a cell wall stained densely by ruthenium red and with fine fibrils extending toward cuticle, and another organism contains many electronlucent granules, probably polysaccharide. Note ruthenium red contrasting external surface of cuticle. This material may derive from salivary glycoprotein and/or bacterial polysaccharide (×19,875) (E = enamel space).

essential amino acids from salivary proteins [42]. Intact salivary protein may be precipitated spontaneously onto the tooth surface [61,111,123, 126,139, 146–151] and oral epithelium [151].

Acquired pellicle reforms within 30 sec of its mechanical removal [112] and is microscopically demonstrable within 2 hr of exposure to saliva of an abraded tooth surface [152–154]. Although it may form and reform rapidly, the processes responsible for the production of mature pellicle seem to require time. The chemical and structural changes that take place over a 7-day period from its initial formation are not clear; however, such a pellicle restricts acid diffusion far more effectively than does a 4-day-old film [155].

On abraded surfaces, carboxyl groups in the glycoprotein molecules of pellicle may bind to calcium ions in enamel HA [61,109,111,147]. Calcium ions enhance adsorption of salivary proteins to HA by promoting aggregation of proteins [111]. This is an example of the effect of some divalent cations in supporting biological adhesion [156].

Bernardi et al. [157] observed that HA may be regarded as an amphoteric substance, which may bind both positively and negatively charged macromolecules due mainly to electrostatic interactions. Calcium ions, for instance, may attract phosphate, carboxyl, or sulfate groups. Acidic calcium-binding proteins are adsorbed selectively to human enamel [133, 154,158] in preference to basic salivary proteins. Similarly, positively charged groups on salivary proteins have been shown to possess an affinity for phosphate groups in tooth mineral [158]. Anions, notably fluoride, inhibit the binding of acidic proteins to HA [158], possibly by decreasing the surface free energy of enamel [159,160]. Lowering of pH to 5 appeared to increase protein adsorption but to abolish the effect of calcium ions [111]. The net charge of apatite crystals depends on the presence or absence of divalent cations such as Ca^{++} and Mg^{++}, which favor adsorption of acidic protein. Recent studies by Arends and his group suggest that enamel HA has a net negative surface charge [161]. Anions such as F^- and HPO_4^{--} favor adsorption of basic macromolecules [128] and may enhance adhesion by increasing the wettability of HA [162,163]. The notion of selective adsorption of salivary proteins to teeth has been confirmed by Belcourt

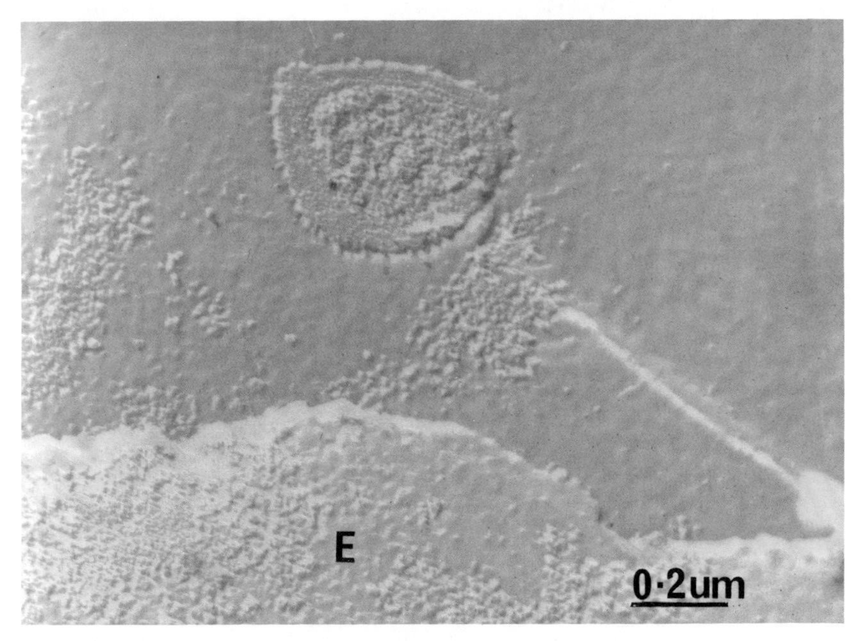

Figure 7.7 Freeze-etch micrograph. Thick-walled organism with short cell-wall fibrils and attached by amorphous matter to cuticle (×66,500) (E = enamel space).

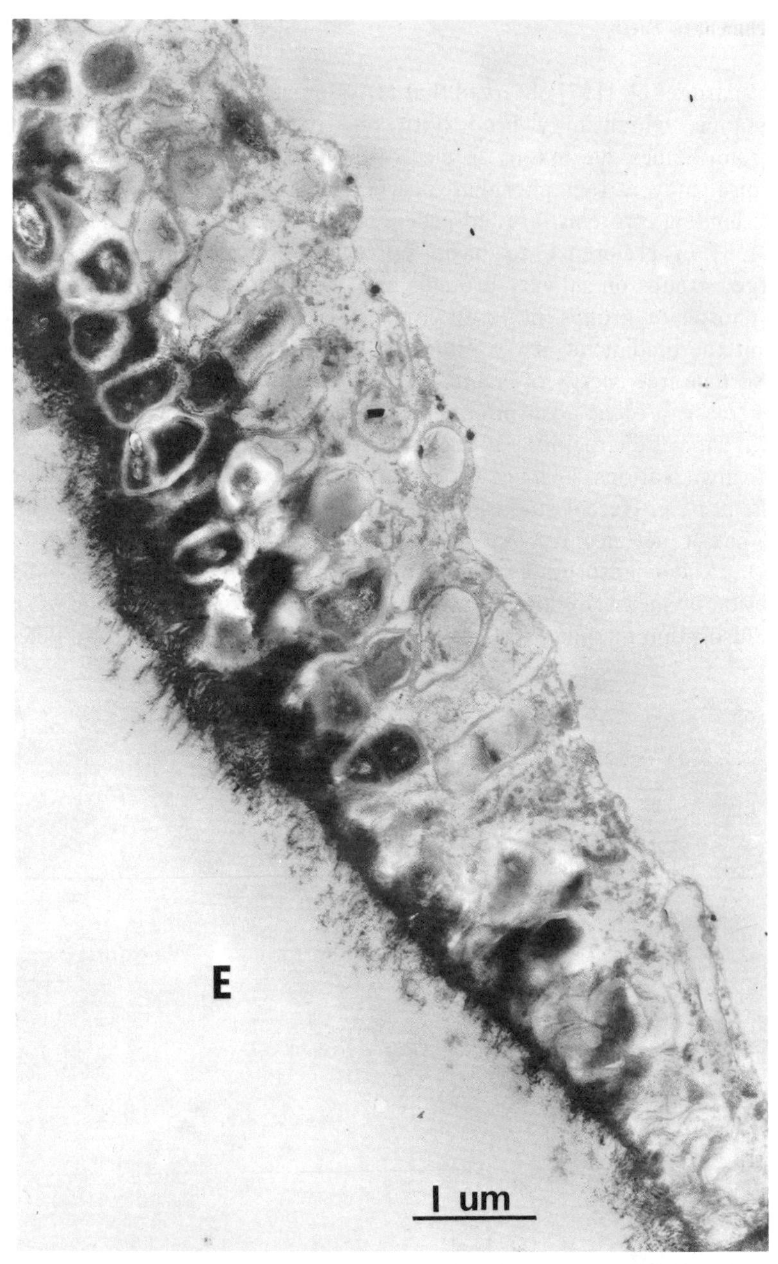

Figure 7.8 Bacteria in relation to electron-dense scalloped pellicle. Cause of this scalloping is not known; it may reflect bacterial metabolism or merely follow outline of organisms. Outer cells at subcontact area apical border are frequently lysed. Avoidance of exposure of plaque to chelator EDTA preserves the more diffuse outer portion of plaque matrix, particularly the discrete outer layer. This layer explains lack of detail of bacterial morphology in SEMS of (undecalcified) approximal plaque. Note prolongations of enamel cuticle continuous with enamel matrix. Note electronlucent gaps left by chelation of surface enamel crystals (×17,875).

[164] and Mayhall [165], who have shown that such proteins are identical with those precipitable from saliva by calcium ions.

Little is known of the physical chemistry of adhesion of pellicle to tooth. It may be related to the known ability of proteins to spread. This appears to be independent of the polar–apolar characteristics of the protein and of denaturation [166]. Leach [40] has suggested that the formation of acquired pellicle may depend on hydrophobic mechanisms and hydrogen bonding rather than electrostatic interaction. He points out that it is entropically favorable for salivary proteins to bond with one another so as to release the otherwise entropically unfavorable water molecules associated with them. Large molecules are held to a surface much more effectively than smaller molecules and are difficult to desorb. As Leach remarked, any large molecule capable of modifying its spacial structure could adsorb nonspecifically to any surface, which would explain the formation of pellicles on a variety of surfaces exposed experimentally in the mouth. Large molecules, as in pellicle, are preferentially surrounded by like molecules instead of being mixed, probably because of the large interaction energy evident between large similar molecules. This would provide an energetically more favorable state than would occur if the molecules moved freely through an aqueous system. The opposite applies to small molecules [40]. It should be stressed that pellicle (and plaque) found on amalgam and other dental-restoration surfaces may differ quantitatively and qualitatively from natural acquired pellicle [167].

7.4.3 Bacterial Attachment to Teeth

Bacteria are rarely attached directly by their cell walls to a host surface, as stressed previously. Almost invariably, the link is effected by means of bacterial polysaccharide or mucilage and a host cuticle or carbohydrate-rich cell coat [38,55,56,90,92,168–170]. Certain properties of the acquired pellicle are relevant to the attachment of bacteria to teeth:

1. Salivary proteins may aggregate oral organisms and thereby reduce their ability to colonize the tooth surface [171]. Sulfur-containing products of bacterial metabolism that occur in the mouth may enhance salivary aggregating factors [172]. Conversely, agglutinated masses of organisms may be incorporated in plaque instead of being swallowed [173].
2. Although bacteria do attach to acquired pellicle, pellicle may reduce attachment since it has a lower surface free energy than does a variety of agents used to study attachment *in vitro* [174–177]. Salivary proteins reduce the adsorption of "cariogenic" streptococci to HA [178]. Meckel [179] observed that the absence of an organic cuticle or pellicle seemed

to delay bacterial attachment for 1 day, the time required in his experiments to form such a layer on a previously cleaned tooth.

3. Salivary pellicle contains proline-rich proteins uniquely sensitive to a bacterial collagenase. Bacterial enzymes, therefore, may act on pellicle and not be available to damage periodontal tissues [180]. Similarly, organisms colonizing gastrointestinal mucin generally use this material as a nutrient source [36].
4. The carbohydrate moieties on glycoprotein molecules may serve as identification markers, enabling correct location of a given protein molecule at the site of function, when sites of secretion and function are distant from one another [181]. Salivary proteins found in dental or epithelial pellicle may be localized by this means and may have as a principal function the competitive inhibition of bacterial attachment [182]. Bacteria may have evolved a partial counter to this protective mechanism by producing surface components, such as those with blood-group antigenicity, which may be recognized by the oral mucosa as self [49]. It is possible that some salivary proteins are designated to bind organisms capable of such mimicry, thus preventing direct attachment of the bacteria to the host surface [30]. If, as Gibbons and Qureshi suggest [183], organisms acquire the blood-group reactivity of their host while proliferating in the mouth, this may influence their transmission between individuals of different blood types. Sönju and Rölla [184] have observed that the erythrocyte surface and the acquired pellicle possess common surface structures since both contain blood-group substances. On this basis, Rölla and Kilian [185] demonstrated a comparable affinity of a microorganism for both red blood cells and acquired pellicle. Strains of *S. sanguis*, *S. mutans*, *S. miteor*, and *A. viscosus* showed a marked affinity for erythrocytes, with that of *S. mutans* dependent on the presence of sucrose in the growth medium. The actual hemagglutinating factor was probably polysaccharide, but not a dextran. Rölla and Kilian [185] stated that there was conflicting evidence as to the possible role of the lipoteichoic-acid content of *S. mutans* EPS in the phenomenon and suggested that hemagglutination might depend on the fibrillar coatings present on many bacterial cell walls [70,186].
5. Biologically active salivary enzymes, (e.g., lysozyme and amylase) and antibodies may be bound to pellicle and react with bacterial cell-surface determinants and thereby influence preferential adhesion of oral strains [137,138]. Parotid secretory IgA can inhibit the adhesion of some oral streptococci to oral epithelial cells [76].

Bacterial colonization of tooth surfaces may involve Ca^{++} ion-mediated interaction between negatively charged bacteria and negatively charged pellicle on the tooth surface; cations that bind to bacteria and pellicle will inhibit bacterial adsorption to teeth [187]. Hydrogen bonding and specific agents such as blood-group-reactive salivary mucins also may be important [154, 183–185, 187]. Grigsby and Sabiston [188] suggested that bacteria may interact with the acquired pellicle through covalent bonds, for example, between an amino-acid side-chain carboxyl or amino group from the pellicle protein and a peptide unit of bacterial cell-wall peptidoglycan. Another possibility they suggest is a covalent link between an hydroxyl group from a serine residue or from the carbohydrate moiety of pellicle to the *N*-acetyl muramic acid or the *N*-acetyl glucosamine portions of the bacterial cell wall. On epithelial cells, the binding sites may be phospholipids of the plasma membrane [189]. The initial stage in adsorption of bacteria may involve EPS (Figs. 7.9 and 7.10), since polymers of glucose have been found on human tooth surfaces 2 hr after the onset of pellicle formation. Rölla [187] has shown that bacterial EPS may adsorb to pellicle, possibly by means of calcium ions, due to the positive charge of these ions interacting with negative surface charges on the polysaccharides and in pellicle (cationic bridging). Sialic-acid-containing glycoprotein may be adsorbed to HA in competition with other salivary proteins. Dextran appears to have a lesser affinity for HA, especially if the latter is already coated with pellicle. The amount of dextran adsorbed is increased at higher salivary phosphate concentrations [131]. It must be emphasized that plaque will form in the absence of extraneous sources of carbohydrate [125,190]. Serotype antigen and cell-surface polysaccharide appear to function in the binding of sucrases, which then synthesize the polysaccharide binding the producer (and possibly other) organisms to the surface [191]. Calcium ions also have been shown to increase cohesion between polysaccharide chains [192]. Negatively charged phosphorylated EPS produced by prominent plaque organisms such as *S. mutans* and *S. sanguis* have a higher affinity for the tooth surface than do similar nonphosphorylated macromolecules [193].

The structure of EPS is related to its solubility in water and appears to affect bacterial adhesion. For example, under identical conditions, *S. sanguis* produces about 10 times as much soluble glucan as does *S. mutans*, but its glucans are more sensitive to hydrolysis by dextranases [194]. Similar hydrolytic enzymes may play some part in regulating EPS metabolism in natural plaque [195,196]. Apart from calcium ions, hydrogen bonding between bacterial EPS and pellicle provides a further mechanism of plaque formation, since these polymers possess many hydroxyl groups [109,128]. Rölla [109,197] and Kelstrup et al. [198] have noted the presence of weak chemical forces in the initial phase of bacterial sorption to teeth. It is not established whether a reversible sorption phase follows initial attachment as observed in the marine habitat [55,56].

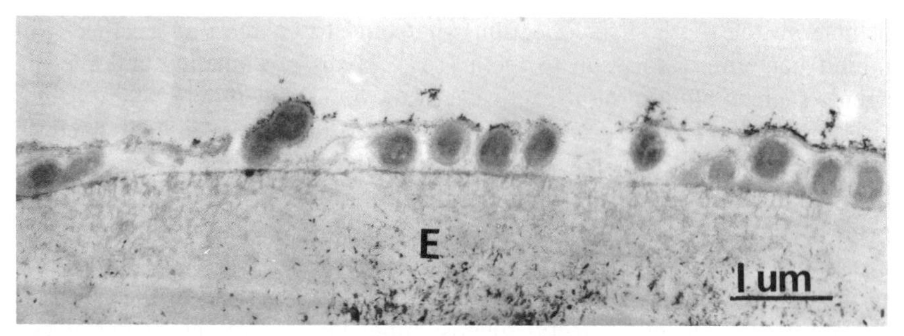

Figure 7.9 Gram-positive, probably coccal monolayer (SEM shows initial layer in this region to be almost invariably coccal). Note positive colloidal iron reaction of layer on outer aspect of organisms (×13,625) (E = enamel space).

Although calcium ions may have a function in the formation of pellicle and plaque, this can only be part of the mechanism because of the failure of chelating agents such as ethylenediaminetetraacetic acid (EDTA) to disaggregate plaque even after long periods of immersion [199]. However, EDTA does seem to remove the most superficial interbacterial matrix regardless of the thickness of the plaque layer [200]. Possibly, it dissociates the salivary glycoprotein component of plaque matrix. Such a "coat" is present at the surface of most, if not all, animal cells [168] and may be removed by EDTA, as in the separation of corium from epithelium [201,202]. It is known that EDTA may remove lipopolysaccharide and a little lipid and protein from some organisms, such as *Pseudomonas aeruginosa* [203] and *E. coli* [204]. This may affect the structure and properties of the cell wall [205]. Undoubtedly, the effect of EDTA would be more severe if plaque bacteria had the metal–oxygen bonded protein cell-wall structure of marine nitrifying bacteria, which is readily disrupted by EDTA [206]. Results on the metabolic effects of this complexing agent on oral organisms are sparse. It lowers the viable count of plaque organisms [207] as of other bacteria [208]. It has little effect on invertase activity of *S. mutans* [209], but levansucrases of *S. mutans* produce only oligosaccharides in its presence as the complexing agent breaks up the enzyme–levan complex [210]. Loss of pellicle Gram positivity appears less in plaques removed from enamel with EDTA than with HCl [124]. There may be a need to distinguish between agglutination of organisms, disrupted by EDTA [211], and adhesion of bacteria to teeth [212,213].

Regarding the possible role of calcium in promoting adhesion, Weiss [47,214] has shown that cells binding calcium peripherally do not exhibit calcium-sensitive adhesion. Chelation of this calcium appears to induce deformability of the cell, indicated by structural weakness. On this basis,

Weiss proposed that a possible function of calcium in short-term contact phenomena is to raise the *cohesive* strength of contiguous cell peripheries, rather than to promote intercellular *adhesion*. This function of calcium remains to be demonstrated in plaque, although it is known that chelation disaggregates the most recently formed (outer) layer of plaque [198]. It does seem that calcium ions may be critical for cell spreading in the presence of serum proteins and may thus affect binding of bacteria to acquired pellicle, which contains such matter [112,215]. Leach [40] concluded from his studies of pellicle and plaque formation that the presence of calcium is fortuitous in terms of maintaining the structure of dental plaque. Coman [216] proposed that calcium ions were probably responsible for forming chemical bridges between carboxyl, phosphate, and other such anionic groups on opposing cell surfaces, so that the ability of EPS to bind cells to a foreign surface should not be affected by calcium removal. In seawater, bacteria may precipitate carbonates as insoluble calcium or magnesium salts [217] and become adsorbed with these crystals to the sides of culture vessels [218]. It is possible that such deposits act as binding agents (universal cements [168]). A similar phenomenon may occur as part of plaque formation, since it is known that calcium salts (phosphates) are deposited in pellicle and plaque.

Several examples of specific interaction between protein and polysaccharide are known [187], including that between salivary glycoprotein and the surfaces of some oral bacteria [219]. However, there is considerable variation in this phenomenon, which adds to the confusion. For instance, *S. sanguis* and *S. miteor* adhere best to saliva-coated HA, whereas *S. salivarius* adheres best to uncoated HA. *Streptococcus mutans* adheres better to dextran-coated HA than to uncoated or saliva-coated HA, whereas the

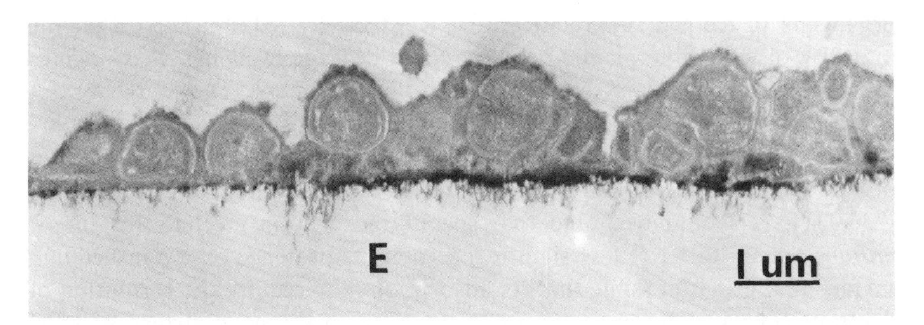

Figure 7.10 Generally large organisms surrounded by matrix contrasted with Alcian blue and lanthanum nitrate. Reactions to colloidal iron, Alcian blue, and ruthenium red suggest that this material may be salivary glycoprotein, which is known to contribute to plaque matrix as well as acquired pellicle (×11,500) (E = enamel space).

adsorption of *S. sanguis* is not enhanced by dextran [220]. It is of interest that *S. mutans*, currently of concern because of its probable involvement in dental caries, may require cell-bound polysaccharide for attachment to dextran-coated HA. This process seems to depend on the type of polymer present in the incubation medium, since low-molecular-weight dextrans block the binding sites involved in the adhesion process [220] and inhibit the ability of organisms to adhere to teeth [221]. Hydrolytic plaque enzymes are also adsorbed to enamel [222]. Whatever the precise mechanisms relating to individual organisms, both ionic effects and hydrogen bonding seem to be of importance in adsorption of bacteria to teeth. Kelstrup and Funder-Nielsen [223] considered plaque to consist of bacteria aggregated largely by EPS, partially dissociable by EDTA. Various cations (Li^{+}, Mg^{++}, Ca^{++}, Mn^{++}, and Ba^{++} but neither Na^{+} nor K^{+}) reaggregated such EPS. Calcium ions also were readily adsorbed by negatively charged polysaccharide on cells. Similarly, protein readily adsorbed to the polysaccharide. Their results also indicated that hydrogen bonding was important in aggregation of organisms. They suggested that sucrose-derived polysaccharide contained negatively charged sites and many free OH–groups, with the latter forming hydrogen bonds leading to aggregation of polysaccharide-coated cells. They confirmed that cations eliminated the counteracting repulsion of the negative charges. They suggested further that protein might be incorporated into plaque by polar bonding to charged polysaccharides or through divalent cation bridges. They subsequently demonstrated [224] that all strains of *S. mutans* except a noninsoluble EPS-producing mutant [225,226] were agglutinated by dextran and showed that the proteinaceous bacterial cell-surface dextran receptors were discrete from glycosyltransferase. However, adhering dental deposits are not invariably associated with detectable EPS synthesis [227], and there seems to be evidence of a clear distinction between such agglutination and polysaccharide-mediated adsorption of organisms to teeth [212]. Incidentally, Kelstrup et al. [227] noted that the majority of cultivable plaque bacteria could degrade carbohydrates, including sucrose, to acid and store glycogen-like polymers. This corroborates suggestions of the adaptability of plaque organisms to excess sugars [54,228,229] and may be explained at least in part by the ability of organisms to transfer such properties to one another by transfection [230,231]. It should be noted that specific adsorption mechanisms for *S. mutans* appear to be of less significance than its presence on a nonshedding surface [232]. It is possible that its adsorption may require the formation of dextrans (glucans) by enzyme molecules close to the host surface and the binding of the cells to the polymers thus formed by means of a cell-surface dextran receptor [72,194]. However, *S. mutans* can bind firmly to HA in the absence of sucrose, the primary substrate for its EPS production [233], nor does it require sucrose to colonize the teeth [234]. Some plaque organisms may attach by means of protein rather than EPS, since trypsin reduces the ability of some streptococci to adhere to saliva-coated HA. Lipid elements on the bacterial cell wall may inhibit adsorption since lipase treatment of *S. sanguis* cells enhanced adsorption to saliva-coated HA [220].

It has been suggested that more bacteria adsorb to saliva-treated rather than untreated HA because the layer of selectively adsorbed salivary components increases the number of binding sites available for the organisms [235]. Conversely, saliva may reduce the ability of organisms to adhere to oral epithelial cells [106]. Pellicle in particular [219] and organic matter in general [236] seem to reduce the ability of organisms to adhere to teeth.

Whereas organisms may colonize an undisturbed tooth surface rapidly, this rate is not maintained [237]. Apparently, a monolayer is formed to which additional organisms do not readily adhere [219]. Unfortunately, it is not possible to draw conclusions for all oral microorganisms from experiments using a few strains. Adsorption and accumulation of bacteria on oral surfaces appear to depend on complex relations between many factors, and it is probably unwise at present to assert that general mechanisms apply invariably to all colonizing organisms. Ørstavik [238] suggested that whereas the formation of an acquired pellicle generally renders the enamel surface less suitable as a substrate for bacterial sorption, specific chemical groups in the pellicle may act as determinants for attachment or lack of attachment of oral bacteria (see also McGaughey et al. [178].

Frank and Houver [66] found that superficial enamel apatite crystals were covered by a 100-nm-thick bacteria-free cuticle within 24 hr. Within 2 days plaque formation was evident as a coccal monolayer interspersed with extracellular matrix with an electron density lower than the acquired pellicle. Many matrix vesicles were found that, they suggested, might mediate attachment. After 6 days there was minimal extracellular matrix between bacteria and acquired pellicle. Extracellular polysaccharide was more abundant near the enamel than near the salivary surface of the plaque [103].

There has been little correlation of retention mechanisms with information available on the bacteriology of developing plaque. Gram-positive cocci, especially streptococci, are common in such plaque [62,239], perhaps due to their pronounced negative surface charge. This may depend on a high content of extracellular lipoteichoic acid in relation to the polysaccharide coat of such bacteria [240]. Aerobes, such as *Neisseria* and *Rothia*, are also found [62–64, 241]. Streptococci and Gram-positive rods appear to be the initial colonizers of occlusal fissures [220,242,243]. Counts of filaments, such as fusobacteria and actinomycetes, and of Gram-negative cocci, particularly veillonellae, increase as counts of the initial colonizers decrease [211]. In a recent study approximately 10^5–10^6 microorganisms were recovered 5 min after exposure of a previously cleaned natural tooth surface in a nonstagnation site. The organisms on this surface were exclusively cocci and rods for up to 16 days after initial cleaning [244]. This does not reflect the situation at stagnant sites of interest in disease. For instance, spirochetes, vibrios, and similar anaerobic forms abound in such sites [241]. *Streptococcus sanguis* and some unidentified organisms appear to be the earliest colonizers, and this correlates with several studies, suggesting that *S. sanguis* is an early colonizer and prominent plaque organism at most sites and for most ages of plaque [241,245,246]. Socransky et al. [244] suggest that the counts obtained indicate a scattered distribution of organisms on an exposed tooth surface 8–24 hr following the initiation of plaque formation. The presence of rods at this early stage correlates with the proximity of rods to cocci at the borders of natural plaque of unknown age on children's teeth [54]. Polysaccharides do not appear to be as important in the attachment of *S. sanguis* as of *S. mutans* [30], although it should be noted that *S. mutans* is feebly retained on tooth surfaces except at relatively undisturbed stagnation sites [232]. The changes noted in the population with advancing plaque age appear to depend largely on increasing anaerobiosis [241,247], apart from various interbacterial and host–parasite interactions.

7.5 POLYSACCHARIDES AND BACTERIAL RETENTION

Extracellular polysaccharides (EPS) are produced by organisms in a variety of habitats, such as the sugarcane rhizosphere [249] and acid mine water [250,251] in which the environment is acidogenic and aciduric or contains much carbohydrate [200]. Polysaccharide production is a basic cellular function, as judged by its early appearance in phylogenesis and ontogenesis [248]. Apart from involvement in attachment to surfaces, these compounds may subserve a number of functions, including protection against host and other bacterial defense mechanisms, and against high extracellular levels of acid and carbohydrate that are either inhibitory or toxic to the growth of most bacteria [252].

It has been suggested that the initial stage in plaque formation involves the adsorption of EPS to pellicle, since polymers of glucose may be found on human tooth surfaces 2 hr after the onset of plaque formation [154]. Serotype antigen and bacterial-cell surface polysaccharide appear to function in binding of sucrase enzymes, which then synthesize the EPS binding the organism to the surface [192]. Electrostatic interaction between negatively charged bacterial EPS and positively charged protein, as well as frequency of branching of EPS molecules and chain ends, may facilitate the adsorption of bacteria to acquired pellicle [109].

Polysaccharides are abundant in plaque [41, 253–257] and are produced by many plaque organisms [54,228,241]. It has been estimated that on average about 30% of the plaque matrix consists of EPS [257]. In the mouth, too, EPS subserve functions in addition to or other than attachment. For example, EPS capsules of oral streptococci have been related to the incapacity of leukocytes to phagocytose these organisms [258]. It is likely that many organisms in plaque, and not merely the "cariogenic" *S. mutans*, can metabolize common dietary carbohydrates and form polymers from them [226]. This was indicated by rapid and relatively even uptake of labeled sucrose or glucose, and their metabolism to polymeric form, by different morphologic plaque types [54,228]. Extracellular polysaccharide production seems to be dependent on a high extracellular carbon:nitrogen ratio. A prominent EPS producer such as *S. mutans* can grow in the absence of some amino acids [259]. Under these growth-limiting circumstances, however, an *S. mutans* culture takes about 2 days to double its mass [260], whereas its growth in rapidly stirred systems is classically logarithmic [261].

Apart from direct bacterial cell wall–cell wall or wall–cuticle or wall–pellicle contacts, a variety of morphological interconnections have been observed between bacteria and their host oral surface. Some of these consist, in part at least, of EPS (Figs. 7.11–7.13). Amorphous, globular, and fibrillar elements are found [262], and it has been shown that some of this material is carbohydrate [263]. As yet, no direct correlation has been made between biochemical analysis of plaque matrix carbohydrate and its morphology, location, and function in natural plaque. This is due largely to the heterogeneity of these EPS and the lack of specificity of available histochemical techniques. The variety of EPS is matched by that of their synthesizing glycosyltransferases [264]. Several types of bacterial interconnection apparent on freeze etched and thus nondehydrated specimens [70] appear to be partly or wholly polysaccharide. Thus "stained" material is demonstrable between contiguous cell walls as fine fibrils, often parallel to palisaded organisms (Fig. 7.13). The matrix is generally less dense in outer plaque. The predominant morphologic component appears to be fibrillar, occurring as single units or as double-unit protofibrils, showing extensive branching [66,213,226,262,265,266] and consisting mainly of glucan [212]. Nests of fibrils are demonstrable around some organisms with an outer "capsule" of parallel fibrils (Fig. 7.13). Loosely associated fibrils on bacterial cell walls appear to be highly hydrated [267], which may explain difficulties in dehydrating *S. mutans* grown in sucrose broth [226]. Freeze-etch micrographs suggest that organisms in deep palisaded plaque are connected

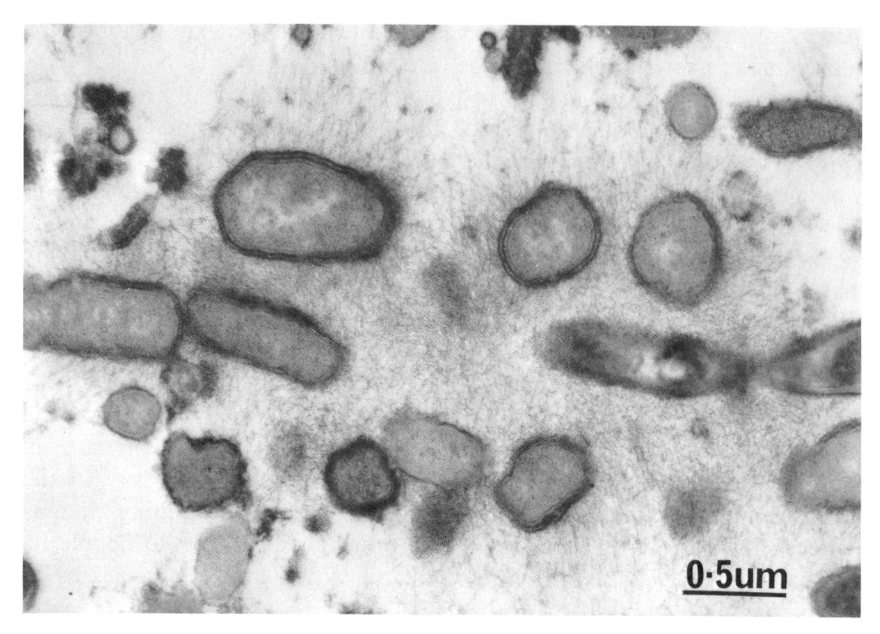

Figure 7.11 Microcolony of organisms with many fine fibrils arising from cells and enveloping organisms. This network may be a relatively intact layer *in vivo*, with fibrillar appearance as a preparation artifact (×28,250).

by short fibrils continuous with the cell walls [70]. Such close contacts may facilitate metabolic interchange, particularly since cytoplasmic membrane may extend through the cell wall, perhaps through these fibrils [267]. As Bayer and Thurow [94] emphasize, the appearance of these various matrix structures is a function of various fixation and dehydration techniques. It may be observed that fibrillar mediation of bacterial attachment common to many mucosal surfaces [268–272] is relatively infrequent on teeth.

The principal structural forms of bacterial attachment to solid surfaces in general include fimbriae or pili, stalks, holdfasts, and capsular polysaccharide [88,92,170,268,271,273,274]. In regard to capsular polysaccharide, a "clear zone" around organisms in several habitats, including dental plaque [275], has been mentioned. Evidence exists suggesting the presence of a water-soluble polysaccharide layer between cell wall and insoluble EPS, removed by conventional dehydration prior to embedding [94], and it is possible that the latter evolves from the soluble form [92]. The junctional zone between bacteria and the organic dental integument is usually either amorphous, or amorphous with some extracellular vesicles present [66]. Whereas the origin and nature of these vesicles is unknown, they are membrane bound and resemble extruded mesosomal material in the electron micrographs due to Ghosh [276]. Vesicles are found in relation to a variety of plaque bacteria *in situ* (Fig. 7.14) and in culture [277]. They are also found in more mature plaque [278,279] and in glucose-containing cultures of *S. mutans* maintained at pH > 6. Such vesicles are not present in other glucose-grown cultures of this organism nor of a mutant strain grown in sucrose at low pH or pH > 6 [226]. Ridgeway and Lewin [280] suggested that similar structures produced by a marine organism might serve to attach the cell to a surface.

Depending on the degree of shelter from cleansing mechanisms, plaque increases in thickness, and a variety of contacts may be observed between organisms [70]. Wall–wall contacts between bacteria on nondehydrated specimens (Fig. 7.15) suggest a degree of proximity producing a strong force of adhesion due to the various chemical and London–van der Waals forces [57,112]. This may explain the difference between *materia alba*, the soft dental accumulations readily removed by thorough mouth rinsing, and established plaque, which resists such treatment [281].

Holdfasts are relatively uncommon in plaque but may be seen more readily in plaque isolates [282] and between coating organisms and central filamentous organisms in some unusual plaque configurations [283]. Polar holdfasts have also been observed in gingival plaque isolates [284]. There

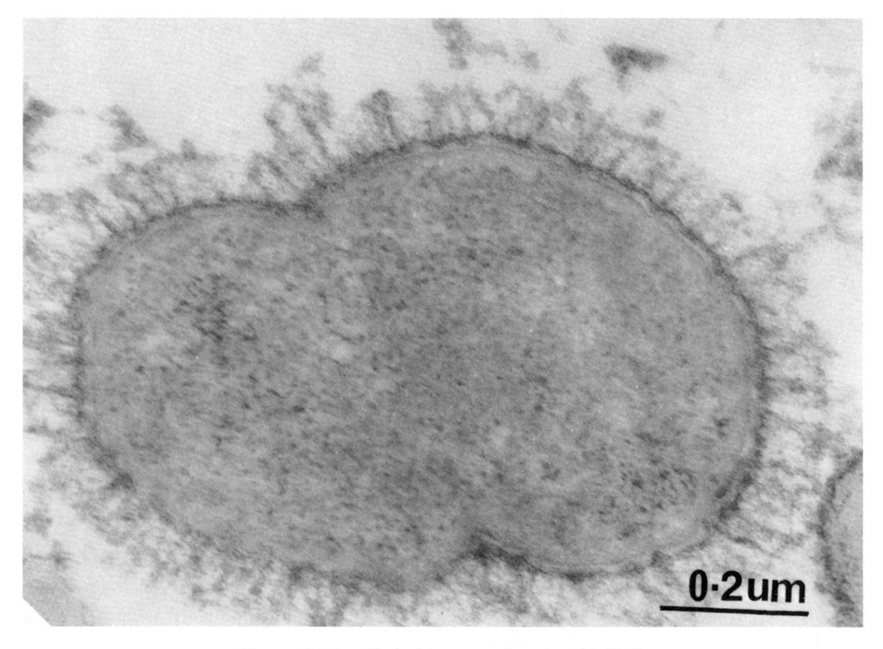

Figure 7.12 Fimbriate organism (×101,250).

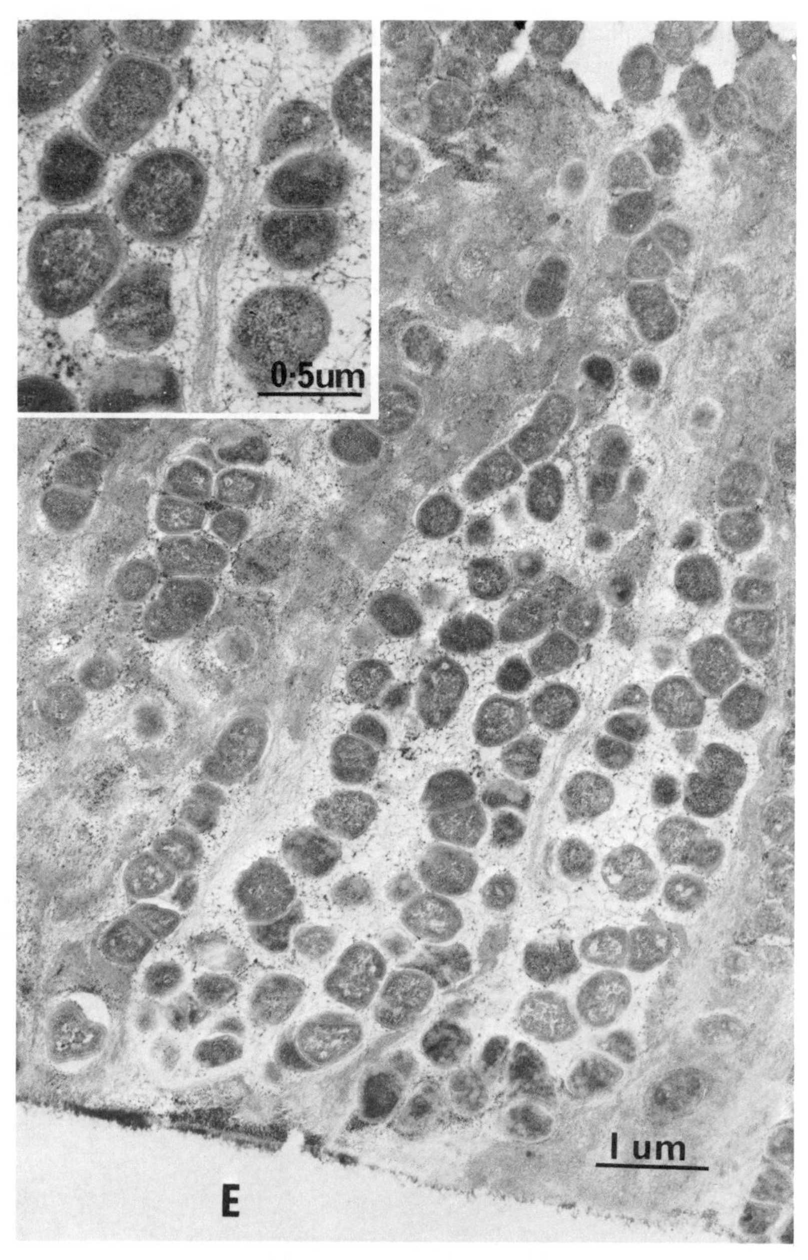

Figure 7.13 Contrasted with Alcian blue and lanthanum nitrate. Microcolonial growth in plaque. Note fine sparse carbohydrate matrix network around organisms, with condensation of fibrils between microcolonies (×15,250 insert at ×29,250).

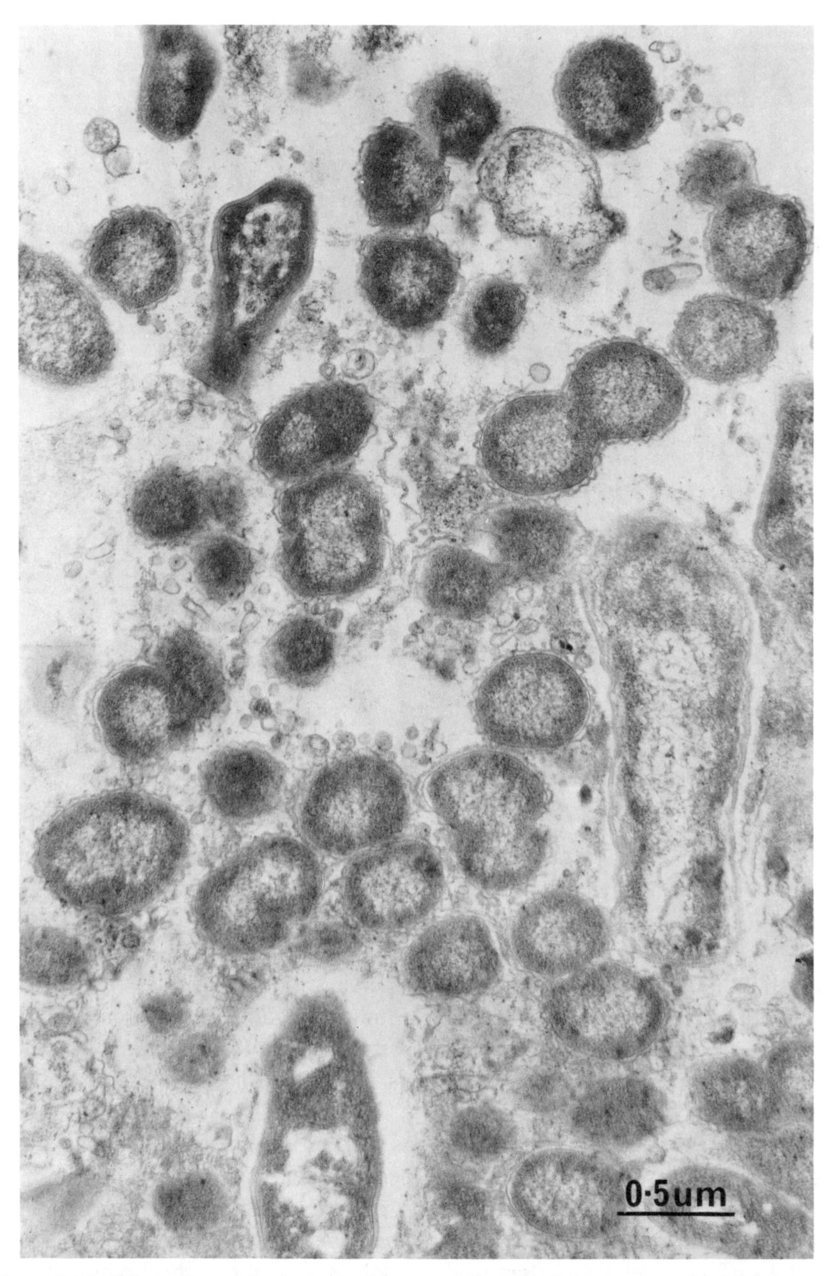

Figure 7.14 Microcolony of organisms with wavy cell walls. Numerous matrix vesicles, some apparently budding off cell wall. Function of these membrane-bound structures is not known (×33,250).

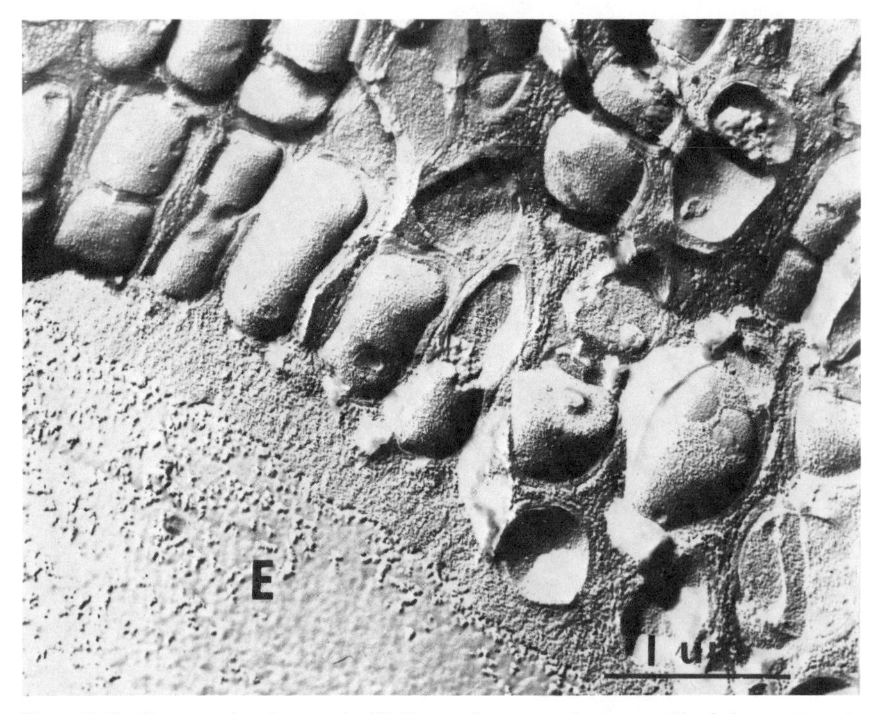

Figure 7.15 Freeze-etch micrograph. Wall-to-wall contacts between palisaded organisms in deep plaque (×25,250). Reprinted from the *J. Periodontol.*, **45,** 478–488 (1974).

are several striking structural relationships between plaque bacteria, notably "corncobs," in which cocci are clustered around a filament tip, and "test-tube brushes," in which bacilli radiate from a central filament [283, 285–290]. Intermicrobial attachment may be maintained even after sonication [200], the principal method of dispersing plaque samples for isolation and culture. This procedure disrupts many organisms, particularly filaments. As spherical organisms are more resistant to sonication, this may explain high counts of these organisms in various investigations of plaque microbiology [291,292].

Because *S. mutans* grows at least as well in the presence of glucose and some other sugars as it does in the presence of sucrose, its ability to colonize teeth may be related to its capacity to adhere to teeth [30]. As stated previously, *S. mutans* has also been described as having a feeble affinity for the tooth surface, and its preferential growth is thought to be favored by its location in stagnant sites where adsorption is not of primary importance [30,232].

Although glycosyltransferases are involved in EPS formation, they are located at different sites to the mainly protein receptors for EPS on the bacterial cell surface, at least in some *S. mutans* strains [293]. Yet another instance of the complexity of the situation on tooth surfaces may be derived from the work of Ellwood et al. [294] indicating that variation in glycosyltransferase activities occurs even between strains of the same serotype of *S. mutans*. There is a similar variety in the plaque enzymes capable of breaking down these polymers [196].

Plaque formation and bacterial agglutination seem to be separate and dissociable traits. Adhesion, but not agglutination, appears related to the synthesis of water-insoluble EPS [212]. The solubility of these compounds may be due largely to their content of α-1:3 glucopyranosyl linkages [295]. It may be observed that prior growth in a sucrose-containing medium does not affect adsorption by several plaque isolates, notably *S. mutans* and *S. sanguis* [236], although colonization by *S. mitis* requires the presence of EPS-synthesizing streptococci [296]. Adhering deposits are not associated invariably with detectable EPS synthesis [227], even though sucrose was found to be necessary for the formation of tenacious "plaques" by monocultures of *S. mutans*, *S. sanguis*, and *S. salivarius* in an artificial mouth system [296].

The EPS-producing enzymes are located mainly extracellularly and are produced principally during the growth phase, with their formation coupled to and closely dependent on protein synthesis. This close relation between cell growth and glycosyltransferase production may confer an adaptive advantage on *S. mutans*, facilitating its colonization of the tooth surface [297]. Walker [298] suggested that EPS formation may be regulated by *exo*- or *endo*-dextranases. Producers of EPS may form dextranases that facilitate growth by lysing polymers in the immediate vicinity of the cell [299]; this may, in turn, relate to the appearance of the clear zone described previously.

7.6 RETENTION ON ARTIFICIAL ORAL SURFACES

The accumulation of bacteria on oral surfaces is enhanced by poorly contoured restorations and orthodontic and prosthetic appliances, which produce additional stagnant loci. Restorations with overhanging or otherwise defective subgingival margins are notoriously difficult to keep clean [298,300,301]. Bacterial accumulation is also affected by the surface properties of the various dental restorative materials. Skjörland [302] has shown that after 8 hr of exposure in the mouth, plaque was most abundant on composite restorations, less on silicate cements, and least on dental silver amalgam. Kaqueler and Weiss [303] found that less plaque formed on porcelain and polished restorations than on enamel. Surface roughness is an important factor in retention, with more plaque forming on rough than on polished gold inlays [304]. Glantz [174] found that polished gold and amalgam had low surface energies comparable to human enamel, whereas nonmetallic materials such as porcelain and silicate had high surface energies. He concluded that it was less difficult to maintain a clean enamel surface than a similar restored surface.

Scaling procedures also may allow more plaque to form by producing surface roughness [305,306]. Polishing may reduce such irregularities [307].

7.7 RETENTION AND PREVENTION

It is now evident that caries and chronic inflammatory periodontal disease, although so widespread, may be countered largely by plaque control and reduction in sucrose intake [200,281,308–311]. Although attempts to immunize against both diseases are currently the subject of intensive research, present methods of prevention and control rely mainly on reducing the microbial volume by mechanical means particularly in stagnation sites prone to disease.

Present understanding of the ecology of the mouth emphasizes that it is neither necessary nor desirable to eliminate oral microorganisms. Their reduction to physiologic proportions may be accomplished by mechanical or antimicrobial means. The use of various antibacterial agents and a number of nonabsorbed antibiotics has been reviewed at length by Loesche [312]. The antiseptic chlorhexidine is the most promising of these since it binds readily and firmly to acquired pellicle and has a wide range of activity. Various side effects preclude its universal use on a permanent daily basis. It is worth stressing that there is no available adequate substitute for regular daily toothbrushing.

Current interest in the retention of bacteria on oral surfaces has generated a desire to understand the mechanisms of bacterial adhesion to effect disruption of the bonds between teeth and plaque. The study of surface properties of enamel from a protective standpoint seems to antedate observations of these properties in relation to plaque formation [313]. Walsh and Green [314] considered that various fats and oils might protect human enamel from carious attack by acid. They found the long-chain amine, hexadecylamine, to be a promising agent because the amine group attached to the tooth surface and the long chain was thought to bind the oil film, displacing saliva. The most successful of all agents tested was found to be the ^{14}C aliphatic amine, tetradecylamine (Armeen 14 D, Armour and Company, Chicago) [315,316]. Various laboratory tests and clinical studies showed that tetradecylamine had a marked affinity for enamel and apatite surfaces and could significantly protect enamel against acid attack [317]. Unfortunately, the compound is not available with an acceptable consistency and taste for use as a marketable dentifrice [318–320]. As water-soluble salts of these amines were found to be equally effective in affording protection, it was evident that it was the surface activity of the amine that gave it its protective properties [313].

Stack and Fletcher [321] were able to reduce enamel solubility with silanes. The most effective of these was the fluorinated derivative of bis(trimethylsilyl)acetamide. Unfortunately, the protective effect induced was transient. Of a number of anticorrosion agents supplied to the author by Armour Hess Chemicals (Leeds, U.K.), an aliphatic nitrogen-derived corrosion inhibitor (Armohib 25) was found to be the most effective, even when compared with hexadecylamine (Armeen 16D) or an amphoteric surfactant $R \cdot NH \cdot CH\ (CH_3)\ CH_2COOH$ (Armeen Z) [321].

Leaver [313] considered that the readily available water-soluble amine hexadecylamine acetate had considerable promise as an anticaries agent, especially if used as a mouthwash. It may be observed that the use of surfactants in this manner is an extension of the phenomenon of natural protection of tooth surfaces from acid decalcification by the natural organic integument [155,322]. Nonbactericidal amine fluorides are more effective than fluoride alone in reducing bacterial adhesion to teeth [323]. A chloromethyl-substituted cationic surfactant reduced human dental plaque formation [324], and Quintana et al. [325] proposed that such action was due to modification of the tooth surface by strong adsorption of this agent. Two recent interesting developments in this aspect of preventive dental research may be mentioned:

1. The possibility that calcium ions may affect the binding of plaque to pellicle prompted Rölla [187] to suggest that displacement of calcium might inhibit bacterial adsorption to teeth. As mentioned previously, calcium chelation does not disrupt plaque but separates plaque from enamel, not from pellicle, and then only on exposure sufficiently prolonged to disrupt the surface enamel.
2. The inhibition of adsorption of *S. mutans* to glass by low-molecular-weight dextrans and maltose led the investigators concerned to suggest that the incorporation of these carbohydrates in food might reduce caries [326].

7.8 CONCLUDING REMARKS

Interest in retention of bacteria on oral surfaces derives mainly from the relationship of portions of that population to the two most prevalent human diseases: chronic inflammatory periodontal disease and dental caries. In the absence of definitive knowledge of the physical chemistry of this phenomenon, it is still possible to make several conclusions from available evidence relevant to the physiology and pathology of this portion of the human bacterial population:

1. Stagnation rather than selective adhesion determines the accumulation of organisms at disease-prone locations.
2. Apparently causative organisms are not specific pathogens but commensals selected for primarily by soft diet texture, which enhances accumulation of organisms and by high carbohydrate content, which favors saccharolytic, acidogenic, and aciduric forms.
3. Motile bacteria are probably attracted to an oral surface by chemotaxis. A reversible attachment phase may exist, but organisms held close to an oral surface for some hours are usually irreversibly adsorbed. It is possible that only viable cells adhere.
4. The most likely bonding mechanisms are short-range (2–3-Å) chemical (electrostatic, covalent, and hydrogen bonds) and longer-range (3–4-Å) van der Waals forces, so that close approximation of organisms to surfaces is essential to effect adhesion of this nature. Extracellular polysaccharide may facilitate this process by linking and then, presumably due to hydrogen-bond cross-linking, approximating host and bacterial surfaces.
5. Cocci may predominate among early tooth colonizers because a diminishing radius of curvature of a cell reduces the potential-energy barrier to contact. Numbers of organisms found at early stages of colonization obviously depend also on availability of nutrients and inhibitory factors.
6. Specific areas of the bacterial cell wall are involved in adhesion. At least some oral streptococci adhere through extracellular-cell-associated carbohydrate receptors that bind probably to polysaccharide or to oligosaccharides in acquired pellicle glycoprotein.
7. Bacteria do not usually attach directly to the host surface, but to an organic layer composed mainly of selectively adsorbed salivary glycoproteins. Carboxyl groups in these proteins may adsorb directly to enamel hydroxyapatite on abraded surfaces. Normally, however, teeth enter the

mouth covered by a cuticle derived from their formative epithelium, to which pellicle binds by a mechanism not yet examined. The important natural adhesion mechanisms are thus of organic (bacteria) to organic (pellicle or cuticle) materials.

8. Acquired pellicle has several important physiological properties, including antibacterial components and the ability to reduce adhesion by some organisms.

Present understanding of oral ecology indicates that it is neither necessary nor desirable to eliminate the oral bacterial population to prevent caries or chronic inflammatory periodontal disease. At present, the only universally applicable long-term regimen for plaque control involves daily toothbrushing. It is probably unrealistic to expect entire populations to attain the necessary level of control required to prevent disease. Therefore, this aspect of preventive medicine must be concerned with restoring the balance of human oral ecology to provide the dental health enjoyed by most other dentate species.

Chronic inflammatory periodontal disease and dental caries are the most widespread human ailments. For many years their prevention has been attempted on the basis of classical theories of infection induced by specific pathogenic microorganisms. As yet no simple answer to these problems has been found with the use of antiseptics, antibiotics, or vaccines. But with an increasing awareness of oral ecology has arisen a need to understand the mechanisms by which bacteria accumulate on teeth. It might eventually be possible to disrupt adhesive or cohesive bonds in dental plaque, thereby maintaining the oral bacteria at clinically insignificant levels; hence the recent concern with surface and colloid phenomena in the oral cavity [327].

7.9 ACKNOWLEDGMENT

I thank Drs. S. A. Leach and A. G. Leaver of the University of Liverpool for their stimulating comments on acquired pellicle formation and surfactants, respectively. To Miss C. Rundle, Mrs. P. Quirke, and Mr. D. Wood of the Institute of Dental Surgery, University of London, I express my appreciation for their assistance with electron microscopy and photography. I gratefully acknowledge the continuing support of the Medical Research Council of the United Kingdom in personal work reported and the facilities made available to me by Professor I. R. H. Kramer, Dean and Director of Studies of the Institute of Dental Surgery, University of London.

7.10 REFERENCES

1. G. W. Burnett, H. W. Scherp, and G. S. Schuster, *Oral Microbiology and Infectious Disease*, 4th ed., Williams and Wilkins, Baltimore, 1976.
2. H. D. Donoghue, *J. Dent. Res.* **53,** 1289 (1974).
3. H. Löe, E. Theilade, and S. Börglum Jensen, *J. Periodont.*, **36,** 177 (1965).
4. G. J. Tucker, R. J. Andlaw, and C. K. Burchell, *Br. Dent. J.*, **141,** 75 (1976).
5. J. S. Wallace, *Br. Dent. J.*, **25,** 861 (1904).
6. F. Colyer, *Abnormal Conditions of the Teeth of Animals in their Relationships to Similar Conditions in Man*, The Dental Board of the United Kingdom, London, 1931, p. 39.
7. A. Baarregaard, *Oral Surg.*, **2,** 995 (1949).
8. P. R. Begg, *Am. J. Orthodont.*, **40,** 298 (1954).
9. T. G. H. Davies and P. O. Pedersen, *Br. Dent. J.*, **99,** 35 (1955).
10. T. R. Murphy, *Br. Dent. J.*, **116,** 483 (1964).
11. R. M. S. Taylor, *J. Polynes. Soc.*, **71,** 167 (1962).
12. H. Beyron, *Acta Odont. Scand.*, **22,** 597 (1964).
13. J. Egelberg, *Odont. Revy.*, **16,** 31 (1965).
14. H. N. Newman, *Br. Dent. J.*, **136,** 491 (1974).
15. T. L. Cleave, G. D. Campbell, and N. S. Painter. *Diabetes, Coronary Thrombosis and the Saccharine Disease*, John Wright, Bristol, U.K., 1969.
16. J. Ainamo, *Scand. J. Dent. Res.*, **80,** 505 (1972).
17. J. Robertson, *Nature (Lond.)*, **238,** 290 (1972).
18. D. P. Burkitt, *Br. Med. J.*, **1,** 274 (1973).
19. M. A. Eastwood, J. R. Kirkpatrick, W. D. Mitchell, A. Bone, and T. Hamilton, *Br. Med. J.*, **4,** 392 (1973).
20. K. W. Heaton and E. W. Pomare, *Lancet*, **7846,** 49 (1974).
21. D. A. T. Southgate, *Plant Foods for Man*, **1,** 45 (1973).
22. H. Trowell, *Plant Foods for Man*, **1,** 11, (1973).
23. J. Lindhe and P. O. Wicén., *J. Periodont. Res.*, **4,** 193 (1969).
24. A. Boyde, in R. W. Fearnhead and M. V. Stack, Eds., *Tooth Enamel II*, John Wright, Bristol, U.K., 1971, p. 39.
25. H. N. Newman and D. F. G. Poole, *Arch. Oral. Biol.*, **19,** 1135 (1974).
26. P. Axelsson and J. Lindhe, *J. Clin. Periodont.*, **1,** 126 (1974).
27. S. Poulsen, N. Agerbaek, B. Melsen, D. C. Korts, L. Glavind, and G. Rölla, *Commun. Dent. Oral Epidemiol.*, **4,** 195 (1976).
28. J. van Houte, R. J. Gibbons, and M. M. O'Gara, *J. Dent. Res.*, **50**(Abstr. 509), 180 (1971).
29. J. van Houte, R. J. Gibbons, and A. J. Pulkkinen, *Arch. Oral Biol.*, **16,** 1131 (1971).
30. R. J. Gibbons and J. van Houte, *Ann. Rev. Microbiol.*, **29,** 19 (1975).
31. H. A. Bartels, H. Blechman, and D. Lorieo, *J. Dent. Res.*, **39**(Abstr. 96), 687 (1960).
32. E. Sanders, *J. Infect. Dis.*, **120,** 698 (1969).
33. K. Sprunt and W. Redman, *Ann. Int. Med.*, **68,** 579 (1968).

34. W. G. Johanson, R. Blackstock, A. K. Pierce, and J. P. Sandford, *J. Lab. Clin. Med.*, **75,** 946 (1970).
35. E. T. Cetin, Ö. Ang, K. Töreci, and R. Berkiten, *Path. Microbiol.*, **37,** 185 (1971).
36. D. C. Savage, *Am. J. Clin. Nutr.*, **23,** 1495 (1970).
37. Monitor, *New Sci.*, **50,** 75 (1971).
38. R. Campbell and A. D. Rovira, *Soil Biol. Biochem.*, **5,** 747 (1973).
39. A. D. Rovira and R. Campbell, *Microbial Ecol.*, **1,** 15 (1974).
40. S. A. Leach, in T. Lehner, Ed., *The Borderland between Caries and Periodontal Disease*, Academic, New York, 1977, p. 79.
41. S. A. Leach, in W. D. McHugh, Eds., *Dental Plaque*, Livingstone, Edinburgh, 1970, p. 143.
42. R. A. Cowman, R. J. Fitzgerald and S. J. Schaefer, in H. M. Stiles, W. J. Loesche, and T. C. O'Brien, Eds., *Microbial Aspects of Dental Caries III*, Information Retrieval, Inc., Washington D. C., 1976, p. 465.
43. T. D. Brock, *Principles of Microbial Ecology*, Prentice-Hall, Englewood Cliffs., N. J., 1966.
44. M. P. Silverman and E. F. Munoz, *Science*, **169,** 985 (1970).
45. J. W. Schopf, *Orgins of Life*, **5,** 119 (1974).
46. J. W. Schopf, *Endeavour*, **34,** 51 (1975).
47. T. D. Luckey, *Am. J. Clin. Nut.*, **23,** 1430 (1970).
48. H. N. Newman, *Microbios*, **9,** 247 (1974).
49. R. J. Gibbons, D. M. Spinell, and Z. Skobe, *Infect. Immun.*, **13,** 238 (1976).
50. L. Weiss, *J. Theor. Biol.*, **6,** 275 (1964).
51. R. E. Baier, E. G. Shafrin, and W. A. Zisman, *Science*, **162,** 1360 (1968).
52. R. Mitchell, in H. M. Stiles, W. J. Loesche, and T. C. O'Brien, Eds., *Microbial Aspects of Dental Caries I*, Information Retrieval, Inc., Washington, D. C., 1976, p. 47.
53. P. S. Meadows, *Arch. Mikrobiol.*, **75,** 374 (1971).
54. H. N. Newman, *Br. Dent. J.* **138,** 335 (1975).
55. K. C. Marshall, R. Stout, and R. Mitchell, *Can. J. Microbiol.*, **17,** 1413 (1971).
56. K. C. Marshall, R. Stout, and R. Mitchell, *J. Gen. Microbiol.*, **68,** 337 (1971).
57. L. Weiss, in R. S. Manly, Ed., *Adhesion in Biological Systems*, Academic, New York, 1970, p.1.
58. J. Olsson, P.-O. Glantz, and B. Krasse, *Scand. J. Dent. Res.*, **84,** 240 (1976).
59. G. Silverman and I. Kleinberg, *Arch. Oral. Biol.*, **12,** 1387 (1967).
60. R. M. Frank and A. Brendel, *Arch. Oral Biol.*, **11,** 883 (1966).
61. C. McGaughey and E. C. Stowell, *Arch. Oral Biol.*, **12,** 815 (1967).
62. H. L. Ritz, *Arch. Oral Biol.*, **12,** 1561 (1967).
63. H. L. Ritz, *Arch. Oral. Biol.*, **14,** 1073 (1969).
64. H. L. Ritz, in W. D. McHugh, Ed., *Dental Plaque*, Livingstone, Edinburgh, 1970, p. 17.
65. W. G. Armstrong and A. F. Hayward, *Caries Res.*, **2,** 294 (1968).
66. R. M. Frank and G. Houver, in W. D. McHugh, Ed., *Dental Plaque*, Livingstone, Edinburgh, 1970, p. 85.
67. A. F. Hayward and W. G. Armstrong, in W. D. McHugh, Ed., *Dental Plaque*, Livingstone, Edinburgh, 1970, p. 187.

68. K. C. Marshall, *Interfaces in Microbial Ecology*, Harvard U. P., Cambridge Mass., 1976.
69. K. C. Marshall and R. H. Cruickshank, *Arch. Mikrobiol.*, **91,** 29 (1973).
70. H. N. Newman and A. B. Britton, *J. Periodont.*, **45,** 478 (1974).
71. M. M. McCabe and E. E. Smith, in W. H. Bowen, R. J. Genco, and T. C. O'Brien, Eds., *Immunologic Aspects of Dental Caries*, Information Retrieval Inc., Washington, D. C., 1976, p. 111.
72. M. M. McCabe, A. V. Haynes, and R. M. Hamelik, in H. M. Stiles, W. J. Loesche, and T. C. O'Brien, Eds., *Microbial Aspects of Dental Caries II*, Information Retrieval, Inc., Washington, D. C., 1976, p. 413.
73. C. F. Schachtele, in H. M. Stiles, W. J. Loesche, and T. C. O'Brien, Eds., *Microbial Aspects of Dental Caries II*, Information Retrieval, Inc., Washington, D. C., 1976, p. 429.
74. A. S. Bleiweis, in H. M. Stiles, W. J. Loesche, and T. C. O'Brien, Eds., *Microbial Aspects of Dental Caries II*, Information Retrieval, Inc., Washington, D. C., 1976, p. 245.
75. V. J. Iacono, M. A. Taubman, D. J. Smith, P. R. Garant, and J. J. Pollock, in W. H. Bowen, R. J. Genco, and T. C. O'Brien, Eds., *Immunologic Aspects of Dental Caries*, Information Retrieval, Inc., Washington, D. C., 1976, p. 75.
76. R. C. Williams and R. J. Gibbons, *J. Dent. Res.*, **51**(Abstr. 879), 267 (1972).
77. B. Rosan and R. J. Eisenberg, *Arch. Oral Biol.*, **18,** 1441 (1973).
78. K. Holmberg and H. O. Hallander, *J. Dent. Res.*, **51,** 588 (1972).
79. K. Holmberg and H. O. Hallander, *Arch. Oral Biol.*, **18,** 423 (1973).
80. F. H. M. Mikx, J. S. Van der Hoeven, K. G. König, A. J. M. Plasschaert, and B. Guggenheim, *Caries Res.*, **6,** 211 (1972).
81. F. H. M. Mikx, J. S. Van der Hoeven, A. J. M. Plasschaert, and K. G. König, *Caries Res.*, **9,** 1 (1975).
82. L. S. Fosdick and G. D. Wessinger, *J. Am. Dent. Assoc. Dent. Cosmos*, **27,** 203 (1940).
83. B. Regolati, B. Guggenheim, and H. R. Mühlemann, *Helv. Odont. Acta*, **16,** 84 (1972).
84. C. H. Miller and J. L. Kleinman, *J. Dent. Res.*, **53,** 427 (1974).
85. S. Selwyn and H. Ellis, *Br. Med. J.*, **1,** 136 (1972).
86. M. D. McMillan, *Arch. Oral Biol.*, **20,** 815 (1975).
87. R. Fuller and A. Turvey, *J. Appl. Bacteriol.*, **34,** 617 (1971).
88. R. C. Wagner and R. J. Barrnett, *J. Ultrastruct. Res.*, **48,** 404 (1974).
89. M. L. Barnett, *J. Dent. Res.*, **52,** 1160 (1973).
90. H. C. Jones, I. L. Roth, and W. M. Sanders, *J. Bacteriol.*, **99,** 316 (1969).
91. K. J. Cheng, R. Hironaka, D. W. A. Roberts, and J. W. Costerton, *Can. J. Microbiol.*, **19,** 1501 (1973).
92. M. Fletcher and G. D. Floodgate, *J. Gen. Microbiol.*, **74,** 325 (1973).
93. R. Fuller, personal communication.
94. M. E. Bayer and H. Thurow, *J. Bacteriol.*, **130,** 911 (1977).
95. D. K. McCallum, *J. Dent. Res.*, **53,** 138 (1974).
96. J. M. Brady, W. A. Gray, and W. Lara-Garcia, *J. Dent. Res.*, **54,** 777 (1975).
97. J. G. G. Ottow, *Annu. Rev. Microbiol.*, **29,** 79 (1975).

98. W. F. Liljemark and R. J. Gibbons, *Infect. Immun.*, **4,** 264 (1971).
99. R. Fuller and B. E. Brooker, *Am. J. Clin. Nutrition*, **27,** 1305 (1974).
100. D. C. Savage and R. V. H. Blumershine, *Infect. Immun.*, **10,** 272 (1966).
101. H. Hoffman and M. E. Frank, *Acta. Cytol.*, **10,** 272 (1966).
102. N. Tinanoff and A. Gross, *J. Dent. Res.*, **55,** 580 (1976).
103. P. Critchley, C. A. Saxton, and A. B. Kolendo, *Caries Res.*, **2,** 115 (1968).
104. H. I. Sussman, H. A. Bartels, and S. S. Stahl, *J. Periodont.* **40,** 210 (1969).
105. H. Takeuchi, M. Sumitani, K. Tsubakimoto, and M. Tsutsui, *J. Dent. Res.*, **53,** 132 (1974).
106. R. J. Gibbons and J. van Houte, *J. Dent. Res.*, **50**(Abstr. 508), 179 (1971).
107. J. van Houte and D. B. Green, *Infect. Immun.*, **9,** 624 (1974).
108. J. van Houte, J. D. Hillman, and R. J. Gibbons, *J. Dent. Res.*, **49,**(Abstr. 158), 88 (1970).
109. G. Rölla, *Arch. Oral Biol.*, **16,** 527 (1971).
110. E. I. F. Pearce, *Arch. Oral Biol.*, **21,** 545 (1976).
111. C. McGaughey and E. C. Stowell, *J. Dent. Res.*, **50,** 542 (1971).
112. R. E. Baier, in R. S. Manly, Ed., *Adhesion in Biological Systems*, Academic, New York, 1970, p. 15.
113. C. Dawes, G. N. Jenkins, and C. H. Tonge, *Br. Dent. J.*, **115,** 65 (1963).
114. P. Pincus, *Br. Dent. J.*, **59,** 372 (1935).
115. A. H. Meckel, *Arch. Oral Biol.*, **10,** 585 (1965).
116. W. H. Bowen, in A. H. Melcher and W. H. Bowen, Eds., *Biology of the Periodontium*, Academic, New York, 1969, p. 485.
117. C. J. Smith, in A. H. Melcher and W. H. Bowen, Eds., *Biology of the Periodontium*, Academic, New York, 1969, p. 157.
118. H. E. Schroeder and M. A. Listgarten, in A. Wolsky, Ed., *Fine Structure of the Developing Attachment of Human Teeth*, Karger, Basel, 1971.
119. A. Weinstock, *J. Dent. Res.*, **51**(Abstr. 136), 82 (1972).
120. H. G. Huxley, *J. Dent. Res.*, **51,** 828 (1972).
121. R. S. Manly, *J. Dent. Res.*, **22,** 479 (1943).
122. M. A. Rushton, *Br. Dent. J.*, **97,** 64 (1954).
123. S. A. Leach, P. Critchley, A. B. Kolendo, and C. A. Saxton, *Caries Res.*, **1,** 104 (1967).
124. W. A. McDougall, *Austral. Dent. J.*, **8,** 261 (1963).
125. W. A. McDougall, *Austral. Dent. J.*, **8,** 398 (1963).
126. C. W. Mayhall, *Arch. Oral Biol.*, **15,** 1327 (1970).
127. K. S. Birdi, *J. Colloid Interface Sci.*, **57,** 228 (1976).
128. G. Rölla, *Caries Res.*, **11,** (Suppl. 1), 243 (1977).
129. S. A. Leach and C. A. Saxton, *Arch. Oral Biol.*, **11,** 1081 (1966).
130. S. A. Leach, *Brit. Dent. J.*, **122,** 537 (1967).
131. G. Rölla and P. Mathiesen, in W. D. McHugh, Ed., *Dental Plaque*, Livingstone, Edinburgh, 1970, p. 129.
132. D. I. Hay, *Arch. Oral Biol.*, **18,** 1517 (1973).
133. D. I. Hay, *Arch. Oral Biol.*, **18,** 1531 (1973).
134. W. G. Armstrong, *Caries Res.*, **5,** 215 (1971).

135. G. Rölla, *J. Periodont. Res.*, **4,** 165 (1969).
136. T. Lie and K. A. Selvig, *Scand. J. Dent. Res.*, **83,** 145 (1975).
137. F. W. Kraus, D. Ørstavik, D. C. Hurst, and C. H. Cook, *J. Oral Pathol.*, **2,** 165 (1973).
138. D. Ørstavik and F. W. Kraus, *J. Oral Pathol.*, **2,** 68 (1973).
139. W. G. Armstrong, *Caries Res.*, **1,** 89 (1967).
140. C. Dawes and G. N. Jenkins, *J. Dent. Res.*, **42**(Abstr. 362), 126 (1963).
141. S. A. Leach and M. L. Hayes, *Caries Res.*, **2,** 38 (1968).
142. S. A. Leach and T. H. Melville, *Arch. Oral Biol.*, **15,** 87 (1970).
143. S. A. Leach and P. Critchley, *Nature* (*Lond.*), **209,** 506 (1966).
144. J. K. Pinter, J. A. Hayashi, and A. N. Bahn, *Arch. Oral Biol.*, **14,** 735 (1969).
145. C.-E. Nord, L. Linder, T. Wadström, and A. A. Lindberg, *Arch. Oral Biol.*, **18,** 391 (1973).
146. T. Ericson, *Caries Res.*, **1,** 52 (1967).
147. G. Rölla, *J. Periodont. Res.*, **2,** 243 (1967).
148. G. N. Jenkins, *Caries Res.*, **2,** 130 (1968).
149. D. I. Hay, *J. Dent. Res.*, **48,** 806 (1969).
150. D. I. Hay, *J. Dent. Res.*, **49,** 71 (1970).
151. P. Cleaton-Jones and L. Fleisch, *J. Periodont. Res.*, **8,** 366 (1973).
152. H. Lenz and H. R. Mühlemann, *Helv. Odont. Acta*, **7,** 30 (1963).
153. T. Sönju and G. Rölla, *Caries Res.*, **7,** 30 (1973).
154. T. Sönju, T. B. Christensen, L. Kornstad, and G. Rölla, *Caries Res.*, **8,** 113 (1974).
155. R. T. Zahradnik, E. C. Moreno, and E. J. Burke, *J. Dent. Res.*, **55,** 664 (1976).
156. A. Attramadal, *J. Periodont. Res.*, **4,** 281 (1969).
157. G. Bernardi, M. G. Giro, and C. Gaillard, *Biochim. Biophys. Acta*, **278,** 409 (1972).
158. G. Rölla and B. Melsen, *Caries Res.*, **9,** 66 (1975).
159. S. Hoffman, G. Rovelstad, W. S. McEwan, and C. M. Drew, *J. Dent. Rest.*, **48,** 1234 (1969).
160. S. Poulsen and I. J. Møller, *Arch. Oral Biol.*, **19,** 951 (1974).
161. J. Arends, personal communication.
162. R. D. Mulholland and D. D. DeShazer, *Angle Orthodont.*, **38,** 236 (1968).
163. P.-O. Glantz and G. Nyquist, *Odont. Revy.*, **17,** 332 (1966).
164. A. Belcourt, *Arch. Oral Biol.*, **21,** 717 (1976).
165. C. W. Mayhall, *J. Periodont.*, **48,** 78 (1977).
166. K. S. Birdi and J. Jeppesen, *Colloid and Polymer Science*, **256,** 261 (1978)
167. T. Sönju and K. Skjörland, in H. M. Stiles, W. J. Loesche, and T. C. O'Brien, Eds., *Microbial Aspects of Dental Caries I*, Information Retrieval, Inc., Washington, D. C., 1976, p. 133.
168. A. Rambourg, *Internat. Rev. Cytol.*, **31,** 57 (1972).
169. F. T. Last and R. C. Warren, *Endeavour*, **31,** 143 (1972).
170. W. A. Corpe, in R. S. Manly, Ed., *Adhesion in Biological Systems*, Academic, New York, 1970, p. 73.
171. D. I. Hay, R. J. Gibbons, and R. M. Spinell, *Caries Res.*, **5,** 111 (1971).
172. S. Kashket and S. R. Hankin, *Arch. Oral Biol.*, **22,** 49 (1977).

173. R. J. Gibbons and D. M. Spinell, in W. D. McHugh, Ed., *Dental Plaque*, Livingstone, Edinburgh, 1970, p. 207.
174. P.-O. Glantz, *Odont. Revy.*, **20**(Suppl. 17) (1969).
175. D. Ørstavik, F. W. Kraus, and L. C. Henshaw, *Infect. Immun.*, **9,** 794 (1974).
176. M. V. Stack and R. P. Fletcher, personal communication.
177. P. Ruangsri and D. Ørstavik, *Caries Res.*, **11,** 204 (1977).
178. G. McGaughey, B. D. Field, and E. C. Stowell, *J. Dent. Res.*, **50,** 917 (1971).
179. A. H. Meckel, *J. Dent. Res.*, **50**(Abstr. 654) 216 (1971).
180. D. I. Hay and F. G. Oppenheim, *Arch. Oral Biol.*, **19,** 627 (1974).
181. P. J. Winterburn and C. F. Phelps, *Nature* (*Lond.*), **236** 147 (1972).
182. R. C. Williams and R. J. Gibbons, *Infect. Immun.*, **11,** 711 (1975).
183. R. J. Gibbons and J. V. Qureshi, in H. M. Stiles, W. J. Loesche, and T. C. O'Brien, Eds., *Microbial Aspects of Dental Caries I*, Information Retrieval, Inc., Washington, D. C., 1976, p. 163.
184. T. Sönju and G. Rölla, *Acta Pathol. Microbiol. Scand. C.*, **83,** 215 (1975).
185. G. Rölla and M. Kilian, *Caries Res.*, **11,** 85 (1977).
186. H. N. Newman, *J. Periodont. Res.*, **7,** 91 (1972).
187. G. Rölla, in H. M. Stiles, W. J. Loesche, and T. C. O'Brien, Eds., *Microbial Aspects of Dental Caries II*, Information Retrieval,/Inc., Washington D. C., 1976, p. 309.
188. W. R. Grigsby and C. B. Sabiston, *J. Oral Pathol.*, **5,** 175 (1976).
189. R. J. Gibbons, J. van Houte, and W. F. Liljemark, *J. Dent. Res.*, **51,** 424 (1972).
190. W. H. Bowen, *Arch. Oral Biol.*, **19,** 231 (1974).
191. H. Mukasa and H. D. Slade, *Infect. Immun.*, **10,** 1135 (1974).
192. G. T. Grant, E. R. Morris, D. A. Rees, P. J. C. Smith, and D. Thom, *FEBS Lett.*, **32,** 195 (1973).
193. K. L. Melvaer, K. Helgeland, and G. Rölla, *Arch. Oral Biol.*, **19,** 589 (1974).
194. C. F. Schachtele, A. E. Loken, and M. K. Schmitt, *Infect. Immun.*, **5,** 263 (1972).
195. C. A. Saxton and P. Critchley, *Caries Res.*, **6,** 254 (1972).
196. R. H. Staat, T. H. Gawronski, and C. F. Schachtele, *Infect. Immun.*, **8,** 1009 (1973).
197. G. Rölla, *Coll. Int. CNRS* (*230*), 459 (1973).
198. J. Kelstrup, T. D. Funder-Nielsen, and E. N. Möller, *Acta Odont. Scand.*, **31,** 249 (1973).
199. H. N. Newman, *Br. Dent. J.*, **135,** 64, 106 (1973).
200. H. N. Newman and D. F. G. Poole, in F. A. Skinner and J. G. Carr, Eds., *The Normal Microbial Flora of Man*, Academic, London, 1974, p. 111.
201. L. J. Scaletta and D. K. McCallum, *J. Dent. Res.*, **49**(Abstr. 574), 192 (1970).
202. I. B. Stern and K. Sekeripataryas, *J. Dent. Res.*, **51**(Abstr. 311), 125 (1972).
203. E. Barrett and A. W. Asscher, *J. Med. Microbiol.*, **5,** 355 (1972).
204. L. Leive, V. K. Shovlin, and S. E. Mergenhagen, *J. Biol. Chem.*, **243,** 6384 (1968).
205. M. E. Bayer and L. Leive, *J. Bacteriol.*, **130,** 1364 (1977).
206. S. W. Watson and C. C. Remsen, *Science*, **163,** 685 (1969).
207. H. D. Seed, J. Spear, and H. N. Newman, *J. Dent. Res.*, **50**(Abstr. 30), 1175 (1971).
208. A. B. Spicer and D. F. Spooner, *J. Gen. Microbiol.*, **80,** 37 (1974).
209. H. K. Kuramitsu, *J. Bacteriol.*, **115,** 1003 (1973).

210. J. Carlsson, *Caries Res.*, **4,** 97 (1970).
211. W. J. Loesche and S. A. Syed, *Caries Res.*, **7,** 201 (1973).
212. M. L. Freedman and J. M. Tanzer, *Infect. Immun.*, **10,** 189 (1974).
213. J. Nalbandian, M. L. Freedman, J. M. Tanzer, and S. M. Lovelace, *Infect. Immun.*, **10,** 1170 (1974).
214. L. Weiss, *J. Cell Biol.*, **35,** 347 (1967).
215. A. C. Taylor, in M. J. Brennan and W. L. Simpson, Eds., *Biological Interactions in Normal and Neoplastic Growth*, Little Brown, Boston, 1962, p. 169.
216. D. R. Coman, *Cancer Res.*, **21,** 1436 (1961).
217. L. H. Greenfield, *Ann. N. Y. Acad. Sci.*, **109,** 23 (1963).
218. W. A. Corpe, unpublished observations from W. A. Corpe, (1969) in R. S. Manly, Ed., *Adhesion in Biological Systems*, Academic, New York, 1970, p. 73.
219. T. Ericson, J. Sandham, and I. Magnusson, *Caries Res.*, **9,** 325 (1975).
220. W. F. Liljemark and S. V. Schauer, *Arch. Oral Biol.*, **20,** 609 (1975).
221. R. J. Gibbons and P. H. Keyes, *Arch. Oral Biol.*, **14,** 721 (1969).
222. I. K. Paunio, *Helv. Odont. Acta*, **15,** 96 (1971).
223. J. Kelstrup and T. D. Funder-Nielsen, *Arch. Oral Biol.*, **17,** 1659 (1972).
224. J. Kelstrup and T. D. Funder-Nielsen, *J. Gen. Microbiol.*, **81,** 485 (1974).
225. J. D. de Stoppelaar, K. G. König, A. J. M. Plasschaert, and J. S. van der Hoeven, *Arch. Oral Biol.*, **16,** 971 (1971).
226. H. N. Newman, H. D. Donoghue, and A. B. Britton, *Microbios*, **15,** 113 (1976).
227. J. Kelstrup, J. Theilade, S. Poulsen, and I. J. Möller, *Caries Res.*, **8,** 61 (1974).
228. C. A. Saxton, *Caries Res.*, **9,** 418 (1975).
229. H. D. Donoghue and H. N. Newman, *Infect. Immun.*, **13,** 16 (1976).
230. M. Higuchi, G. H. Rhee, S. Araya, and M. Higuchi, *Infect. Immun.*, **15,** 938 (1977).
231. M. Higuchi, G. H. Rhee, S. Araya, and M. Higuchi, *Infect. Immun.*, **15,** 945 (1977).
232. J. van Houte, in H. M. Stiles, W. J. Loesche, and T. C. O'Brien, Eds., *Microbial Aspects of Dental Caries I*, Information Retrieval, Inc., Washington, D.C., 1976 p. 4.
233. W. Clark, R. J. Gibbons, and Z. Skobe, *J. Dent. Res.*, **53,** (spec. issue Abstr. 749), 242 (1974).
234. J. van Houte, R. C. Burgess, and H. Onose, *Arch. Oral Biol.*, **21,** 561 (1976).
235. R. J. Gibbons, E. C. Moreno, and D. M. Spinell, *Infect. Immun.*, **14,** 1109 (1976).
236. F. Pourdjabbar and C. Russell, *J. Dent. Res.*, **56**(Abstr.) D108, abs. 77 (1977).
237. S. S. Socransky and S. D. Manganiello, *J. Periodont.*, **42,** 485 (1971).
238. D. Ørstavik, *Acta Pathol. Microbiol. Scand. B*, **85,** 47 (1977).
239. E. Theilade and J. Theilade, in W. D. McHugh, Ed., *Dental Plaque*, Livingstone, Edinburgh, 1970, p. 27.
240. G. Rölla, *Swedish Dent. J.*, **1.** 241 (1977).
241. J. M. Hardie and G. H. Bowden, in F. A. Skinner and J. G. Carr, Eds., *The Normal Microbial Flora of Man*, Academic, London, 1974, p. 47.
242. E. Theilade, R. H. Larson, and T. Karring, *Caries Res.*, **7,** 130 (1973).
243. L. E. A. Folke, O. B. Sveen, and E. K. Thott, *Scand. J. Dent. Res.*, **81,** 411 (1973).

244. S. S. Socransky, A. D. Manganiello, D. Propas, V. Oram, and J. van Houte, *J. Periodont. Res.*, **12,** 90 (1977).

245. C. Lai, M. Listgarten, and B. Rosan, *Infect. Immun.*, **11,** 193 (1975).

246. C. Lai, M. Listgarten, and B. Rosan, *Infect. Immun.*, **11,** 200 (1975).

247. J.-O. Berg and C.-E. Nord, *Acta Odont. Scand.*, **30,** 503 (1972).

248. L. C. U. Junquiera and F. Fava-de-Moraes, in W. Bothermann, Ed., *Sekretion und Exkretion. 2*, Springer, Berlin, Heidelberg, 1965, p. 36.

249. G. Pitts and B. B. Keele, *J. Dent. Res.*, **52,** 1303 (1973).

250. P. R. Dugan, C. B. MacMillan, and R. M. Pfister, *J. Bacteriol.*, **101,** 973 (1970).

251. P. R. Dugan, C. B. MacMillan, and R. M. Pfister, *J. Bacteriol.*, **101,** 982 (1970).

252. J. F. Wilkinson, *Bacteriol. Rev.*, **22,** 46 (1958).

253. R. J. Gibbons and S. S. Socransky, *Arch. Oral Biol.*, **7,** 73 (1962).

254. R. J. Gibbons and S. S. Socransky, *Arch. Oral Biol.*, **8,** 319 (1963).

255. J. van Houte, *Arch. Oral Biol.*, **9,** 91 (1964).

256. F. Bramstedt, R. Naujoks, and I. Benedict, *Adv. Fluorine Res.*, **2,** 173 (1964).

257. J. M. Wood and P. Critchley, *Arch. Oral Biol.*, **11,** 1039 (1966).

258. J. H. J. Huis, in 't Veld and J. M. N. Willers, *Antonie van Leeuwenhoek*, **39,** 281 (1973).

259. T. Ikeda and H. J. Sandham, *Arch. Oral Biol.*, **16,** 1237 (1971).

260. J. M. Tanzer, W. I. Wood, and M. I. Krichevsky, *J. Gen. Microbiol.*, **58,** 125 (1971).

261. J. M. Tanzer, A. T. Brown, and M. F. McInerney, *J. Bacteriol.*, **116,** 192 (1973).

262. B. Guggenheim and H. E. Schroeder, *Helv. Odont. Acta*, **11,** 131 (1967).

263. P. Critchley, J. M. Wood, C. A. Saxton, and S. A. Leach, *Caries Res.*, **1,** 112 (1967).

264. B. Guggenheim, *Internat. Dent. J.*, **20,** 657 (1970).

265. E. Newbrun, R. Lacy, and T. M. Christie, *Arch. Oral Biol.*, **16,** 863 (1971).

266. M. C. Johnson, J. J. Bozzola, and I. L. Schechmeister, *J. Bacteriol.*, **118,** 304 (1974).

267. G. D. Shockman, M. L. Higgins, L. Daneo-Moore, S. J. Mattingly, J. R. Dipersio, and B. Terleckyj, *J. Dent. Res.*, **55,** (Spec. Issue A), A10 (1976).

268. J. P. Duguid, *J. Gen. Microbiol.*, **21,** 271 (1959).

269. K. B. Pedersen, L. O. Fröholm, and K. Bövre, *Acta Pathol. Microbiol. Scand. B*, **80,** 911 (1972).

270. G. A. Wistreich and R. F. Baker, *J. Gen. Microbiol.*, **65,** 167 (1971).

271. M. E. Ward and P. J. Watt, *Br. Med. J.* **1,** 485 (1973).

272. B. E. Brooker and R. Fuller, *J. Ultrastruct. Res.*, **52,** 21 (1975).

273. C. C. Brinton, in B. D. Davis and L. Warren, Eds., *The Specificity of Cell Surfaces*, Prentice-Hall, Englewood Cliffs, N. J., 1965, p. 37.

274. J. P. Duguid, E. S. Anderson, and I. Campbell, *J. Pathol. Bacteriol.*, **92,** 107 (1966).

275. H. N. Newman, in R. Fuller and D. W. Lovelock, Eds., *Microbial Ultrastructure. The Use of the Electron Microscope*, Academic, New York, 1976, p. 223.

276. B. K. Ghosh, *Sub-cell. Biochem.* **3,** 311 (1974).

277. H. N. Newman, in T. Lehner, Ed., *The Borderland between Caries and Periodontal Disease*, Academic, New York, 1977, p. 79.

278. B. Kérébel, S. Clergeau-Guérithault, and P. Forlot, *Ann. Microbiol.* (*Inst. Pasteur*), **126A,** 203 (1975).

279. J. Theilade, O. Fejerskov, and M. Hörsted, *Arch. Oral Biol.*, **21,** 587 (1976).
280. H. F. Ridgeway and R. A. Lewin, *J. Gen. Microbiol.*, **79,** 119 (1973).
281. W. D. McHugh, Ed., *Dental Plaque*, Livingstone, Edinburgh, 1970.
282. P. M. C. James, personal communication.
283. M. A. Listgarten, *J. Periodont.*, **47,** 1 (1976).
284. N. Halhoul and J. R. Colvin, *Can. J. Microbiol.*, **20,** 1307 (1974).
285. F. Vincentini, memoir read before the Medical Academy of Naples, March 1890.
286. F. Vincentini, *Dent. Cosmos*, **45,** 701 (1903).
287. J. L. Williams, *Dent. Cosmos*, **41,** 317 (1899).
288. S. J. Jones, *Arch. Oral Biol.*, **17,** 613 (1972).
289. H. N. Newman and G. S. McKay, *Microbios*, **8,** 117 (1973).
290. C. Mouton, Thesis, University, Louis Pasteur de Strasbourg, France, 1976.
291. S. A. Robrish, S. B. Grove, R. S. Bernstein, P. T. Marucha, S. S. Socransky and B. Amdur, *J. Dent. Res.*, **54,** (Spec. Issue A, Abstr. 236), 103 (1975).
292. S. A. Robrish, S. B. Grove, R. S. Bernstein, P. T. Marucha, S. S. Socransky, and B. Amdur, *J. Clin. Microbiol.*, **3,** 474 (1976).
293. J. Kelstrup and T. D. Funder-Nielsen, *J. Biol. Buccale*, **2,** 347 (1974).
294. D. C. Ellwood, J. K. Baird, J. R. Hunter, and V. M. C. Longyear, *J. Dent. Res.*, **55**(Spec. Issue C), C42 (1976).
295. T. Nisizawa, S. Imai, M. Hinoide, and S. Araya, *Arch. Oral Biol.*, **21,** 207 (1976).
296. C. Russell and W. A. Coulter, *J. Appl. Bacteriol.*, **42,** 337 (1977).
297. W. M. Janda and H. K. Kuramitsu, *Infect. Immun.*, **14,** 191 (1976).
298. G. J. Walker, *J. Dent. Res.*, **51,** 409 (1972).
299. B. Guggenheim and J. J. Burckhardt, *Helv. Odont. Acta*, **18,** 101 (1974).
300. E. Romine, *J. Am. Dent. Assoc.*, **44,** 742 (1952).
301. H. Zander, *J. Am. Dent. Assoc.*, **55,** 11 (1957).
302. K. Skjörland, *Scand. J. Dent. Res.*, **81,** 538 (1973).
303. J. C. Kaqueler and M. B. Weiss, *J. Dent. Res.*, **49**(Abstr. 615), 202 (1970).
304. W. Mörmann, B. Regolati, and H. H. Renggli, *J. Clin. Periodont.*, **1,** 120 (1974).
305. J. Waerhaug, *J. Dent. Res.*, **35,** 323 (1956).
306. R. Rosenberg and M. Ash, *J. Periodont.*, **45,** 146 (1974).
307. J. A. Clayton and E. Greene, *J. Prosthet. Dent.*, **23,** 407 (1970).
308. E. Newbrun, *Odont. Revy.*, **18,** 373 (1967).
309. G. B. Winter, *Br. Dent. J.*, **124,** 407 (1968).
310. G. N. Jenkins, *Internat. Dent. J.*, **22,** 350 (1972).
311. K. K. Mäkinen, *Internat. Dent J.*, **22,** 363 (1972).
312. W. J. Loesche, *Oral Sci. Rev.*, **9,** 65 (1976).
313. A. G. Leaver, *N. Z. Dent. J.*, **67,** 99 (1971).
314. J. P. Walsh and R. W. Green, *J. Dent. Res.*, **29,** 270 (1950).
315. R. W. Green and J. P. Walsh, *J. Dent. Res.*, **30,** 218 (1951).
316. R. M. King, *J. Dent. Res.*, **30,** 399 (1951).

317. T. J. Roseman, W. I. Higuchi, B. Hodes, and J. J. Hefferren, *J. Dent. Res.*, **48,** 509 (1969).

318. R. B. Nevin, J. P. Walsh, and R. M. King, *N. Z. Dent. J.*, **49,** 237 (1953).

319. M. Irwin, A. G. Leaver, and J. P. Walsh, *J. Dent. Res.*, **36,** 166 (1957).

320. T. G. Ludwig, *N. Z. Dent. J.*, **59,** 220 (1963).

321. M. V. Stack and R. P. Fletcher, *J. Dent. Res.*, **50**(Abstr. 121). 693 (1971).

322. E. C. Dobbs, *J. Dent. Res.*, **12,** 581 (1932).

323. J. Olsson and B. Krasse, *Scand. J. Dent. Res.*, **84.** 20 (1976).

324. S. Turesky, N. D. Gilmore, and I. Glickman, *J. Periodont.*, **41,** 41 (1970).

325. R. P. Quintana, A. Lasslo, J. W. Clark, and R. E. Baier, in A. Lasslo and R. P. Quintana, Eds., *Surface Chemistry and Dental Integuments*, Thomas, Springfield, Ill., 1970, p. 392.

326. E. Newbrun, F. Finzen, and M. Sharma, *Caries Res.*, **11,** 153 (1977).

327. "Surface and Colloid Phenomena in the Oral Cavity," *Swedish Dent. J.*, **1,** 205 (1977).

CHAPTER 8

Adsorption of Microorganisms to Roots and Other Plant Surfaces[1]

FRANK B. DAZZO[2]

Departments of Microbiology and Public Health and Crop and Soil Sciences, Michigan State University, East Lansing, Mich. 48824, U.S.A.

CONTENTS

[1] Dedicated to Drs. E. B. Fred, I. L. Baldwin, and E. McCoy, pioneers in the study of plant root-microorganism interactions.

[2] Formerly, Department of Bacteriology and Center for Studies of Nitrogen Fixation, University of Wisconsin, Madison

8.1 INTRODUCTION

Plants come into constant contact with microorganisms, and the events that ensue have profound influences on plant morphogenesis, nutrition, and pathogenesis. This chapter deals with the interface that forms a common boundary between microorganisms and living plant tissue, with particular emphasis on roots. A fully integrated approach is utilized at the molecular, subcellular, cellular, histological, and community levels of organization. Examples have been selected that illustrate similarities and differences between noninvasive and invasive interactions that benefit and/or harm the plant host. Several of the areas discussed are controversial issues. All interactions between the microorganism and plant interface are considered symbiotic in the broadest sense: the intimate living together of two or more dissimilar organisms [1]. Bracker and Littlefield [2] indicated that all structures between the plasma membranes of both symbionts are considered components of the "interface," and they include normal cell-surface components and modified or newly formed structures as a result of the interaction. They emphasize that the "interface" is the site of high cytoplasmic and metabolic activity.

The plant root functions as an absorptive organ for water and mineral nutrients, a mechanical support for aerial plant parts, a reserve organ for food, and it accomplishes a number of biosynthetic functions for the plant as a whole [3]. The root surface comes into contact with mineral particles, air, water, solutes, and organic matter including living and dead organisms. In 1904 Hiltner [4] recognized that soil near the roots harbored a more dense microbial population than did soil devoid of roots. Hiltner called the region of the soil under the immediate influence of living roots the *rhizosphere*. Later, Clark [5] suggested the term *rhizoplane* to include the external surface of plant roots together with closely adhering particles of soil (Fig. 8.1). The intense microbial activity associated with plant roots has been attributed largely to root exudation of energy-rich and growth-promoting organic compounds [6]; indeed, it has been estimated that some cereals release 12–18% of their photosynthetically fixed carbon as organic com-

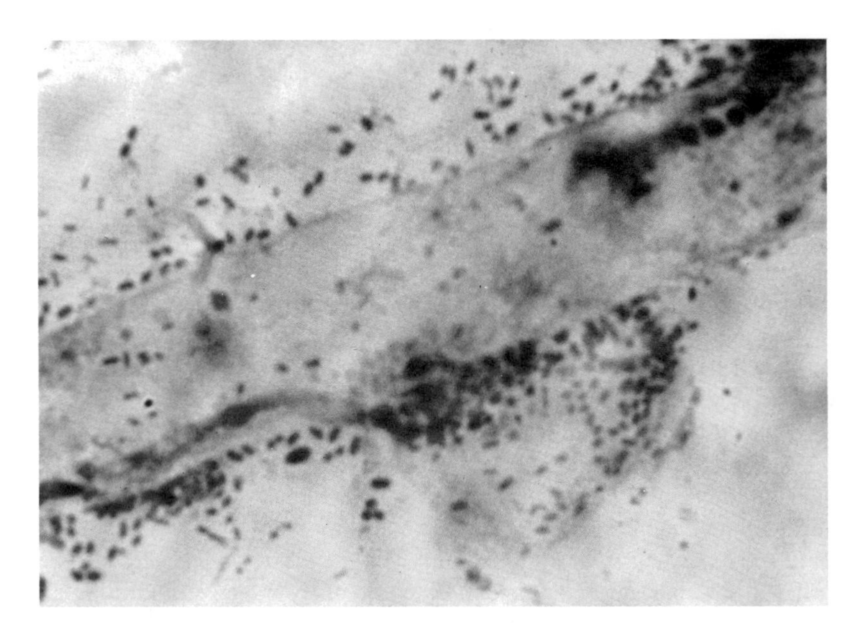

Figure 8.1 Adsorption of bacteria to a millet root hair.

pounds in root exudate [7]. Interactions between microorganisms attached to dead or sloughed-off root cells in the rhizosphere soil, and "root colonization" studies that do not document the physical contact of microorganisms to plant roots have been discussed in detail elsewhere [6, 8–14] and fall outside the scope of this chapter.

8.2 CHEMOTAXIS TOWARD ROOTS

The chemotactic response of microorganisms to roots may be limited by adsorption of chemotactic substances to soil particles. Movement of motile bacteria in soil is limited more by moisture content than by the soil pore-neck diameter [15]. The movement of *R. trifolii* in soil [15] ceases on a macroscale when the soil water tension is higher than the amount that produces discontinuous water-filled pores. This water tension value corresponds to 1.25%, 3.5%, and 2% water content for a coarse sand, fine sand, and silty loam soil, respectively. The steep moisture gradient close to the root during days with high transpiration rates would also tend to restrict bacterial movement to roots [16].

There is no evidence that bacteria swim toward legume roots in nature. Currier and Strobel [17,18] have proposed that chemotaxis of rhizobia toward the legume root surface may be the first step in the complex interaction that leads to nodulation and nitrogen fixation. Rhizobia exhibit a positively chemotactic response to root exudates collected aseptically from a wide variety of legumes. The chemotactic response is not restricted to homologous *Rhizobium*–legume systems and, therefore, chemotaxis is not a major determinant of host specificity. A glycoprotein ("chemotactin") from birdsfoot trefoil can maintain rhizobial cells within capillary tubes, and Currier and Strobel proposed that chemotactin elicited a chemotactic response in rhizobia [18]. An equally plausible alternative explanation is that "chemotactin" possesses no chemotactic activity but is merely a lectin (see Section 8.5.2) that binds to and immobilizes the bacteria so that they cannot swim out of the capillary tube.

Field studies have shown that good nodulation of legume roots is achieved following inoculation with rhizobia directly below the seed, but not following lateral inoculation, despite high rainfall activity [19]. These results indicate that chemotaxis of rhizobia to legume root surfaces in the field would be important, at most, only on a microscale. Furthermore, nonflagellated mutants of *Rhizobium* retain their infective and nodulating capabilities [20]. The chemotactic growth of legume root hairs *toward* rhizobia remains to be examined, and experiments with immobilized rhizobial flocs suggest that this phenomenon may occur (see Fig. 7 in Napoli et al. [21]). Time-lapse cine-photomicroscopy of inoculated roots in hydroponic culture may answer this important question.

8.3 ACCUMULATION OF MICROORGANISMS NEAR ROOTS

The R:S ratio is a measure of the *rhizosphere effect* and represents the number of microorganisms around roots (R) divided by the number of microorganisms in nonrhizosphere soil (S) [22]. The R:S ratio of *Rhizobium* for legumes generally falls within the range 10^2 to 10^6, with preferential stimulation of *Rhizobium* representing 1–10% of the total rhizosphere microorganisms [23]. The occurrence of compounds likely to be used as carbon, nitrogen, and energy sources for *Rhizobium* does not explain host specificity [24]. However, rhizobia that nodulate alfalfa and birdsfoot trefoil specifically accumulate in the rhizosphere of their associated hosts and are stimulated by the root exudates [24–26]. Also, pea roots release proportionally larger amounts of homoserine [27,28] utilized selectively by the symbiont *R. leguminosarum* [27]. Although other *Rhizobium* species are not inhibited by homoserine and are able to utilize other compounds released from pea roots, it was suggested that homoserine might selectively stimulate the growth of *R. leguminosarum* on the pea rhizoplane [27]. Munns [26] found that higher numbers of *R. meliloti* accumulate on symbiont alfalfa roots in the first few hours after inoculation than can be accounted for only by multiplication of the bacteria. He suggested that the rhizoplane population develops in hydroponic culture as a result of an initial adsorption followed by some bacterial multiplication at the root surface.

The motility of fungal zoospores aids their accumulation on the surface of roots [27,29,30]. Maximum accumulation of zoospores on the rhizoplane of different plants occurs immediately behind the root tip, over wounds, or on the root-elongation region [30]. The most active attractants for *Pythium aphanidermatum* zoospores in pea-root exudate are mixtures of aspartic acid, arginine, glutamine, homoserine, and serine [27]. With few exceptions, the chemotactic responses are nonspecific, and the zoospores of fungal pathogens accumulate along the rhizoplane of host and nonhost plants. Interestingly, chemotropic responses of germ tubes of *Phytophthora cinnamomi* zoospores oriented toward the region of elongation of susceptible avocado roots may play an important role in root infection [30]. Following adsorption of the germ tube to the avocado rhizoplane, penetration and infection of the fungus takes place rapidly.

8.4 THE MUCIGEL

Jenny and Grossenbacher [31] described *mucigel* as a mucilaginous gel substance extending from the surface of root epidermal cell walls into the rhizosphere. The mucigel was revealed under the electron microscope by adding positively charged colloidal iron hydroxide particles that bound to negatively charged carboxyl groups on the outer surface of the gel. Other studies have demonstrated a firm and distinct boundary of mucigel when bacteria were present [32–34] (Fig. 8.2). The effect of this outer boundary on nutrient uptake or microbial adsorption to the root is unknown. Ultrastructural evidence has been presented suggesting digestion of mucigel by attached bacteria [35]. More mucigel accumulates when certain bacteria colonize roots that when roots are grown axenically [33]. The distribution of mucigel on grass roots is varied, ranging from none to a complete covering that obscures bacterial colonies (Fig. 8.3) and junctions between adjacent epidermal cells. Areas exist where root hairs penetrate the mucigel and also where scattered threads of mucigel adhere to cell surfaces [36]. Sand grains in the rhizosphere near some plants may be covered with mucigel [37]. The mucigel is highly hydrated [38] and stains with ruthenium red, a stain specific for acidic polysaccharides [33]. The mucigel is undoubtedly complex in composition, and analytical studies are complicated by its ability to absorb contaminating plant and microbial products. The majority of bacteria entrapped within the mucigel of field-grown sorghum are Gram-negative bacilli [39]. Actinomycetes adsorbed to root hairs may be enclosed within mucigel [40].

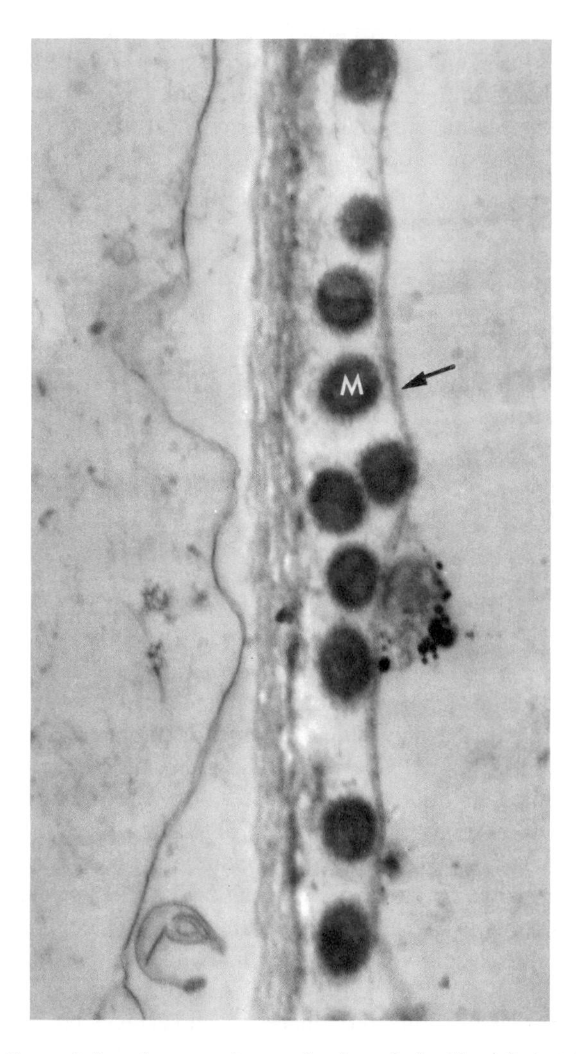

Figure 8.2 Transmission electron micrograph of mucigel with a distinct boundary (arrow) embedding microorganisms (M) on root-hair zone of onion. Courtesy of M. P. Greaves and J. F. Darbyshire and Pergamon Press, New York [33].

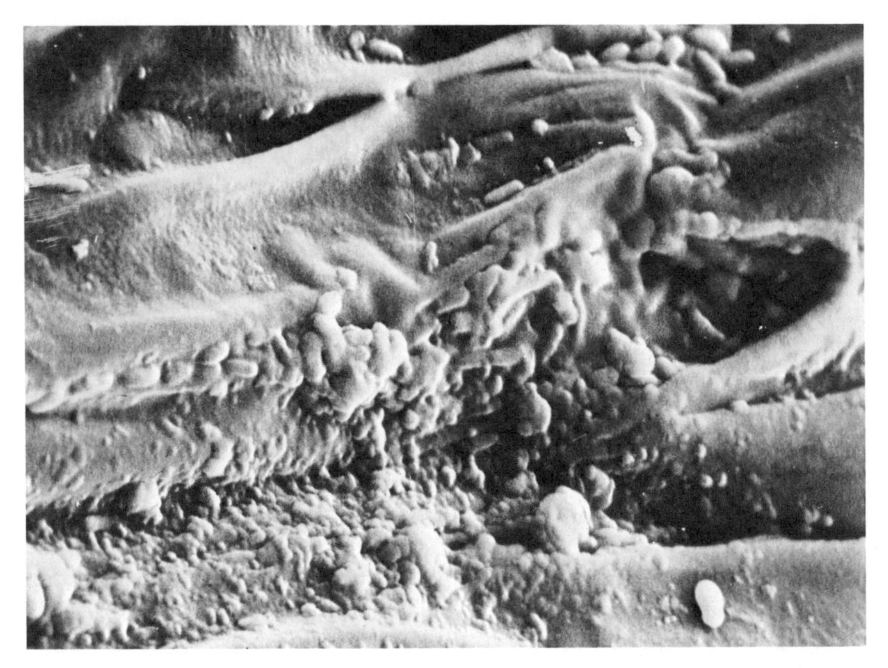

Figure 8.3 Scanning electron micrograph of a bacterial colony partially embedded with mucigel on a sand-dune grass. Courtesy of K. M. Old and T. H. Nicolson.

8.5 ADSORPTION TO ROOT SURFACE

Morphological and biochemical studies have been undertaken to describe the adsorption of microorganisms to plant root surfaces. This section has been subdivided arbitrarily to distinguish unidentified or noninvasive microorganisms and microorganisms that infect roots after they adsorb.

8.5.1 Noninvasive or Undefined Interactions

Studies that identify the rhizoplane microbial communities as saprophytic, or leave them undefined, will overlap in discussion with root-infecting plant pathogens, especially those where invasive microbial activities on old roots are suggested by extensive plant wall digestion.

Starkey [41] used the Rossi–Cholodny buried slide technique [42] to study the development of microorganisms on roots growing in soil. Slides taken from soil free of roots had well-defined and regularly scattered single microbial cells and colonies. In contrast, microorganisms on roots were distributed in many irregular aggregates and colonies, with bacteria, fungal hyphae, actinomycetes, and even nematodes adsorbed to root surfaces. Areas of the rhizoplane uncolonized by microorganisms were also noted. Some root hairs had chains of attached bacteria.

Root hairs of young oat plants were colonized by large numbers of bacteria, but tomato root hairs surprisingly were devoid of microorganisms [43]. Bacteria colonized the middle lamella along the longitudinal junctions between adjacent epidermal cells on roots of tomato [6], sand-dune grasses [36], and wheat [34,44]. In contrast, bacteria develop as discrete colonies on *Phalaris roots* [6]. Accumulation of bacterial cells between the longitudinal epidermal junctions may reflect the relative susceptibility of this region to bacterial decomposition, enchanced nutrient availability [36] and a more continuous and substantial volume of moisture and mucigel [13]. Experiments to test these possibilities remain to be performed. Pectic substances in the middle lamella may influence bacterial adsorption (see Sections 8.5.4 and 8.6.3). Root regions which release certain nutrients can be identified with defined auxotrophic mutants of *Neurospora crassa* [45]. The root tip of many plants often is devoid of microbial cover [34,37,46,47]. Colonization of bean roots by saprophytic fungi [46], especially *Fusarium oxysporum*, occurred rapidly on the hypocotyl which is above the oldest portion of the root. As roots increased in length, the fungi were isolated more frequently near the root tip. Taylor and Parkinson [46] (Table 8.1) indicated that some fungi confined to the surface of young roots penetrate and colonize internal root tissue of older roots.

Newman and Bowen [48] quantitated the pattern of distribution of bacteria on the rhizoplane of several grasses. Bacteria were aggregated nonrandomly on root surfaces in contact with soil. Plants had 4–10% of the root surface covered with bacteria [49]. In studies of microbial colonization of pine roots [13] Bowen and Rovira found that there was a high association between microcolonies and organic matter particles, suggesting that colonized organic fragments were a major source of inoculum. "Total" bacterial growth-rate studies during colonization of pine-seedling roots showed generation times of 7.5 hr in the apical centimeter, 9.1 hr in the root-elongation region, and 6.6 hr in the proximal portion (10 cm from the tip) of the root [50]. These studies exclude the many dividing cells that are not washed off of root surfaces. The data agree with calculated generation times of 9–12 hr for *Rhizobium japonicum* on soybean roots [13]. The localization of microcolonies and their embedment in mucigel suggest that dispersal of bacteria on roots would probably be a minor consideration. However, dispersal of fungal hyphae along roots may be extensive and sometimes involves rhizomorphs as dispersal units. These quantitative studies suggest that microorganisms on roots behave as a collection of spatially separate com-

Table 8.1 Fungi associated with dwarf bean roots showing estimated relative importance of each species in each of four recognizable habitats[a,b]

Fungus	Root surface	Cortex	Outer stele	Inner stele
Mucor spp.	+	+		
Mortierella vinacea	+	+		
Trichoderma viride	++	+		
Fusarium sambucinum	++	+		
Penicillium spp.	++	++		
Penicillium lilacinum	++	++		
Mortierella spp.	++	++	+	
Gliocladium spp.	+++	++	+	+
Fusarium oxysporum	+++++	+++++	+	+
Cylindrocarpon radicicola	+++	+++	+++++	++++
Sterile dark forms	+	+	++++	+++++
Varicosporium elodea		+	++	++
Sphaeropsidales spp.		+	+	
Botrytis cinerea		+	+	
Sterile hyaline forms			+	++

[a] From Taylor and Parkinson [46] and courtesy of Martinus Nijhoff.
[b] Symbols: +, usually present but only low frequency of isolation; +++++, dominant form with high frequency of isolation.

munities, determined largely by the organisms that contact the root in the soil, rather than as one large interacting community [13].

Ultrastructural studies have illustrated the spatial relationships of rhizoplane microorganisms and the "interface" between the plant root and microorganisms. Foster and Rovira [34,44] found that epidermal cells and most of the outer cortex cells of healthy wheat were collapsed, devoid of cytoplasm, and decayed by bacteria at the time of flowering. Lysis of epidermal cells and decomposition of the root was progressive inward from the soil–root "interface" and was accomplished by the combined action of various morphologically distinguishable microorganisms, many of which were encapsulated. Walls and middle lamellae showed signs of erosion following the contours of the invading bacteria, eventually forming extensive lysis troughs and holes. These results illustrate the occurrence of low-level infections in the field. Clay minerals were aligned with the broken edges of platelets attached to the root surface, and the authors suggested [34] that this arose by mechanical reorientation of randomly dispersed particles as the wheat root expands into the soil during growth. Fibrous material often surrounded bacteria adsorbed to epidermal cell walls [47].

Marchant [51] combined light microscopy, SEM, and cultural plating techniques to study the root surface of *Ammophila arenaria* as a substrate for microorganisms. Consistent with other studies, no microorganisms were found on the root tip. Microbial colonization began 4–5 mm from the root tip. Some rhizoplane microorganisms could not be removed by washing, suggesting firm adhesion to the root surface. Tests for cellulolytic activity suggested that most fungi, and to a lesser extent bacteria, eroded epidermal cells and exposed underlying cortical cell walls (Fig. 8.4). Roots from plants dominant in different stages of ecological succession had the same rhizoplane microtopography.

Old and Nicolson [36] selected sand-dune grasses to study the ultrastruc-

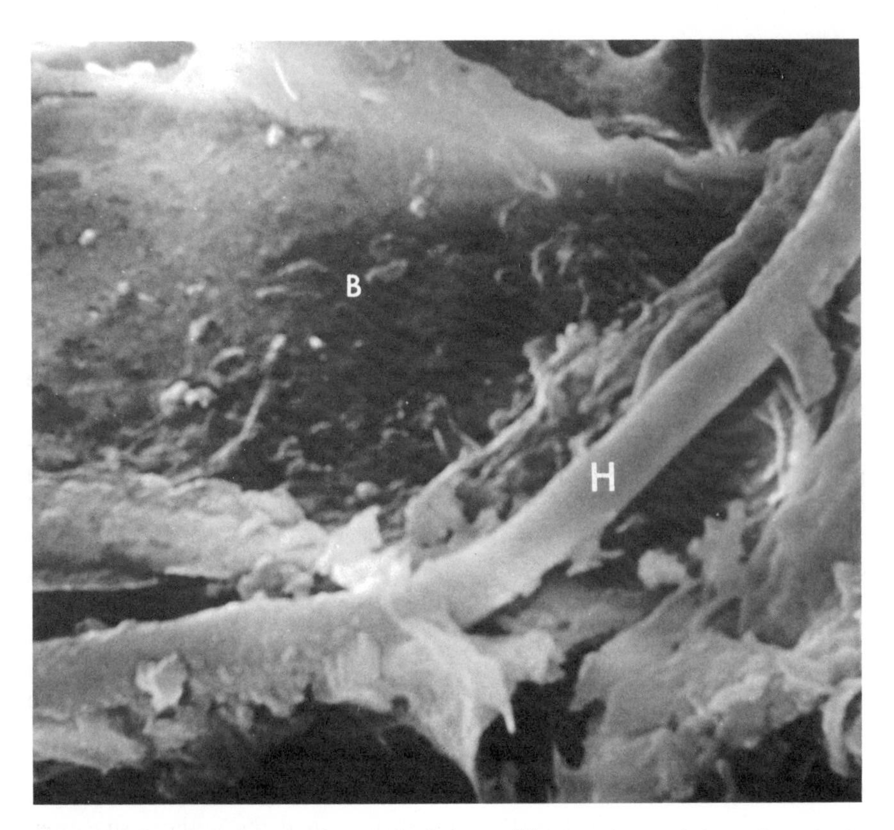

Figure 8.4 Scanning electron micrograph of *Ammophila arenaria* root 4–5 mm from the tip, showing a fungal hypha (H) associated with cell wall debris and bacteria (B) on the smooth wall surface. Courtesy of R. Marchant and Cambridge University Press [51].

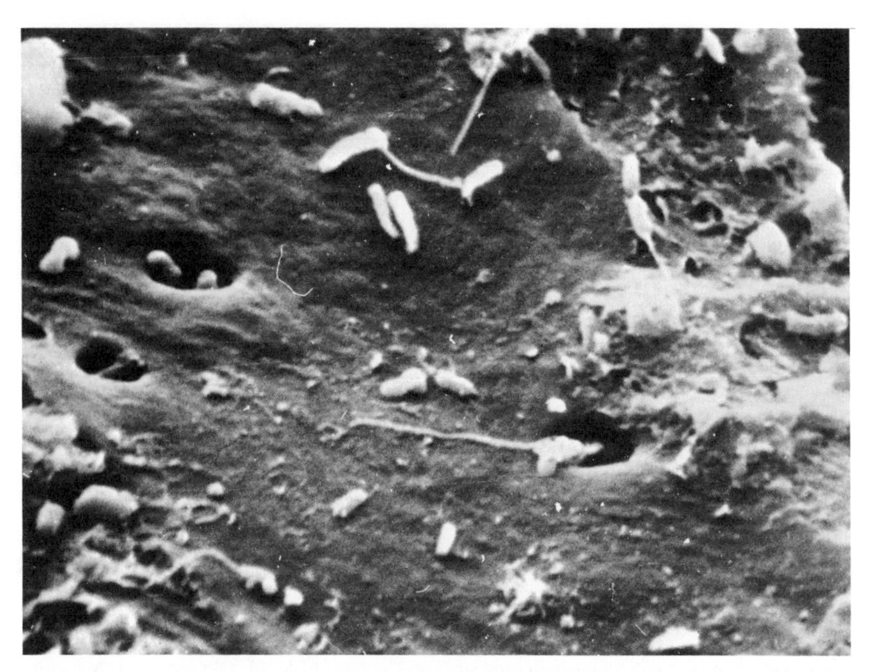

Figure 8.5 Scanning electron micrograph of *Ammophila arenaria* root 4 cm from the tip, showing bacterial cells with filamentous appendages forming bridges between adjacent cells and cells in pits. Courtesy of K. M. Old and T. H. Nicolson and Blackwell Scientific Publications, Oxford, England [36].

ture of the rhizoplane. This environment lacks colloidal organic matter and clays that otherwise complicate interpretations of electron micrographs. Many of the root epidermal cells were perforated with holes of various dimensions that harbor bacteria (Fig. 8.5). Bacteria were also embedded in spaces between adjacent epidermal cells. Microorganisms apparently colonized underlying cortical tissue after epidermal cells were sloughed off. Bacteria may have entered roots by direct perforation of host cell walls and then penetrated along the middle lamella between adjacent epidermal cells. Microcolonies of dividing cells occurred on the epidermal root surface (Fig. 8.6), and some were embedded in mucigel (Fig. 8.3). Bacteria, fungal hyphae, actinomycetes, and structures resembling budding yeasts were found. Some bacteria adsorbed to the rhizoplane were encapsulated (Fig. 8.7) or had filamentous appendages (Fig. 8.8). Dart [52] and Dazzo (in press) observed similar appendages on rhizobia attached to legumes after extended incubation.

Nissen [53,54] found that adsorption of *Pseudomonas tolaasii* to barley roots was dependent on divalent cations (Ca^{++}, Mg^{++}, and Sr^{++}), but not

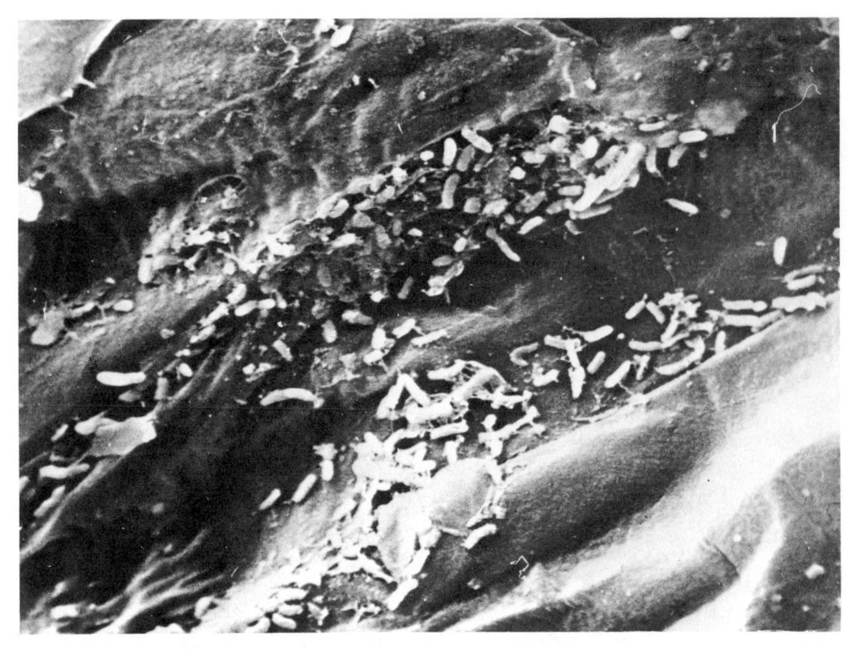

Figure 8.6 Bacterial colony on epidermal surface. Courtesy of K. M. Old and T. H. Nicolson.

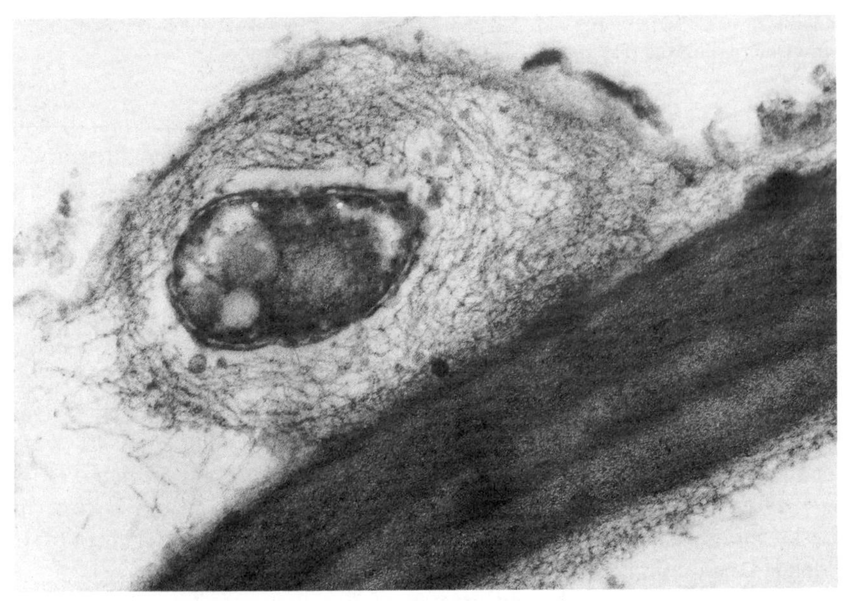

Figure 8.7 Encapsulated bacterium adsorbed on the cell wall of *Festuca rubra* var. *arenaria*. Courtesy of K. M. Old and T. H. Nicolson and Blackwell Scientific Publications, Oxford, England [36].

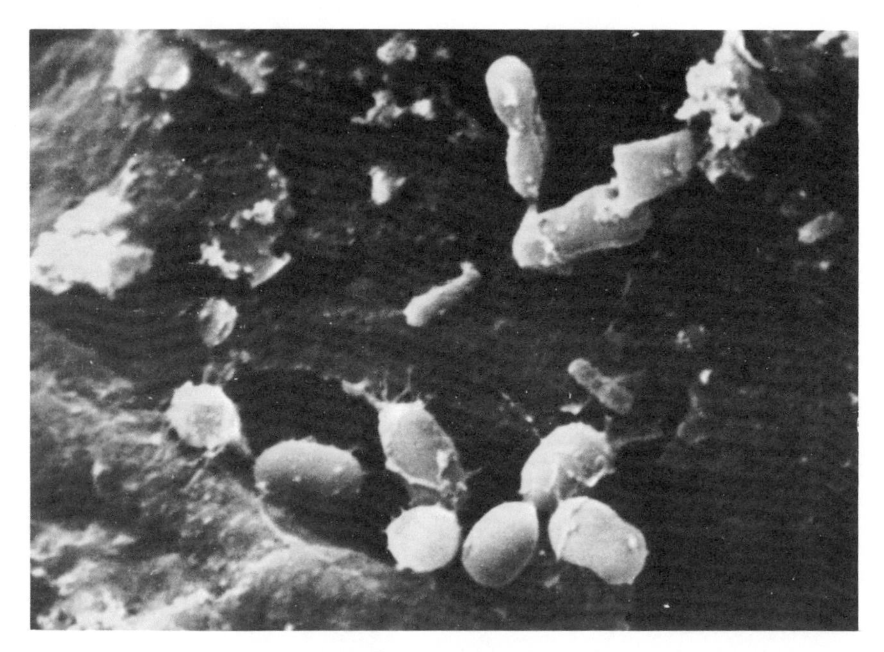

Figure 8.8 Yeast-like cells with appendages on rhizoplane of *Ammophila arenaria*. Courtesy of K. M. Old and T. H. Nicolson and Blackwell Scientific Publications, Oxford, England [36].

on monovalent ions (Figs. 8.9*a*,*b*). Desorption and release of the bacteria occurred immediately after transfer of roots to distilled water. Two mechanisms for the adsorption of bacteria to roots were described. One mechanism does not require divalent metal cations or energy, is operational at acid pH, and is not specific for bacteria. The other mechanism requires divalent cations, is only operative near neutral pH, requires plant-derived energy, and is very specific for certain bacteria. Nissen interpreted this latter mechanism as the bridging of negatively charged groups on the bacterial and plant root surfaces by divalent cations, but he had no explanation for the energy requirement and the specificity displayed by certain Gram-negative bacteria. Nissen speculated that Ca^{++}-mediated adsorption may constitute the first step in the invasion of plant tissues by phytopathogenic and root nodule bacteria. The ability of the bacteria to become adsorbed to roots may be an important factor in determining their prevalence in the rhizosphere and rhizoplane. This report constitutes one of the first efforts to unravel the biochemical components of the bacterial–plant root "interface."

8.5.2 *Rhizobium*–Legume Associations

Species of *Rhizobium* infect, nodulate, and enter into a nitrogen-fixing symbiosis with roots of various leguminous plants. With many temperate

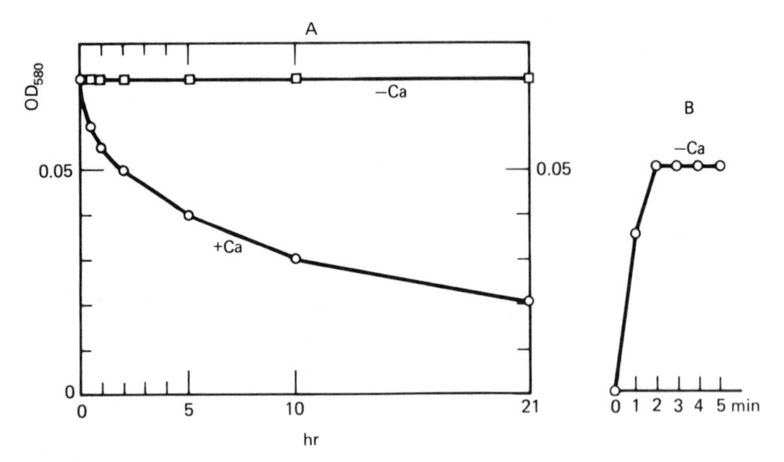

Figure 8.9 Calcium-dependent adsorption (A) and release (B) of *Pseudomonas tolaasii* from barley roots as measured by optical density of the bacterial cell suspension. Courtesy of P. Nissen and North-Holland Publishing Company, Amsterdam [53].

legumes as hosts, infectiveness is defined as the ability to form infection threads in root hairs. Figure 8.10 shows a clump of infective *R. trifolii* cells at the tip of an infected and curled root hair containing the refractile infection thread. The high degree of *Rhizobium*–host specificity is displayed prior to the formation of root-hair infection threads [55,56].

The most useful technique to study the interaction of *Rhizobium* and small-seeded legume roots is the plant-slide culture devised by Fahraeus [57]. This procedure involves aseptic germination and growth of a small-seeded legume (e.g., clover) in a thin layer of nitrogen-free, mineral nutrient agar on a glass slide. Following inoculaton with *Rhizobium*, the roots are covered with a cover glass and the seedling slide assembly is inserted in a sterile tube containing the mineral nutrient solution. The slide assembly can be removed, blotted, and the roots examined microscopically at any time.

The Fahraeus slide technique is artificial since roots are in a hydroponic environment and may behave differently in soil. For instance, some plants produce secondary water roots that are anatomically different from roots grown in natural soil. Studies with this technique, however, are usually conducted on the seedlings' primary roots. Two other considerations are illumination of roots and crowding of seedlings. Nutman [58] found that clover root hair infection by *R. trifolii* was enhanced by some light and that the number of infections per plant dropped significantly if four or more seedlings were present per vessel. Despite these shortcomings, the Fahraeus slide technique has been instrumental to our understanding of the *Rhizobium*–legume symbiosis at the molecular, subcellular, cellular, histological, and community levels. An elegant time-lapse movie of the root-hair infection process using Fahraeus slides is available [59].

The role of cell-wall-degrading enzymes in infection [60] has remained controversial. Some of the difficulty in reproducing the data is explained by inhibitors from seeds and seedlings that inactivated the enzymes [61]. Pectolytic activity is detected with rhizobia in pure culture [62,63], and so the role of wall-degrading enzymes can now be studied at the molecular level. Infection threads are formed by invagination of the root-hair cell wall [64–66]. The root-hair nucleus [57,59], as well as dictyosome and rough endoplasmic reticulum [21,65], are in close proximity with the growing root-hair tip of the infection thread.

Some tropical legumes (peanut and joint vetch) are infected by rhizobia entering through intercellular spaces created as epidermal cells are sloughed off at lateral root emergence [67,68]. Ultrastructural studies show erosion of wall material at the point of rhizobial attachment [67,69] (Fig. 8.11). Interestingly, this erosion is associated with one cell pole.

Dazzo et al. [70] modified the Fahraeus slide technique [57] to quantitate the adsorption of rhizobial cells to target clover root hairs. Seedlings (24 hr old) were inoculated with 10^7 cells/plant (direct count), incubated for 12 hr in a growth chamber, rinsed with saline or Fahraeus solution, and then

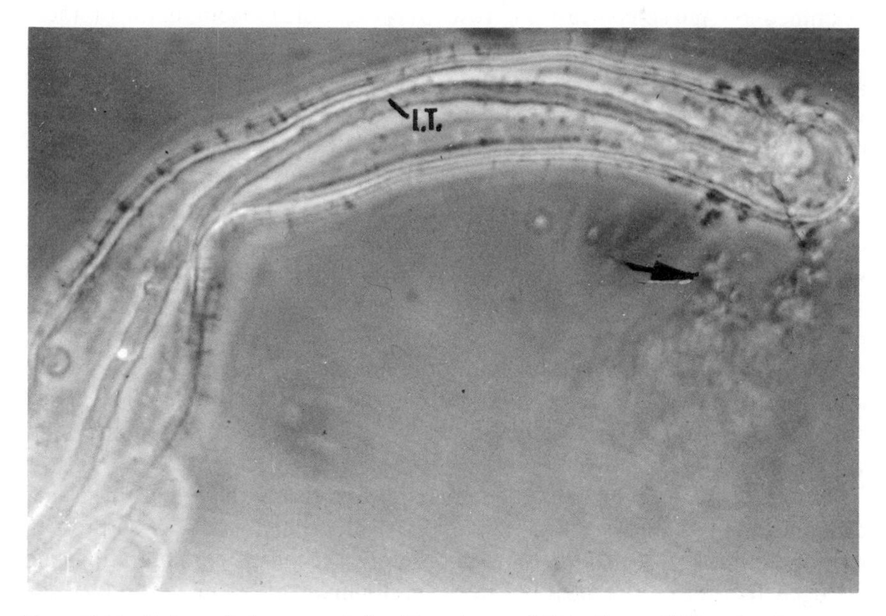

Figure 8.10 Deformed clover root hair with a clump of *Rhizobium trifolii* at the tip (arrow) and an intracellular refractile infection thread (IT). Note the polarly attached bacteria.

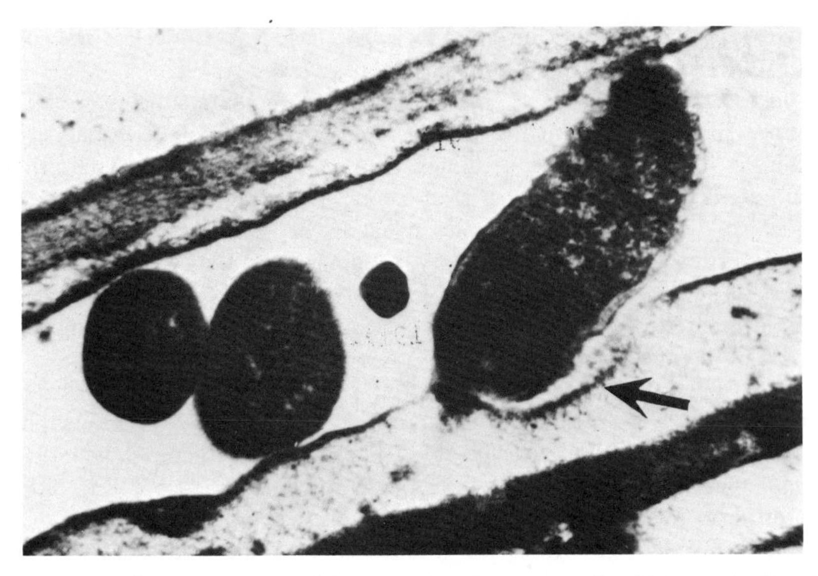

Figure 8.11 *Rhizobium* in intercellular spaces between cortex cells of joint vetch. Note the contour of the host cell-wall cuticle that conforms to the shape of the *Rhizobium* at the attached cell pole (arrow). Courtesy of C. A. Napoli [67].

examined by phase-contrast microscopy (Fig. 8.12). Root hairs of these seedlings were straight; adsorbed bacteria were refractile and easily distinguishable (Fig. 8.13). The numbers of bacteria adsorbed to root hairs approximately 200 μm in length were determined, and the data were compared statistically (Table 8.2). Only root hairs along the optical median plane were examined. This assay allows comparisons of bacterial and plant variability, and the effect of metabolites, antimetabolites, and sugar inhibitors on rhizobial attachment to target cells with a standard surface area and with a uniform state of physiological development. The direct microscopic assay is reproducible and scores only bacteria with their cell surface in firm physical contact with root-hair cell walls. The assay eliminates sources of error in quantitative adsorption studies by not scoring bacterial cells or flocs adsorbed to undifferentiated epidermal (nontarget) host cells. The assay also takes into account the variability of root-surface areas among individual plants.

Host specificity in the root–nodule symbiosis forms the basis for species differentiation within the genus *Rhizobium* [71,72] and is expressed prior to root infection [55,56] (days to weeks before the onset of nodulation and nitrogen fixation). Is it possible that development of a unique "interface" between the microbe and the plant root may be required for recognition? The evidence is growing that plant proteins called *lectins* (from the Latin *legere*, to choose) may be an integral part of the initial *Rhizobium*-legume "interface." Lectins are proteins or glycoproteins that have carbohydrate-binding specificity. The *lectin-recognition* hypothesis [70, 73–98] is based on the possibility that recognition at infection sites involves the binding of specific legume lectins to unique carbohydrate structures found exclusively

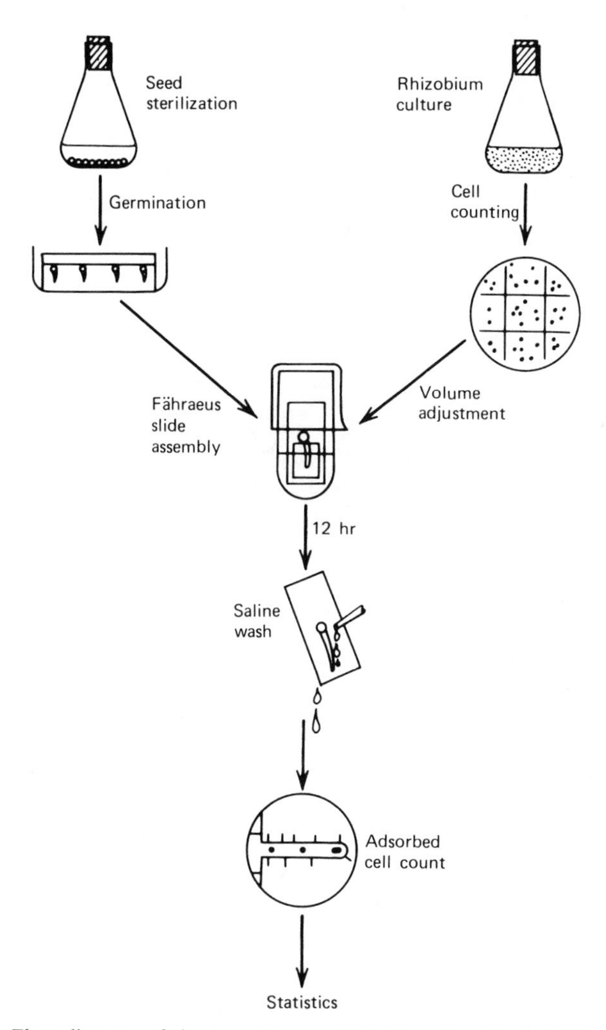

Figure 8.12 Flow diagram of the assay to quantitate the adsorption of rhizobia on legume seedling root hairs. Taken from Dazzo et al. [70].

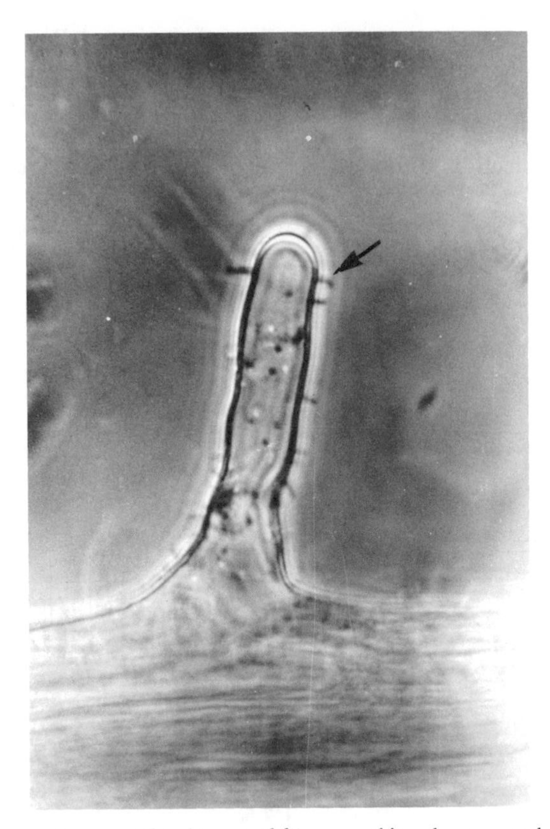

Figure 8.13 Adsorption of *Rhizobium trifolii* to a white clover root hair. Note polarly attached bacteria (arrow).

on the surface of the appropriate rhizobial symbiont. This working hypothesis is being tested. Hamblin and Kent [73] found that an agglutinin from kidney bean could bind to *R. phaseoli*, the symbiont of red kidney beans. A lectin from kidney beans binds to the blood group A substance on human erythrocytes. Consequently, these authors used human group A erythrocytes to detect an agglutinin on epidermal root surfaces and in saline extracts of seeds, nodules, and roots below the nodules [73]. Bohlool and Schmidt [74] introduced the possible role of lectins in host specificity by demonstrating that a fluorescein isothiocyanate (FITC)-labeled soybean lectin preparation bound only to strains of *R. japonicum* capable of nodulating soybean. Bauer et al. [84,85] have confirmed these observations with highly purified FITC- and ^{3}H-labeled soybean lectin and have shown that the binding to *R. japonicum* is reversible by *N*-acetylgalactosamine, a potent specific inhibitor of binding by soybean lectin. Wolpert and

Albersheim [78] reported similar lectin-binding specificities with rhizobial lipopolysaccharide preparations, although no hapten-reversible interactions were demonstrated in this study. Other observations of the soybean lectin–*Rhizobium japonicum* interaction include: the bacterial biphasic lectin-binding curves suggestive of multiple carbohydrate receptors [84,85]; transient appearance of lectin receptors on the *R. japonicum* cell surface [84,85]; binding of lectin to polar tips of some *R. japonicum* strains [90,91]; rapid decline of lectin levels in developing soybean seedling roots [86]; induction of lectin-binding surface receptors on *R. japonicum* in the soybean rhizosphere [95]; and the reports by Albersheim and Wolpert [92,93] that lectins, including one from soybean, are enzymes that specifically degrade the lipopolysaccharide (LPS) of their symbiont rhizobia.

A major criticism of many *Rhizobium*–lectin studies is that there exists no *a priori* reason why erythrocytes and rhizobia should have the same carbohydrate receptors that bind the lectin. Many studies have considered only "classical" lectins that agglutinate erythrocytes and, therefore, overlooked hitherto unrecognized legume proteins that may bind to and distinguish nodulating and nonnodulating strains of *Rhizobium* [73,78,81,83,94,97] but do not agglutinate erythrocytes. Plants may contain several lectins, each with different carbohydrate specificities [99,100]. Furthermore, sugar-inhibited controls were not included in some studies to rule out nonspecific binding of lectin.

Table 8.2 Adsorption of *Rhizobium* spp. to *Trifolium repens* root hairs[a,b]

Species	Number of strains tested	Infective on *T. repens*	Adsorbed cells per root hair (means ± SD)[c]
Rhizobium trifolii	5	+	23.48 ± 2.03
Rhizobium trifolii	5	—	5.36 ± 2.25
Rhizobium meliloti	3	—	2.63 ± 0.24
Rhizobium japonicum	3	—	2.20 ± 0.71
Rhizobium leguminosarum	1	—	1.90 ± 0.85
Rhizobium phaseoli	4	—	3.59 ± 0.79
Rhizobium lupini	1	—	3.80 ± 0.28
Rhizobium sp.	1	—	2.45 ± 0.64
Rhizobium trifolii[d]	1	—	3.50 ± 1.05

[a] From Dazzo et al. [70] and courtesy of the American Society for Microbiology.
[b] Total 15–20 root hairs (ca. 200 μm in length) examined per strain.
[c] Standard deviation.
[d] Heat-killed infective *R. trifolii* 0403 cells (80°C for 10 min).

Dazzo and colleagues [21,70,76,77,80,88,89, 96–98] have tested the lectin-recognition hypothesis in the *Rhizobium trifolii*–clover symbiosis. Preliminary studies showing antigenic differences between infective and noninfective strains of *R. trifolii* [76] led to the demonstration that the surfaces of infective encapsulated *R. trifolii* and clover epidermal cells contain an immunochemically unique polysaccharide that is antigenically cross-reactive [77]. This unique cross-reactive antigen (CRA) contains receptors that bind to a clover lectin called *trifoliin*, which has recently been purified and characterized [88]. Electron microscopy shows that bacterial attachment sites consist of the fibrillar capsule of *R. trifolii* in physical contact with electron-dense globular aggregates lying on the outer periphery of the fibrillar clover root-hair cell wall [77] (Fig. 8.14). Dazzo and Hubbell [77] have proposed a model to explain this early recognition step occurring at the root-hair surface (Fig. 8.15). Trifoliin recognizes these unique surface receptors and cross-bridges them to form the correct molecular structure that allows for specific adsorption of the bacteria to the root-hair surface.

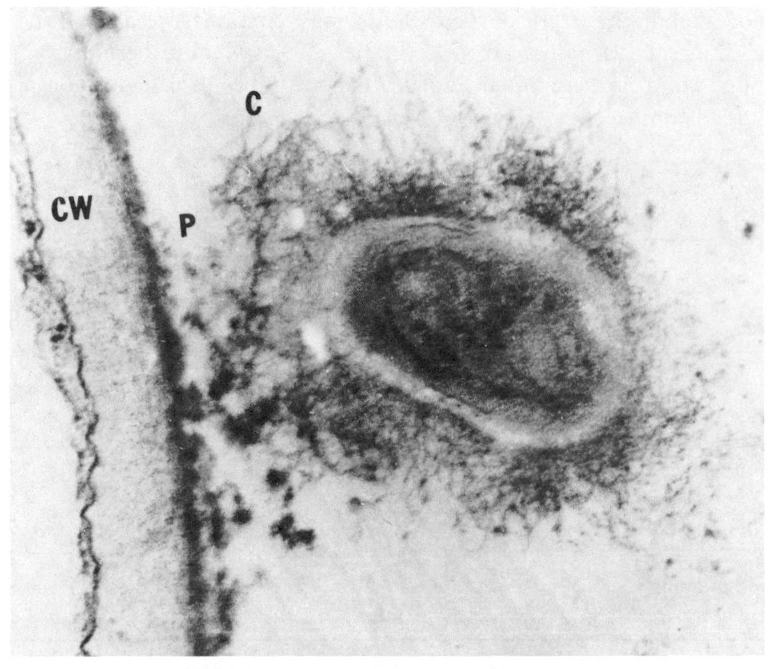

Figure 8.14 Transmission electron micrograph of infective *Rhizobium trifolii* NA30 in association with a strawberry clover root hair. Fibrillar capsule (C) of the bacterium is in contact with globular particles (P) on the outer periphery of the root-hair cell wall (CW). Courtesy of C. A. Napoli and the American Society for Microbiology [77].

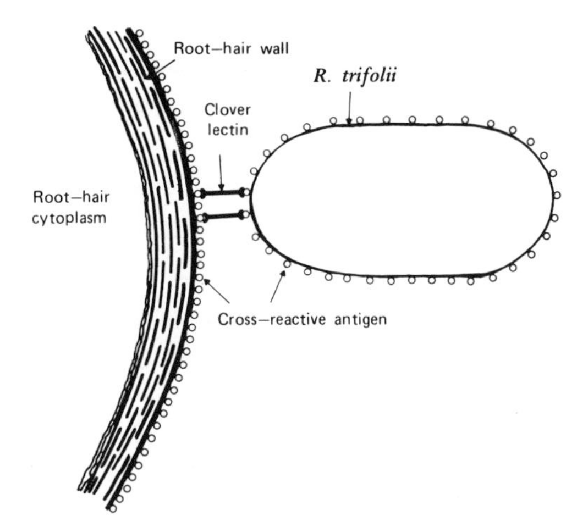

Figure 8.15 Schematic diagram showing proposed cross-bridging of cross-reactive antigen receptors of *Rhizobium trifolii* and clover root hairs with clover lectin trifoliin. Courtesy of the American Society for Microbiology [77].

This cross-bridging model is a modification of an earlier hypothesis by Hamblin and Kent [73] and Bohlool and Schmidt [74], who proposed the role of lectin in adsorption and specific recognition of rhizobia on the root surface. However, it proposes two other necessary elements of the specificity-determining complex: the unique CRA on the rhizobia and the plant cell, which bind the multivalent lectin. There is much experimental evidence to support this model. Only rhizobia that infect clover have the surface CRA [77,89,96] and the trifoliin receptor [88]. Clover root hairs preferentially adsorb infective *R. trifolii* [70] and its CRA [80]. The receptor sites that bind *R. trifolii* and its CRA match the distribution of immunologically detectable trifoliin on clover root surfaces [88]. These receptor sites accumulate at root-hair tips and diminish toward the base of the root hair [80,88]. By contrast, the CRA is uniformly distributed on both root hairs and undifferentiated epidermal cell walls in the root-hair region of the clover seedlings [89]. Trifoliin is multivalent in binding to and specifically agglutinating *R. trifolii* [77,80,88]. The sugar 2-deoxyglucose specifically inhibits agglutination of *R. trifolii* by trifoliin and the anti-clover root CRA [77,80,96], the specific binding of *R. trifolii* or its capsular polysaccharide to clover root hairs [70,80] (Table 8.3), and specifically elutes trifoliin from intact clover roots or trifoliin-coated *R. trifolii* [80,88] (Table 8.4). As a negative control, 2-deoxyglucose does not inhibit adsorption of *R. meliloti* or its capsular polysaccarides to alfalfa root hairs [70,80] (Table 8.3). This *Rhizobium*–host combination constitutes a different cross-inoculation group [101]. Rhizobia precoated with

Table 8.3 Inhibition of adsorption of *Rhizobium* spp. to legume root hairs[a]

Strain	Treatment[b]	Adsorbed bacteria (means ± SD)
Rhizobium trifolii strain 0403–*T. repens*	None	24.70 ± 1.80
	2-Deoxyglucose	2.60[c] ± 1.06
	2-Deoxygalactose	23.55 ± 2.34
	α-D-Glucose	24.00 ± 2.75
Rhizobium meliloti strain F28–*M. sativa*	None	35.50 ± 4.11
	2-Deoxyglucose	34.25 ± 4.79

[a] From Dazzo et al. [70] and courtesy of the American Society for Microbiology.
[b] Final sugar concentration was 30m*M*.
[c] Significantly less than untreated control at 99% level.

Table 8.4 Agglutination of *Rhizobium* spp. by protein eluted from intact clover roots[a]

Bacterium	Additions to root-washing solution[b]	Specific activity (agglutination units/mg of protein)
Rhizobium trifolii	2-Deoxyglucose	1240
Rhizobium trifolii	α-D-Glucose	42
Rhizobium trifolii	—	59
Rhizobium trifolii	2-Deoxyglucose, protease[c]	<5
Rhizobium meliloti	2-Deoxyglucose	<5
Rhizobium japonicum	2-Deoxyglucose	<5

[a] From Dazzo and Brill [80] and courtesy of the American Society for Microbiology.
[b] Concentration of sugar was 30m*M* in 0.01*M* phosphate buffered saline (pH 7.2); sugar was removed by dialysis prior to assay.
[c] Root washings were incubated with 2-deoxyglucose or protease (1 mg/ml) 4 hr prior to assay.

their corresponding plant lectin adsorb in greater numbers to host-specific roots [70,75]. Only clover roots have the *R. trifolii*-specific CRA on their surface [89].

Competition experiments show that trifoliin and anti-clover root CRA bind to the same or similar overlapping determinants on the surface of encapsulated *R. trifolii*, and exposure of this polysaccharide antigenic determinant is necessary for attachment of the bacteria to clover root hairs [89]. This was shown by three experiments, two of which utilized anti-CRA Fab monovalent fragments. An IgG antibody molecule has two protein Fab regions that constitute the antigen-combining sites and that can be purified after papain digestion of IgG. Because of their monovalency, Fab fragments bind immunospecifically to cell-surface antigens but cannot cross-bridge antigens on neighboring cells to agglutinate them. Thus Fab fragments of IgG anti-clover root CRA are specific reagents that bind to and mask the CRA on the *R. trifolii* surface. In the first experiment, these anti-clover root CRA Fab fragments specifically blocked the agglutination of *R. trifolii* by purified trifoliin. In the second experiment the anti-CRA Fab fragments specifically blocked the ability of *R. trifolii* to bind to clover root hairs. In both of these experiments Fab fragments purified from normal rabbit preimmune IgG were without effect, demonstrating that the blocking observed with immune anti-CRA was very specific. In the third experiment, intergeneric transformation was conducted with *Azotobacter vinelandii* nitrogenase minus (*nif*$^-$) mutants and *R. trifolii* donor deoxyribonucleic acid (DNA) [98]. The genetic hybrids were selected on nitrogen-free medium for correction of *nif*. The transfer of *nif* was 100-fold above the spontaneous reversion rate. The *nif*$^+$ hybrids were examined for other *Rhizobium*-specific traits, and 13% of them expressed the clover root CRA on their cell surface. Only the same hybrids expressed the trifoliin receptor, indicating that the genetic cotransformation frequency of the genes that control the synthesis of the trifoliin receptor and the CRA on *R. trifolii* is 100%, suggesting that they are indeed the same determinants [98]. Only the *A. vinelandii–R. trifolii* hybrids that carry the trifoliin receptor bind to clover root hairs [89], and this specific binding is inhibited by 2-deoxyglucose.

Trifoliin is a carbohydrate-containing protein with subunit molecular weight ca. 50,000 daltons, and aggregates of trifoliin at pH 7 are 10 nm in diameter [88]. It is interesting to note the same dimensions of soybean lectin aggregates [84] and the presence of electron-dense aggregates [77] (Fig. 8.14) and subcellular particles (see Fig. 5C in Dart [52]) on clover root-hair walls at *R. trifolii* adsorption sites.

Solheim has proposed a similar model (Fig. 8.16) of rhizobial adsorption. His model is the product of studies on the curling factor [102] (see section 8.7.1) and the effect of *Vicia* lectin on rhizobial attachment to *Vicia* roots

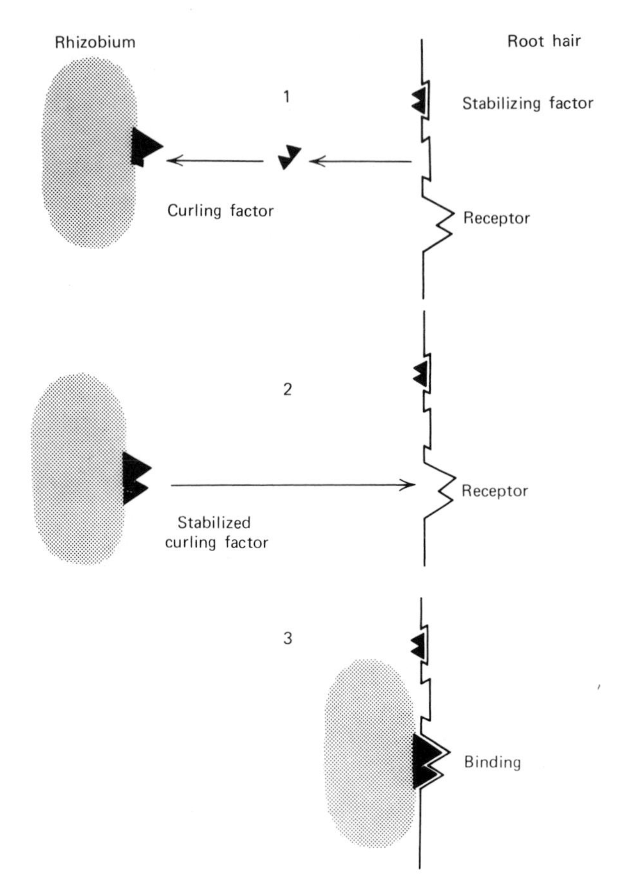

Figure 8.16 Model of host recognition by *Rhizobium* on legume roots. Courtesy of B. Solheim [75].

[75] (Figs. 8.17*a,b*). His model indicates that the root-hair wall has a "stabilizing factor" (presumably a lectin), which is secreted from the root and binds to a curling factor on the surface of *Rhizobium*. The bacteria containing the plant "stabilizing factor" then acquire a strong affinity for receptor sites on the root. The difference between Solheim's model and that of Dazzo and Hubbell [77] is based on whether the lectin is released or remains attached to the roots. Detection of trifoliin in root washings with isotonic buffer [80] (Table 8.4) and agglutination of homologous rhizobial cells in hydroponic culture suggests that some lectin may indeed be released. Both models may be operative, the second one suggesting a salvage mechanism for the eluted multivalent lectin. Preparations of capsular polysaccharides

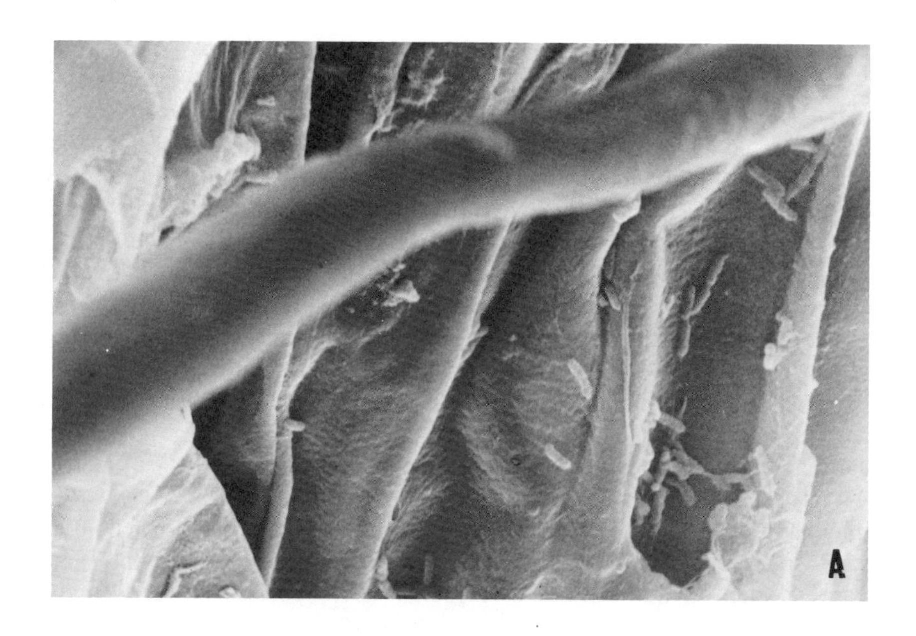

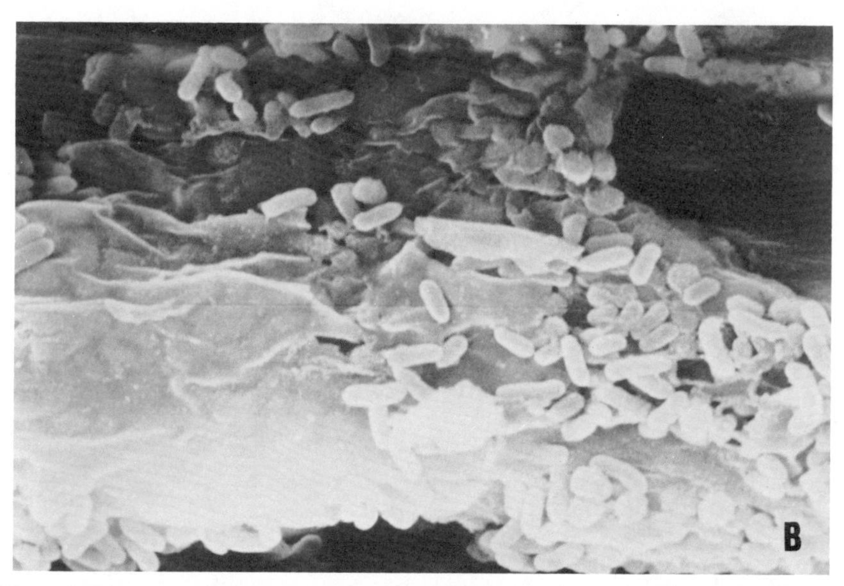

Figure 8.17 Scanning electron micrograph of *Vicia* rhizoplane after 4-hr incubation period with: (A) *Rhizobium leguminosarum* (infective on *Vicia*) or (B) *R. leguminosarum* coated with the *Vicia* lectin. Courtesy of B. Solheim [75].

have curling activity [77] (also Dazzo and Brill, unpublished observation) and lectin-coated rhizobia have a mechanism for specifically recognizing binding sites on roots of their homologous legume host [70,75,102].

A major controversy is the nature of the carbohydrate receptor for the lectin on the *Rhizobium* cell. The lectin receptors identified as capsular polysaccharide [77], lipopolysaccharide [78], or a glycan [87] are all surface acidic heteropolysaccharides [77,87,89,103]. Electron-microscopic examination of *R. trifolii* revealed a very prominent capsule that stains positive for acidic polysaccharide with ruthenium red [89] (Fig. 8.18). Fluorescence microscopic techniques indicate greater reactivity of anti-CRA and purified trifoliin with encapsulated as compared with unencapsulated *R. trifolii* cells separated by differential centrifugation [77,80,88,89]. The same observation has been made with *R. japonicum* and soybean lectin [85,104]. A purified acidic heteropolysaccharide from capsular preparations of *R. trifolii* [89] interacts specifically with purified trifoliin and anti-clover root CRA, although it lacks heptose, 2-keto, 3-deoxyoctonic acid, and endotoxic lipid A, which are chemical markers present in lipopolysaccharide from the same strain. Under certain conditions capsular acidic polysaccharides can be

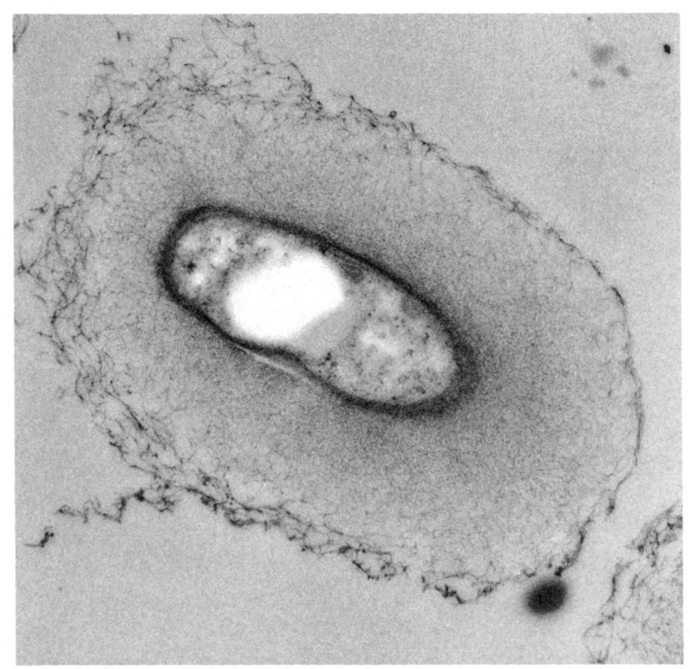

Figure 8.18 Transmission electron micrograph of encapsulated *Rhizobium trifolii* 0403 stained with ruthenium red to show the acidic polysaccharide.

transferred to the core-lipid A and replace the neutral O-antigen polysaccharide [89].

Nodulating *Rhizobium* strains that produce small, nongummy colonies, nevertheless, have a "microcapsule" on their cell surface that stains positive for acidic polysaccharide with ruthenium red [89,104,105]. A low-molecular-weight "exopolysaccharide" from *Rhizobium* does not interact with legume lectins [78,93]. Maier and Brill [105] showed that the antigenic difference between nodulating and nonnodulating strains of *R. japonicum* lies in the immunochemistry of their surface acidic heteropolysaccharide. Saunders et al. [106] showed that mutant strains of *R. leguminosarum* failed to nodulate the pea host, and they proposed that these strains were defective in exopolysaccharide synthesis. An equally plausible explanation is that these mutant strains have a defective O-antigen polymerase that functions in LPS biosynthesis, since their LPS has less than half the carbohydrate content of the LPS from the wild-type strain (106). Only careful structural and immunochemical analyses will reveal the identity of the lectin receptors on both the *Rhizobium* and the legume host.

The data in Tables 8.2–8.4 [70,80] demonstrate that there are multiple mechanisms of rhizobial adsorption to clover root hairs. There exists a nonspecific mechanism that allows all rhizobia (dead or alive) to attach at low background levels (2–5 cells/200 μm of root hair). In addition, there is a specific mechanism of adsorption for infective *R. trifolii* cells (22–27 cells/200 μm of root hair). The differences were evaluated statistically after the square-root transformation of the means and found to be significant at the 99.5% level [70]. The data on kinetics of bacterial adsorption to pea roots [94] (Fig. 8.19) also suggests two mechanisms, the first one very rapid and the second one slower. However, it should be emphasized that these and other similar data [24] are generated from studies that count unadsorbed bacteria that are not removed by washing as well as bacteria adsorbed onto undifferentiated epidermal cells. Furthermore, Kotarski and Savage [108] have shown that use of radiolabeled cells for adsorption studies is complicated by the narrow range of linearity in relation of counts per minute versus cell number and the transfer of radiolabeled metabolites from the bacterium to the host tissue.

The Ca^{++}-dependent specific attachment of *Pseudomonas tolaasii* to barley roots [53,54] may possibly be mediated by root lectins, which require divalent cations for activity. By contrast, the nonspecific mechanism is insensitive to pH and Ca^{++} and may represent the mechanism for background attachment of noninfective *R. trifolii* and heterologous rhizobia, as well as heat-killed infective *R. trifolii* on clover root hairs [70]. Multiple mechanisms of cell adsorption to legume roots are also suggested by their different sensitivities to dissociation by high salt concentrations [107].

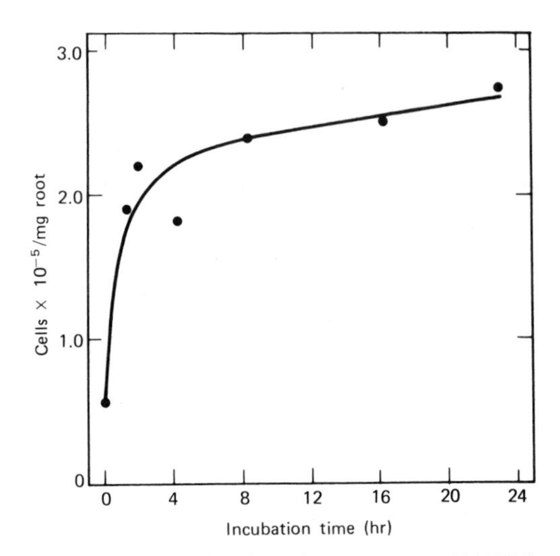

Figure 8.19 Binding kinetics of ^{32}P-*Rhizobium leguminosarum* 129C53 by pea root segments. Courtesy of A. P. Chen and D. H. Phillips and the Scandanavian Society for Plant Physiology [94].

A third mechanism that functions in adsorption of rhizobia to clover root hairs is suggested since many bacterial cells cannot be desorbed after 12-hr incubation. Many *Rhizobium* species, including noninfective mutants, produce extracellular cellulose microfibrils in culture [21,109]. It has been proposed that cellulose microfibrils may serve to anchor the bacteria firmly to the root surface once recognition has been established [21]. This maintenance of an intimate physical contact between the bacterium and the root hair could permit the localized biochemical interactions to trigger successful infection [21]. Cellulose microfibrils clearly cannot account for specific recognition of rhizobia by legumes [21]. Napoli [110] reported that fibrillar appendages appear on *R. trifolii* growing in aseptically collected clover root exudate. Similar appendages have been observed on *R. meliloti* or *R. trifolii* adsorbed to barrel medic [32] or clover root hairs [52] (also Dazzo, in press), respectively, and on unidentified bacteria in soil [111].

Presumably, adsorption of infective cells to target root hairs is required for infection but is insufficient by itself. Heterologous rhizobia can *attach* in low numbers to root hairs but do not infect [70]. Very few root hairs containing adsorbed infective rhizobia eventually become infected. Genetic hybrids of *A. vinelandii* transformed with *R. trifolii* DNA attached to clover root hairs but do not infect [89].

The signals that trigger infection once infective bacteria attach to root hairs are undefined. Only small numbers of *R. japonicum* and *R. leguminosarum* cells adsorb to roots of nonnodulating soybean and pea-plant varieties, respectively, as compared to roots of their normal nodulating host varieties [112,113]. The host genetic block possibly may create defective recognition events on the root surfaces. This hypothesis remains to be tested.

Equally important to understanding the recognition process is the basis for differentiation of root epidermal cells into root hairs. Binding studies [70,80,88,94] clearly show the importance of root hairs containing receptor sites that bind rhizobia and their capsular polysaccharides. This selective-binding effect can be observed very early in root hair differentiation and development [80]. Studies with tissue culture may help to explain the mechanism of root-hair differentiation and how it relates to rhizobial infection [114,115]. In culture, soybean root cells selectively attach homologous *R. japonicum* [114], and attachment can be inhibited by pectinase. *Rhizobium japonicum* contains galacturonic acid in its extracellular polysaccharides [116], and the inhibition by polygalacturonase possibly may be exerted at the bacterial, rather than the plant surface (compare with Section 8.6.3). Studies with suspension-culture techniques [115] may provide a useful approach to study details of root-hair biogenesis and the infection process in soybean. Care must be exercised to distinguish infection threads from plasmolyzed root hairs.

Rod-shaped *Rhizobium* cells adsorb end on to legume root surfaces grown in hydroponic culture [21,32,52,62,64,67,69,70,77,110,114,115, 117–119] (Figs. 8.10, 8.13, and 8.14). The physiochemical mechanism for this striking adsorption habit is unknown. The polar orientation is not unique to the rhizoplane but is also characteristic of microbial attachment to other biological and nonbiological surfaces.

The genetics and biochemistry of fimbriae on soil bacteria and their role in star formation and bacterial conjugation have been studied [120–128]. These fimbriae are inserted at one pole of the bacterial cell (Fig. 8.20) and are involved in the initial development of star formation in rhizobia. Fimbriae can contract [122,126] and consist of a low-molecular-weight protein [127]. Some bacteria adsorb by their fimbriated pole to red blood cells [122]. The hemadsorption is inhibited by mannose [122], suggesting that the proteinaceous fimbriae terminate with lectins that bind to mannose-containing determinants on the erythrocyte membranes. The involvement of fimbriae in the adsorption of bacteria to root surfaces remains to be examined.

Some *R. japonicum* strains growing exponentially have cell-surface polarity [90,91], as illustrated by star formation (Fig. 8.21) and binding of antibody or soybean lectin. Electron microscopy shows accumulation of reserve polymer and exopolysaccharide at opposite cell poles [91]. Bohlool and Schmidt [90] speculated that the polar attachment of rhizobia to root surfaces may occur on the legume rhizoplane as a prelude to nodulation.

The polar antigens common to many rhizobia [90] could reflect Braun lipoproteins of the outer envelope, which are exposed at poles just after cell division and separation [129,130]. These lipoproteins are convalently attached to peptidoglycan through peptide linkage and extend through the periplasmic space and associate with the outer envelope membrane by noncovalent, nonpolar interactions. Ballou [131] has argued that the polar binding of soybean lectin to *R. japonicum* may reflect only an enhanced exposure of surface receptors owing to a disorganization of wall structure in the area of the rapidly growing cell extension. Whether these polar receptors of rhizobia occur on the rhizoplane remains to be determined. The low

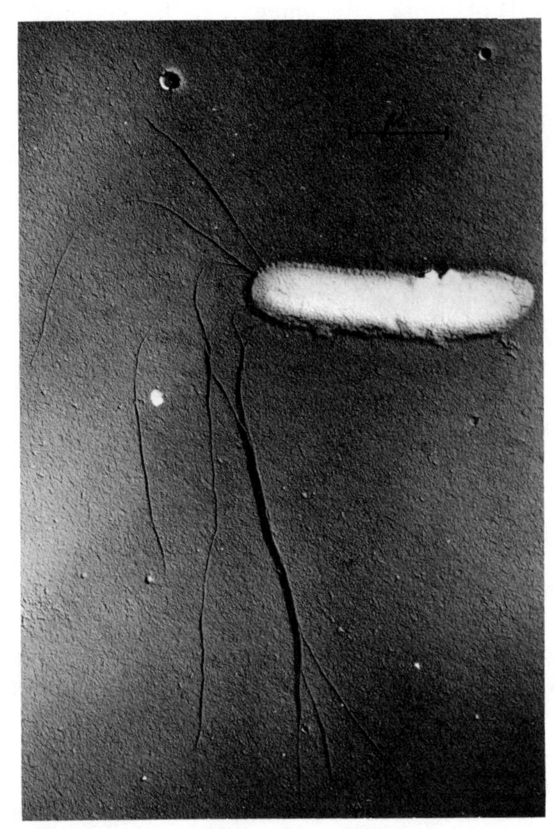

Figure 8.20 Polar insertion of fimbriae into the soil bacterium *Pseudomonas echinoides*. Courtesy of W. Heumann and R. Marx and Springer-Verlag, New York [122].

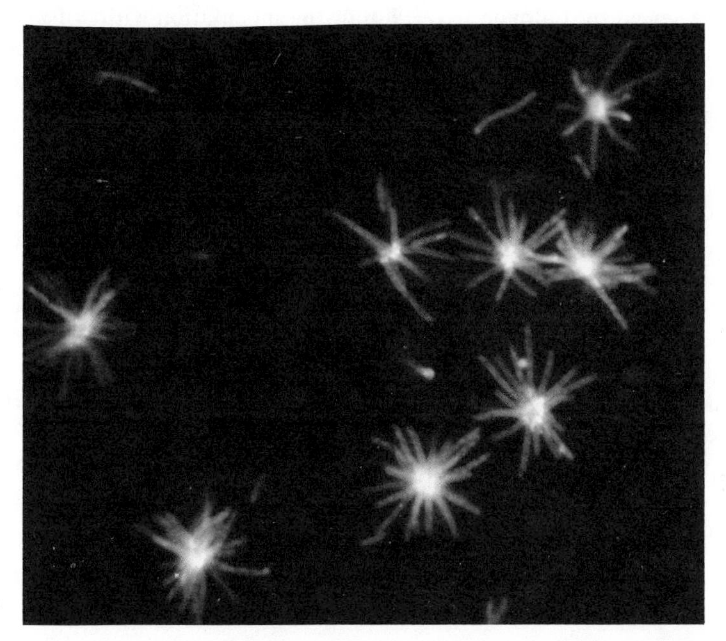

Figure 8.21 Starlike clusters of *Rhizobium japonicum* USDA31 on contact slide in sterilized soil. Courtesy of B. Bohlool and E. L. Schmidt, and the American Society for Microbiology [90].

background adsorption of heterologous rhizobia to legume root hairs is polar [70,117], and thus the role of polar rhizobial attachment in specific recognition and in the infection process is open to question [90,91,117,119].

Marshall et al. [119,132] have examined cell-surface hydrophobicity and polar orientation of microorganisms to surfaces. They propose that polar orientation at interfaces results from a relatively hydrophobic cell pole that is rejected from the aqueous environment. The requirement for attachment of sulfur granules to microorganisms that degrade them is accomplished by interactions of their common hydrophobic surfaces [133–135]. One must not assume that polar attachment of microorganisms to biological and nonbiological surfaces is mediated by the same mechanism. This point is emphasized by the specific polar attachment of rhizobial bacteriophage tails to the host cell where the bacterial pili (phage receptors) are inserted [136].

8.5.3 *Frankia*–Nonleguminous Angiosperm Association

The adsorption of the actinomycete-like endophyte, *Frankia* sp., to root hairs of its host *Alnus* and subsequent events in the infection process have been reviewed by Lalonde [137]. Following inoculation with crushed nodule inoculum, the endophyte proliferates in the rhizosphere. The organism adsorbs to root hairs, forming a polar thread (Figs. 8.22 and 8.23). Endophytic bacterial cells are enveloped with membrane-bound "blebs" containing electron-dense materials (Fig. 8.24). Blebs are observed only on inoculated plants. The deposition of blebs forms an "*exo*-encapsulation thread" containing endophyte cells arranged end to end on a deformed root hair. The endophyte penetrates the root-hair wall and assumes the more customary filamentous–hyphal morphology. The dictyosome is involved in deposition of the endophyte's capsule. The encapsulated endophyte grows to the base of the root-hair cell wall, perforates the wall, branches to invade neighboring cortical cells of the roots, and finally produces septate vesicles.

Callaham et al. [138] have isolated in pure culture the nitrogen-fixing actinomycete endophyte from *Comptonia* and have fulfilled Koch's postulates with it. This discovery opens new avenues to the study of recognition phenomena at the root surface and the regulation of nitrogen fixation by this root–nodule symbiosis.

8.5.4 *Azospirillum* (*Spirillum*)–Grass Association

Dobereiner and Day [139] found an associative symbiosis between *Spirillum lipoferum* (now *Azospirillum*) [140] and roots of various grasses, which they considered as a primitive nitrogen-fixing symbiosis. Bacteria on the roots

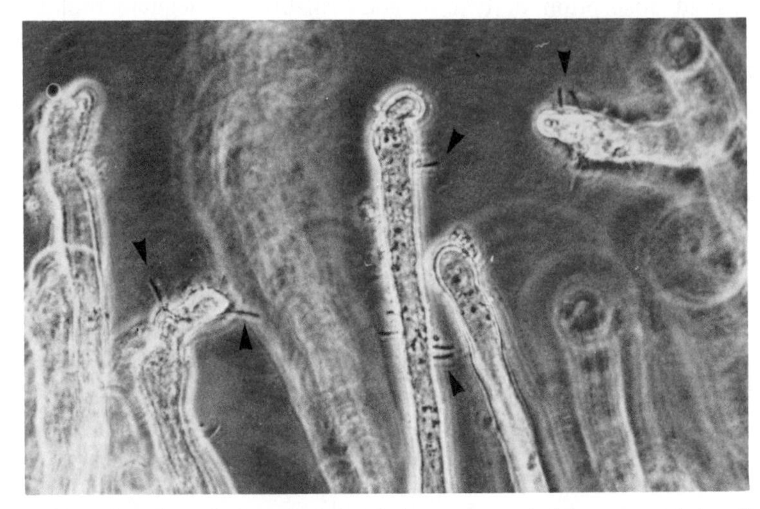

Figure 8.22 Adsorption of nitrogen fixing actinomycete endophyte to root hairs of *Alnus glutinosa*. Courtesy of M. Lalonde and Academic Press, New York [137].

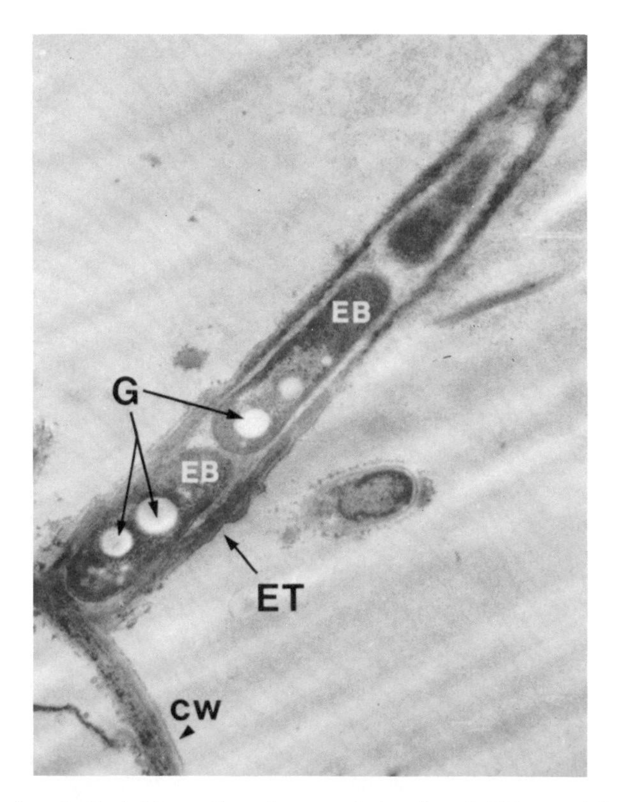

Figure 8.23 Longitudinal thin section of encapsulation thread (ET) containing endophytic bacterial cells (EB) aligned end-to-end and adsorbed in the cell wall (CW) of the deformed root hair of host. Cytoplasmic translucent granules (G) are seen. Courtesy of M. Lalonde and Academic Press, New York [137].

were examined microscopically after their reduction of tetrazolium salts to the red formazan chromophore [139]. These areas were interpreted as active sites of nitrogen fixation [139], but subsequent investigations showed that tetrazolium reduction was not specific for nitrogen-fixing organisms [141].

Umali-Garcia et al. [142,143] have investigated infection of guinea grass and pearl millet roots by *Azospirillum brasilense* using the Fahraeus slide technique with nitrogen-free medium. Young inoculated roots of pearl millet and guinea grass had more mucigel, root hairs, and lateral roots than did uninoculated controls. The bacteria were found within mucigel accumulating on the root cap and along the root axis. *Azospirillum brasilense* also adsorbed firmly to root hairs and undifferentiated epidermal cells. Some of the bacteria were oriented in a polar fashion on root-hair cell walls. Supplementing the medium with fixed nitrogen ions suppressed the firm adsorption

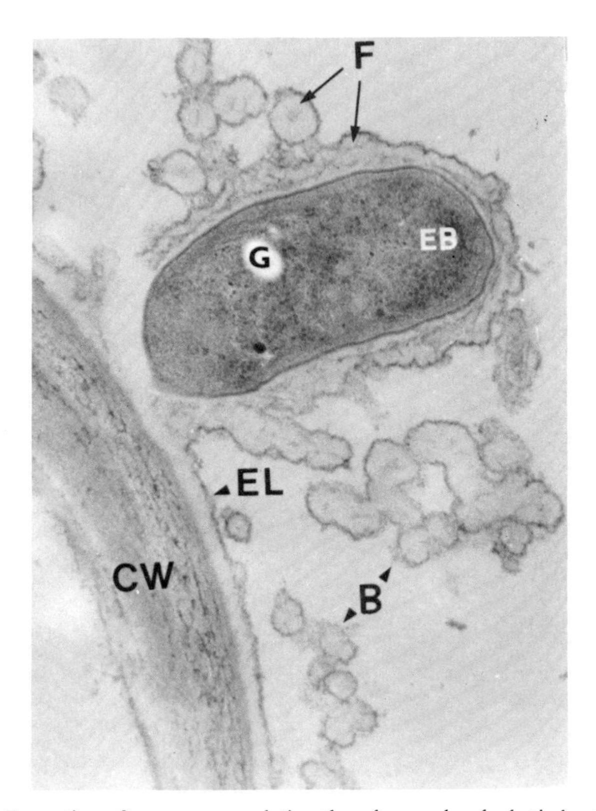

Figure 8.24 Formation of an exoencapsulation thread around endophytic bacterial cell (EB). Note the continuity of external layer (EL) covering root-hair cell wall (CW), bordering the developing encapsulation thread. Membrane-bound "blebs" (B) fuse with thread-bordering layer and discharge fibrillar material (F) on the endophytic bacterial cell. Note the translucent granule (G) in bacterial cytoplasm. Courtesy of M. Lalonde and Academic Press, New York [137].

of the bacteria to the roots. Ultrastructural examination of the pearl-millet rhizosphere revealed *Azospirillum* cells enclosed within a common slime layer that had a high affinity for electron-dense materials normally associated with the plant cell walls. Preliminary studies revealed that root exudate from pearl millet contained substances that bound to *Azospirillum* and promoted their firm adsorption to the root epidermal cells of this host. These substances were nondialyzable and were inactivated by protease. *Azospirillum brasilense* entered root tissue through void spaces created by profuse epithelial desquamation and lateral root emergence. The bacteria invaded the middle lamella and showed signs of hydrolysis that followed the contour of the bacterial cell walls. However, adjacent cortical cells retained active cytoplasm, and the bacteria were not found within living host cells. Pectolytic activity (pectin lyase and endopolygalacturonase) was detected in pure cultures of *A. brasilense* induced with purified pectin. Cortex invasion by the bacteria has been confirmed with the tetrazolium reduction technique [141]. These studies show that *A. brasilense* is invasive on these grass roots and that there is intimate but limited colonization of the middle lamella.

8.5.5 Mycorrhizal Associations

Ectotrophic and *endotrophic* mycorrhizal associations are distinguished by their rhizoplane colonizing habit. The infection process of ectotrophic mycorrhizas begins with a weft of fungal hyphae on the root surface prior to intercellular colonization [144]. Dispersal of the fungus along the root is accomplished by rhizomorph growth. The distinctive features of ectotrophic mycorrhizas are a dense collection of hyphae encasing the root surface (the mantle) and fungal colonization of the intercellular spaces in the cortex, which resemble a net in cross sections (the Hartig net). Diatoms, fungi, bacteria, and actinomycetes interact in close association with the mantle. This microbial environment created by the symbiotic fungus on the root has been termed the mycorrhizosphere [145].

Endotrophic mycorrhizal fungi infect primarily root hairs of lateral roots [146]. Fungal penetration begins about 0.5 mm behind the lateral root cap and extends back into older parts of the root system. Enzymes and mechanical pressure presumably are involved in the penetration process [147]. A papilla forms opposite to the appressorial pressure point. The host cell reacts to invasion with an accumulation of dictyosome, mitochondria, and endoplasmic reticulum cisternae in the vicinity of the parasite [148]. The intracellular fungal hyphae have an encasement layer that becomes even more electron dense as the hyphae disintegrate. The layer is identical to and continuous with host wall material [2]. Intercellular hyphae have no encasement layer.

Many authors have documented the suppression of mycorrhizal infection under high soil phosphorous regimes [149–152], but the absolute biochemical mechanism of this suppression is undefined. Mosse suggests that reduction of mycorrhizal infection may be attributed to phosphate toxicity [149]. Saunders used foliar-applied phosphate to eliminate fungal toxicity in the soil and found that mycorrhizal infection was still reduced [151]. Marx et al. [152] found that high levels of nitrogen and phosphorus in soil decrease sucrose content of short roots of loblolly pine and decrease their susceptibility to ectomycorrhizal development by *Pisolithus tinctorius*.

8.5.6 Root-infecting Plant Pathogens

A distinguishing feature of the microorganisms considered in this section is whether they induce sufficient physiological disfunction in the plant host following invasion so that symptoms of disease are expressed.

Light and fluorescence microscopy have been used to study fungal attachment sites and points of root infection [153–158] (Fig. 8.25). At the ultrastructural level, the appressorium and haustorium are dominant structures of the "interface" [2]. The appressorium is a hyphal structure that attaches the germ tube to the host specifically during early stages of infection. The haustorium is a distinctive fungal branch formed inside living host

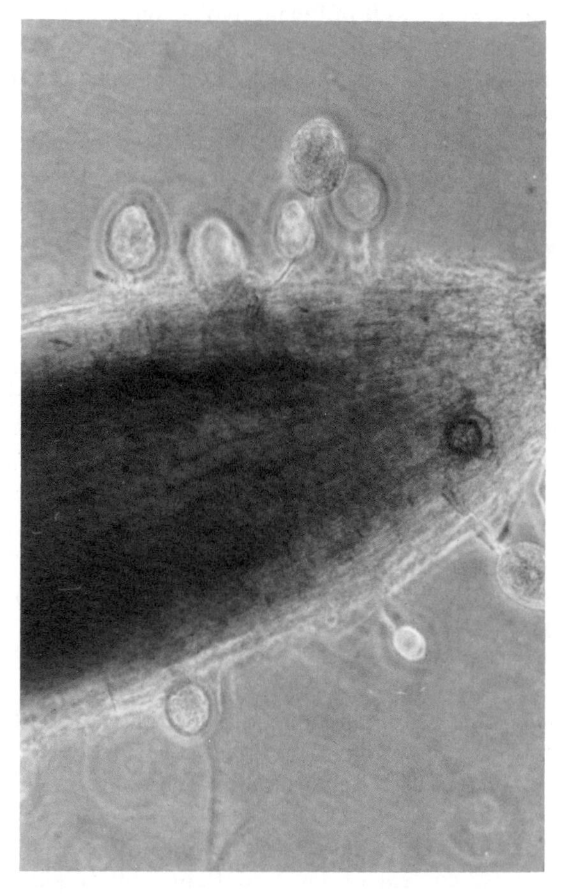

Figure 8.25 *Phytophthora megasperma* var. *sojae* sporangia adsorbed to susceptible soybean seedling root. Courtesy of J. B. Sinclair and the American Phytopathological Society [157].

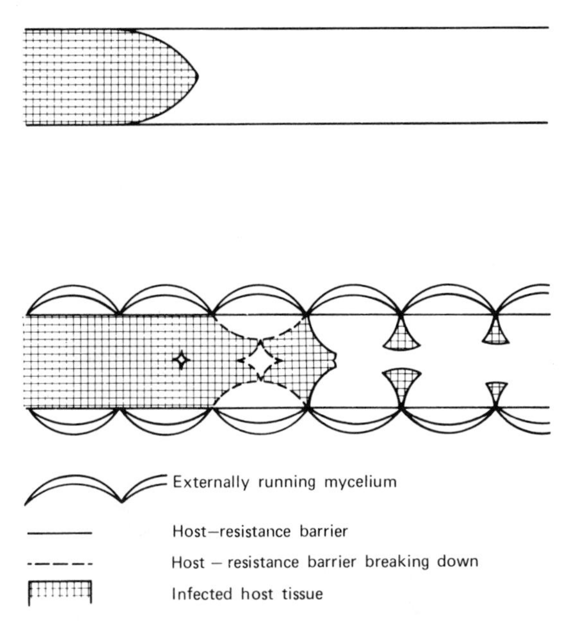

Figure 8.26 Diagram illustrating a possible mechanism whereby ectotrophic infecting habit synergistically enables pathogenic root-infecting fungus to overcome host resistance. Courtesy of S. D. Garrett and Cambridge University Press [159].

cells and probably plays a role in the interchange of substances between host and pathogen. Generalized infection by obligate parasites is similar to endomycorrhizal fungi and proceeds by the following sequence of events. The fungal wall contacts the host cell wall, and then an amorphous layer is deposited between them. This material may aid in the adhesion of the pathogen to the host-cell wall. Fungal appressoria and host papillae form at the penetration site. A penetration peg erodes the host cell wall. Very frequently, host wall dissolution is limited to the immediate penetration area, so enzymes active in wall hydrolysis are probably limited to the surface of the penetration peg. Penetration is believed to result from a combination of enzymatic activity and physical force. A flurry of cytoplasmic activity by the host cell generally occurs at the site of penetration. The fungal wall is thinnest where it passes through the host cell wall. The invading intracellular pathogen is surrounded by an additional encapsulation or apposition layer that is peripheral to the fungal cell wall.

The ectotrophic-infection habit described by Garrett [159] applies to specialized root pathogens like *Gaumannomyces graminis*, which causes the take-all disease of cultivated cereals (Fig. 8.26). This fungus progresses over the host root system as a sparse network with hyaline branches penetrating

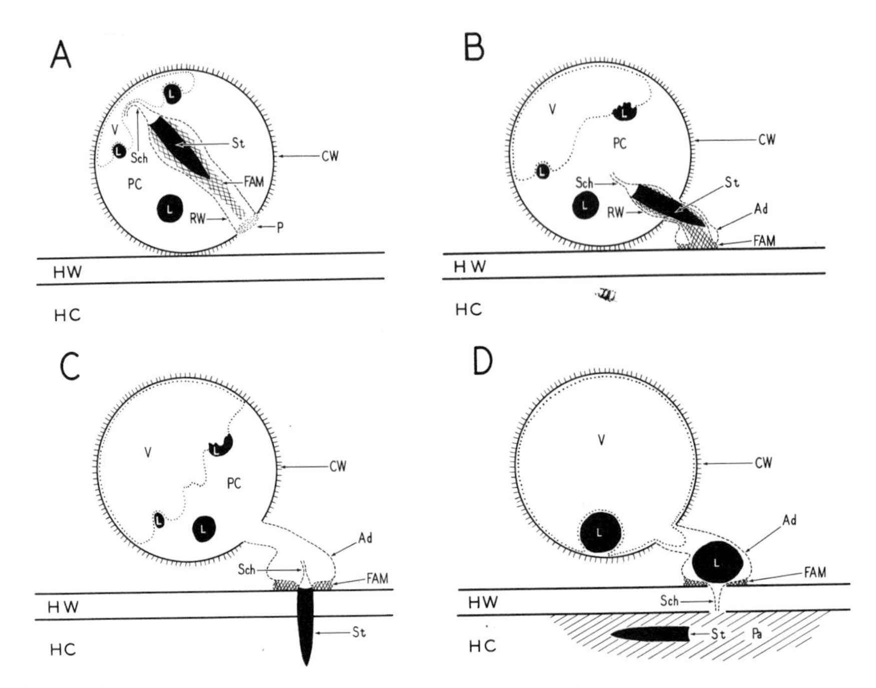

Figure 8.27 Diagrammatic summary of penetration process of *Plasmodiophora brassicae* on cabbage root hairs: (A) cyst vacuole before enlargement; (B) vacuole enlargement and appearance of adhesorium; (C) stachel punctures host cell wall; (D) penetration occurs and host protoplast deposits papilla at penetration site. Symbols: L, lipid body; V, vacuole; Sch, Schlanch; St, stachel; CW, zoospore cell wall; RW, Rohr wall; P, plug; FAM, fibrillar adhesorium material; Ad, adhesorium; HW, host wall; HC, host cytoplasm; Pa, protozoa. Courtesy of J. R. Aist and P. H. Williams and the Canadian Research Council [155].

the epidermis, infecting the underlying cortex and the vascular cylinder. Occasionally, the fungus abandons the ectotrophic habit on the distal part of the root and produces runner hyphae that intercellularly penetrate a few cells deep within the cortex. Garrett postulated that the ectotrophic-infection habit allowed the pathogen to initiate a series of rapid successive infections, which act synergistically along the length of the root. This habit was imposed on the specialized pathogen by the action defense response of roots in possession of their tissue resistance. The hypothesis was confirmed later in the field [160].

Aist and Williams [155] illustrated the adsorption and penetration of cabbage root hairs with *Plasmodiophora brassicae*, the clubroot pathogen (Figs. 8.27*a,b*). Zoospores collide with root hairs several times before they become quiescent and attach on the side opposite the inserted flagellum. The flagellum coils around the zoospore body, which becomes slightly flattened against the host cell wall. The flagellar axoneme retracts within 1 min of attachment. Within 2–3 min the parasite rounds up and encysts, with breakdown of flagella and development of Golgi body, rough endoplasmic reticulum, a large vacuole, inclusion bodies, and a large tubular cavity called the *Rohr*. The Rohr contains a dense-staining sharp-pointed rod called the *Stachel* and is occluded with a light-staining plug oriented toward the host cell wall. The Rohr erects and then swells at the point of contact with the host to form the adhesorium, which grows within 1 min to about one-third the diameter of the cyst. A thick layer of adhesive fibrillar material accumulates between the adhesorium and the host cell wall. The Stachel penetrates the host wall, and the parasitic amoeba is injected

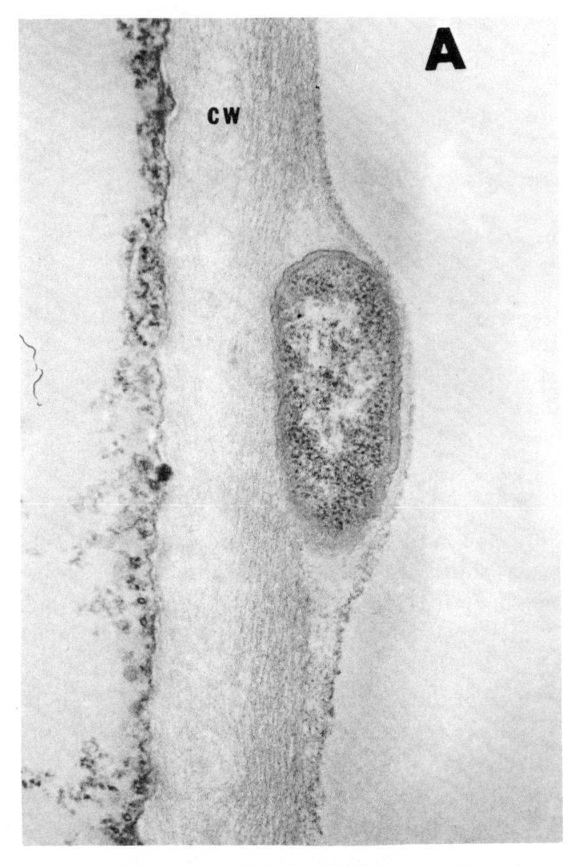

Figure 8.28 Bacterial cell with (A) capsule (ca) and (B) particles (arrow) presumed to be condensing phage embedded in the root epidermal cell wall (CW) of *Panicum virgatum*. Courtesy of R. F. Lewis and W. J. Crotty and the Botanical Society of America [161].

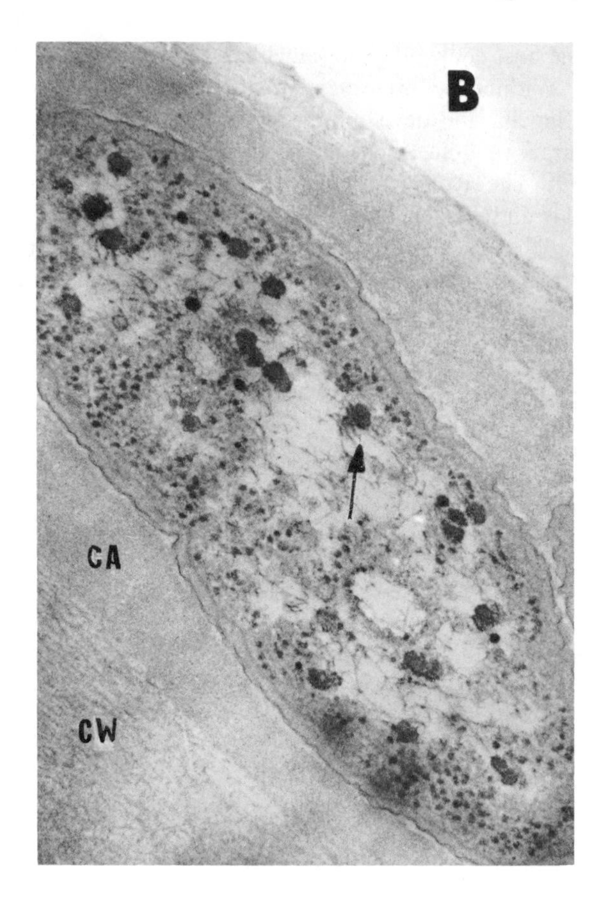

immediately into the host. The young ameoba remains attached to the adhesorium by a short strand for a few seconds before dislodgement. Nothing is known of the molecular components of the adhesorium–root-hair cell-wall "interface" or the biochemistry of infection (see addendum to this chapter).

In *Panicum viratum* the loss of cytoplasmic inclusions in basal elongating root epidermal cells is correlated with the acquisition of surface-associated bacteria (tentatively identified as *Erwinia herbicola*) and the appearance of dense-cored particles in the cytoplasms of the bacterium and plant host [161] (Figs. 8.28*a*,*b*). The bacteria were adsorbed to the surface and embedded in the walls of elongating epidermal cells but were never found within epidermal cytoplasm or in contact with host plasma membrane. The bacteria isolated from roots swell and lyse when incubated with homogenized root extracts, and particles resembling bacteriophage heads appear in the lysates. Lewis and Crotty [161] suggested that virulent bacteriophage lysed the bacteria and prevented their intracellular invasion of epidermal cells.

8.6 ADSORPTION OF MICROORGANISMS TO OTHER PLANT TISSUES

Advancing the understanding of surface interactions between microorganisms and plants is a prerequisite for developing methods to control plant diseases [162]. This need has provided the impetus for research discussed in this section.

8.6.1 The Phylloplane

The occurrence of microorganisms on aerial plant parts is well known. The leaf surface, known as the *phylloplane*, is an area of intensive microbiological investigation [163,164]. In the humid tropics the diurnal supply of dew and rain regularly leaches plant substances from the leaf, which develops an extensive phylloplane microbial community [165]. The high carbon:nitrogen ratio of leaf leachates selects for nitrogen-fixing bacteria, which may become important to productivity in the tropics where temperature, moisture, and light are not limiting [165]. *Cladosporium* species are some of the most abundant fungi in the earth's atmosphere, and thus they are very common inhabitants of the phyllosphere [166]. Under favorable conditions of extended wet periods and deposition of pollen during flowering, the phyllosphere mycoflora become very active [166]. Pathogenic fungi with complex life cycles may produce distinctive reproductive structures adsorbed to the phylloplane of specific plants. Examples of obligate parasites that display a high degree of host specificity after they adsorb to leaves are *Sclerospora graminicola* (downy mildew) and the telial stage of *Puccinia graminis* (rust) on wheat (Fig. 8.29*a*,*b*). The molecular basis for this biological specialization awaits discovery.

8.6.2 Mesophyll Cell Walls

The adsorption to and envelopment of microorganisms by mesophyll cell walls have recently received considerable attention [167–176]. After infiltration into leaves, only avirulent or nonpathogenic cells agglutinate, adsorb to the plant cell walls, and are enveloped by electron-dense vesicles and fibrillar materials (Fig. 8.30*a;* cf. Fig. 8.24). In contrast, virulent cells remain

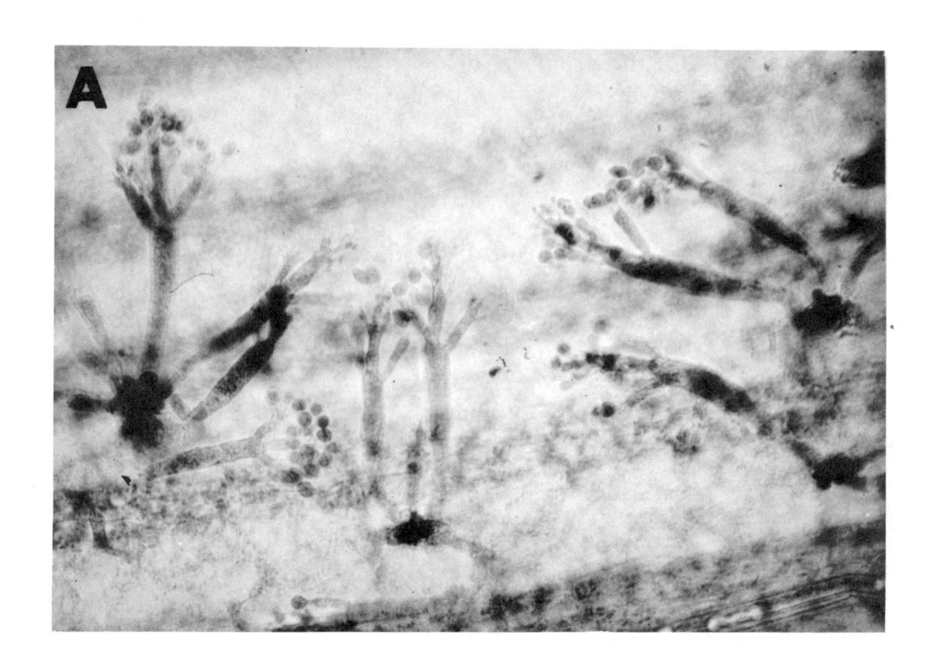

Figure 8.29 Adsorption of obligate parasitic fungi to phylloplanes of their plant hosts: (A) *Sclerospora graminicola* sporangiopores on millet; (B) *Puccinia graminis* teliospores on wheat.

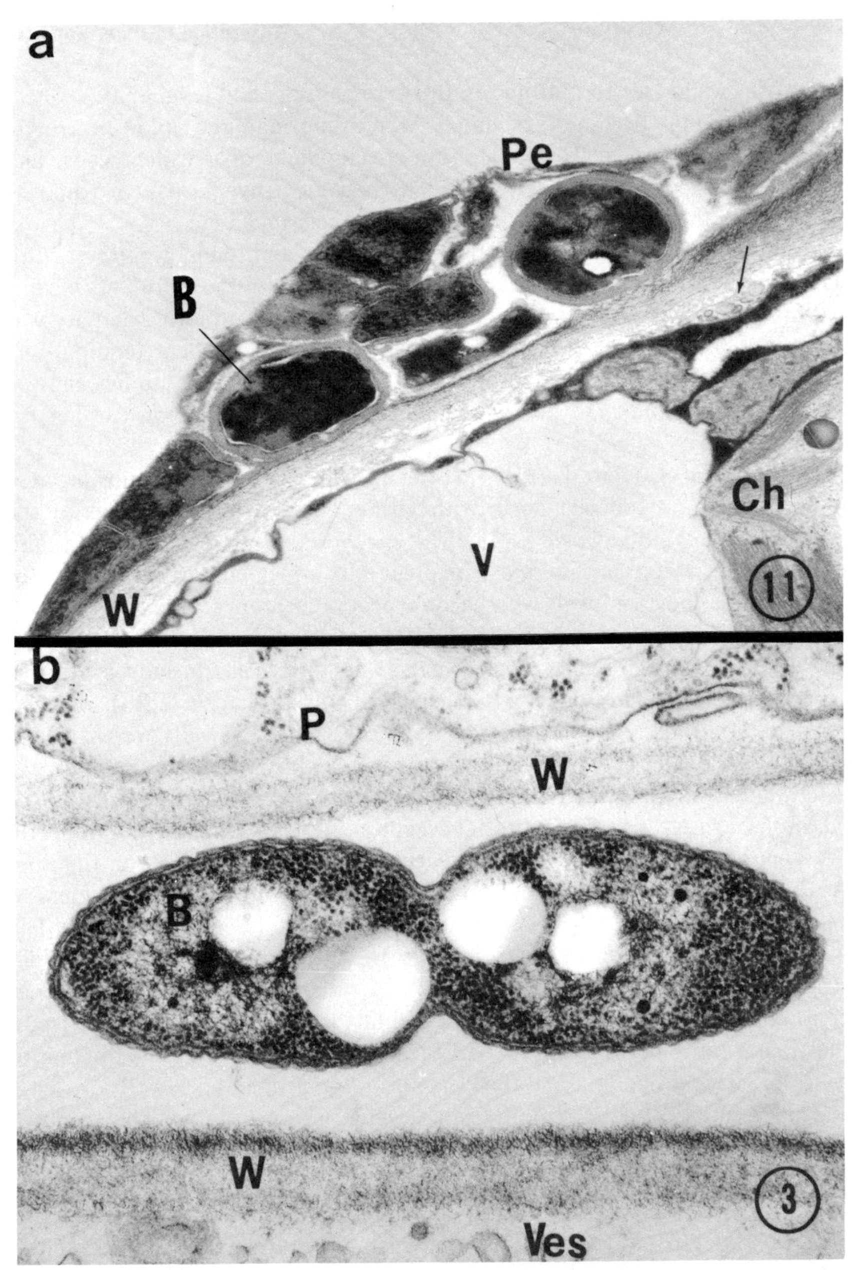

Figure 8.30 Interactions of *Pseudomonas solanacearum* with mesophyll cell walls of tobacco leaves: (a) attachment of *P. solanacearum* S_{210} incompatible bacteria to cell walls (W) at 7 hr after infiltration—note pellicle (Pe) encasement around immobilized bacteria (B), vesicle accumulation (arrow), vacuole (V), and chloroplast (Ch); (b) virulent *P. solanacearum* K_{60} dividing in intracellular space—note separation of plasmalemma (P) and accumulation of vesicles (Ves) next to the cell wall. Courtesy of L. Sequeira, G. Gaard, and G. A. De Zoeten and Academic Press, New York [170].

unadsorbed and free to multiply in the intercellular fluid (Fig. 8.30*b*), where they eventually become systemic. Membrane damage and electrolyte leakage of the host cells soon follow envelopment of avirulent cells, and these cytological changes are viewed as part of a hypersensitive response that plays a fundamental role in plant disease resistance.

Goodman et al. [171] extracted tobacco leaves and found greater agglutinating activity in the leaves previously infiltrated with the avirulent (incompatible) organism, *Pseudomonas pisi*. Little or no agglutinating activity was recovered from leaves previously infiltrated with the virulent (compatible) organism *P. tabaci* or the saprophyte *P. fluorescens*. Their studies showed that the agglutinin was nondialyzable and required divalent cations for activity.

Graham and Sequeira [173,174] studied the interactions of lectins isolated from potato and tobacco hosts with virulent and avirulent strains of the pathogen *P. solanacearum*. Their hypothesis is that adsorption and host–pathogen recognition may involve a complementary interaction of lectins on the host cell wall with carbohydrate receptors on the bacterial cell surface. Their studies showed that lectins purified from these hosts agglutinate only avirulent cells. Failure of bacterial cells to bind lectin was correlated with the presence of extracellular polysaccharide (EPS), which is formed by virulent but not by avirulent cells. Virulent cells were agglutinated by potato lectin if their EPS were removed by washing, and avirulent cells would not agglutinate if EPS from virulent cells was added prior to the addition of the lectin. The bacterial receptor sites that bind the potato lectin are apparently internal repeating *N*-acetylglucosamine residues in the lipid A:R core moieties of LPS, and inhibition studies indicate that oligomers of *N*-acetylglucosamine (e.g., chitotriose) are potent hapten inhibitors. The basis behind inhibition of lectin binding to cells by the EPS from virulent strains is currently under investigation. Graham and Sequeira (personal communication) propose that the EPS binds to the same site on the lectin that is normally occupied by the LPS, and thus EPS blocks the binding of lectin to virulent bacteria and these bacteria to mesophyll cell walls.

The lipid A:R core region from several Gram-negative bacteria will induce resistance in potatoes if applied 24 hr prior to inoculation with virulent *P. solanacearum* [173,175]. Intact LPS from *P. solanacearum* binds strongly to tobacco mesophyll cell walls [175]. Graham and Sequeira [173] propose that this induction of disease resistance may involve the interaction between the host cell-wall lectin and the same components of the bacterial cell wall involved in recognition and attachment. The application of induced-disease resistance to curtail infection may well be of agricultural value.

There are both striking similarities and differences emerging from studies with *Rhizobium* and plant pathogens in relation to lectins isolated from their corresponding hosts (see Section 8.5.2). These comparisons should be viewed in context with the consequences of bacterial adsorption to host-plant cell walls where the lectins are found. Specific lectins on legume root hairs presumably ensure that the right *Rhizobium* bacteria will be concentrated at the right time and location for successful symbiotic infection at discrete foci on the host. On the other hand, lectins on mesophyll cell walls presumably recognize invading microorganisms, immobilize them, and prevent their further systemic spread in the intercellular spaces. It is not known whether a hypersensitive reaction in legume roots is induced by heterologous or noninfective rhizobia.

8.6.3 *Agrobacterium* Adsorption to Host Cell Walls

The adsorption of virulent *Agrobacterium* cells to a specific wound site on the host cell wall is an essential early step of the infection process in crown-gall tumor formation [177–187]. Whole cells of some avirulent strains [177–179] or hot phenol–water extracts containing LPS from "site-binding" virulent or avirulent strains [180] were found to be potent inhibitors of tumor formation. Lippincott and Lippincott [181,182] have reviewed the nature of adsorption. Site competition occurred only when the total concentration of cells exceeded 10^8/ml, added either with or prior to inoculation with virulent cells. Inhibition was dependent on the absolute numbers of each cell type and follows a one-particle hit curve. Competition was specific for infective agrobacteria. More recent studies showed that the O-antigen:R core (but not the lipid A) of LPS from site-binding strains and cell-wall preparations from bean leaves [182,183,187] could inhibit tumor initiation by virulent agrobacteria. At high concentrations, polygalacturonic acid, pectin, and *arabino*-galactan were also inhibitory to tumor initiation [182,187]. Cell-wall preparations from bean embryos and monocotyledonous plants were without effect [182,187]. These results may account in part for the curious ability of *Agrobacterium* to infect dicotyledonous plants and gymnosperms but not monocotyledons or meristematic tissues in general. These resistant tissues are high in pectin methyl groups, and pectin methylesterase treatment of cell walls of corn (a monocot) and bean embryos converted them to an inhibitory state. These data suggest that *Agrobacterium* cannot adhere to cell walls that contain heavily methylated pectic substances. Lippincott and Lippincott [182] propose that the adhesion of *Agrobacterium* to wounded host cells is initiated by interactions between bacterial and plant polysaccharides. The possible involvement of lectins in this adsorption process is being investigated [186]. As in the case with

Rhizobium, these studies should not be limited to considering only "classical" lectins that agglutinate erythrocytes.

8.7 CONSEQUENCES OF MICROBIAL ADSORPTION TO PLANT ROOTS

All physical, chemical, and biological components of the soil ecosystem that affect plant physiology and pathology are subject to microbial modification on the rhizoplane. Lynch [188] has reviewed the products of soil microorganisms that affect plant growth. These include plant growth regulators (e.g., cytokinins, auxins, ethylene, organic acids, phenolics, cyanogenic glycosides, amino-acid derivatives, and volatiles, including alcohols, HCN, and H_2S) [189,190] and soil enzymes (e.g., cellulases, hemicellulases, proteases, and pectinases) [191]. The active concentration supplied to the absorbing root may be inhibitory at one concentration yet stimulatory at another.

Microbial metabolites may be toxic to plants. Rhizobitoxine is an amino acid derivative made by certain strains of *R. japonicum*, which adsorb to and infect soybean roots [192]. This microbial toxin produces a dramatic yellowing or chlorosis in soybean leaves, and the mode of action involves interference with the metabolism of cystathione [193].

8.7.1 Rhizoplane Morphogenesis

Rhizoplane microorganisms influence plant morphogenesis and development. Very dramatic morphological modifications include the formation of proteoid roots [194], the branching and curling of host legume and nonlegume root hairs with subsequent infection and nodulation by nitrogen-fixing symbionts [40,72,118,195], the induction of adventitious roots by *Agrobacterium rhizogenes* [196], *Azospirillum brasilense* [142,143,197], mycorrhizal fungi [198], and gametophore biogenesis in mosses [199–203].

The biochemical agents responsible for the morphologic alterations on legume root surfaces induced by *Rhizobium* are undefined. Root hairs of axenically grown clover seedlings are straight. Root hair curling and branching is induced specifically by rhizobia and their culture filtrates [55,77,102, 204–208]. The active fraction from *R. trifolii* have been partially purified by ethanol precipitation [206], ion-exchange precipitation, and chromatography [77,102]. Attempts have failed to demonstrate curling and branching of clover root hairs by a wide concentration range of indoleacetic acid [207,208] or isopentenyladenine (Dazzo and Brill, unpublished observation). These are plant hormones produced by *Rhizobium* [209,210]. The

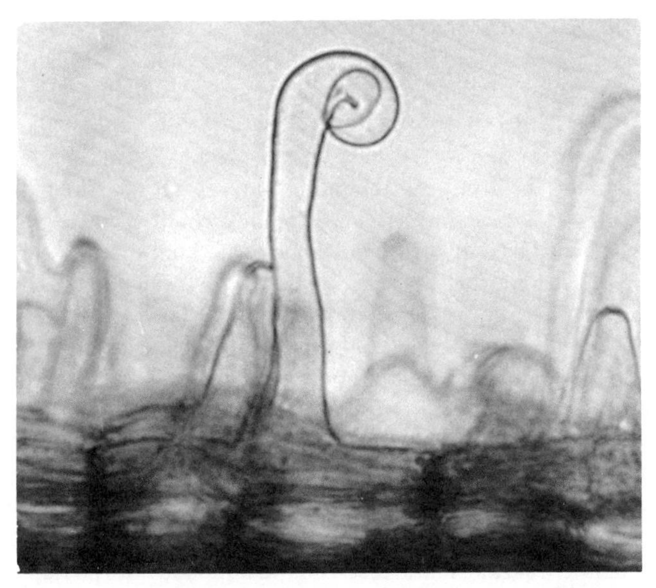

Figure 8.31 Clover root hair shepherd's crook induced by infective *Rhizobium trifolii.*

very tight curling at root-hair tips, known as the "shepherd's crook" (Fig. 8.31), is induced only by viable rhizobia specifically capable of infecting that legume host [205,208]. Shepherd's crooks are not induced if homologous viable rhizobia are separated from legume roots by a dialysis barrier, although root-hair branching is abundant under these conditions [208]. These data suggest that shepherd's crook formation requires attachment or close proximity of viable, homologous, infective rhizobia to the root surface. Serial thin sectioning reveals that rhizobia always are enclosed in the center overlap of the "knot" created by the shepherd's crook [65]. The root-hair deformation factor is inactivated by trypsin [208] and ribonuclease [102]. The curling factor was heat stabilized when incubated with clover root exudate [102]. Of potential importance is the recent report by Hubbell et al. [62,63] that *Rhizobium* release an active pectinase. The successful transfer of symbiotic genes from *Rhizobium trifolii* and *R. japonicum* to *Azotobacter vinelandii* by Bishop et al. [98] may be a useful approach in unraveling the genetic and biochemical control of legume root-hair deformation induced by *Rhizobium*.

Major gaps exist in understanding the biogenesis of legume root nodules. Torrey [211] reported that kinetin triggers mitosis in polyploid callus tissue but not in diploid cells, suggesting that the polypoid cell is induced to produce nodular growth. Studies using autoradiography of *R. trifolii* labeled with ^{3}H-leucine or ^{3}H-proline indicated that rhizobial metabolites are transferred to polyploid nuclei of young clover nodules [212]. Hybridization *in situ* by autoradiography indicated that *Rhizobium* DNA is detected in cortical cells of its specific plant host shortly after the roots are inoculated [213]. In pea roots, cortex cells opposite the xylem radii are stimulated to divide rapidly in front of the advancing infection thread containing rhizobia [214]. This root mitotic activity may result from a response to a hormone gradient in front of the advancing infection thread [215]. Kijne et al. [216,217] have reported that rhizobial LPS can replace kinetin in induction of cell proliferation of pea root explants. Their hypothesis for root nodule initiation is that cell proliferation begins with an LPS-induced enhancement of endogenous cytokinin production [217]. Syoño et al. [218] have found that mitotic activity of pea nodule meristem parallels cytokinin levels. The cytokinins in nodules are identified as zeatin and its riboside and lesser amounts of isopentenyladenine and isopentenyladenosine. The symbiotic capabilities of cytokinin-minus or auxin-minus mutants of *Rhizobium* and revertants of these mutations remain to be examined.

Adsorbed microorganisms affect root length and morphology and thus influence uptake of plant nutrients. Plant hormones are likely candidates as inducers of root morphological changes in ectomycorrhizas. In pure culture mycorrhizal fungi may produce indoleacetic, indolepropionic, and indolebutyric acids [219], and these hormones are abundant in mycorrhizal roots. Ectomycorrhizas may have simple monopodal, dicotomously branched (coralloid), and nodular morphologies. Root structures resembling each morphology can be induced by auxin at critical concentrations. When the symbiotic relationship terminates, mycorrhizal morphology disappears in a manner similar to the auxin-induced root structure when the auxin is discontinued. These observations suggest, but do not prove, that growth hormones produced by the fungus are released continuously in sufficient quantities to elicit the mycorrhizal morphology on the plant root.

The developmental changes from the filamentous moss protonema to bud formation and then into a leafy gametophore provide an excellent system for studying control of plant morphogenesis. Speiss et al. [199–203] have shown that gametophore induction was enhanced by the adsorption of certain Gram-negative bacteria (agrobacteria and rhizobia included) to special cell sites on the surface of spore buds (Fig. 8.32). Gametophore induction by site-binding strains follows a one-particle hit curve. The order of addition of competing viable and nonviable cells is important, and furthermore, the carbohydrate portion of LPS is identified as the binding component necessary for attachment of agrobacteria in order to induce gametophores [203]. Parabiotic chamber experiments revealed that the bacterial product(s) responsible for gametophore induction either are larger

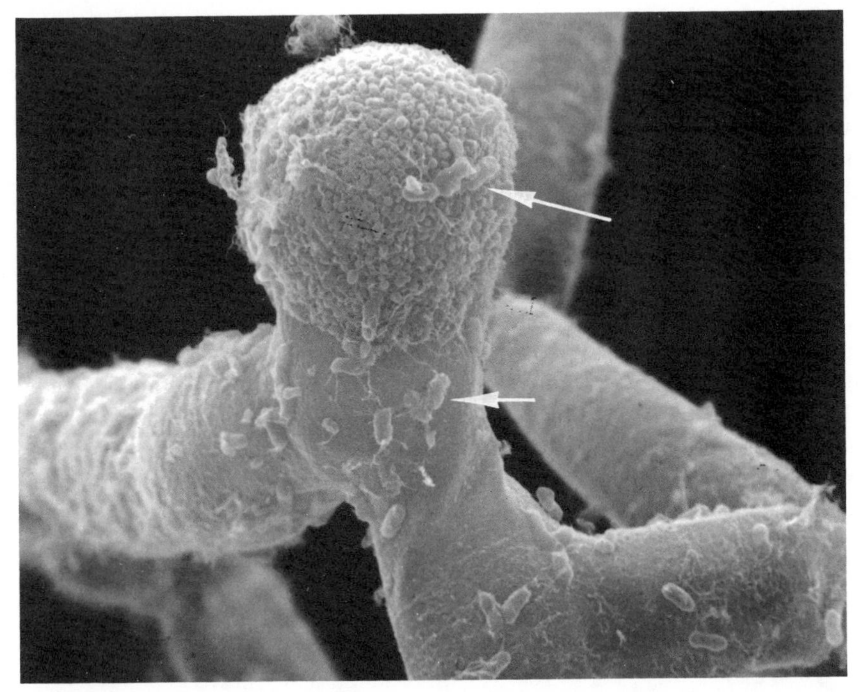

Figure 8.32 Scanning electron micrograph of adsorption of *Agrobacterium tumefaciens* B_6 to spore initial and filament cells of the moss *Pylaisiella selwynii*. Courtesy of L. D. Speiss, B. B. Lippincott, J. A. Lippincott, P. Mahlberg, and J. Turner and the Hattori Botanical Laboratory [201].

than 0.22 μm or are never released directly into the medium in sufficient quantity to be effective, as the combination of bacteria plus moss in one chamber fails to produce developmental changes of moss in the opposite chamber [200]. It has been suggested that a cytokinin (ribosyl-transzeatin) produced by the bacterium and transferred directly to the moss may be involved in bud formation [201,202].

8.7.2 Plant Nutrition

Hypothetical models of nutrient uptake by plants must consider the rhizoplane microorganisms as the normal case to be accurate. Microorganisms on the rhizoplane influence substances moving into and out of roots, and their own metabolites may also enter roots. Metabolic activities of rhizoplane microorganisms influence rhizoplane pH, which, in turn, influences solubility and availability of ions for plants.

Extremes that illustrate the dependence of higher plants on rhizoplane microorganisms include mycorrhizas that provide nutrients for the achlorophyllous angiosperm *Monotropa* [221]and some orchids [222]. The success of both terrestrial and aquatic mycorrhizal plants, which fluorish in nutrient-poor soils and sediments, illustrates the important influences of these rhizoplane microorganisms in plant nutrient uptake [223,224]. In infertile soils, plants deficient in root hairs have a greater dependence on mycorrhizas for growth than do plants with finely branched root systems and copious root hairs [225]. A major exception occurs with ferns, which have long, persistent root hairs and well-developed root systems but in nature are consistently mycorrhizal [150]. A bidirectional transport of ^{14}C from $^{14}CO_2$ between the fungus and the roots of pine and ericoid mycorrhizae and certain orchids has been demonstrated [226,227]. Microscopic autoradiography has indicated that much of the increase in plant uptake of phosphate by endomycorrhizas is due to uptake and translocation by fungal hyphae external to the root, rather than to fungal stimulation of ion uptake by the infected root cells [228]. The improved performance of agriculturally important crops (wheat, corn, and potatoes) inoculated with mycorrhizal fungi has been reported [229–231].

Noninvasive microorganisms adsorbed to the rhizoplane may influence

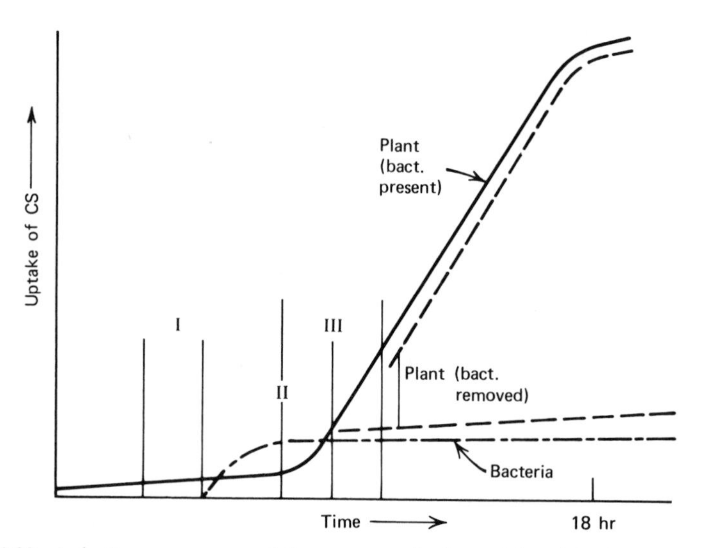

Figure 8.33 Induction processes and time courses for bacterial and barley root uptake of choline sulfate. Bacterial uptake is induced during phase I with a specific permease. Plant uptake remains at low constitutive level during phase II. Plant uptake increases during phase III when roots adsorb bacteria. Bacteria can eventually be removed with no change in uptake rates. Courtesy of P. Nissen and North-Holland Publishing Company, Amsterdam [53].

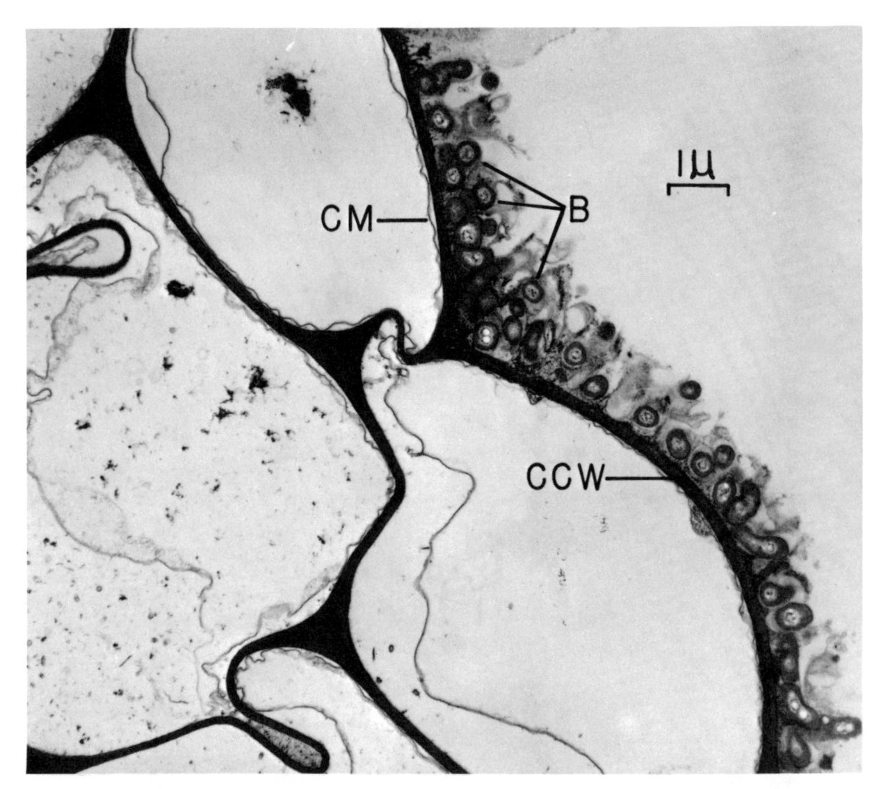

Figure 8.34 Bacteria (B) adsorbed to the rhizoplane of the Florida mangrove *Lemna minor*. Note cortical cell wall (CCW) and cell membrane (CM) of root cells and alignment of rhizoplane microorganisms. Courtesy of D. Zuberer and W. Silver.

ion uptake by roots in many ways. These include the concentrations of ions at the root surface, the uptake efficiency and metabolism of plants, root-hair growth, and root length and morphology [232].

Rhizoplane microorganisms may also produce substances that affect nutrient permeability in roots [233]. Nissen made pioneering efforts to elucidate the biochemical mechanism behind the ability of noninvasive rhizoplane microorganisms to enhance plant nutrient uptake [53,54] (Fig. 8.33). He found that the uptake of choline sulfate by barley roots is increased after Gram-negative bacteria possessing an inducible choline–sulfate permease, adsorbed to the root surface. Uptake kinetics of choline sulfate were characteristic of the plant, and not the bacteria. Nissen interpreted these data as a possible transfer of information from bacteria to plant that regulates choline sulfate uptake.

The rhizospheres of many terrestrial and aquatic plants have associated

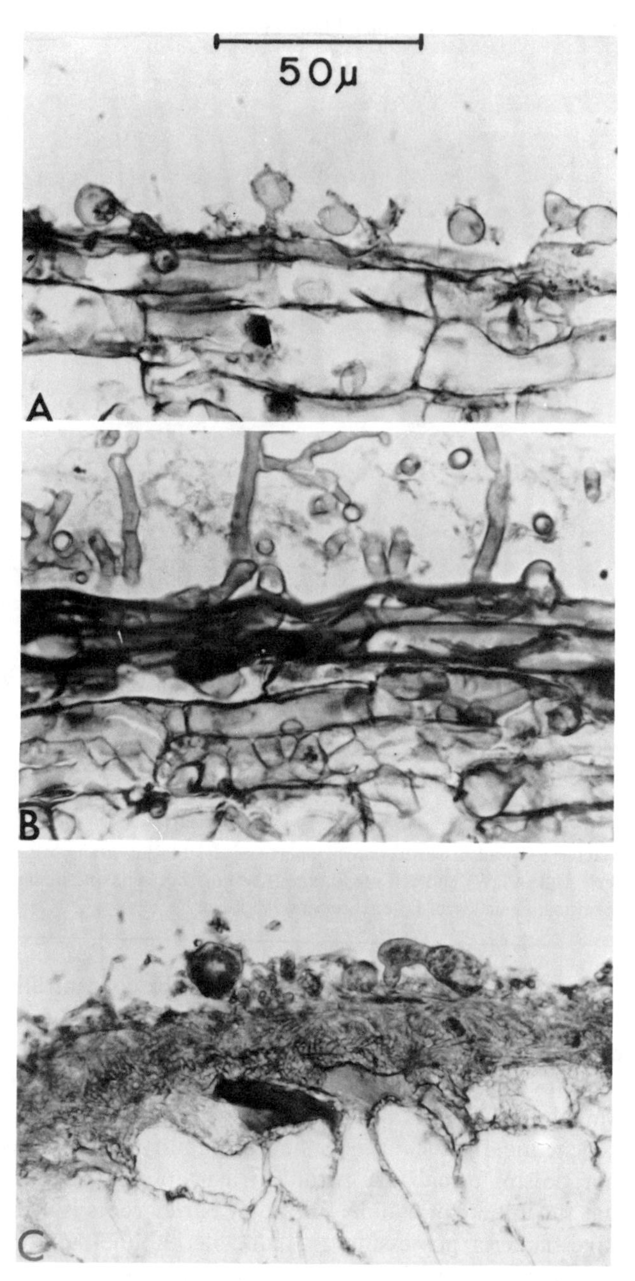

Figure 8.35 Susceptibility and resistance of nonmycorrhizal and mycorrhizal feeder roots of pine to *Phytophthora cinnamomi*. Direct penetration of zoospores (A) and vegetative hyphae (B) of *P. cinnamomi* into nonmycorrhizal roots with intracellular cortex infection. Vesicles and hyphae of *P. cinnamomi* are restricted to mantle barrier of mycorrhizas in (C). Courtesy of D. H. Marx and C. B. Davey and Academic Press, New York, and the American Phytopathological Society [237, 238].

nitrogenase activity, as measured by acetylene-dependent ethylene production [234].The amount of nitrogen fixed in the rhizosphere of corn grown in the arid tropics falls well below the level of nitrogen required by the plant [235].

Mangrove vegetation lines 75% of the coastal areas along the tropical and subtropical countries. These habitats are very productive but often nitrogen limited. Zuberer and Silver [236] found considerable nitrogen-fixing activity and extensive microbial colonization on the rhizoplane of three Floridian mangrove species (Fig. 8.34). They propose that the bacteria used root exudates and sloughed cell debris as energy sources for nitrogen fixation.

8.7.3 Plant Protection

Marx [237] has shown the potential of ectomycorrhizal fungi in the biological control of soil-borne root pathogens. Extensive mycorrhizal tissue leaves little susceptible (nonmycorrhizal) root surface accessible to infection by pathogens. The susceptibility and resistance of nonmycorrhizal and mycorrhizal feeder roots of pine to infection by *Phytophthora cinnamomi* [238] is shown in Figs. 8.35*a–c*.

Several mechanisms may be responsible for the protection, although it is not known which is more important. In general, the protection is not systemic, and nonmycorrhizal sections of the same root system can be infected. However, mycorrhizal fungi were found to produce a potent antibiotic that was translocated to adjacent nonmycorrhizal roots and protected them effectively against a feeder root pathogen [237]. Other possible mechanisms for disease resistance include: (1) a mechanical barrier of the fungal mantle growing on the root surface, (2) utilization of carbohydrates otherwise stimulatory to the pathogen, (3) induction of host resistance (e.g., phytoalexins) as a result of mycorrhizal infection, (4) stimulation of a mycorrhizosphere microflora antagonistic to the soilborne plant pathogen, and (5) possible suppression of the chemotactic responses of pathogens toward roots [237].

Binding of lipopolysaccharide to mesophyll cell walls of tobacco leaves results in systemically induced disease resistance and protection against various pathogens [175]. The ultrastructural changes in the host are similar to those resulting from attachment of whole heat-killed bacterial cells.

8.8 CONCLUDING REMARKS

This chapter has dealt with the adsorption of microorganisms to plant tissues in an integrated approach at the molecular, subcellular, cellular, histological, and community levels. Both the bacterial and the root surfaces have net negative surface charges at the pH values of soil solutions [239]. The mechanisms whereby microorganisms and plants overcome the repulsion energy barriers, which tend to prevent direct contact, are currently an area of intensive investigation. In some cases the adsorption steps constitute specific recognition phenomena of fundamental importance to plant–microorganism interactions. At the molecular level studies implicate the involvements of hydrophobic interactions and complementary recognition of unique cell-surface polysaccharides and carbohydrate-binding proteins. Complementary carbohydrate–carbohydrate interactions may also be important. Genetic control of cell surface components involved in microbial adsorption to plant surfaces is being studied [89,98,180]. At the subcellular level these components may exist as electron-dense particles, outer envelope polysaccharides, capsules, slime layers, microfibrils, fimbriae, and mucigel. At the cellular level cell differentiation dictates the distribution of receptor binding sites [80]. At the histological level the rhizosphere concept requires modification to include organisms colonizing not only the rhizosphere and rhizoplane, but also plant epidermal tissue and underlying cortical tissue [34,36]. Several mechanisms of microbial adsorption to plants have been recognized; some are very specific and others, nonspecific. All contribute at the community level to the development of spatially separated, independent, and interacting microbial communities on plant surfaces. These communities may have profound influences on plant morphogenesis, nutrition, growth, pathogenesis, and disease resistance.

Many questions concerning the adsorption of microorganisms to plant root surfaces remain to be answered: the importance of microbial activity in cortical tissue to plant nutrition and root decomposition, the extent to which rhizoplane microorganisms of different species coexist despite overlapping niches, the essential factors controlling growth rates of rhizoplane microbes, the responses of microbe and plant to the rhizoplane–rhizosphere "interface", the mechanisms that alter plant ion uptake; the hormonal communications between microorganisms and plants, the basis for ecological dominance and succession of rhizoplane populations, and the energetic relationships and balances for growth and maintenance of the rhizoplane microorganisms. Biophysical aspects of the adsorption phenomenon on rhizoplanes have received very little consideration. Efforts by Zvyagintsev et al. [240] to measure the adhesive forces of microorganisms to solid surfaces may possibly be modified to study the plant-surface habitat. Major gaps in understanding events of adsorption to roots have been emphasized.

Possible benefits from studies of adsorption of microorganisms to roots and other plant tissues include the provision of basic knowledge to plant physiology, the improvement of the symbiotic performance of rhizobial and mycorrhizal inoculants, and the curtailment of pathological processes in roots and other plant tissues by pathogens.

8.9 ACKNOWLEDGMENTS

This research was supported by the College of Agriculture and Life Sciences, University of Wisconsin, Madison; the Michigan Agricultural Experiment Station, Michigan State University, East Lansing, Michigan; and by grants No. PCM 76-24271 and AER 77-00879 from the National Science Foundation. I thank the many scientists who contributed figures to this chapter; R. Heinzen and S. Vicen for technical photographic assistance; and Drs. W. Brill, T. Brock, S. Ela, D. Hubbell, E. McCoy, L. Sequeira, and G. Zeikus for helpful suggestions.

ADDENDUM

ADSORPTION OF MICROORGANISMS TO ROOTS AND OTHER PLANT SURFACES

E. Shimshick and R. Herbert [*Biochem. Biophys. Res. Commun.*, **84,** 736 (1978)] found that attachment of *Rhizobium japonicum* to wheat and rice seedling roots for periods of up to 6 days is an equilibrium process that follows a Langmuir adsorption isotherm. This model predicts that there is a maximum, fixed number of adhesion sites on the root surface and that there is a slow, time-dependent dissociation of the rhizobia from wheat roots.

S. Pull, S. Pueppke, T. Hymowitz, and J. Orf [*Science*, **200,** 1277 (1978)] found that the major soybean lectin is not detected in seeds in a few varieties that nevertheless develop seedling roots that can be nodulated by rhizobia. However, it was not determined whether the lectin was present on soybean root hairs at infection sites or whether these mutant varieties of soybean lost their selectivity for rhizobia.

Recent studies have shown that the selective ability of *R. trifolii* to adhere to clover root hairs is influenced by conditions that affect the accumulation of trifoliin on the host root surface and the saccharide receptor on the bacterium. In one study fixed nitrogen ions (e.g., NO_3^- and NH_4^+) in the rooting medium were found to regulate the levels of trifoliin on clover root hairs and the concurrent ability of *R. trifolii* to adhere to these surfaces [F. Dazzo and W. Brill, *Plant Physiol.*, **62,** 18 (1978)]. In a subsequent study it was found that the appearance of trifoliin receptors on *R. trifolii* is transient under certain growth conditions and only during this period were the bacteria able to attach to clover root hairs [F. Dazzo, M. Urbano, and W. Brill, *Curr. Microbiol.*, **2,** 15 (1979)].

M. Lapp and W. Skoropad [*Transact. Br. Mycol. Soc.*, **70,** 221 (1978)] found that the mucilage of the appressoria of *Colletotrichum graminicola*, a fungal leaf pathogen of oats, contains hemicellulose as an adhesive factor.

8.10 REFERENCES

1. H. Hanson, *Dictionary of Ecology*, Philosophical Library, New York, 1962, p. 336.
2. C. Bracker and L. Littlefield, "Structural Concepts of Host–Pathogen Interfaces," in R. Byrde and I. Cutting, Eds., *Fungal Pathogenicity and the Plant's Response*, Academic, New York, 1973, pp. 159–313.
3. J. G. Torrey and D. T. Clarkson, Eds., *The Development and Function of Roots*, Academic, New York, 1975.
4. L. Hiltner, *Arb. Deut. Landwirtsch. Ges.*, **98,** 59 (1904).
5. F. E. Clark, *Adv. Agron.*, **1,** 241 (1949).
6. A. D. Rovira, "Plant Root Exudates and Their Influence upon Soil Microorganisms," in K. F. Baker and W. C. Snyder, Eds., *Ecology of Soil-Borne Plant Pathogens*, California U. P. Berkeley, 1970, pp. 170–184.
7. D. A. Barber, and J. K. Martin, *New Phytol.*, **76,** 68 (1976).
8. R. Starkey, *Bacteriol. Rev.*, **22,** 154 (1958).
9. N. A. Krasil'nikov, *Soil Microorganisms and Higher Plants*, Academy of Sciences, U.S.S.R., 1958.
10. A. D. Rovira, *Annu. Rev. Microbiol.*, **19,** 241 (1965).
11. A. D. Rovira and B. M. McDougall, in H. McLaren and B. Peterson, Eds., *Soil Biochemistry*, Vol. 1, Marcel Dekker, New York, 1967, p. 417.
12. A. D. Rovira and C. B. Davey, "Biology of the Rhizosphere," in E. Carson, Ed., *The Plant Root and Its Environment*, Virginia U. P., Charlottesville, Va., 1974, pp. 153–175.
13. G. D. Bowen, and A. D. Rovira, *Annu. Rev. Phytopathol.*, **14,** 121 (1976).
14. T. Hattori, *Microbial Life in the Soil*, Marcel Dekker, New York, 1973, p. 327.
15. Y. A. Hamdi, *Soil Biol. Biochem.*, **3,** 212 (1971).
16. R. I. Papendick and G. S. Campbell, "Water Potential in the Rhizosphere and Plant and Methods of Measurement and Experimental Control," in G. W. Bruel, Ed., *Biology and Control of Soil-Borne Plant Pathogens*, American Phytopathological Society, St. Paul, Minn., 1975, pp. 39–49.
17. W. W. Currier and G. A. Strobel, *Plant Physiol.*, **57,** 820 (1976).
18. W. W. Currier and G. A. Strobel, *Science*, **196,** 434 (1977).
19. W. T. Scudder, "Improving the efficiency of Inoculation with *Rhizobium japonicum*," in *Proceedings of Sixth American Rhizobium Conference, University of Florida, Gainesville*, 1977, p. 19.
20. C. A. Napoli and P. Albersheim, *Abstracts of Annual Meeting of American Society for Microbiology*, 1978, p. 169(N43).
21. C. A. Napoli, F. Dazzo, and D. H. Hubbell, *Appl. Microbiol.*, **30,** 123 (1975).

22. M. I. Timonin, "Interaction of Higher Plants and Soil Microorganisms," in C. M. Gilmour and O. N. Allen, Eds., *Microbiology and Soil Fertility*, Oregon State U. P., Corvallis, 1964, pp. 135–158.
23. P. S. Nutman, *Ann. Bot.* (*Lond.*), **21,** 321 (1957).
24. R. J. Peters and M. Alexander, *Soil Sci.*, **102,** 380 (1966).
25. A. C. Robinson, *J. Austral. Inst. Agric. Sci.*, **33,** 207 (1967).
26. D. N. Munns, *Plant Soil*, **28,** 129 (1968).
27. Y. Chang-Ho and C. J. Hickman, "Some Factors Involved in the Accumulation of Phycomycete Zoospores on Plant Roots," in T. Tousoun, R. Bega, and P. Nelson, Eds., *Root Diseases and Soil-Borne Pathogens*, California U. P., Berkeley, 1970, pp. 103–108.
28. A. W. Van Egeraat, *Plant Soil*, **42,** 381 (1975).
29. C. J. Hickman and H. H. Ho, *Annu. Rev. Phytopathol.*, **4,** 195 (1966).
30. G. Zentmyer, "Tactic Responses of Zoospores of *Phytophthora*," in T. Toussoun, R. Bega, and P. Nelson, Eds., *Root Diseases and Soil-Borne Pathogens*, California U. P., Berkeley, 1970, pp. 109–111.
31. H. Jenny and K. Grossenbacher, *Proc. Soil Sci. Soc. Am.*, **27,** 273 (1963).
32. P. J. Dart and F. V. Mercer, *Arch. Mikrobiol.*, **47,** 344 (1964).
33. M. P. Greaves and J. F. Darbyshire, *Soil Biol. Biochem.*, **4,** 443 (1972).
34. R. C. Foster and A. D. Rovira, *New Phytol.*, **76,** 343 (1976).
35. A. Guchert, H. Breisch, and O. Reisinger, *Soil Biol. Biochem.*, **7,** 241 (1975).
36. K. M. Old and T. H. Nicolson, *New Phytol.*, **74,** 51 (1975).
37. R. Campbell and A. D. Rovira, *Soil Biol. Biochem.*, **5,** 747 (1973).
38. D. H. Northcote and J. D. Pickett-Heaps, *Biochem. J.*, **98,** 159 (1966).
39. F. B. Dazzo, "The Microbial Ecology of Cultivated Soil Receiving Cow Manure Waste," M.S. thesis, University of Florida, Gainesville, 1972.
40. J. H. Becking, "Root Nodules in Nonlegumes," in J. G. Torrey and D. L. Clarkson, Eds., *The Development and Function of Roots*, Academic, London, 1975, pp. 507–566.
41. R. L. Starkey, *Soil Sci.*, **45,** 207 (1938).
42. N. Cholodny, *Arch. Mikrobiol.*, **1,** 620 (1930).
43. A. D. Rovira, *J. Appl. Bacteriol.*, **19,** 72 (1956).
44. R. C. Foster and A. D. Rovira, *Bull. Ecol. Res. Commun.* (*Stockholm*), **17,** 93 (1973).
45. B. Frenzel, *Planta*, **55,** 169 (1960).
46. G. S. Taylor and D. Parkinson, *Plant Soil*, **15,** 261 (1961).
47. A. D. Rovira and R. Campbell, *Microb. Ecol.*, **1,** 15 (1974).
48. E. I. Newman and H. J. Bowen, *Soil Biol. Biochem.*, **6,** 205 (1974).
49. A. D. Rovira, E. I. Newman, H. J. Bowen, and R. Campbell, *Soil Biol. Biochem.*, **6,** 211 (1974).
50. G. D. Bowen and A. D. Rovira, *Bull. Ecol. Res. Commun.* (*Stockholm*), **17,** 443 (1973).
51. R. Marchant, *Transact. Br. Mycol. Soc.*, **54,** 479 (1970).
52. P. J. Dart, *J. Exp. Bot.*, **22,** 163 (1971).
53. P. Nissen, "Choline Sulfate Permease: Transfer of Information from Bacteria to Higher Plants? II. Induction Processes," in L. Ledoux, Ed., *Information Molecules in Biological Systems*, North Holland, Amsterdam, 1971, pp. 201–212.

54. P. Nissen, *Sci. Rep. Agric. Univ. Norway*, **52,** 1 (1973).
55. E. McCoy, *Proc. Roy. Soc. (London) Ser. B.*, **110,** 514 (1932).
56. D. Li and D. H. Hubbell, *Can. J. Microbiol.*, **15,** 1133 (1969).
57. G. Fahraeus, *J. Gen. Microbiol.*, **16,** 374 (1957).
58. P. S. Nutman, "The Relation between Nodule Bacteria and the Legume Host in the Rhizosphere and the Process of Infection," in K. F. Baker and W. C. Snyder, Eds., *Ecology of Soil-Borne Pathogens*, California U. P., Berkeley, 1970, pp. 231–247.
59. P. S. Nutman, C. C. Doncaster, and P. J. Dart, "Infection of Clover by Root-nodule Bacteria," film available from the British Film Institute, London, 1973.
60. H. Ljunggren and G. Fahraeus, *J. Gen. Microbiol.*, **26,** 274 (1961).
61. G. Fahraeus and K. Sahlman, *Ann. Acad. Reg. Sci. Upsaliensis*, **20,** 103 (1977).
62. D. H. Hubbell, "Association of Nitrogen Fixing Bacteria with Plant Roots," in H. M. Vines, Ed., *Physiology of Root–Microorganisms Associations*, Georgia U. P., Athens, 1977, p. 11–25.
63. D. H. Hubbell, V. M. Morales, and M. Umali-Garcia, *Appl. Environ. Microbiol.*, **35,** 210 (1978).
64. K. Sahlman and G. Fahraeus, *J. Gen. Microbiol.*, **33,** 425 (1963).
65. C. A. Napoli and D. H. Hubbell, *Appl. Microbiol.*, **30,** 1003 (1975).
66. P. S. Nutman, *Biol. Rev. Cambridge Phil. Soc.*, **3k,** 109 (1956).
67. C. Napoli, F. Dazzo, and D. Hubbell, "Ultrastructure of Infection and Common Antigen Relationships in *Aeschynomene*," *Proceedings of Fifth Australian Conference*, Brisbane, Australia, 1975.
68. O. N. Allen, and E. K. Allen, *Bot. Gaz.*, **102,** 121 (1940).
69. D. Raveed, M. Reporter, and G. Norris, "Invasion of Soybean Root Cells in Culture by *Rhizobium japonicum*," *Proceedings of 33rd Electron Microscope Society of America*, 1975, pp. 582–583.
70. F. B. Dazzo, C. A. Napoli, and D. H. Hubbell, *Appl. Microbiol.*, **32,** 166 (1976).
71. I. Baldwin and E. B. Fred, *J. Bacteriol.*, **17,** 141 (1929).
72. E. B. Fred, I. Baldwin, and E. McCoy, *The Root Nodule Bacteria and Their Leguminous Plants*, Wisconsin U. P., Madison, 1932.
73. J. Hamblin and S. P. Kent, *Nature New Biol.*, **245,** 28 (1973).
74. B. B. Bohlool and E. L. Schmidt, *Science*, **185,** 269 (1974).
75. B. Solheim, "A Model of the Recognition-reaction between *Rhizobium trifolii* and *Trifolium repens*," *Proceedings of NATO Conference on Specificity in Plant Diseases*, Advance Study Institute, Sardinia, 1975.
76. F. B. Dazzo and D. H. Hubbell, *Appl. Microbiol.*, **30,** 172 (1975).
77. F. B. Dazzo and D. H. Hubbell, *Appl. Microbiol.*, **30,** 1017 (1975).
78. J. S. Wolpert and P. Albersheim, *Biochem. Biophys. Res. Commun.*, **70,** 729 (1976).
79. R. J. Maier and W. J. Brill, *J. Bacteriol.*, **127,** 763 (1976).
80. F. B. Dazzo and W. J. Brill, *Appl. Environ. Microbiol.*, **33,** 132 (1977).
81. T. S. Brethauer and J. D. Paxton, "The Role of Lectin in Soybean–*Rhizobium japonicum* Interactions," in B. Solheim and J. Raa, Eds., *Cell Wall Biochemistry Related to Specificity in Host–Plant Pathogen Interactions*, Universitetsforlanget, Oslo, Norway, 1977, pp. 381–387.

82. J. W. Kijne, "Pea Lectin Purification," in B. Solheim and J. Raa, Eds., *Cell Wall Biochemistry Related to Specificity in Host–Plant Pathogen Interactions*, Universitetsforlaget, Oslo, Norway, 1977, pp. 403–406.
83. I. J. Law and B. W. Strijdom, *Soil Biol. Biochem.*, **9,** 79 (1977).
84. W. D. Bauer, *Basic Life Sci.*, **9,** 283 (1977).
85. T. V. Bhuvaneswari, S. G. Pueppke, and W. D. Bauer, *Plant Physiol.*, **60,** 486 (1977).
86. S. G. Pueppke, K. Keegstra, A. L. Ferguson, and W. D. Bauer, *Plant Physiol.*, **61,** 779 (1978).
87. K. Planqué and J. W. Kijne, *FEBS Lett.*, **73,** 64 (1977).
88. F. B. Dazzo, W. E. Yanke, and W. J. Brill, *Biochim. Biophys. Acta*, **539,** 276 (1978).
89. F. B. Dazzo and W. J. Brill, *J. Bacteriol.* **137,** 1362 (1979).
90. B. B. Bohlool and E. L. Schmidt, *J. Bacteriol.*, **125,** 1188 (1976).
91. H. C. Tsien and E. L. Schmidt, *Can. J. Microbiol.*, **23,** 1274 (1977).
92. P. Albershiem and J. S. Wolpert, *Plant Physiol. Suppl.*, **57,** 79 (1976).
93. P. Albershiem and J. Wolpert, "Molecular Determinants of Symbiont–Host Selectivity between Nitrogen-fixing Bacteria and Plants," in B. Solheim and J. Raa, Eds., *Cell Wall Biochemistry Related to Specificity in Host–Plant Pathogen Interactions*, 1977, Universitetsforlaget, Oslo, Norway, pp. 373–376.
94. A. P. Chen and D. H. Phillips, *Physiol. Plant*, **38,** 83 (1976).
95. T. V. Bhuvaneswari and W. D. Bauer, *Plant Physiol.*, **62,** 71 (1978).
96. F. B. Dazzo, "Cross-reactive Antigens and Lectin as Determinants of Host Specificity in the *Rhizobium*-legume Symbiosis," in B. Solheim and J. Raa, Eds., *Cell Wall Biochemistry Related to Specificity in Host–Plant Pathogen Interactions*, Universitetsforlaget, Oslo, Norway, pp. 389–400.
97. F. B. Dazzo and D. H. Hubbell, *Plant Soil*, **43,** 713 (1975).
98. P. E. Bishop, F. B. Dazzo, E. R. Appelbaum, R. J. Maier, and W. J. Brill, *Science*, **198,** 938 (1977).
99. H. Kauss and C. Glaser, *FEBS Lett.*, **45,** 304 (1974).
100. P. N. Shankar Iyer, K. D. Wilkinson, and I. J. Goldstein, *Arch. Biochem. Biophys.* **177,** 330 (1976).
101. J. C. Burton and P. W. Wilson, *Soil Sci.*, **47,** 293 (1939).
102. B. Solheim and J. Raa, *J. Gen. Microbiol.*, **77,** 241 (1973).
103. R. Carlson, R. E. Saunders, C. Napoli, and P. Albersheim, *Plant Physiol.* **62,** 912 (1978).
104. A. K. Bal, S. Shantharam, and S. Ratnam, *J. Bacteriol.*, **133,** 1393 (1978).
105. R. J. Maier, and W. J. Brill, *J. Bacteriol.*, **133,** 1295 (1978).
106. R. E. Saunders, R. W. Carlson, and P. Albersheim, *Nature*, **271,** 240 (1978).
107. D. Werner, J. Wilcocson, and E. Zimmerman, *Arch. Microbiol.*, **105,** 27 (1975).
108. S. F. Kotarski and D. C. Savage, *Abstracts of Annual Meetings of American Society for Microbiology*, 1978, p. 15(B11).
109. M. H. Dienema, and L. P. Zevenhuizen, *Arch. Mikrobiol.*, **78,** 42 (1971).
110. C. A. Napoli, "Physiological and Ultrastructural Aspects of the Infection of Clover (*Trifolium fragiferum*) by *Rhizobium trifolii* NA30, Ph.D. dissertation, University of Florida, Gainesville, 1976.
111. D. L. Balkwill and L. E. Casida, Jr., *J. Bacteriol.*, **114,** 1319 (1973).

112. G. H. Elkan, *Can. J. Microbiol.*, **8,** 79 (1962).
113. T. L. Degenhardt, T. A. LaRue, and E. A. Paul, *Can. J. Bot.*, **54,** 1633 (1976).
114. M. Reporter, D. Raveed, and G. Norris, *Plant Sci. Lett.*, **5,** 73 (1975).
115. N. Hermina, and M. Reporter, *Plant Physiol.*, **59,** 97 (1977).
116. W. F. Dudman, *Carbohydr. Res.*, **46,** 97 (1976).
117. G. Menzel, H. Uhig, and G. Weischsel, *Zent. Bakteriol. Parasitenkd. Infektionski. Hyg. Abt.* **2,** 127, 348 (1972).
118. P. J. Dart, "The Infection Process," in A. Quispel, Ed., *The Biology of Nitrogen Fixation*, North Holland, Amsterdam, 1974, pp. 381–429.
119. K. C. Marshall, R. H. Cruickshank, and H. V. Bushby, *J. Gen. Microbiol.*, **91,** 198 (1975).
120. W. Heumann, *Arch. Microbiol.*, **24,** 362 (1956).
121. R. Marx and W. Heumann, *Arch. Mikrobiol.*, **43,** 245 (1962).
122. W. Heumann and R. Marx, *Arch. Mikrobiol.*, **47,** 325 (1964).
123. F. Mayer, *Arch. Mikrobiol.*, **68,** 179 (1969).
124. F. Mayer, *Zeit. Allg. Mikrobiol.*, **10,** 329 (1970).
125. W. Heumann, *Molec. Gen. Genet.*, **102,** 132 (1968).
126. F. Mayer, *Arch. Mikrobiol.*, **76,** 166 (1971).
127. F. Mayer and R. Schmitt, *Arch. Mikrobiol.*, **79,** 311 (1971).
128. F. Mayer, S. Kall, and R. Schmitt, *Zeit. Allg. Mikrobiol.*, **14,** 221 (1974).
129. V., Braun, K. Rehn, and H. Wolff, *Biochem.*, **9,** 5041 (1970).
130. S. Halegona, A. Hirashima, and M. Inouye, *J. Bacteriol.*, **120,** 1204 (1974).
131. C. Ballou, "Cell Wall Structure and Recognition in Yeasts," in B. Solheim and J. Raa, Eds., *Cell Wall Biochemistry Related to Specificity in Host–Plant Pathogen Interactions*, Universitetsforlaget, Oslo, Norway, 1977, pp. 364–371.
132. K. C. Marshall and R. H. Cruickshank, *Arch. Mikrobiol.*, **91,** 29 (1973).
133. K. G. Vogler and W. W. Umbreit, *Soil Sci.*, **51,** 331 (1941).
134. W. Schaeffer, P. Holbert, and W. Umbreit, *J. Bacteriol.*, **85,** 137 (1963).
135. J. L. Beebe and W. W. Umbreit, *J. Bacteriol.*, **108,** 612 (1971).
136. W. Lotz and H. Pfister, *J. Virol.*, **16,** 725 (1975).
137. M. Lalonde, "The Infection Process of *Alnus* Root Nodule Symbiosis," in E. Newton, J. Postgate, and C. Rodriquez-Barrueco, Eds., *Recent Developments in Nitrogen Fixation*, Academic, London, 1977, pp. 569–589.
138. D. Callaham, P. D. Tredici, and J. G. Torrey, *Science*, **199,** 899 (1978).
139. J. Dobereiner and J. M. Day, *Proc. Internat. Symp. Nitrogen Fixation*, **2,** 518 (1976).
140. N. R. Kreig, *Basic Life Sci.*, **9,** 463 (1977).
141. D. G. Patriquin and J. Dobereiner, "Bacteria in the Endorhizosphere of Maize in Brazil," in *International Symposium on the Limitations and Potentials of Biological Nitrogen Fixation in the Tropics*, University of Brazilia, Brazil, 1977, p. 76.
142. M. Umali-Garcia, D. H. Hubbell, and M. H. Gaskins, *Bull. Ecol. Res.* (*Stockholm*), **26,** 373 (1978).
143. M. Umali-Garcia, D. H. Hubbell, M. H. Gaskins, and F. B. Dazzo, "Adsorption and Infection of Grass Roots by *Azospirillum brasilense* sp. 7," in W. Orme-Johnson, and W.

Newton, Eds., *Steenbock-Kettering 3rd International Symposium on Nitrogen Fixation*, University of Wisconsin Press, p. 24 (B-47).

144. G. C. Marks and R. C. Foster, "Structure, Morphogenesis, and Ultrasturcture of Ectomycorrhizae," in G. C. Marks and T. T. Kozlowski, Eds., *Ectomycorrhizae, Their Ecology and Physiology*, Academic, New York, 1973, pp. 1–41.
145. R. C. Foster and G. C. Marks, *Austral. J. Biol. Sci.*, **20,** 915 (1967).
146. D. J. Read and P. P. Stribley, "Some Mycological Aspects of the Biology of Mycorrhizae in the *Ericaceae*," in F. Saunders, B. Mosse, and P. Tinker, Eds., *Endomycorrhizae*, Academic, London, 1975, pp. 105–117.
147. H. Kaspari, "Fine Structure of the Host–Parasite Interface in Endotrophic Mycorrhiza of Tobacco," in F. Saunders, B. Mosse, and P. Tinker, Eds., *Endomycorrhizae*, Academic, London, 1975, pp. 325–334.
148. G. Hadley, "Organization and Fine Structure of the Orchid Mycorrhiza," in F. Saunders, B. Mosse, and P. Tinker, Eds., *Endomycorrhizae*, Academic, London, 1975, pp. 335–351.
149. B. Mosse, *New Phytol.*, **72,** 127 (1973).
150. K. M. Cooper, "Growth Response to the Formation of Endotrophic Mycorrhizas in *Solanum leptospermum* and New Zealand Ferns," in F. Saunders, B. Mosse, and P. Tinker, Eds., *Endomycorrhizae*, Academic, London, 1975, pp. 392–407.
151. F. Saunders, "The Effect of Foliar Applied Phosphate on the Mycorrhizal Infections of Onion Roots," in F. Saunders, B. Mosse, and P. Tinker, Eds., *Endomycorrhizae*, Academic, London, 1975, pp. 261–276.
152. D. H. Marx, A. B. Hatch, and J. F. Mendicino, *Can. J. Bot.*, **55,** 1569 (1977).
153. H. E. Wilcox, *Am. J. Bot.*, **55,** 688 (1968).
154. J. M. Phillips and D. S. Hayman, *Transact. Br. Mycol. Soc.*, **55,** 158 (1970).
155. J. R. Aist and P. H. Williams, *Can. J. Bot.*, **49,** 2023 (1971).
156. A. A. Holland and R. G. Fulcher, *Austral. J. Biol. Sci.*, **24,** 819 (1971).
157. R. L. Slusher and J. B. Sinclair, *Phytopathol.*, **63,** 1168 (1973).
158. E. L. Schmidt, S. A. Biersbrock, B. B. Bohlool, and D. H. Marx, *Can. J. Microbiol.*, **20,** 137 (1974).
159. S. D. Garrett, *Pathogenic Root-Infecting Fungi*, Cambridge U. P., London, 1970, pp. 82–89.
160. G. W. Wallis, *Can. J. Bot.*, **39,** 109 (1961).
161. R. F. Lewis and W. J. Crotty, *Am. J. Bot.*, **64,** 190 (1977).
162. B. Solheim and J. Raa, *Cell Wall Biochemistry Related to Specificity in Host-Plant Pathogen Interactions*, 1977, Universitetsforlaget, Oslo, Norway (Preface, p. 5).
163. T. F. Preece and C. H. Dickinson, Eds., *Ecology of Leaf Surface Microorganisms*, Academic, London, 1971.
164. C. H. Dickinson and T. F. Preece, Eds., *Microbiology of Aerial Plant Surfaces*, Academic, London, 1976.
165. J. Ruinin, "The Grass Sheath as a Site for Nitrogen Fixation," in T. F. Preece and C. H. Dickinson, Eds., *The Ecology of Leaf Surface Microorganisms*, Academic, London, 1971, pp. 567–579.
166. H. G. Diem, *J. Gen. Microbiol.*, **80,** 77 (1974).
167. R. N. Goodman, *Phytopathol.*, **62,** 1327 (1972).

168. P. Y. Huang, J. S. Huang, and R. N. Goodman, *Physiol. Plant Pathol.*, **6,** 283 (1975).
169. R. N. Goodman, P. Y. Huang, and J. A. White, *Phytopathol.*, **66,** 754 (1976).
170. L. Sequeira, G. Gaard, and G. A. DeZoeten, *Physiol. Plant Pathol.*, **10,** 43 (1977).
171. R. N. Goodman, P. Y. Haung, J. S. Haung, and V. Thaipanich, "Induced Resistance to Bacterial Infections," in Y. Tomiyama, Ed., *Biochemistry and Cytology of Plant-Parasite Interactions*, Kodansha, Tokyo, 1977, pp. 35–42.
172. R. N. Goodman, D. J. Politis, and J. A. White, "Ultrastructural Evidence of an Active Immobilization Process of Incompatible Bacteria in Tobacco Leaf Tissue: A Resistance Reaction," in B. Solheim and J. Raa, Eds., *Cell Wall Biochemistry as Related to Specificity in Host–Plant Pathogen Interactions*, Universitetsforlaget, Oslo, Norway, 1977, pp. 423–437.
173. T. L. Graham and L. Sequeira, "Interaction between Plant Lectins and Cell Wall Components of *Pseudomonas solanacearum:* Role in Pathogenicity and Induced Disease Resistance," in B. Solheim and J. Raa, Eds., *Cell Wall Biochemistry Related to Specificity in Host-Plant Pathogen Interactions*, Universitetsforlaget, Oslo, Norway, 1977, pp. 417–422.
174. L. Sequeira and T. L. Graham, *Physiol. Plant Pathol.*, **11,** 43 (1977).
175. T. L. Graham, L. Sequeira, and J. R. Haung, *Appl. Environ. Microbiol.*, **34,** 424 (1977).
176. V. O. Sing, and M. N. Schroth, *Science*, **197,** 259 (1977).
177. B. B. Lippincott and J. A. Lippincott, *J. Bacteriol.*, **97,** 620 (1969).
178. R. A. Schilperoort, "Investigations on Plant Tumors Crown Gall: On the Biochemistry of Tumor Induction by *Agrobacterium tumefaciens*," Ph.D. thesis, University of Leiden, Wageningen, 1969.
179. R. Beiderbeck, *Z. Naturforsch.*, **28,** 198 (1973).
180. M. H. Whatley, J. S. Bodwin, B. B. Lippincott, and J. A. Lippincott, *Infect. Immun.*, **13,** 1080 (1976).
181. J. A. Lippincott and B. B. Lippincott, *Encycl. Plant Physiol.*, **4,** 356 (1976).
182. J. A. Lippincott and B. B. Lippincott, "Nature and Specificity of the Bacterium–Host Attachment in *Agrobacterium* Infection," in B. Solheim and J. Raa, Eds., *Cell Wall Biochemistry Related to Specificity in Host–Plant Pathogen Interactions*, Universitetsforlaget, Oslo, Norway, 1977, pp. 439–451.
183. B. B. Lippincott, M. H. Whatley, and J. A. Lippincott, *Plant Physiol.*, **59,** 388 (1977).
184. A. G. Matthysse, R. M. Wyman, and H. L. Miller, *Plant Physiol. Suppl.*, **59,** 109 (1977).
185. W. Glogowski and A. Galsky, *Plant Physiol. Suppl.*, **59,** 108 (1977).
186. V. K. Anad, S. G. Pueppke, and G. T. Heberlein, *Plant Physiol. Suppl.*, **59,** 109 (1977).
187. J. A. Lippincott and B. B. Lippincott, *Science*, **199,** 1075 (1978).
188. J. M. Lynch, *CRC Rev. Microbiol.*, **5,** 67 (1976).
189. J. W. Einset, E. M. Greene, F. Skoog, E. Doyle, and R. S. Hansen, *Plant Physiol. Suppl.*, **59,** 109 (1977).
190. T. M. McCalla and F. A. Haskins, *Bacteriol. Rev.*, **28,** 181 (1964).
191. J. J. Skujins, "Soil Enzymes," in A. McLaren and B. Peterson, Eds., *Soil Biochemistry*, Vol 1, 371, Marcell Dekker, New York, 1967.
192. L. Owens, *Science*, **165,** 18 (1965).
193. L. D. Owens, S. Guggenhaeim, and J. Hilton, *Biochem. Biophys. Acta*, **158,** 219 (1968).

194. N. Malajczuk and G. D. Bowen, *Nature* (*Lond.*), **251,** 316 (1974).

195. J. G. Torrey, *Am. J. Bot.*, **63,** 335 (1976).

196. D. L. Hopkins and R. D. Durbin, *Can. J. Microbiol.*, **17,** 1409 (1971).

197. T. M. Tien, M. H. Gaskins, and D. H. Hubbell, *Appl. Environ. Microbiol.*, **37,** 1016 (1979).

198. J. L. Hartley, "Mycorrhiza," in K. F. Baker and W. C. Snyder, Eds., *Ecology of Soil-Borne Plant Pathogens*, California U. P., Berkeley, 1970, pp. 218–230.

199. L. D. Speiss, B. B. Lippincott, and J. A. Lippincott, *Am. J. Bot.*, **58,** 726 (1971).

200. L. D. Speiss, B. B. Lippincott, and J. A. Lippincott, *Am. J. Bot.*, **63,** 324 (1976).

201. L. D. Speiss, B. B. Lippincott, and J. A. Lippincott, *J. Hattori Bot. Lab.*, **41,** 185 (1976).

202. L. D. Speiss, B. B. Lippincott, and J. A. Lippincott, *Bot. Gaz.*, **138,** 35 (1977).

203. M. H. Whatley and L. D. Speiss, *Plant Physiol.*, **60,** 765 (1977).

204. A. Haack, *Zentr. Bakteriol. Parasitk.*, **117,** 343 (1964).

205. P. Y. Yao and J. M. Vincent, *Aust. J. Biol. Sci.*, **22,** 413 (1969).

206. D. H. Hubbell, *Bot. Gaz.*, **131,** 337 (1970).

207. K. Sahlman and G. Fahraeus, *Kung. LantbrHogsk. Ann.*, **28,** 261 (1962).

208. P. Y. Yao and J. M. Vincent, *Plant Soil*, **45,** 1 (1976).

209. J. Dullaart, *Acta Bot. Neerlandica*, **19,** 573 (1970).

210. D. Phillips and J. G. Torrey, *Plant Physiol.*, **49,** 1115 (1972).

211. J. G. Torrey, *Exp. Cell Res.*, **23,** 281 (1961).

212. M. L. Brueing and L. H. Wullstein, *Physiol. Plant*, **27,** 244 (1972).

213. A. A. Lepidi, M. P. Nuti, G. Bernacchi, and R. Neglia, *Plant Soil*, **45,** 555 (1976).

214. K. R. Libbenga and P. A. Harkes, *Planta*, **114,** 17 (1973).

215. K. R. Libbenga, K. R. VanIren, R. Bogers, and M. F. Schraag-Lamers, *Planta*, **114,** 29 (1973).

216. J. Kijne, "Cell Proliferation in Pea Root Explants Supplied with Rhizobial LPS," in B. Solheim and J. Raa, Eds., *Cell Wall Biochemistry as Related to Host Specificity in Host–Plant Pathogen Interactions*, Universitetsforlaget, Oslo, Norway, 1977, pp. 415–416.

217. J. Kijne, S. D. Adhin, and K. Planqué, "Effect of Rhizobial LPS on Cell Division in Pea Root Explants," *Proceedings of Second International Congress on Nitrogen Fixation*, Varona, Salamanca, Spain, 1976, p. B6.

218. K. Syoño, W. Newcomb, and J. G. Torrey, *Can. J. Bot.*, **54,** 2155 (1976).

219. V. Slankis, "Hormonal Relationships in Mycorrhizal Development," in G. C. Marks, and T. Kozlowksi, Eds., *Ectomycorrhizae: Their Ecology and Physiology*, Academic, New York, 1973, pp. 231–298.

220. L. D. Speiss, *Plant Physiol.*, **55,** 583 (1975).

221. W. D. Gray, *The Relation of Fungi to Human Affairs*, Henry Holt, New York, 1959, pp. 62–63.

222. S. Purves and G. Hadley, "Movement of Carbon Compounds between Partners in Orchid Mycorrhiza," in F. Saunders, B. Mosse, and P. Tinker, Eds., *Endomycorrhiza*, Academic, London, 1975, pp. 175–194.

223. M. Sondergaard and L. Laegaard, *Nature* (*Lond.*), **268,** 232 (1977).

224. P. P. Kormanik, W. C. Bryan, and R. C. Schultz, "The Role of Mycorrhizae in Plant Growth and Development," in H. M. Vines, Ed., *Physiology of Root–Microorganism Associations*, Georgia U. P., Athens, 1977, pp. 1–10.
225. G. T. Baylis, "The Magnoloid Mycorrhiza and Mycotrophy in Root Systems Derived from It," in F. Saunders, B. Mosse, and P. Tinker, Eds., *Endomycorrhizas*, Academic, London, 1975, pp. 373–389.
226. D. H. Lewis, "Comparative Aspects of the Carbon Nutrition of Mycorrhizas," in F. Saunders, B. Mosse, and P. Tinker, Eds., *Endomycorrhizas*, Academic, London, 1975, pp. 119–148.
227. G. Cox, F. Saunders, P. Tinker, and J. Wild, "Ultrastructural Evidence Relating to Host–Endophyte Transfer in a Vesicular–Arbuscular Mycorrhiza," in F. Saunders, B. Mosse, and P. Tinker, Eds., *Endomycorrhizas*, Academic, London, 1975, pp. 297–312.
228. G. Bowen, D. Bevege, and B. Mosse, "Phosphate Physiology of Vesicular–Arbuscular Mycorrhizas," in F. Saunders, B. Mosse, and P. Tinker, Eds., *Endomycorrhizas*, Academic, London, 1975, pp. 241–260.
229. B. Mosse, D. S. Hayman, and G. J. Ide, *Nature* (*Lond.*), **267,** 510 (1975).
230. A. G. Kahn, *Ann. Appl. Biol.*, **80,** 27 (1975).
231. R. L. Black and P. B. Tinker, *Nature* (*Lond.*), **267,** 510 (1975).
232. A. D. Rovira and G. D. Bowen, "Microbial Effects on Nutrient Uptake by Plants," in R. M. Samish, Ed., *Recent Advances in Plant Nutrition*, Vol. 2, Gordon and Breach, New York, 1971, pp. 307–320.
233. A. Norman, *Arch. Biochem. Biophys.*, **58,** 461 (1955).
234. F. B. Dazzo and D. H. Hubbell, *Proc. Fla. Soil Crop Sci. Soc.*, **34,** 71 (1974).
235. J. Tjepkema and P. V. Berkum, *Appl. Environ. Microbiol.*, **33,** 626 (1977).
236. D. Zuberer and W. Silver, *Appl. Environ. Microbiol.*, **35,** 567 (1978).
237. D. H. Marx, "Mycorrhizae and Feeder Root Disease," in G. C. Marks and T. Kozlowski, Eds., *Ectomycorrhizae: Their Ecology and Physiology*, Academic, New York, 1973, pp. 351–382.
238. D. H. Marx and C. B. Davey, *Phytopathology*, **59,** 559 (1969).
239. K. C. Marshall, *Interfaces in Microbial Ecology*, Harvard U. P., Cambridge, Mass., 1976.
240. D. Z. Zvyagintsev, A. F. Pertsovskaya, E. D. Yakhin, and E. I. Averbakh, *Mikrobiologia*, **40,** 889 (1972).

CHAPTER 9

Adsorption of Microorganisms to Soils and Sediments

K. C. MARSHALL

School of Microbiology, The University of New South Wales, Kensington, N. S. W., Australia

CONTENTS

9.1 INTRODUCTION

Microbial recycling of organic materials and minerals in soils and sediments is a process vital to the continued functioning of all biological communities. These soils and sediments are complex microbial habitats, as they are dominated by particulate components of almost infinitely variable type and structural organization. The increasing interest in studies on interactions between microorganisms and particulates is reflected in the number of recent reviews on various aspects of this subject [1–8]. Rather than provide a repetitive review of literature, this chapter endeavors to present a conceptual view of adhesion processes as they occur in soils and sediments, to provide a sound basis for future studies on microorganism–particulate interactions and on the significance of such interactions in these ecosystems.

Inorganic particulates in soils and sediments are classified, on the basis of size, into coarse sands (diameter 200–2000 μm), fine sands (20–200 μm), silts (2–20 μm), and clays ($<2\mu$m). A significant feature of these particulates, from a microbiological viewpoint, is the surface area available for nutrient accumulation and for microbial adhesion. The surface area ranges from approximately 20 $cm^2\ g^{-1}$ for a coarse sand to in excess of 20,000 $cm^2\ g^{-1}$ for certain clays. A knowledge of the particle size distribution in soils and sediments, however, rarely provides any predictive value in relation to the growth of and interactions between microorganisms in these systems. It is the structural organization of the various particulate materials that defines the type of habitat available to the microorganisms. Fluctuating wet and dry conditions in soils influence the availability of organic matter, the degree of weathering, the leaching and redeposition of soluble materials, and the translocation of smaller particulate fractions, thereby leading to the development of structured layers (horizons) within a single soil profile. The type and extent of horizonation varies according to the climate, topography, and parent material. Sediments in aquatic systems develop under more constant conditions, but distinct layering of sediment materials does result where regular fluctuations occur in the inorganic or organic particulate load of the overlying water column.

9.2 SOILS AS MICROBIAL HABITATS

Attention is beginning to focus on a more precise definition of soil microhabitats in relation to microbial activity [9]. Marshall and Marshman [10] suggest that a soil microhabitat represents a restricted but variable volume of soil wherein specific or unique reactions directly affect microorganisms or are directly affected by the microorganisms. The microhabitat is not a particular site, but it represents the sum of all environmental variables (physical, chemical, and biological) with which the organisms interact [11]. These variables can be represented as axes of an m-dimensional coordinate system (with 1,2,3,4, . . . , m axes), wherein each of the m dimensions represents one of the environmental variables with which the species interacts. The m dimensions can be expressed on an m dimensional figure that defines the *habitat hyperspace* or *hypervolume* [12]. The particulate material of a soil contributes to the actual shape of this habitat hypervolume [11].

The three-dimensional organization of soil particulates (the *soil fabric*) [13] determines the nature and extent of the solid–liquid interfaces in soils. The concept of soil fabric includes the voids (pore spaces) that are surrounded by solid-phase components. Voids may be partly or completely filled with water, thereby providing the aqueous medium required for microbial activity. The solid phase of soil is divided into the *skeleton grains* and the *plasma*. Skeleton grains are relatively stable and are not readily translocated, concentrated, or reorganized by soil-forming processes, whereas plasma is the portion of the soil that has been or is capable of being moved, reorganized, or concentrated by the soil-formation processes [13]. It is the arrangement of the skeleton grains and plasma that determines the micropedological features and, to some extent, the nature of microbial habitats in various soils.

The distribution of microorganisms in soils is not homogeneous [14–18]; an extreme instance is sandy soils where the majority of bacteria are associated with a minor fraction of the soil—the organic matter [18,19]. Clays present in more complex soils may be distributed as clay skins (cutans) around sand grains [7,10,13], giving rise to a microbial habitat in which the clay dictates the properties of most solid–water interfaces. Microorganisms were observed readily by scanning electron microscope (SEM) on such cutans [7]. In organic matter and clay-rich soils, however, only filamentous microorganisms growing into the void space could be recognized by SEM. Bacteria in such soils may have been so completely enveloped in colloidal material as to be unrecognizable [7].

9.3 SEDIMENTS AS MICROBIAL HABITATS

Microhabitats in sediments may be defined in the same manner as in soils, except that fabric development and modification are less extensive. The physical environment of sediments is more constant than that of soils, especially with regard to water availability, temperature and O_2 levels, and there tends to be a more consistent input of organic matter into sediments [20]. If the organic-matter content of sediments is high, anaerobic conditions are rapidly established, and the extent of anaerobiosis depends on conditions existing in the water column. Oxidizing conditions exist at the sediment–water interface in well-aerated waters, but the subsurface sediments rapidly become anaerobic as a result of microbial activity. In stratified lakes or marine sites of high organic input, the lower portion of the water column contains little or no dissolved oxygen and the sediment is wholly anaerobic [20].

The relative amounts of sand, silt, clay, and organic matter, and the way in which these are distributed, will again dictate the shape of the habitat hypervolume in sediments of various origin. Little or no work has been done to define the exact microhabitats existing in such sediments.

9.4 MICROORGANISM—PARTICULATE ASSOCIATIONS

Zvyagintsev [16] differentiated between the types of interactions observed between bacteria and particulates in terms of the size of the particulates relative to that of the bacteria. The categories proposed were: (1) adsorption of bacteria on the surface of larger soil particulates, (2) coflocculation (aggregation) of bacteria and soil particles of similar size, and (3) adsorption of colloidal particles at the bacterial surface. Hattori [4] has developed these ideas in relation to bacterial–clay (*BC*) complex formation as follows:

$$mB + nC = B_m C_n$$

Zvyagintsev's first two categories fit into Hattori's type $m > 1$, $n = 1$, which describes situations where many bacteria adhere to single clay particles equal or greater in size to the bacteria [21]. Hattori also described an unusual association between *Escherichia coli* and pyrophyllite to give a complex of type $m = 1$, $n = 1$, as demonstrated in the results presented in Fig. 9.1. The third category described by Zvyagintsev is represented by Hattori's type $m = 1$, $n > 1$, in which many colloidal particles sorb at the surface of a single bacterium [22–25].

Aggregate stability in soils and the degree of coflocculation between microorganisms and soil particles is very dependent on the valency of the dominant cations in the system. Peele [26] demonstrated a sharp increase in sorption of *Azotobacter* to soil particulates as the valency of the saturating cation increased from the mono-, to the di-, and to the trivalent state. In a study of aggregation between bacteria and clay particles of comparable size, Santoro and Stotzky [21] found that the degree of aggregation increased with the valency of the cation as predicted by the Schultz–Hardy rule. This rule states that flocculation of sols is caused by ions of opposite charge, the flocculating power increasing with the valency of the appropriate ions. Soil destabilization results from a lowering of the electrokinetic potential of the particles or, under some conditions, an actual charge reversal in the presence of trivalent ions. Hattori [4] has described similar effects of cation valency.

A bacterial surface provides a significant solid–water interfacial area on which colloidal materials can adhere. Neihof and Loeb [27,28] have provided evidence for the adsorption of natural organic materials to the surface

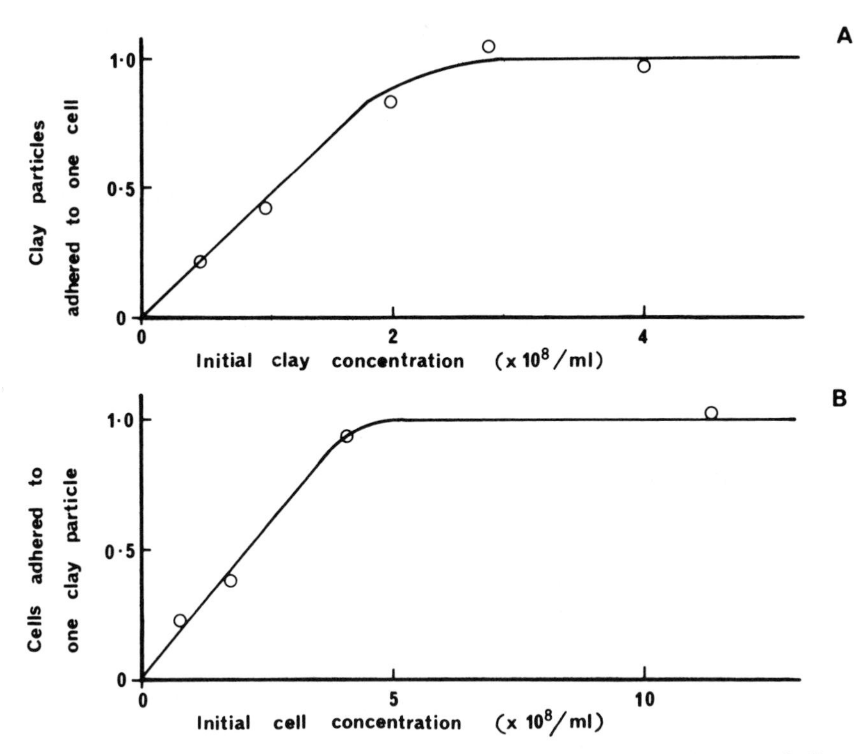

Figure 9.1 Adhesion between cells of *Escherichia coli* and particles of sodium pyrophyllite as functions of concentrations of (A) clay and (B) bacteria. Cells and clay particles had effective mean diameters of 0.8 μm and 0.9 μm, respectively. Redrawn from Hattori [4].

of bacteria and other particles in seawater. Earlier, Lahav [22] had reported the adsorption of colloidal clay platelets to bacterial surfaces and suggested possible modes or orientation of the clay platelets at the bacterial surface. Similar results were described by Marshall [23–25], who presented evidence for a predominantly edge-to-face orientation of clay platelets at negatively charged (carboxyl) sites on bacterial cells and for a degree of face-to-face orientation on those bacteria possessing some positively charged (amino) surface ionogenic sites.

9.5 REVERSIBLE AND PERMANENT ADHESION

The term *reversible sorption* [29] is used to describe the situation where microorganisms are attracted to surfaces of like charge under conditions where van der Waals attraction energies exceed the electrical double-layer repulsion energies (Fig. 9.2). Under these conditions the microorganisms are held at a small but finite distance from the surface. Any further decrease in interparticle distance results in a condition where the resultant energy is a substantial repulsion energy. When microorganisms are sorbed reversibly at a surface, they are not anchored in any way to the surface. Reversible sorption of microorganisms to surfaces can be overcome by the application of some shear force, by flagellar motion of motile microorganisms, or by reducing the electrolyte concentration to a point where the organisms are repelled from the surface [29]. In relatively static, nutrient deficient systems, such as soils and sediments, reversibly sorbed microorganisms do

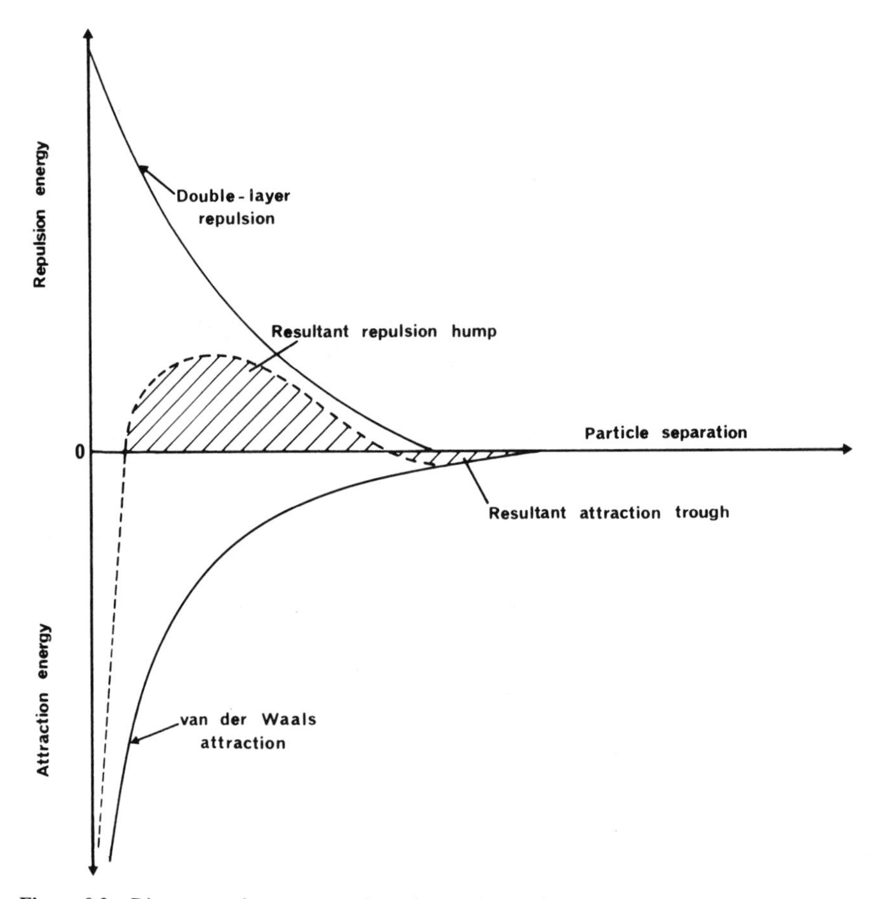

Figure 9.2 Diagrammatic representation of repulsion and attraction energies with increasing particle separation (i.e., between bacterium and solid surface), showing high resultant repulsion energy at small interparticle distances and significant resultant attraction energy at somewhat larger interparticle distances.

benefit from the enriched nutrient status existing at solid–liquid interfaces. An interesting adaptation to this situation is the observation of dinoflagellates grazing along surfaces using one flagellum for propulsion, with the second flagellum held at the point of attraction existing near the surface and acting as a trailing anchor [8].

Permanent adhesion of microorganisms to surfaces appears to be a common feature in all natural ecosystems. However, definitive evidence for permanent adhesion of microorganisms to soil and sediment particulates is lacking. Permanent adhesion, in this sense, refers to the anchoring of microorganisms to solid surfaces by means of polymer bridging [29], ensuring that the organisms remain attached even when substantial shear forces are applied. In some microbial ecosystems the colonizable surfaces are suitable for embedding, ultrathin sectioning, and, finally, visualization under a transmission electron microscope (TEM) of the polymeric material responsible for anchoring bacteria to the surface (see Chapters 4, 5, 7, and 8). Direct observation of microorganisms in opaque soils and sediments is difficult, and it is almost impossible to determine whether the organisms are attached firmly to surfaces, even using a combination of techniques such as phase, fluorescence, and scanning electron microscopy. Especially with SEM, it is dangerous to assume that all microorganisms seen on solid surfaces were firmly adhering to the surfaces prior to specimen drying. Many of the organisms may have been in the bulk aqueous phase or superficially attracted to the surfaces and merely deposited on the surfaces during the drying process.

Reversible sorption to saline sediment particulates by natural and artificially added microorganisms was demonstrated by Roper and Marshall [30] using a stepwise washing technique. Following each washing and centrifugation of the sediment, numbers of microorganisms in the aqueous phase (supernatant) were determined (Fig. 9.3). Except in the case of bacteriophage (Fig. 9.3*b*), desorption of microorganisms at relatively high salinities was negligible. A sudden appearance of high numbers of microorganisms in the supernatant coincided with a dispersion of colloid materials, giving a significant increase in supernatant turbidity (Figs. 9.3*a–c*). At high salinities, the colloids were in a flocculated state and the organisms reversibly sorbed at particulate surfaces. Dilution of the electrolyte beyond a critical concentration resulted in the repulsion of colloidal particles and associated microorganisms from each other (dispersion). The recovery of the phage and *E. coli* was almost complete, but it was assumed in the case of the indigenous sediment population that many microorganisms still remained firmly attached to the sediment particulates throughout these washing treatments.

Many investigations on the study of adsorption of microorganisms to soil

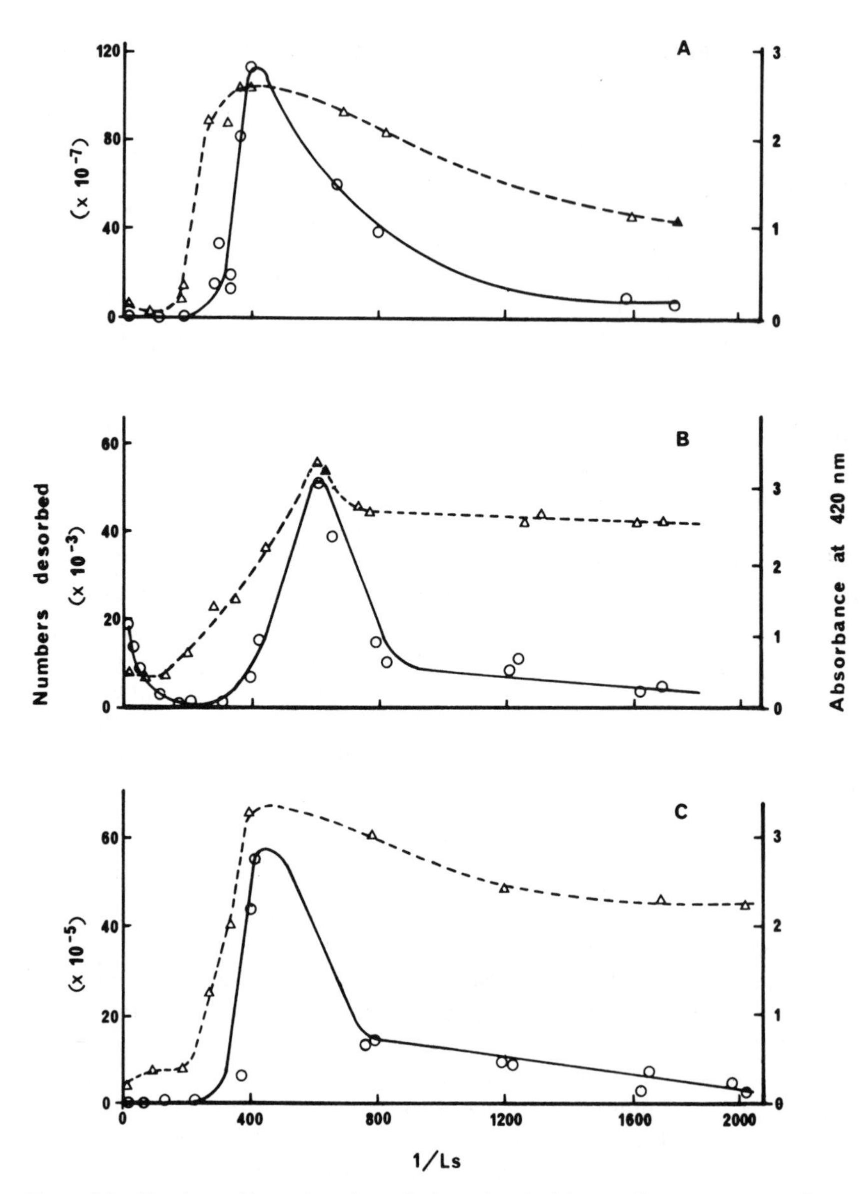

Figure 9.3 Numbers of bacteria or bacteriophage desorbed from sediment or montmorillonite as function of decreasing salinity (expressed as reciprocal of specific conductivity, L_s): (A) *Escherichia coli* + montmorillonite, input = 2.0×10^9 ml^{-1}, recovery = 2.1×10^9 ml^{-1}; (B) phage + sediment, input = 5×10^7 PFU, recovery = 1.3×10^5 PFU; (C) indigenous bacteria from sediment; O——O, numbers desorbed; Δ----Δ, absorbance at 420 nm (indicating state of dispersion of colloids). Redrawn from Roper and Marshall [30].

[6] have involved the use of pure cultures added to either sterile or unsterile soils. The percentage of microorganisms sorbed to the soil is given by the following equation [26]:

$$\text{Percent sorption} = \frac{A - (B - C) \times 100}{A}$$

where A is the count of the untreated bacterial suspension, B is that of the bacterial suspension plus soil, and C is that of the uninoculated soil suspension. This method does not differentiate between reversible and permanent adhesion of the added microorganisms. Attempts to increase the efficiency of release of microorganisms from soils [31,32] for viable count determinations have not provided any clear indication of the true relationship between the microorganisms and soil particulates. Although sonication [31] may break the bridging polymers responsible for permanent microbial adhesion, it is more likely that the main effect of this treatment is in aggregate disruption. In any natural soil it is likely that microorganisms are found in the bulk aqueous phase (with or without colloidal material associated with the microbial surface), physically entrapped within aggregates, reversibly sorbed to particulates of variable size, or permanently attached to such particulates by means of the polymer bridging process. The ecological significance of these various microorganism–particulate associations are considered in Section 9.6.

It should be obvious that there is a real need for more suitable techniques for assessing the extent of permanently attached microorganisms in sediments and soils. Until such techniques are available, the full ecological significance of permanent attachment of microorganisms in such systems cannot be appreciated.

9.6 ECOLOGIC CONSIDERATIONS

The water phase adjacent to the organized solid components in sediments and soils provides the major medium for microbial growth and development within such ecosystems. There is no doubt, however, that the nature and organization of the solid phase in such systems has profound effects on the ecology of the microorganisms. The various interactions between microorganisms and solid components described in Sections 9.4 and 9.5 are summarized in Fig. 9.4 with respect to individual microorganisms. All or some of these conditions may exist within any finite site in a soil or sediment.

With the exception of the plant root zone (rhizosphere) or following seasonal addition of organic matter (crop debris or leaf litter), most soils

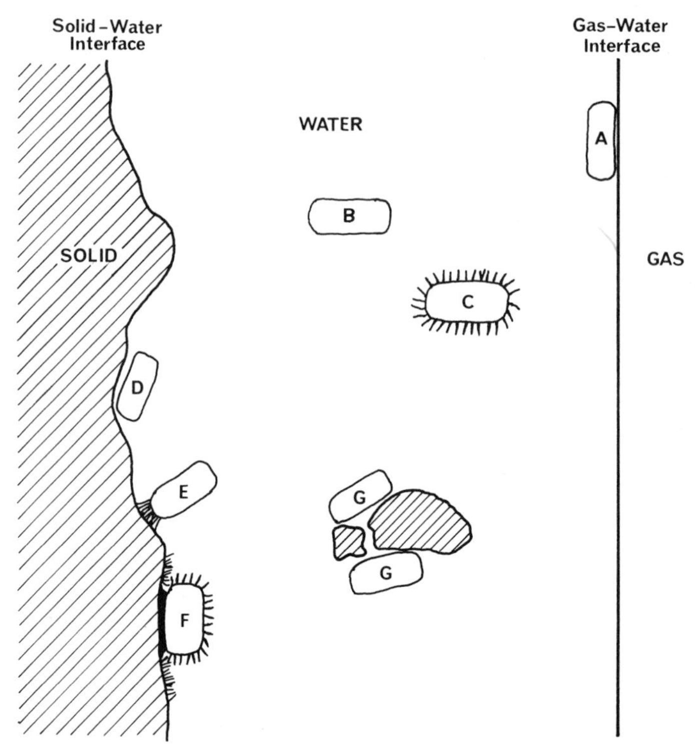

Figure 9.4 Diagrammatic representation of bacteria in aqueous phase of soils or sediments: (A) bacterium at the gas–liquid interface; (B) bacterium in bulk aqueous phase; (C) bacterium in aqueous phase where colloidal material is present; (D) bacterium reversibly sorbed at solid surface; (E) bacterium irreversibly sorbed at solid surface; (F) bacterium as in (E) but enveloped in colloidal material; (G) coflocculation of bacteria and particulates of comparable size.

are in a state of chronic nutrient deficiency with respect to microbial growth. Extremely small bacteria are common in soils [33], which suggests the development of dwarf forms [34,35]. The situations depicted in Fig. 9.4 can be related to such a nutrient-deficient habitat. Organism A should benefit from nutrients accumulated at the gas–liquid interace [8], provided that the gas phase is suitable for the particular organism. The gas phase may be air or, in restricted microenvironments and in waterlogged conditions where oxygen levels are low or zero, other gases such as H_2, CO_2, and CH_4 might be present. This latter condition is the norm in anaerobic sediments where gases accumulate. Microorganisms in the bulk aqueous phase (organism B) should be nutrient deficient and, possibly, present as dwarf forms. If colloidal materials such as clays or organic matter are suspended in the aqueous phase, these may adhere to microbial surfaces (organism C) and alter nutrient input, metabolite output, buffering, and other conditions that may stimulate or inhibit growth [6,8]. Reversible or irreversible sorption to the solid surface (organisms D and E) should allow the microorganisms to benefit from any nutrients accumulating at or near the solid–liquid interface. Firm adhesion would provide a selective advantage only where high shear forces exist near the surface. In soils or sediments containing high levels of colloidal material (particularly clays), some microorganisms may be so effectively enveloped by the colloidal material (organism F) that normal metabolism is inhibited [6,8]. Organic matter may accumulate under such conditions because of the reduced rate of microbial degradation. Coflocculation of microorganisms and particulates of comparable size (organism G) may lead to sedimentation and certainly modifies the behavior of the microorganisms significantly [2].

Clays have a dominant role in the ecology of soil and sediment microorganisms. The swelling clay montmorillonite, in particular, provides a buffering effect in soils that favors the growth of acid-sensitive bacteria [36,37]. This characteristic appears to be correlated with the inability of *Fusarium oxysporum* f. *cubense* (a plant pathogen) and *Histoplasma capsulatum* (a human pathogen) to colonize montmorillonite-containing soils [38]. It appears that the maintenance of high bacterial activity in such soils results in the competitive exclusion of these fungi [37,39]. Clays also modify the survival of microorganisms by providing protection from UV and X-irradiation [40,41], desiccation [42–44], antibiotics [45], and predator–prey interactions [30,46,47].

In estuarine situations the particulate loading of river water can influence the distribution and activity of microorganisms in the water. Both the microorganisms and the suspended particulates carried in freshwater streams tend to flocculate and sediment as the salinity increases. There is some evidence that coliform bacteria survive longer in sediments than in the overlying water [48,49]. Various microbial parasites and predators are partly responsible for the kill of alien bacteria [i.e., coliform bacteria] in aquatic systems [50,51]. The presence of colloidal and particulate materials reduces the efficiency of these parasites and predators [30,46,47,51].

The dumping of raw sewage into river estuaries probably leads to the deposition of enteropathogenic bacteria and viruses in sediments as a result of flocculation of microorganisms and particulates at the higher salinities found in the lower reaches of the estuaries. Such pathogens may be protected from biological-control agents, and the saline sediments may be a reservoir of such pathogens. Desorption of these microorganisms following a significant decrease in salinity, as would occur in flood conditions, may constitute a serious public-health hazard [30,46,49].

9.7 CONCLUSIONS

Microbial habitats in soils and sediments are established in part by the types and the three-dimensional organization of the particulates contained in these media. In view of the complex range of associations between microorganisms and particulates described in this chapter, I have attempted to emphasize the need for a clearer understanding of the role of particulates in determining the distribution and functioning of microorganisms in soils and sediments. These interactions include the reversible and permanent adhesion of microorganisms to particulate surfaces, as well as the adsorption of colloidal particulates to microbial surfaces.

9.8 REFERENCES

1. G. Stotzky, *Transact. N. Y. Acad. Sci., Series II*, **30,** 11 (1967).
2. G. Stotzky, *Crit. Rev. Microbiol.*, **2,** 59 (1972).
3. Z. Filip, *Folia Microbiol.*, **18,** 56 (1973).
4. T. Hattori, *Microbial Life in the Soil: An Introduction*, Marcel Dekker, New York, 1973.
5. G. Bitton, *Water Res.*, **9,** 473 (1975).
6. K. C. Marshall, in A. D. McLaren and J. J. Skujins, Eds., *Soil Biochemistry*, Vol. 2, Marcel Dekker, New York, 1971, p. 409.
7. K. C. Marshall, *Annu. Rev. Phytopathol.*, **13,** 357 (1975).
8. K. C. Marshall, *Interfaces in Microbial Ecology*, Harvard U. P., Cambridge, Mass., 1976.
9. T. R. G. Gray, *Symp. Soc. Gen. Microbiol.*, **26,** 327 (1976).
10. K. C. Marshall and N. A. Marshman, in W. Krumbein, Ed., *Environmental Biogeochemistry and Geomicrobiology*, Ann Arbor Science Publications, Ann Arbor, Mich., 1978, p. 611.
11. N. A. Marshman, "The Ecology of Soil Microorganisms as Influenced by Particulate Matter," Ph.D. thesis, University of New South Wales, 1978.
12. R. H. Whittaker, S. A. Levin, and R. B. Root, *Am. Natur.*, **107,** 321 (1973).
13. R. Brewer, *Fabric and Mineral Analysis of Soils*, Wiley, New York, 1964.
14. E. Burrichter, *Z. Pflanzenernaer. Dung. Bodenk.*, **63,** 154 (1953).
15. N. A. Krasil'nikov, *Soil Microorganisms and Higher Plants*, Academy of Sciences U.S.S.R., Moscow (Engl. transl.). 1978.
16. D. G. Zvyagintsev, *Soviet Soil Sci.*, **1962,** 140.
17. T. R. G. Gray, *Science*, **155,** 1668 (1967).
18. T. R. G. Gray, P. Baxby, I. R. Hill, and M. Goodfellow, in T. R. G. Gray and D. Parkinson, Eds., *The Ecology of Soil Bacteria*, Liverpool U. P., Liverpool, U.K., 1968, p. 171.
19. A. Siala, I. R. Hill, and T. R. G. Gray, *J. Gen. Microbiol.*, **81,** 183 (1974).

20. G. E. Hutchinson, *A Treatise on Limnology*, Vol. 1, Wiley, New York, 1957.
21. T. Santoro and G. Stotzky, *Can. J. Microbiol.*, **14,** 299 (1968).
22. N. Lahav, *Plant Soil*, **17,** 191 (1962).
23. K. C. Marshall, *Biochim. Biophys. Acta*, **156,** 179 (1968).
24. K. C. Marshall, *J. Gen. Microbiol.*, **56,** 301 (1969).
25. K. C. Marshall, *Biochim. Biophys. Acta*, **193,** 472 (1969).
26. T. C. Peele, Cornell University Agriculture Experiment Station Memoir No. 197, 1936.
27. R. A. Neihof and G. I. Loeb, *Limnol. Oceanogr.*, **17,** 7 (1972).
28. R. A. Neihof and G. Loeb, *J. Mar. Res.*, **32,** 5 (1974).
29. K. C. Marshall, R. Stout, and R. Mitchell, *J. Gen. Microbiol.*, **68,** 337 (1971).
30. M. M. Roper and K. C. Marshall, *Microbial Ecol.*, **1,** 1 (1974).
31. D. G. Zvyagintsev, *Sov. Soil Sci.*, **1966,** 811.
32. T. Hattori, *Bull. Inst. Agr. Res. Tohoku Univ.*, **18,** 159 (1967).
33. L. E. Casida, *Can. J. Microbiol.*, **23,** 214 (1977).
34. H. W. Jannasch, *J. Gen. Microbiol.*, **18,** 609 (1958).
35. J. A. Novitsky and R. Y. Morita, *Appl. Environ. Microbiol.*, **32,** 617 (1976).
36. G. Stotzky, *Can. J. Microbiol.*, **12,** 831 (1966).
37. G. Stotzky and L. T. Rem, *Can. J. Microbiol.*, **12,** 547 (1966).
38. G. Stotzky and R. T. Martin, *Plant Soil*, **18,** 317 (1963).
39. G. Stotzky and A. H. Post, *Can. J. Microbiol.*, **13,** 1 (1967).
40. G. Bitton, Y. Henis, and N. Lahav, *Appl. Microbiol.*, **23,** 870 (1972).
41. H. P. Müller and L. Schmidt, *Arch. Mikrobiol.*, **54,** 70 (1966).
42. K. C. Marshall, *Austral. J. Agric. Res.*, **15,** 273 (1964).
43. H. V. A. Bushby and K. C. Marshall, *Soil Biol. Biochem.*, **9,** 143 (1977).
44. Y. Dommergues, *Sci. Sol*, **1964.** 141.
45. F. A. Skinner, *J. Gen. Microbiol.*, **14,** 393 (1956).
46. M. M. Roper and K. C. Marshall, *Microbial Ecol.*, **4,** 279 (1978).
47. G. Bitton and R. Mitchell, *J. Environ. Eng. Div. Proc. Am. Soc. Civil Eng.*, **100,** 1310 (1974).
48. D. J. Van Donsel and E. E. Geldreich, *Water Res.*, **5,** 1079 (1971).
49. M. M. Roper and K. C. Marshall, *Geomicrobiol. J.*, **1,** 103 (1979).
50. R. Mitchell, *Nature* (*Lond.*), **230,** 257 (1971).
51. M. M. Roper and K. C. Marshall, *Austral. J. Mar. Freshwater Res.*, **29,** 335 (1978).

CHAPTER 10

Adsorption of Viruses to Surfaces: Technological and Ecological Implications

GABRIEL BITTON
Department of Environmental Engineering Sciences, University of Florida, Gainesville

CONTENTS

10.1 INTRODUCTION

Among a wide variety of microorganisms known to man, viruses are the smallest intracellular parasites, with a size range of 20–500 nm. From a public-health point of view, the viruses of greatest concern are the enteric viruses that are shed into feces and, subsequently, may initiate a wide variety of diseases ranging from skin rash to paralysis [1,2]. Most of these infectious agents fall within the size range (20–2000 Å) of colloidal particles (Fig. 10.1). Thus viruses are biocolloids with electrical properties governing their sorptive behavior toward biological (host-cell surfaces) and nonbiological surfaces [3–5]. Since their nucleic acid is encased in a proteinaceous capsid, they behave as amphoteric colloids with surface charges resulting from the ionization of surface groups, such as carboxyl– ($-COO^-$) and amino– groups ($-NH_3^+$).

Once in the cell-free state, viruses may encounter a solid surface and this interaction is governed by Brownian motion, electrostatic forces, and electrical double-layer phenomena [4,6]. Even the interaction of viruses with the surface of their host cells is influenced by the presence of specific chemical groups, cations, and appropriate pH [7,8]. The phage–bacterial cell interaction has been the most studied system, but less is known about the adsorption of animal viruses to their host cells. Similar principles govern the attachment of viruses to surfaces other than host-cell surfaces. This subject has been summarized by Bitton [3], and details are given later

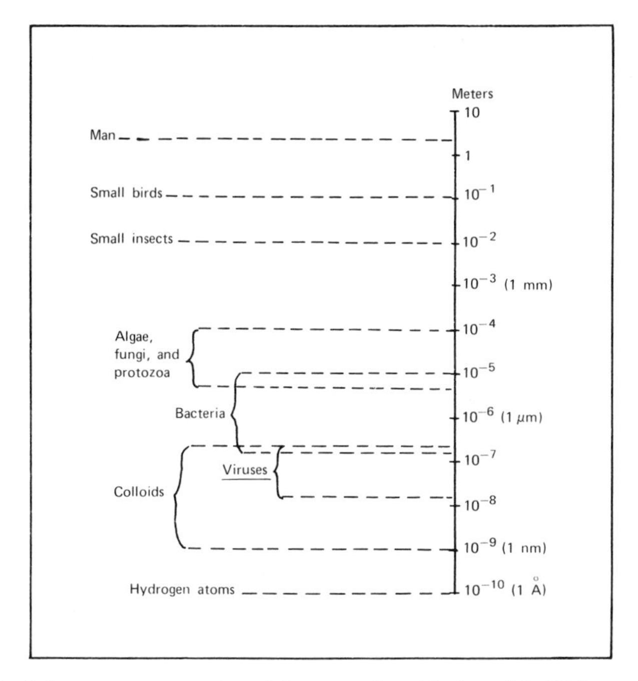

Figure 10.1 Schematic representation of size range (logarithmic scale) of living organisms and colloidal particles [4]. Courtesy of Harvard University Press, Cambridge, Mass.

(Section 10.2) on the sorptive interactions between viruses and particular surfaces.

Desorption of viruses from surfaces may occur following changes in the physicochemical properties of the suspending medium. These changes are brought about by lowering the ionic strength of the medium [9,10] or by adding proteinaceous materials at high pH values [11–15]. Other eluents include "humic substances" (Bitton, Gifford, Lanni, and Allinson, unpublished), sodium dodecyl sulfate [16], EDTA [15,17], Tween 80 [18], phosphate buffer [19], saturated $NaHCO_3$ [20], glycine buffer [17], and many others. Moreover, the efficiency of elution of viruses from surfaces can be improved by exposing the samples to acoustic energy [21].

The topic of virus desorption from surfaces is undoubtedly of utmost importance in environmental virology. Understanding this phenomenon will help us improve the virus detection methodology (e.g., desorption from membrane filters) and better assess some public-health problems, such as those associated with the release of viruses from sediments and soils.

The purpose of the present chapter is to address the topic of virus adsorption onto surfaces and its implications in removal and detection techniques used in the rapidly developing field of environmental virology. In addition, the impact of these sorptive phenomena on virus survival in the host-cell free state is examined.

10.2 TECHNOLOGICAL IMPLICATIONS OF SORPTIVE INTERACTION BETWEEN VIRUSES AND SURFACES

10.2.1 Virus Adsorption Phenomena in Wastewater-treatment Processes

Virus association with solids plays a tremendous role in the removal of viruses by waste treatment operations, such as activated sludge or oxidation ponds. This association and its impact on virus removal from wastewater are described.

Virus Association with Solids in Wastewater and Sludge. Domestic wastewater contains a variety of solids of biological and mineral origin [22]. During waste-treatment processes more solids are produced due to the conversion of soluble organic substances to cellular material. Virus association with wastewater solids originates from their affinity for fecal solids. Many investigators have attempted to evaluate the proportion of viruses attached to wastewater solids. Duff [23] found that 25–70% of viruses were bound to solids. Using a sonication technique to free attached and embedded viruses, Wellings et al. [21] reported that 16–100% of the total viruses were associated with solids. In another study [24] they found that 68–90% of viruses were bound to solids in influent and effluent samples from a package-treatment plant in Gainesville, Florida. The solid-associated viruses in mixed liquor samples with a suspended solids concentration ranging 980–2400 mg/liter are shown in Fig. 10.2. The levels of solid-bound viruses were 5–10 fold higher than those of "free" viruses [13,25]. The kinetics of viral adsorption to suspended solids in sewage were studied by Moore et al. [13]. Sewage samples containing 2600 ppm of suspended solids were inoculated with ^{3}H-tagged poliovirus. Batch experiments showed that most of the viruses were adsorbed within a 30-min period (Table 10.1). A comparison of counts per minute (cpm) to plaque-forming units (PFU) revealed that some inactivated viruses were released after 30 min of contact time. In field studies other authors reported the association of viruses with solids in activated sludge, primary sludge, and wastewater [26–28].

It is important to take solids-associated viruses into account when attempting to monitor wastewater for levels of indigenous viruses. Failure to do so would result in a false sense of security. The solid-bound viruses are not inactivated in the sorbed state (see Section 10.3).

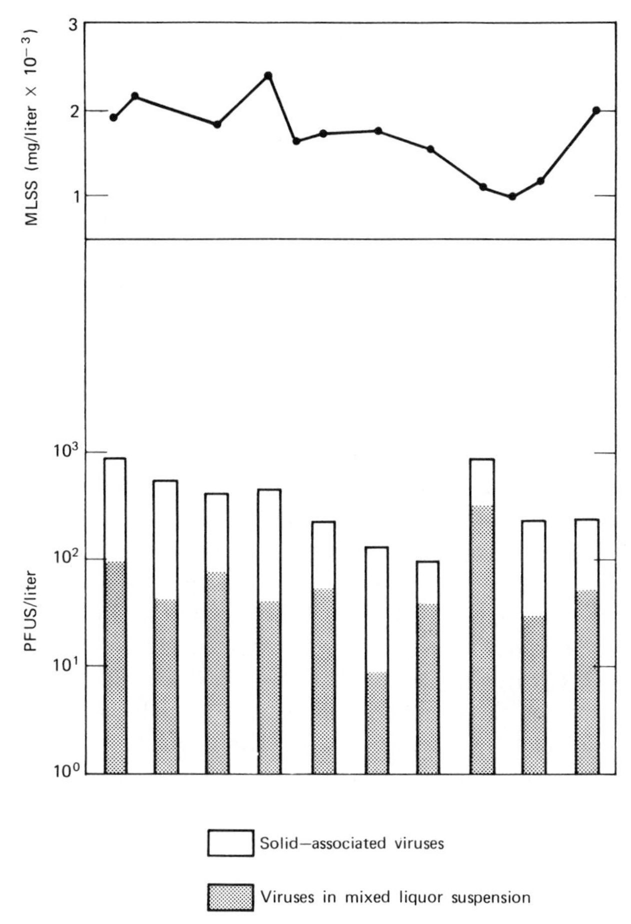

Figure 10.2 Virus association with wastewater solids [13]. Courtesy of the Center for Research in Water Resources, University of Texas at Austin.

Removal of Viruses by Activated Sludge Process. Following the primary clarification step, which removes some solid-associated viruses by sedimentation, the domestic sewage is generally subjected to biological treatment to convert dissolved organics into cell mass. Viruses are adsorbed to or embedded in flocs composed of microbial cells and inorganic or organic debris. The activated sludge process is probably the most efficient way to remove viruses during biological treatment [29–31]. Virus removal results from biological inactivation by bacteria [32] or protozoa [33] or by the adsorption of viruses onto sludge particles. This latter removal mechanism probably accounts for a 10-fold reduction in virus PFUs and was first

Table 10.1 Association of ^{3}H-poliovirus with mixed liquor suspended solids (2600 ppm)[a]

Contact time (min)	Virus in supernatant (%)	
	CPM	PFU
1	100	100
5	28	24
10	20	18
30	17	2
60	20	1
120	32	1

[a] Adapted from Moore et al. [13].

demonstrated by Clarke and his collaborators [34]. They reported that the pattern of virus removal by activated sludge conformed to the Freundlich adsorption isotherm. This phenomenon is shown for Coxsackie virus A9 in the presence of 2320 ppm and 4960 ppm of suspended solids at 4°C (Fig. 10.3). A close examination of Fig. 10.3 shows that the slopes of the straight lines were close to 1, indicating that the percent adsorption was constant over the range of virus levels tested. However, these authors were not successful in desorbing the virus from the sludge particles and concluded that the virus–sludge complex was very stable. Undoubtedly, the adsorption pattern is complicated by the fact that viruses also may be inactivated by bio-

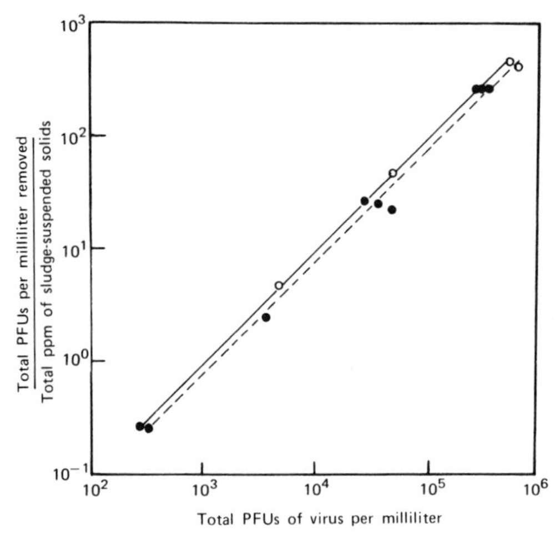

Figure 10.3 Freundlich isotherm describing adsorption of viruses to sludge-suspended solids [34]. Courtesy of the American Public Health Association.

logical or chemical components of wastewater [35]. Ranganathan et al. [36] also demonstrated that the reduction of poliovirus by activated sludge was above 98% and that the attachment process accounted for an approximately 10-fold reduction. However, a more recent work showed that a large proportion of the sludge-associated virus can be recovered in the infective state [37]. Thus the activated sludge process is merely a transfer of infective viruses from wastewater to the sludge [27], and this must be taken into account when one contemplates the disposal of sludge onto land.

Virus Removal in Oxidation Ponds. Viruses are not removed completely by passage of wastewater through oxidation ponds [29]. Malherbe and Strickland-Cholmley [38] showed reovirus and enterovirus reduction after passage through these ponds. Three basic factors are known to contribute to the removal of viruses: solar radiation [38], inactivation by pond bacteria [39], and virus adsorption to suspended solids with subsequent settling to the bottom of the pond. Sobsey and Cooper [39] studied the latter phenomenon and found that approximately 25–30% of the added viruses became associated with the pond solids within only 3 min of contact time (Table 10.2). These attached viruses also were found to be infective, indicating that the sorption process was not irreversible. Moreover, as in the case of activated sludge, the adsorption process conformed to the Freundlich isotherm.

10.2.2 Adsorption of Viruses to Soils: Implications in the Disposal of Sewage Effluents and Residuals onto Land

The U.S. Water Pollution Control Act Amendment of 1972 (PL 92-500) states that "Waste treatment management plans and practices should

Table 10.2 Association of viruses with stabilization pond solids[a]

Contact time (min)	Viruses associated with solids (%)
0	—
3	30
11	24
22	18
35	25
95	30
122	30

[a] Adapted from Sobsey and Cooper [39].

provide for the application of the best practicable waste treatment technology before any discharge into receiving water." In this context, land application of wastewater effluents and residuals appears as a viable alternative [40,41] and is discussed here in more detail with reference to the fate of viruses. The practice of land application is hardly a novelty and has been known to mankind since antiquity [42]. Within the Western world, England was probably the first country to dispose of sewage onto land, and the United States adopted this practice in the 1870s [43]. However, the public-health aspect of sewage application on land was addressed much later, and this triggered some research on the fate of pathogens applied to soil; but the virus question was ignored, probably due to the lack of adequate methodology.

Land spreading of wastewater effluents and residuals has many advantages, such as water conservation, supply of nutrients to crops, enrichment of the soil with organic matter, and protection of surface waters from pollution. There is a growing concern, however, over the contamination of groundwater with heavy metals and nitrates and with microbial pathogens [42,44,45]. Foster and Engelbrecht [44] have estimated the number of pathogens applied to soil when chlorinated effluents were used at a rate of 2 in./week. It was calculated that the pathogen density was not greater than one viable microorganism/4 ft^2 day^{-1}. As a result of land disposal, four categories of public-health hazards are in sight: aerosol generation by spray irrigation systems, contamination of surface waters by runoff, crop contamination, and movement of microbial pathogens through soils.

Of most concern is the transport pattern of virus particles through soils. This topic was reviewed by Bitton [3] and Gerba et al. [46] and is discussed in light of recent findings. A close look at field studies on wastewater application to soils shows that viruses may or may not be present in groundwater supplies. Epidemics of infectious hepatitis were connected with groundwater supplies and it was found on at least two occasions that the virus traveled through fissures and fractures within the substratum [47,48]. Mack et al. [49] isolated poliovirus from a 100-ft-deep well located near a wastewater drainfield. In Hawaii, following the application of chlorinated sewage effluents, virus was isolated from a 5-ft-deep well [50]. More recent studies by Wellings et al. [24,51,52] demonstrated the presence of enteric viruses in groundwater from a spray irrigation site and from a cypress dome site treated with wastewater effluents. However, wastewater reclamation projects in California [53] and Arizona [54] did not show any virus movement through the soil. The most quoted work is probably the Santee Recreation project in Santee, California, where viruses were retained in less than 200 ft of travel through sandy soil [55]. These conflicting reports prompted research on the mechanism of virus removal by soils.

Sorptive Phenomena in Land Application of Sewage Effluents. Straining and sieving are important removal mechanisms for relatively large particles [56] of the size of an amoeboid cyst or even a bacterial cell. These mechanisms become negligible when particles are of colloidal size. Therefore, the major mechanism of virus removal is probably adsorption [3,57,58].

The adsorption of viruses to soils includes a *transportation process* (virus is carried to the vicinity of the soil particle) and an *attachment process* [57]. Batch and column experiments were undertaken to better understand this attachment step. It is known that this process is controlled by the type of soil, flow rate, degree of saturation of pores, cations, organic materials, and type of virus.

Type of Soil. It is assumed that the presence of clay minerals enhances the adsorption of bacteria and viruses to soils. This assumption is based on the fact that pure clay minerals were found to display a high adsorptive capacity toward viruses [9], as a result of the significant surface area and ion-exchange capacity of these minerals. A study of nine soils from Arkansas and California showed that phage adsorption increased with the clay content and the specific surface area of the soil [58,59]. Soils are more adsorbent than sand toward viruses [60]. Bitton and his collaborators (unpublished) assessed the sorptive capacity of sandy clay loam containing 28% kaolinite. It was shown that this soil displayed a good sorptive capacity toward viruses. No poliovirus breakthrough was observed even after the application of eight pore volumes (4.37 cm) of groundwater (Fig. 10.4). After percolation of 58 cm of water, more than 99% of applied viruses were adsorbed to the soil.

It is still debated whether the organic-matter fraction of soils increases or decreases the attachment of viruses to soils. Other soil components, including iron oxides (hematite, goethite, and magnetite), may actively participate in the virus sorption process in soils, but their precise role remains to be determined.

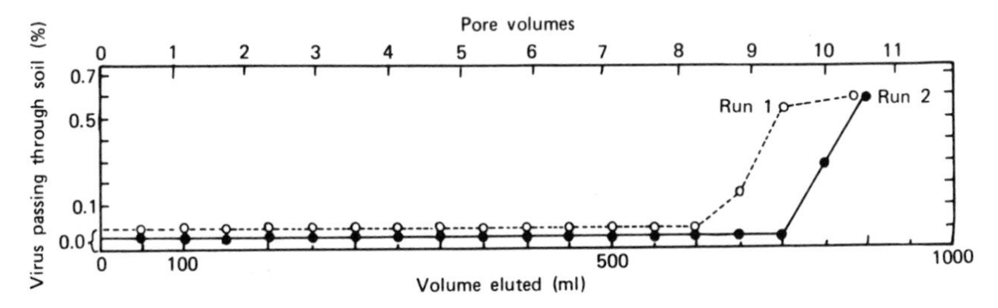

Figure 10.4 Movement of poliovirus through sandy clay loam following application of groundwater [68].

Flow Rate. It is agreed that low flow rates promote virus adsorption to soils or to pure sand. Eliassen et al. [61] demonstrated a better removal of T2 phage at lower flow rates, and green [60] showed that poliovirus suspended in septic tank effluents was retained better at low flow rates by sand mounds.

Degree of Saturation of Pores. The opportunity for virus contact with soil particles is reduced when the degree of saturation of the pores within the soil matrix is increased [60]. This phenomenon has practical implications; drying of the soil between sewage applications could prevent the desorption of viruses by rainwater, and this may be a good management practice in land application of sewage effluents [62].

Level of pH. As the pH of the soil solution increases, viruses and soil particles behave as anions and will repel each other. Therefore, a low pH will result in a decrease of the net negative charge on the particles and will promote virus adsorption to soils [58,59, 63–66]. Studies on the effect of pH on the adsorption of T1 phage to five soils (Fig. 10.5) have revealed that the adsorption could decrease to less than 40% when the pH value was around 8 [59].

Ionic Strength of Percolating Water. The concentration and species of cations present in percolating water have a marked influence on the degree of adsorption of viruses to soil particles. Adsorption of viruses to soils seems to follow the Schultz–Hardy rule; that is, the degree of retention of viruses increases with the valency of the cation. The effect of cation concentration on adsorption of phage T2 to various soils is shown in Fig. 10.6 [59]. Viruses are generally retained in the top inches of soil in the presence of an appropriate cation concentration [67]. This phenomenon is illustrated (Table 10.3) by poliovirus, which was concentrated in the two top fractions of a sand column [67].

In field situations, Wellings et al. [51,52] have reported a "burst" of viruses after a heavy rainfall, indicating that rainwater is an important factor in the redistribution of viruses within the soil matrix. Furthermore, laboratory experiments showed that the leaching of soil columns with distilled water (to simulate rainwater) brings about a decrease in ionic strength of the soil solution and subsequent desorption of soil-bound viruses [62,63]. When the soil columns were leached with distilled water, maximum poliovirus breakthrough was observed before the minimum conductivity was reached. Total organic carbon (TOC) followed a similar pattern (Fig. 10.7). The desorption phenomenon was less pronounced when using rainwater to elute attached virus from a sandy clay loam containing 28% kaolinite [68]. Only 0.0004% of total applied poliovirus was eluted after application of 7

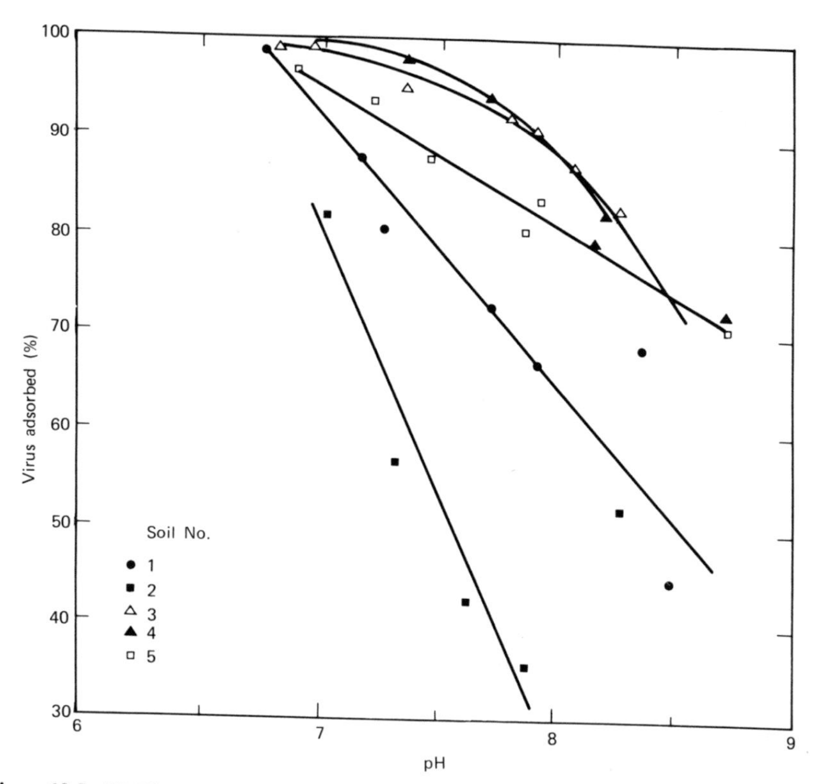

Figure 10.5 T1 Phage adsorption to soils as function of pH [59]. Courtesy of the Water Pollution Control Federation.

Table 10.3 Effect of 0.01*N* $CaCl_2$ or $MgCl_2$ on retention of poliovirus in sand columns.

Column Fraction	Number of PFU/Samples	
	$CaCl_2$	$MgCl_2$
Total influent	3.93×10^6	6.10×10^6
1 + 2	3.00×10^6	1.78×10^7
3 + 6	3.32×10^2	7.99×10^2
5	0	0

[a] Adapted from Lefler and Kott [67].

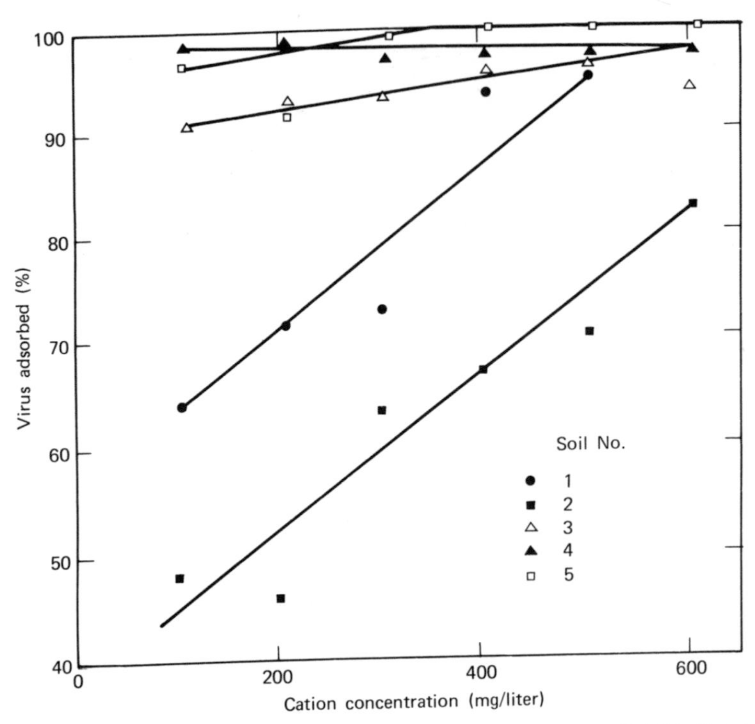

Figure 10.6 Effect of cation concentration on T2 phage adsorption to soils [59]. Courtesy of the Water Pollution Control Federation.

cm of rainwater. Rainwater has a low conductivity (~20 μmhos/cm) and its application on the soil surface results in a decrease of the ionic strength of the soil solution, and thus, it is expected to play a major role in the redistribution of viruses within the soil profile.

Effect of Soluble Organic Materials. In general, soluble organic materials compete with viruses for the adsorption sites on the soil particles. Many examples drawn from the literature illustrate this phenomenon [3]. In this context, interest is focused on the effect of organics present in sewage effluents on the movement pattern of viruses. The amount of organics present in a typical secondary sewage effluent probably does not hinder the adsorption of viruses to the soil matrix. Poliovirus, suspended in activated sludge effluent, did not move appreciably through a Florida sandy soil (Bitton, Pancorbo, Gifford, and Overman, unpublished). Moreover, typical secondary sewage effluents have conductivities ranging 500–700 μmhos/cm, and this is enough to promote the retention of viruses by the soil matrix.

The presence of "humic substances" in water may interfere with the adsorption of viruses to soils. This particular situation was encountered when secondary sewage effluents were applied to a cypress "dome" in Gainesville, Florida. (The trees in the center of this wetland are the tallest and form a "dome".) The sewage effluent became highly colored by the decomposition products of cypress needles ("humic substances"). Soil-column experiments showed that these colored materials competed with poliovirus for the adsorption sites in the soil [69]. Results on the movement of poliovirus 1 through the cypress dome soil are displayed in Fig. 10.8. The adsorption was higher in the presence of tap water than in the presence of "dome water" (standing water, which had a color content ranging 420–750 platinum units). Leachates from the black sediments in the cypress dome are highly colored (2000–6000 units) and repeatedly promote the movement of viruses through the soil profile. The removal of the "humic substances" by activated carbon treatment led to more suitable conditions for virus retention by soils (Fig. 10.9). Colored substances were also found to displace soil-bound viruses [69] and to lower the adsorption of viruses to membrane filters [70].

Types of Virus. Viruses differ in size, shape, structure, and isoelectric points. As far as adsorption pattern is concerned, it is difficult to compare the relatively large-tailed DNA *T*-phages to small tailless RNA particles

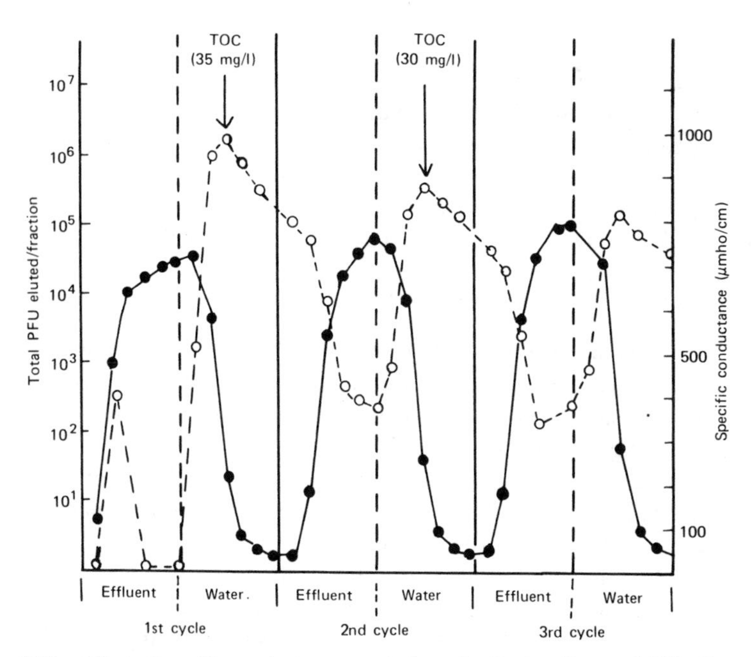

Figure 10.7 Effect of specific conductance on elution of poliovirus from soil [63]. Courtesy of the American Society for Microbiology.

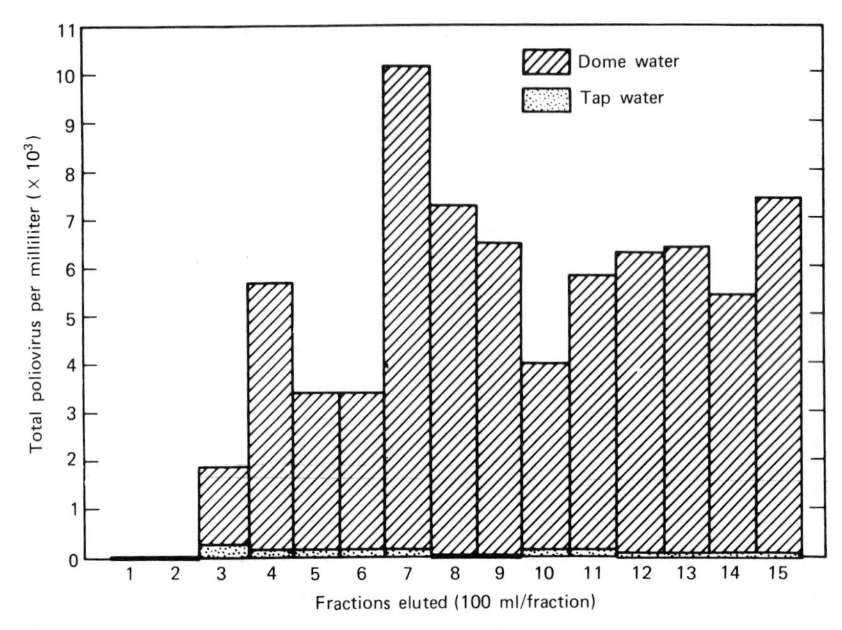

Figure 10.8 Effect of "dome water" (standing colored water in cypress dome) on movement of poliovirus through soil [69]. Courtesy of the American Society of Agronomy.

like picornaviruses in general and to poliovirus in particular. The behavior of enteric viruses could be simulated more effectively through the use of small and tailless RNA phages, such as f2 or MS2. The problem of viral indicator is not resolved yet since there are questions concerning the use of phages to simulate the behavior of animal viruses.

Some authors have looked at the nature of the adsorption process, and it was found that the sorption of viruses to soils conform to the Freundlich isotherm [59,65,71], as illustrated in Fig. 10.10 [59].

Sorptive Phenomena in Land Application of Sludge. Land application of sludge is probably the most promising disposal practice in the future. This method has the same basic advantages and disadvantages already discussed at the beginning of Section 10.2.2. However, there are three basic differences between sludge and sewage effluents: (1) sludge contains higher amounts of heavy metals; (2) microbial pathogens are more concentrated in sludge; and (3) a greater portion of the microbial pathogens are solids associated. We have evaluated the adsorption of poliovirus to liquid digested sludge and found that 50–70% of the virus became solids associated (Bitton et al., unpublished). When virus-laden sludge is applied onto land, the "free" viruses are transported through the soil according to a pattern governed by the physicochemical parameters discussed at the beginning of Section 10.2.2. However, the sludge-bound viruses are retained, along with the sludge particles, in the top portion of the soil profile and are inactivated with time. We have studied the movement of sludge-associated virus through undisturbed columns of an Orangeburg soil from Jay, Florida. One inch of seeded liquid digested sludge was applied on top of the column, mixed with the first inch of soil, air dried for 24 hr, and then eluted with rainwater. The results of two runs revealed that only a small amount of viruses passed through the soil (Table 10.4), indicating that there is probably little risk of virus breakthrough when sludge is properly applied to land.

Movement of Viruses through Soils Leached with Septic Tank and Sanitary Landfill Effluents. Septic tank effluents are disposed of by applying them to soils. There is concern over the contamination of groundwater sup-

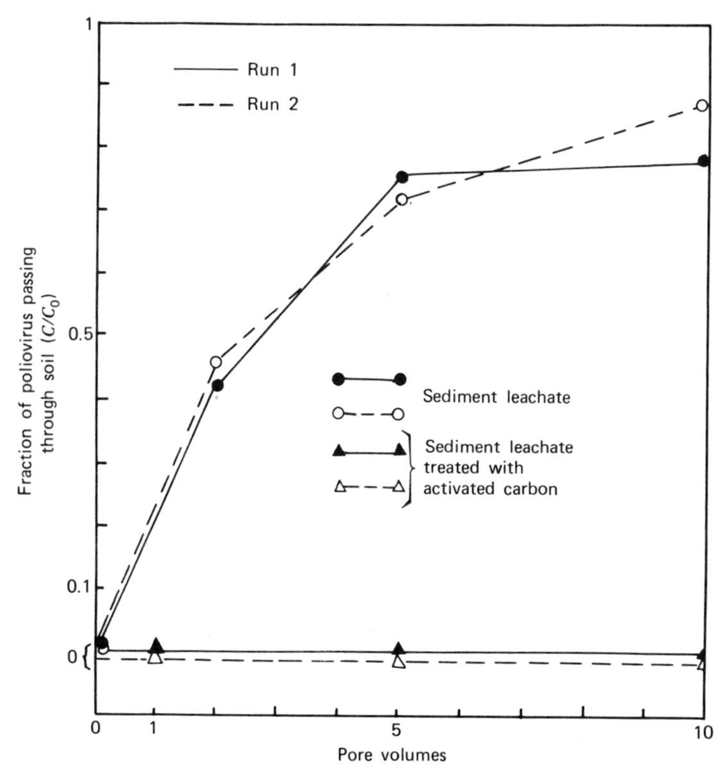

Figure 10.9 Effect of sediment leachates on movement of poliovirus through brown–red sandy soil [68].

Table 10.4 Poliovirus type 1 (Sabin) movement through an undisturbed soil core[a] of an Orangeburg soil (Jay, Florida) following the application of 1 in. of seeded liquid digested sludge[b] (air dried for 24 hr) and the subsequent elution with rainwater[c]

Solution applied	Number of pore volumes eluted	Volume eluted (ml)	Poliovirus eluted (PFU/ml)	Percent of total PFU applied (cumulative)	Conductivity of pore volume collected (μmho/cm at 25°C)	Level of pH of pore volume collected
Run 1						
Rainwater[d]	1/3	157	0	0	245	4.4
	2/3	314	0	0	195	4.7
	1	471	0	0	125	4.8
	1 1/3	628	0	0	106	5.1
	1 2/3	785	6.7	0.2	80	5.2
	2	942	0	0.2	82	5.1
	2 1/3	1099	0	0.2	77	5.1
	2 2/3	1256	0	0.2	72	5.0
	3	1413	0	0.2	66	5.0
	3 1/3	1570	0	0.2	66	5.0

Run 2						
Rainwater[d]	1/3	157	3.3	0.1	232	4.8
	2/3	314	0	0.1	177	5.2
	1	471	0	0.1	157	5.5
	1 1/3	628	0	0.1	115	6.0
	1 2/3	785	0	0.1	98	5.6
	2	942	0	0.1	87	5.4
	2 1/3	1099	0	0.1	75	5.6
	2 2/3	1256	0	0.1	69	5.5
	3	1413	0	0.1	62	5.4
	3 1/3	1570	0	0.1	58	5.5
	3 2/3	1727	0	0.1	54	5.6
	4	1884	0	0.1	54	5.7
	4 1/3	2041	0	0.1	449	6.4

[a] Soil cores were 21.5 in. in length and 2 in. inner diameter; consists of Ap, B1t, and B21t horizons.

[b] The liquid digested sludge was obtained from the City of Gainesville's wastewater-treatment plant; it was anaerobically digested (2% solids). One inch (51.6 ml) of the sludge seeded with a total of 5.1×10^5 PFU of poliovirus was applied to the core, placed on the roof of the Environmental Engineering Sciences building and allowed to air dry for 24 hours, and then worked under 1 in. Elution with rainwater was subsequently undertaken. This solution was applied from an inverted, self-regulated, 1-liter Erlenmeyer flask set to maintain a 1-in. hydraulic head on the column. The flow rate through the column was 2.5 ml/min.

[c] Adapted from Bitton, Pancorbo, Farrah, and Overman (unpublished).

[d] Rainwater was collected next to the Environmental Engineering Sciences building at the University of Florida.

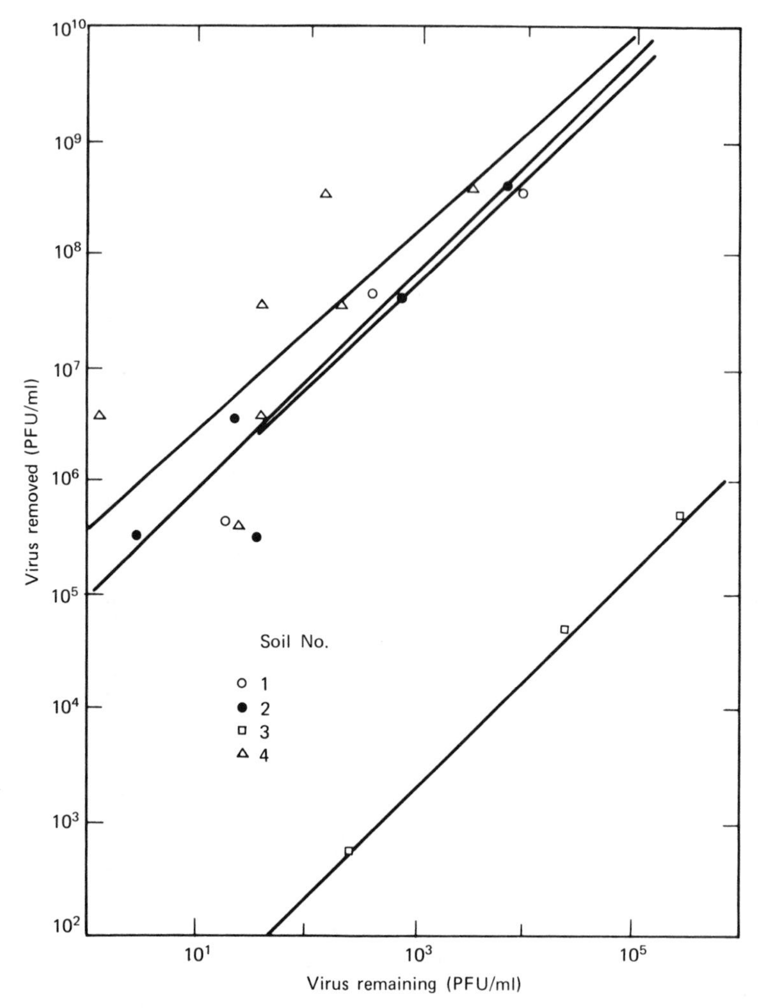

Figure 10.10 Freundlich isotherm describing adsorption of f2 phage to soils [59]. Courtesy of the Water Pollution Control Federation.

plies by these effluents. Field monitoring of coliform bacteria in soils treated with septic tank effluents showed that there was little risk of contamination of groundwater [72]. However, little information is available on the movement of viruses suspended in septic tank effluents. If the soil is shallow and the water table high, viruses may not be removed completely. Green [60] and Green and Cliver [73] studied the feasibility of using sand mounds for treating septic tank effluents. They showed that a properly designed sand mound (sand depth, dose rate, temperature, and conditioning) would be effective in removing viruses from these effluents.

Sanitary landfills are used for the disposal of municipal solid wastes. Among the solid-waste components, diapers are a major source of viruses [74]. Temperature is a major cause of virus die-off in sanitary landfill leachates [75,76]. The soil beneath the sanitary landfill must be effective in adsorbing virus to avoid groundwater contamination. The only study found in the literature is the one performed by Novello [77], who showed that 10 cm of "gravelly silty sand" removed 80–98.6% of poliovirus.

So far, we have seen that adsorption is the major mechanism involved in virus removal in soils systems. A few feet of continuous agricultural soil can effectively remove viruses from wastewater secondary effluents. Virus breakthrough can be expected, however, in situations where the substratum is fissured or fractured. Meanwhile, it appears that we do not have enough data to draw some general conclusions. More work is needed on the modelling of virus transport through soil, and equations describing the transport of pesticides through soils [78] might be adapted to predict the depth of penetration of virus particles.

10.2.3 Virus Adsorption during Water-treatment Operations

The production of safe drinking water by treatment plants involves a series of physicochemical processes for the removal of turbidity, color, and microorganisms, including viruses. The two categories of water-treatment plants are the conventional filter plants and the softening plants. Filter plants treat water by flocculation followed by clarification, filtration (sand or diatomaceous earth filtration), and final disinfection. Softening plants include lime softening to remove hardness, settling, filtration, and disinfection. In general terms, water-treatment plants remove viruses by physical separation, including sedimentation, coagulation, filtration, water-softening precipitation, adsorption, and inactivation processes, including high pH during lime softening and disinfection [79,80]. The virus adsorption capacity of some of the operations involved in both categories of water treatment plants is assessed in the paragraphs that follow.

Sand Filtration. The removal of viruses by sand filtration may be due to both adsorption and filtration processes [80]. As an adsorbent, sand generally displays a low affinity toward viruses due to a limited number of active sites on the sand particles. The virus–sand interaction is governed by the physicochemical properties of the suspending medium. It is enhanced by divalent cations [67] and inhibited by proteinaceous materials [81]. Viruses, when associated with larger particles or when embedded within flocs, are removed more efficiently by filtration through sand beds [82,83]. Coagulation of viruses prior to sand filtration brings about a greater than 99% removal of viruses [83].

Diatomaceous earth (diatomite). Diatomite is composed of remains of siliceous diatoms and is used for treating community supplies or swimming-pool water. Diatomite is inherently a poor adsorbent toward viruses [84]. However, the adsorption process may be enhanced significantly when the surface characteristics of the diatomite are modified following coating with a cationic polyelectrolyte or with ferric or aluminum hydrates [84–86].

Water Softening. Water softening aims at the removal of hardness due to calcium and magnesium compounds. This can be done by using either the lime-soda ash process or ion-exchange resins [87].

Lime–Soda Ash Process. Calcium hydroxide (lime), when used as a softening chemical, removes only carbonate hardness. Noncarbonate hardness (due to calcium and magnesium chlorides and sulfates) necessitates the use of calcium hydroxide in combination with sodium carbonate, commonly named *soda ash*. Carbonate harness is removed according to the following reactions:

$$Ca(HCO_3)_2 + Ca(OH)_2 \rightarrow 2CaCO_3 + 2H_2O \tag{10.1}$$

$$Mg(HCO_3)_2 + 2Ca(OH)_2 \rightarrow Mg(OH)_2 + CaCO_3 + 2H_2O \tag{10.2}$$

Some investigators have looked at the fate of virus during the lime softening process. The addition of lime leads to the generation of high pH values (>11) that contribute to the inactivation of viruses [82,88]. However, other mechanisms contribute to the removal of viruses. Reactions (10.1) and (10.2) show that water softening results in the precipitation of $CaCO_3$ and $Mg(OH)_2$. Wentworth et al. [89] demonstrated that viruses were better removed in the presence of $Mg(OH)_2$ than in the presence of $CaCO_3$. Electrophoretic studies [90,91] show that $CaCO_3$ possesses a net negative charge and $Mg(OH)_2$ a net positive charge throughout the entire pH range generally encountered in softening processes (Fig. 10.11). These results may explain data on the virus removal reported by Wentworth et al. [89]. The negatively charged viruses would adsorb more efficiently to positively charged $Mg(OH)_2$ precipitates than to negatively charged $CaCO_3$.

Ion-exchange Resins. Cation-exchange resins saturated with Na^+ soften the water by exchanging the Na^+ for Ca^{2+} and Mg^{2+}. The binding of viruses to ion-exchange resins was examined in the 1950s [92–95]. A comparison of cation- and anion-exchange resins showed that the latter readily adsorb virus, whereas the former fail to do so [95]. Furthermore, anion-exchange resins

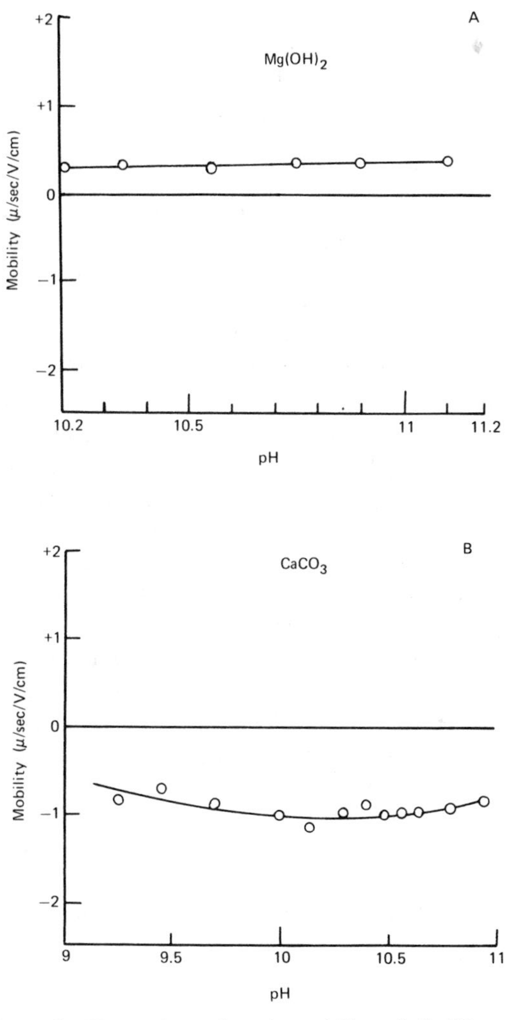

Figure 10.11 Effect of pH on electrophoretic mobility of $CaCO_3$ and $Mg(OH)_2$ [90]. Reprinted from the *Journal of American Water Works Association*, Vol. 53, by permission of the American Water Works Association.

retained viruses more efficiently than activated carbon, but the presence of soluble organics interfered with the adsorption process [96,97]. These research findings show that water softening by cation-exchange resins does not lead to a significant removal of viruses.

Activated Carbon. Activated carbon derives from a number of sources, including petroleum based residues, wood, lignite, and bituminous coal. In water-treatment plants it is used to remove taste, odor, and organic substances. It can be used at any stage of the water-treatment sequence, but beds of granular carbon are normally placed after sand filtration [87]. The removal of organic materials by activated carbon makes the disinfection process more efficient because it removes nitrogenous compounds that might combine with chlorine. Activated carbon offers a large surface area (400–1000 m^2 g^{-1}) for the adsorption of various substances. However, much of this surface area is inside pores that are not large enough to enable virus adsorption. The retention of a variety of viruses onto activated carbon has been examined, and it is generally agreed that carbon fails to retain significant amounts of virus particles, especially in the presence of organics occurring in natural water and wastewater effluents [96, 98–101]. Cookson [102–104] and Cookson and North [105] examined in detail the adsorption of phage T4 onto activated carbon and found that the process was governed essentially by the pH and the ionic strength of the suspending medium. The maximum adsorption was noted at pH 7, and Cookson [103] suggested that there was an electrostatic attraction between carboxyl groups on the activated carbon and amino groups on the virus surface. Following esterification of the carboxyl groups with acid alcohol, it was found that the attachment of T4 to activated carbon was suppressed almost completely (Fig. 10.12). The optimum ionic strength for attachment was 0.08. Above that optimum, T4 adsorption decreased due to changes in configuration of the phage tail fibers. It is known that the phage fibers attach to the tail sheath at high ionic strengths and thus are not available for adsorption to a nonbiological surface [106]. This phenomenon is peculiar to tailed viruses and was

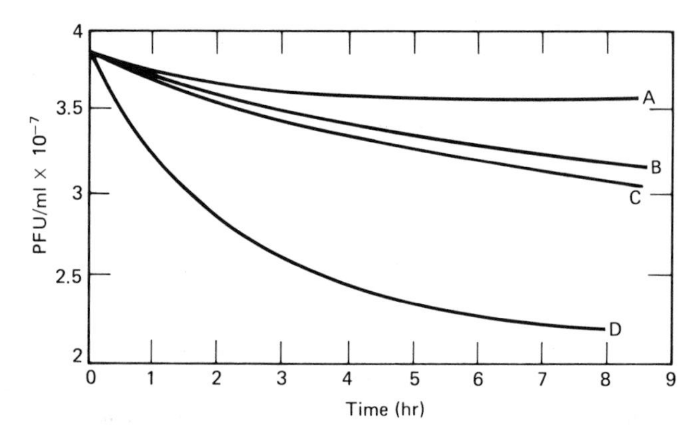

Figure 10.12 Adsorption of phage T4 on treated and untreated carbon [103]: (A) acid alcohol; (B) propylene oxide; (C) acid propylene oxide; (D) untreated carbon. Reprinted from *Journal of the American Water Works Association*, Vol. 61, by permission of the American Water Works Association.

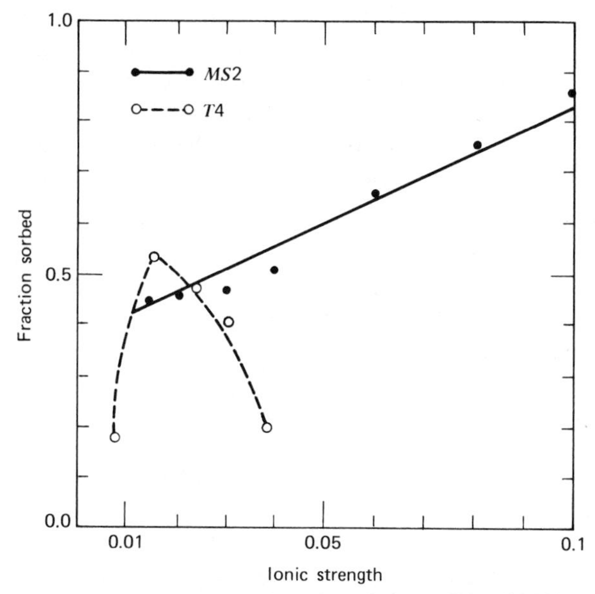

Figure 10.13 Effect of ionic strength on adsorption of phages T4 and MS2 to coal [107,108]. Courtesy of the American Society of Civil Engineering.

confirmed by the data of Oza and Chaudhuri [107,108] on the sorption of T4 and MS2 to coal. For MS2, an RNA tailless phage, the adsorption to coal increased with the ionic strength of the suspending medium, whereas there was a peak of T4 adsorption at an ionic strength of 0.015 followed by a drop in sorption with a further increase in ionic strength (Fig. 10.13). The MS2 data probably better simulate the behavior of animal viruses that do not have any tail.

10.2.4 Sorptive Phenomena in Concentration of Viruses from Large Volumes of Water

In general, raw municipal wastewater contains approximately 10^3 PFU/per liter, with exceptional cases as high as 10^6 PFU/liter [109]. The virus concentration, nevertheless, is lower in treated wastewater effluents and in most natural waters. Therefore, unlike bacterial contaminants that occur in higher concentrations in water, viruses can be detected only after being concentrated by a suitable method. Many different techniques are available, and the subject has been well reviewed [110–113]. Viruses are concentrated from water by gauze pads [114–116], hydroextraction [117,118], two-phase separation [115, 119–121[, ultracentrifugation [122], freeze concentration

[123,124], ultrafiltration [125], reverse osmosis [126], and adsorption to and subsequent elution from surfaces. Only those techniques that make use of adsorption process to concentrate viruses are covered in this chapter. These adsorbents include membrane filters, polyelectrolytes, salt precipitates, iron oxides, and clay minerals [3]. Membrane filters are the most widely used adsorbents for virus concentration in laboratory and field conditions.

Adsorption to Membrane Filters. Membrane filters are available in various pore sizes, chemical structures, and configurations. They are available also as flat or cartridge-type filters and are composed of cellulose acetate, cellulose nitrate, mixtures of cellulose nitrate and fiberglass, epoxy–fiberglass, and other substances. Cellulose nitrate membranes have more affinity for viruses than cellulose acetate type [127]. The mean diameter of viruses (<30 mμ for all the enteroviruses) is much smaller than the pore size of membrane filters (0.45 μm generally used). This means that viruses are not retained by an entrapment process, but rather by adsorption to the membrane filters. However, viruses adsorbed to solid particles or virus aggregates may be entrapped by membrane filters.

Some mechanisms have been suggested to explain the virus–membrane interaction [128]: (1) hydrophobic bonding between nonpolar regions on both viruses and membrane filters, (2) hydrogen bonding between polar groups, and (3) electrostatic forces. The interactions depend on pH, salt concentrations, and organic materials present in the water. The pH is brought to a low value (3.5–4.5) to bring about adsorption [129]. The binding of viruses to membranes increases with the concentration and valency of the cation in the suspending medium [129–131]. The effect of the molarity of two electrolytes ($MgCl_2$ and $AlCl_3$) on the adsorption of poliovirus to a membrane filter is shown in Fig. 10.14 [129]. At all salt concentrations $AlCl_3$ was always more efficient than $MgCl_2$. This, of course, has practical implications when one contemplates the concentration of viruses from large volumes of water under field conditions. For example, the processing of 500 gal in the field will necessitate 0.24 kg and 20.3 kg of $AlCl_3 \cdot 6H_2O$ and $MgCl_2 \cdot 6H_2O$, respectively. The cations present in the suspending medium lead to a reduction of the repulsion forces between the surfaces of both the membrane filter and the viruses, allowing short-range London–van der Waals forces to operate between the two surfaces. Organics present in water generally decrease the sorption of viruses to membrane surfaces [127,131,132]. These organics are called *membrane-coating components* (MCC). Recently, it has been found that the MCC are "humic-like" substances [133]. Once viruses are adsorbed to a filter, they may then be desorbed from the membrane surface by a small volume of eluent generally consisting of a proteinaceous fluid adjusted to high pH.

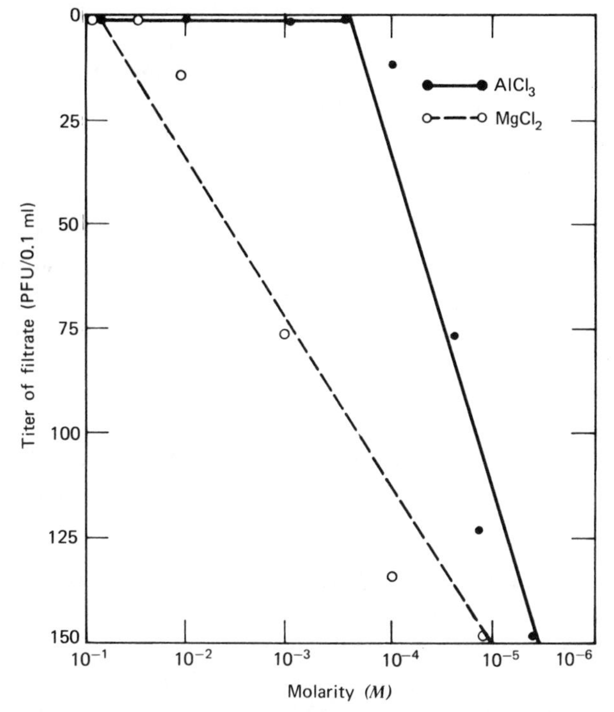

Figure 10.14 Effect of $MgCl_2$ and $AlCl_3$ on adsorption of viruses to membrane filters [129]. Courtesy of the American Society for Microbiology.

The membrane-filtration technique is actually the most commonly used method for the concentration of viruses from large volumes of tap water [134–139], estuarine water [140], seawater [141], wastewater [142], groundwater [143], and sanitary landfill leachates [144]. However, this relatively simple and sensitive method is limited by problems related to clogging and to the presence of MCCs in the water.

Adsorption to Polyelectrolytes, Salt Precipitates, and Iron Oxides. In 1967 Johnson et al. [145] demonstrated the adsorption of tobacco mosaic virus (TMV) and poliovirus to insoluble polyelectrolytes. Since then, other enteric viruses have been studied and were found to be adsorbed at low pH (3–4.5) and easily eluted at higher pH values [146]. Cationic polyelectrolytes generally display a higher adsorptive capacity than do anionic or nonionic polyelectrolytes [147,148]. Negatively charged viruses are probably attracted to positively charged amino groups on the surface of the cationic polyelectrolyte [148]. According to Cookson [149], the virus–polyelectrolyte interaction may be due to hydrogen bonding between the virus and chemical groups (carboxyl or amino groups) on the polymer surface and to electrostatic forces that depend on pH and salt concentration. It has been reported that the adsorption process is reduced at low and high salt concentrations [148]. The lower retention capacity at high ionic conditions is probably due to the contraction of the polyelectrolyte molecules [150]. A cross-linked copolymer of isobutylene maleic anhydride (PE 60) has been widely used as a polymer in the concentration of viruses from water [151] and wastewater [152]. It may be used as a suspension or as a layer sandwiched between two filter pads. The viruses are then eluted from the polymer with proteinaceous fluid at high pH. Despite its good ability to recover viruses from water, this polyelectrolyte was reported to be variable in its performance. It was not recommended as a standard adsorbent for virus concentration [113] since it is no longer available commercially.

Viruses also can be concentrated by adsorption to preformed flocs of aluminum hydroxide, aluminum phosphate, or calcium phosphate [153,154]. Aluminum hydroxide retains viruses more efficiently than do the other precipitable salts. Virus may be recovered from the flocs by dissolution of the flocs or by an elution process. According to Cookson [149], the adsorption of viruses to $Al(OH)_3$ flocs could result from coordinated hydroxyl groups or from electrostatic attraction between negatively charged viruses and positively charged $Al(OH)_3$. The latter mechanism is dependent on pH and salt concentration of the suspending medium. Aluminum phosphate flocs lack the surface charge and the OH^- ion bridges that assist virus adsorption. The concentration of viruses by adsorption to preformed flocs is applicable only to small volumes of water [115] and thus is used as a reconcentration technique.

Attention has been given recently to the use of iron oxides (hematite or magnetite) as adsorbents in virus concentration from water and wastewater. A report published in 1966 described the strong adsorption of influenza virus to hematite [20]. Since then, various studies have shown that magnetic iron oxides (Fe_3O_4) have a strong affinity for phages and enteric viruses [12,19,155,156]. A recent study [156] has concluded that adsorption of poliovirus type 1 to magnetite increased with increasing concentration and valency of the cations in the water but remained unchanged when the pH was varied from 5 to 9. Moreover, the virus removal was not affected by the organics present in wastewater effluents. Attempts to concentrate viruses by adsorption to iron oxides have resulted in three different approaches: (1) column technique, where the viruses are concentrated while passing through a column packed with iron oxide [20], (2) "sandwich" technique, where the iron oxide is "sandwiched" between two prefilter pads [157], [3] magnetic filtration technique, where the test water is "seeded" with magnetite at a concentration of 100–300 ppm and, in the presence of an appropriate salt concentration, the virus adsorbs efficiently to the magnetite [12]. The mixture is then poured through a filter packed with stainless-steel wool placed in a background magnetic field. The magnetite, along with the viruses, is efficiently retained in the filter and can be eluted after turning off the magnetic filter. This method has the advantage of high flow rates and does not suffer from clogging as in the other two methods. Whatever the method adopted, viruses adsorbed to iron oxides can be eluted with a proteinaecous material (10% fetal calf serum or 0.1% casein) at pH 9.

10.3 ECOLOGICAL IMPLICATIONS OF SORPTIVE INTERACTIONS BETWEEN VIRUSES AND SURFACES

10.3.1 Association of Viruses with Suspended Solids in Aquatic Environment

In the past few years studies in environmental virology have put more emphasis on the association of viruses with particulate matter in water and wastewater. This association is of great concern because it plays a significant role in the survival and distribution of pathogenic viruses in the aquatic environment. Viruses may become attached to either inorganic (clay materials or silts) or organic solids present in water and wastewater. The association of viruses with wastewater solids has been examined (see Section 10.2.1) and thus is not discussed further.

Marine and river silts adsorb substantial amounts of enteric viruses [158–160]. Due to their relatively large surface area and wide distribution, clay minerals have received most of the attention in the last few years [3,9,15,37, 161–163]. The attachment of viruses to these minerals depends on the type of clay under consideration. Montmorillonite, an expanding three-layer type of clay, is a better adsorbent than illite, a nonexpanding three-layer type of clay, or kaolinite, a nonexpanding two-layer type of clay [9,37]. The clay–virus interaction depends, furthermore, on the concentration and valency of the cation in the suspending medium [9,15,37]. This is consistent with the Schultz–Hardy rule since, for equivalent virus adsorption, it was found that the required NaCl concentration was approximately 100 times greater than $CaCl_2$ concentration [162]. The kinetics of virus adsorption varies with the type of virus, clay concentration, and the physicochemical properties of the suspending medium. Schaub and Sagik [161] found that a contact time of 1 min was sufficient for the adsorption of 92% of encephalomyocarditis (EMC) virus (Fig. 10.15). However, for poliovirus type 1 [162] the contact time was longer. The effect of water quality on the

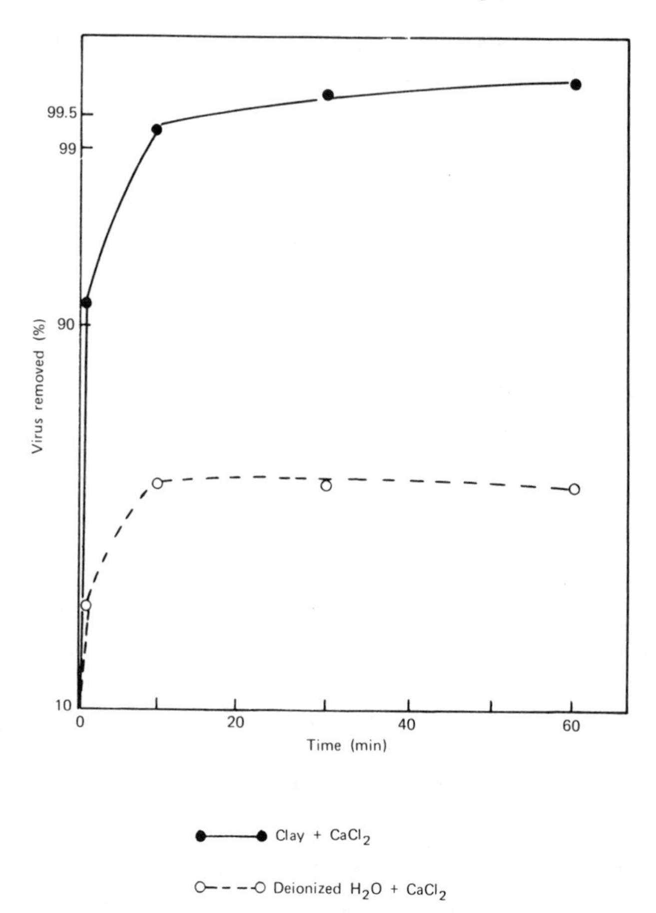

Figure 10.15 Kinetics of encephalomyocarditis (EMC) virus adsorption to montmorillonite [161]. Courtesy of the American Society for Microbiology.

adsorption kinetics is shown in Fig. 10.16. The adsorption rate was lower in the presence of primary sewage effluents than in the presence of tap or distilled water [162]. Although a low pH generally leads to a higher adsorption of viruses to clays, the attachment of EMC virus to these minerals was high (>90%) in the pH range 3.5–9.5 [161]. As for other adsorbents, the virus–clay association is inhibited by the presence of competitive organic materials (proteinaceous substances) present in water [9,161]. From the preceding discussion it appears that the clay–inorganic solids association is greatly dependent on the physicochemical properties of the water and that some dissociation would be expected when these conditions change.

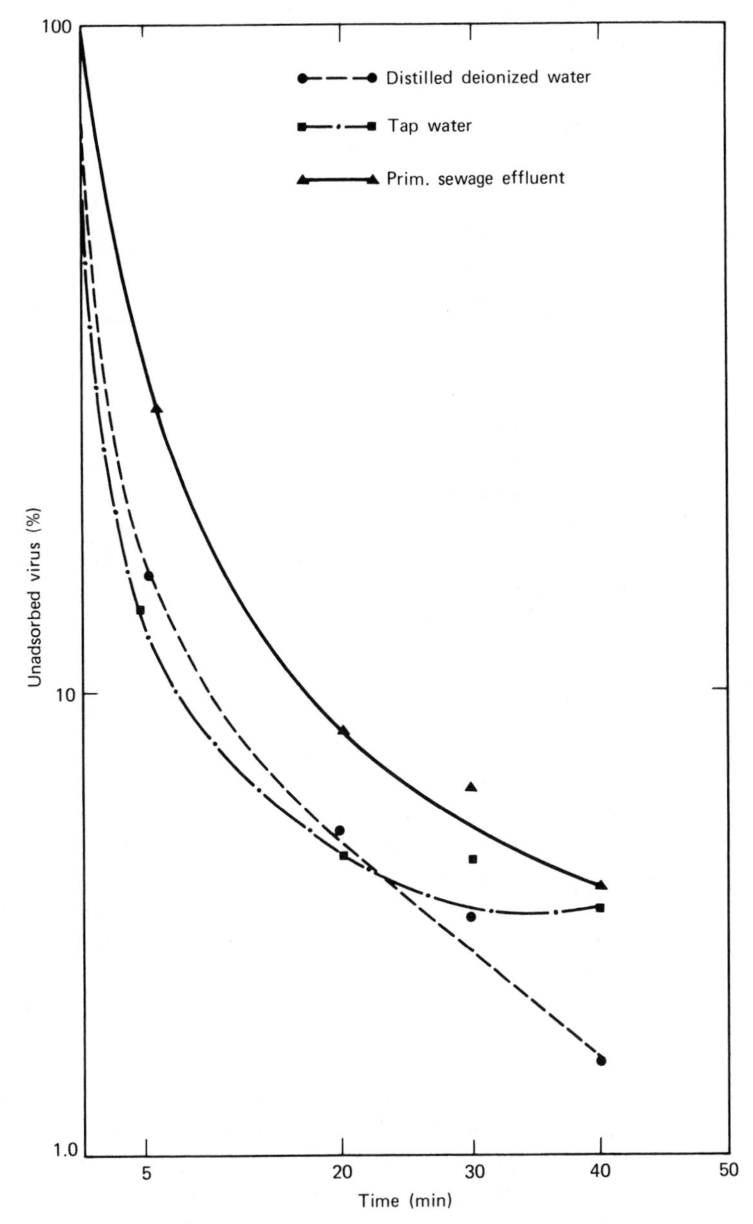

Figure 10.16 Effect of water quality on poliovirus adsorption to montmorillonite [162]. Courtesy of the Center for Research in Water Resources, the University of Texas at Austin.

Table 10.5 Adsorption of encephalomyocarditis (EMC) virus to natural suspended solids[a]

Treatment of lake water	pH	Adsorption (%)
Lake water	7.9	15
Lake water—acidified with HCl	6.0	59
Lake water + 10^{-3} M $CaCl_2$	7.9	64
Lake water + 10^{-3} M $CaCl_2$ and acidified with HCl	6.0	72

[a] Adapted from Schaub et al. [162].

Dry dog food was used to stimulate organic suspended solids present in natural water [37,161]. It was observed that, in the presence of 0.01M $CaCl_2$, organic solids possessed a high affinity for EMC virus, poliovirus, and T7 phage but did not display any affinity for T2 and f2 bacterial phages. Other experiments undertaken to evaluate the adsorption of virus to natural suspended solids in bodies of water showed that suspended solids in a man-made lake adsorbed 5–35% of virus (Table 10.5) and that this adsorption process was enhanced by addition of salts or by a decrease in pH [161].

The virus–solids association, apart from its possible implications in the ecology of viruses, has two important consequences in the field of public health: (1) concentration of viruses; any attempt to concentrate viruses from natural waters must take into consideration not only the "free" viruses, but also the solid-associated ones, as failure to do so would result in error and in a false sense of security [143] and (2) accumulation of viruses in sediment; virus particles associated to suspended solids do settle to the bottom sediments, where they may reach relatively high levels. They may later be released from the solids by resuspension of the sediment material following increased stream velocity.

10.3.2 Infectivity of Viruses in Adsorbed State

For many years it was implied that solids-associated viruses actually were inactivated, that is, loss of the ability to replicate in a host cell. However, recent investigations do show the contrary. It is now accepted that solid-bound viruses are as infective as free virions. A rodent enterovirus, EMC virus, when adsorbed to clay minerals or to natural solids, was as infective in the adsorbed as in the free state [161]. These findings prompted more research on this important topic. Bentonite-associated poliovirus and bacterial phages T2 and T7 were found to retain most of their infectivity when inoculated into host cells. However, f2, an RNA tailless phage, lost its infectivity when adsorbed to that particular clay [37]. The loss of infectivity was also observed when f2 was associated with another three-layer clay, nontronite (Bitton and Fraxedas, unpublished). The infectivity of viruses adsorbed to magnetite has been studied by Bitton et al. [12]. It was found that 57.8–93.5% of adsorbed phage T2 were infective, whereas phage MS2 and poliovirus were infective in totality. It is imperative to investigate whether other enteric viruses behave similarly when in the adsorbed state. The mechanism governing this phenomenon is rather obscure, but it is possible that the virion may be desorbed from the solid surface once it is plated on the host-cell monolayer. This question is best illustrated by a comment made by Dr. Ebba Lund [112]:

> We all have the feeling that viruses bound to soils do not enter cells, but from my own work I must conclude that when viruses are bound to soils, they may enter cells and replicate, i.e., infect. What does that really mean? Does the virus disentangle itself at the cellular membrane or does a phagocytotic process take place?

029

10.3.3 Virus Survival in Aquatic Environment

We have previously examined the association of viruses with particulate matter in the aquatic environment. This association exerts a marked influence on virus survival in water. The topic of viral persistence in the aquatic environment has been thoroughly reviewed elsewhere [164–167], and the major factors influencing viral survival are temperature, salinity, chemical composition of the water, biological inactivation, solar radiation, type of virus, level of pollution, sedimentation, and particulates.

Since viruses are now known to be associated with particulate matter in water (microbial cells, cell debris, silt, clay minerals, fecal material, etc.), some limited research has been undertaken to assess the impact of these sorptive interactions on their survival. Mitchell and Jannasch [168], while examining the rate of decline of phage $\Phi\chi174$ in natural seawater, suggested that suspended organic matter protected the virus against inactivation. Colloidal montmorillonite ($<0.2\ \mu$), when added to seawater at a concentration of 75–750 μg/ml, exerted a protective effect on phage T7 [169]. The phage survival was approximately four orders of magnitude higher in seawater amended with montmorillonite than in seawater alone. Similar experiments described the protective effect of kaolinite (50–500 μg/ml) toward bacterial phage T2 in seawater [170]. Some mechanisms have been proposed [169,170] to explain the viral protection by clays: adsorption of the virus to clay minerals and preservation of its structural integrity and adsorption and then inactivation of antiviral chemicals on the colloid surface. Clay minerals may even modify the interaction between microorganisms. Hence montmorillonite was found to protect *E. coli* from phage attack [10]. An envelope of colloidal clay around the bacterial cell probably prevented the phage from attaching to its host. It also has been observed that adsorption of viruses to clays and other solid particles afforded them protection from high pH and low ionic strength [171]. Schiffenbauer and Stotzky (unpublished data) reported that clay minerals protected coliphages from thermal inactivation. Montmorillonite was found to be more protective than kaolinite at 25°C (Fig. 10.17). The protective effect imparted by suspended solids functions also in the case of viral exposure to light. Bitton (unpublished) observed a protective effect of three different clays toward phage T7 subjected to ultraviolet (UV) irradiation but found no detectable viable virus in control samples (0.85% NaCl) after 30 sec of exposure to UV light (Fig. 10.18). Poliovirus type 1, suspended in groundwater, was exposed to solar radiation in the presence or absence of 500 ppm of nontronite, and it was shown that this clay retarded viral inactivation (Bitton, Fraxedas, and

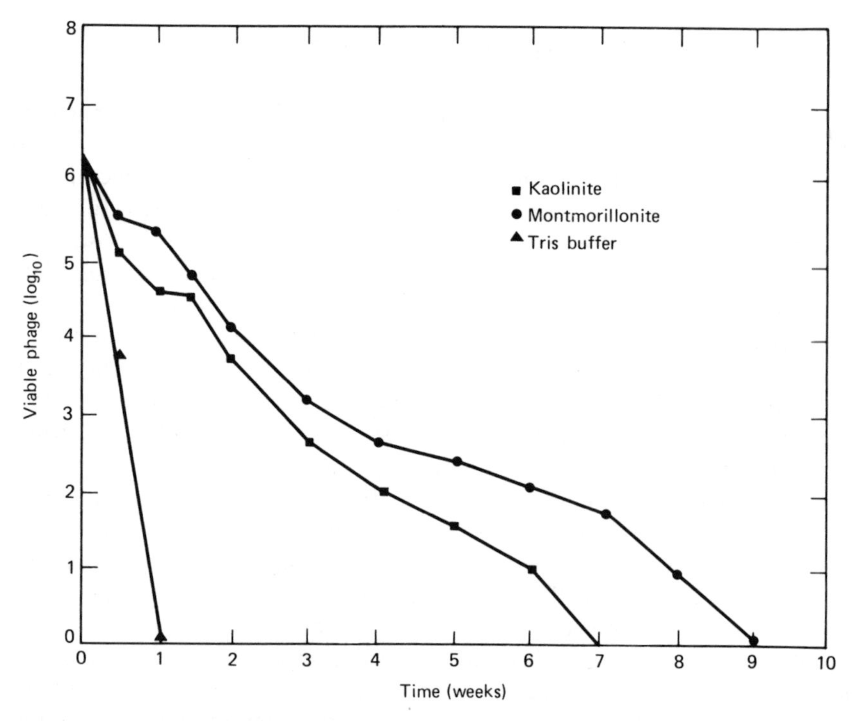

Figure 10.17 Protection of phage T7 from thermal inactivation by kaolinite and montmorillonite (from Schiffenbauer and Stotzky, unpublished).

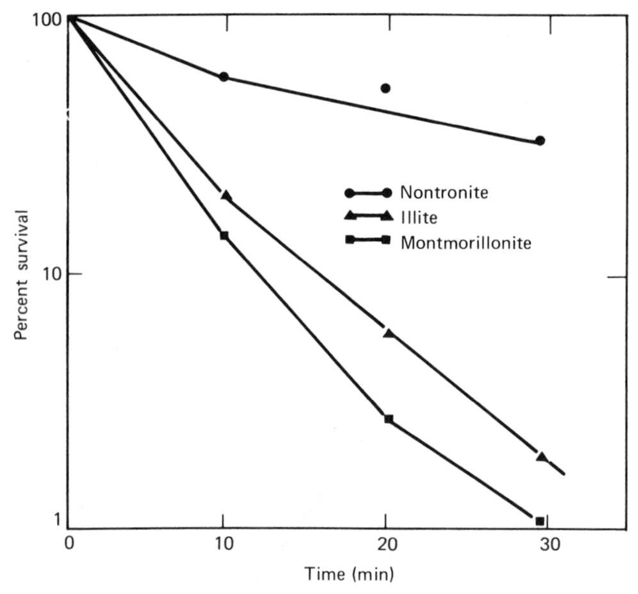

Figure 10.18 Influence of clay minerals on survival of phage T7 exposed to UV irradiation (from Bitton, unpublished).

Gifford, unpublished). Algal cells also may protect viruses from the effect of sunlight. After 60 min of exposure to solar radiation, Bitton, Fraxedas, and Gifford (unpublished) observed a protective effect of algae, predominantly *Microcystis* sp. and *Anabaena* sp. toward poliovirus (Fig. 10.19).

It is reasonable to conclude that, within the aquatic environment, viruses attach readily to particulate matter that modifies their survival following environmental stresses.

10.3.4 Virus Survival in Sediments and Soils

We have seen the protective effect of particulate matter toward viruses in the water column. Following their association with solid particles, viruses may settle and accumulate in the bottom sediments of both freshwater and marine environments. This phenomenon has been well described for enteric bacteria [172–174], but not as thoroughly for viruses due to the lack of adequate methodology. Improvements in detection methods have allowed the monitoring of enteric viruses in sediments where they occur in higher numbers than in the water column [172,175,176]. Gerba et al. [172], studying the distribution of bacterial and viral pathogens in coastal canal communities, found higher numbers of these pathogens in the bottom sediments than in

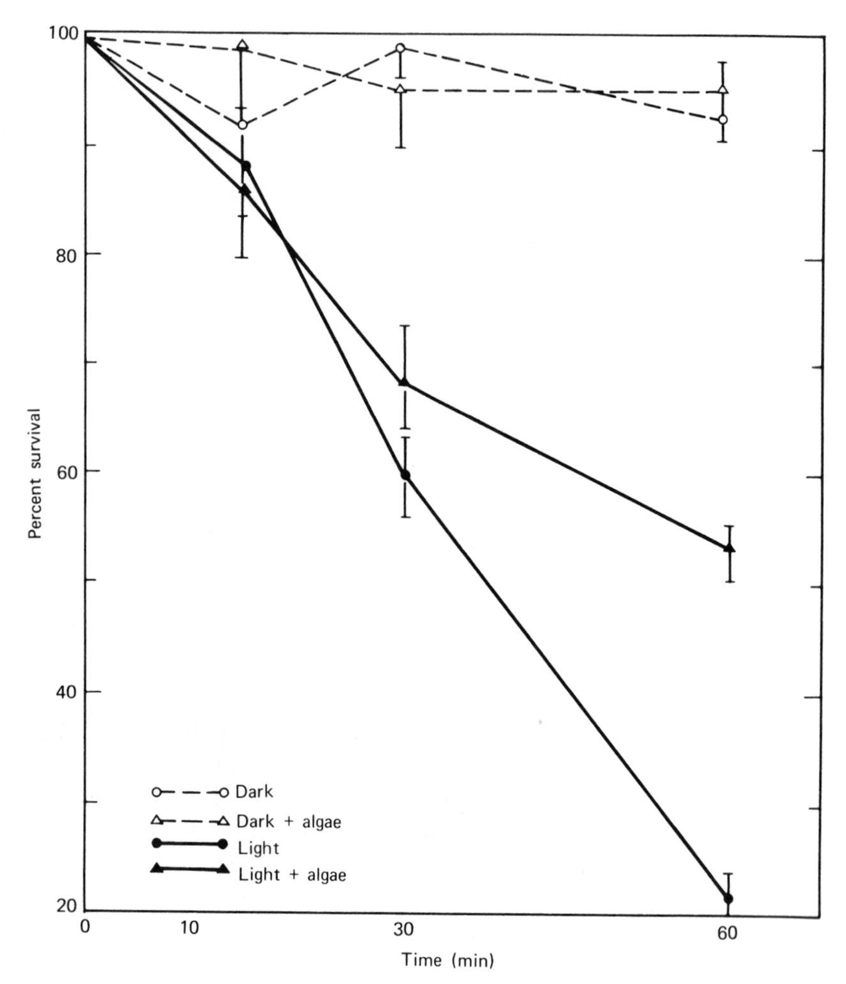

Figure 10.19 Protection of poliovirus from sunlight by algal cells (Bitton, Fraxedas, and Gifford, unpublished).

the water column. Seventy percent of the enteric viruses were poliovirus type 1. DeFlora et al. [176] also observed an accumulation of enteric viruses in marine sediments in the vicinity of sewage outfalls. The association of viruses with sediments depends on the nature of the sediment, fine-textured sediments (rich in clay) adsorbing higher numbers of viruses than coarser ones. This phenomenon is illustrated in Table 10.6 [170]. Once viruses have accumulated in sediments, their survival should be of particular interest to public health workers. Enteric bacteria are known to survive better in sediments than in overlying water [177]. DeFlora et al. [176] investigated the

Table 10.6 Virus adsorption to estuarine sediments (virus T2 input was 10^9 PFU)[a]

Location	Sediment concentration (mg/100 ml)	Adsorption (%)	Nature of sediment
Turkey point	50	91	Clay
	5	43	
Rickenbacker Causeway	1000	9	Sand
	100	0	

[a] Adapted from Gerba and Schaiberger [171].

survival of enteric viruses in marine sediments and observed that the inactivation of type 1 poliovirus was 4.5-fold faster in seawater than in the sandy material (Fig. 10.20). These findings suggest that accumulation of viruses will be favored by their longer survival in sediments. Thus viable viruses are concentrated in sediments and may be resuspended following increases in organic-matter content [170], decreases in salinity [10], wave action, or dredging operations. These observations point to the need for monitoring sediments for the presence of enteric viruses in any public-health survey.

Soil is utilized increasingly for the disposal of partially treated human wastes, and data on the survival of enteric viruses in that particular environment are badly needed. The subject of viral persistence in soil systems has been reviewed by Bitton [166] and by Gerba et al. [46]. Since the solid phase is the major component of a soil matrix, one has to gain knowledge about the survival of viruses in the sorbed state. Some contributions have been made by phytopathologists interested in the degree of persistence of plant viruses in soils. It was found that this category of viruses adsorbs to the clay fraction of the soil and remains viable for many years [178–181]. Enteric viruses may behave similarly with respect to their interactions with soil colloids. Green [60] reported that the inactivation rate of poliovirus in the sorbed state was lower than in suspension. What is the mechanism(s) of this protective effect? Are soil colloids involved in protection of viruses against environmental stresses such as desiccation? These are still some unresolved questions concerning virus survival in soil systems.

10.3.5 Fate of Solids-associated Viruses Following Disinfection

Among the various factors (pH, time, disinfectant concentration, and organic matter) influencing viral disinfection, turbidity has received the least attention. Berg [182] has hypothesized that "free" viruses would be less

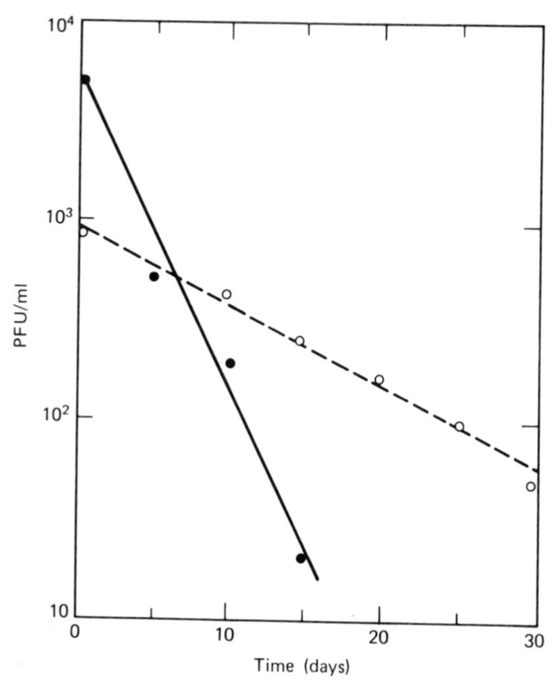

Figure 10.20 Inactivation of poliovirus in seawater and quartz sand [176]. Courtesy of the American Society for Microbiology.

resistant to disinfection than would solids-associated viruses. Culp [183] further suggested that the turbidity of the water should be no more than 1 JTU (Jackson turbidity unit) if good disinfection was the goal. This prompted some limited research on the effect of solid particles on disinfection efficiency. Boardman and Sproul [171] studied the effect of kaolinite, hydrated aluminum oxide, and calcium carbonate on phage T7 subjected to chlorination (0.006–0.036 ppm free chlorine). Hydrated aluminum oxide and $CaCO_3$ were chosen because these are products of coagulation and softening processes in water-treatment plants. It was found that none of the adsorbents studied afforded protection to T7 from chlorine inactivation. However, these particulates protected the virus to some extent against high pH level (10.1) and low ionic strength. These authors suggested that protection from disinfection would occur only after total encapsulation of the virus within the particulate matter. A later study by Stagg and his collaborators [18] showed that bentonite (particle size 2 nm) retarded the rate of inactivation of RNA phage MS2 by hypochlorous acid at pH 6 and at a temperature of 22° ± 1°C. Following exposure to 0.12 ppm of HOCl, the

t_{99} (time to inactivate 99% of the virus) was 17 sec for "free" virus and 30 sec for bentonite-associated virus (turbidity of 2–4 JTU). A study of this phenomenon under field conditions [184] confirmed that solids-associated phages were also more resistant to chlorination than freely suspended ones, although most of the viruses were adsorbed to, rather than embedded in, particulate matter.

Another phenomenon worth mentioning is the effect of viral aggregation on disinfection efficiency [185]. In classical laboratory experiments, inactivation curves describe the fate of populations of single virus particles. Under natural conditions, however, viruses may undergo aggregation, and those infectious particles inside the aggregates have more chance to survive the disinfection process.

Clearly, these investigations show that the factors responsible for turbidity exert an important effect on the disinfection process. One must take into consideration the solids-associated viruses when one develops a disinfection method. A reduction of turbidity to less than 0.1 JTU could be a preventive measure to counterbalance the protective effect of particulates during disinfection.

10.4 CONCLUDING REMARKS

This chapter has dealt with the adsorption of viruses to surfaces and with the technological and ecological implications of this phenomenon.

Sorptive phenomena play a role in the removal of enteric viruses by primary (physical), secondary (biological), and tertiary (physicochemical) treatment of domestic wastewater. This removal process does not lead necessarily to viral inactivation (loss of infectivity) since viruses remain infective in the adsorbed state. Consequently, these treatment processes merely transfer at least some of the viruses to sludges, and this should be an important consideration when the disposal of these sludges is contemplated. In recent years land application has become an attractive alternative for the disposal of sewage effluents and sludges. We have seen that this practice requires a detailed knowledge of the sorptive capacity of soils toward bacterial and viral pathogens. The adsorption of viruses to the soil matrix should be well understood if one seeks to optimize their removal by this system and thus avoid the contamination of groundwater supplies.

Sorptive phenomena also play a role in the removal of viruses by water-treatment plants. Adsorption is involved in the removal of viruses by sand filtration, diatomaceous earth, activated carbon, and water softening agents [adsorption to $CaCO_3$ and $Mg(OH)_2$ precipitates].

Since viruses occur in low numbers in water, efforts have been directed toward finding a suitable concentration method. The membrane-filtration technique is presently the most commonly used in the recovery of viruses from large volumes of water. Viruses are retained on membrane surfaces by adsorption, and research is still heavily directed toward finding the most suitable conditions governing this process.

Viruses are commonly associated with suspended solids in the aquatic environment, and this phenomenon has some bearing on their survival pattern. Adsorbed viruses survive better the rigors of some environmental stresses, namely temperature, light, and some yet undefined factors. Particulates also protect viruses from the inactivating effect of chlorine.

10.5 REFERENCES

1. C. Andrews and H. G. Pereira, *Viruses of Vetebrates*, Williams and Wilkins, Baltimore, 1967.
2. J. T. Cookson, *J. Am. Water Works Assoc.*, **66,** 707 (1974).
3. G. Bitton, *Water Res.*, **9,** 473 (1975).
4. K. C. Marshall, *Interfaces in Microbial Ecology*, Harvard U. P., Cambridge, Mass., 1976.
5. E. C. Pollard, *The Physics of Viruses*, Academic, New York, 1953.
6. R. C. Valentine and H. C. Allison, *Biochem. Biophys. Acta*, **34,** 10 (1959).
7. T. T. Puck and L. J. Tolmach, *Arch. Biochem. Biophys.*, **51,** 229 (1954).
8. T. T. Puck, A. Garen, and J. Cline, *J. Exp. Med.*, **97,** 807 (1951).
9. G. F. Carlson, Jr., F. E. Woodart. D. F. Sproul, and O. J. Sproul, *J. Water Pollut. Control Fed.*, **40,** R89 (1968).
10. M. M. Roper and K. C. Marshall, *Microbial. Ecol.*, **1,** 1 (1974).
11. J. J. Bertucci, C. Lu-Hing, D. R. Zenz, and S. J. Sedita, 34 p (1975). Department of Resources Development, Metropolitan Sanitation District of Greater Chicago, report No. 75-21, 1975.
12. G. Bitton, G. E. Gifford, and O. C. Pancorbo, Water Resources Research Center, University of Florida, Gainesville, Fla., publication No. 40, 1976.
13. B. E. Moore, L. Funderburg, B. P. Sagik, and J. F. Malina, Jr., in J. F. Malina, Jr., and B. P. Sagik, Eds., *Virus Survival in Water and Wastewater Systems*, Texas U. P., Austin, 1974, pp. 3–15.
14. S. A. Sattar, S. Remia, and J. C. N. Westwood, *Can. J. Pub. Health*, **67,** 221 (1976).
15. S. A. Schaub and C. A. Sorber, in G. Berg, H. L. Bodily, E. H. Lennette, J. L. Melnick, and T. G. Metcalf, Eds., *Viruses in Water*, American Public Health Association, Washington, D. C., 1974, pp. 128–138.
16. R. L. Ward, C. S. Ashley, and R. H. Moseley, *Appl. Environ. Microb.*, **32,** 339 (1976).
17. C. P. Gerba, E. M. Smith, and J. L. Melnick, *Appl. Environ. Microb.*, **34,** 158 (1977).
18. C. H. Stagg, C. Wallis, and C. H. Ward, *Appl. Environ. Microbiol.*, **33,** 385 (1977).
19. V. C. Rao, R. Sullivan, R. B. Read, and N. A. Clarke, *J. Am. Water Works Assoc.*, **60,** 1288 (1968).

20. J. Warren, A. Neal, and D. Rennels, *Proc. Soc. Exp. Biol. Med.*, **12,** 1250 (1966).
21. F. M. Wellings, A. L. Lewis, and C. W. Mountain, *Appl. Environ. Microbiol.*, **31,** 354 (1976).
22. D. O. Cliver, *Environ. Lett.*, **10,** 215 (1975).
23. M. F. Duff, *Appl. Microbiol.*, **19,** 120 (1970).
24. F. M. Wellings, A. L. Lewis, and C. W. Mountain, in F. M. d'Itri, Ed., *Wastewater Renovation and Reuse*, Marcel Dekker, New York, 1977, pp. 453–478.
25. B. E. Moore, B. P. Sagik, and C. A. Sorber, *Third National Conference on Sludge Management, Disposal and Utilization*, Miami, Fla., 1976.
26. H. H. Bloom, W. N. Mack, B. J. Krueger, and W. L. Mallman, *J. Infect. Dis.*, **105,** 61 (1961).
27. E. Lund, in G. Berg, H. L. Bodily, E. H. Lennette, J. L. Melnick, and T. G. Metcalf, Eds., *Viruses in Water*, American Public Health Association., Washington, D. C., 1974, pp. 196–205.
28. W. N. Mack, J. R. Frey, B. J. Riegle, and W. L. Mallman, *J. Water Pollut. Control Fed.*, **34,** 1133 (1962).
29. G. Berg, in V. L. Snoeyink, Ed., *Viruses and Water Quality: Occurrence and Control*, 13th Water Quality Conference, University of Illinois, Urbana-Champaign, 1971, pp. 126–136.
30. B. B. Berger, F. A. Butrico, H. J. Dunsmore, V. C. Lischer, H. F. Ludvig, F. Nevins, G. W. Reid, H. Romer, and O. J. Sproul, *J. San. Eng. Div.* **96,** 111 (1970).
31. J. F. Malina, Jr., in L. B. Baldwin, J. M. Davidson, and J. F. Gerber, Eds., *Virus Aspects of Applying Municipal Wastes to Land*, Florida U. P., Gainesville, 1976, pp. 9–23.
32. S. Kelly, W. W. Sanderson, and C. Neidl, *J. Water Pollut. Control Fed.*, **33,** 1056 (1961).
33. S. L. Chang, *J. San. Eng. Div.*, **96,** 111 (1970).
34. N. A. Clarke, R. E. Stevenson, S. L. Chang, and P. W. Kabler, *Am. J. Pub. Health*, **51,** 1118 (1961).
35. J. F. Malina, Jr., K. R. Ranganathan, B. P. Sagik, and B. E. Moore, *J. Water Pollut. Control Fed.*, **47,** 2178 (1975).
36. K. R. Ranganathan, J. F. Malina, Jr., and B. P. Sagik, *Seventh International Conference on Water Pollution Research*, Paris, September 9–13, 1974.
37. B. E. Moore, B. P. Sagik, and J. F. Malina, Jr., *Water Res.*, **9,** 197 (1975).
38. H. H. Malherbe and M. Strickland-Cholmley, in G. Berg, Ed., *Transmission of Viruses by Water Route*, Wiley-Interscience, New York, 1967.
39. M. D. Sobsey and R. C. Cooper, *Water Res.*, **7,** 669 (1975).
40. G. J. Thabaraj, *Fla. Sci.*, **38,** 222 (1975).
41. R. L. Wright, *Fla. Sci.*, **38,** 207 (1975).
42. M. A. Benarde, *J. Am. Water Works Assoc.*, **65,** 432 (1973).
43. S. E. Chase, *Water Works Waste Eng.*, **1,** 48 (1964).
44. D. H. Foster and R. S. Engelbrecht, in W. E. Sopper and L. T. Kardos, Eds., *Recycling Treated Municipal Wastewater and Sludge Through Forest and Cropland*, Pennsylvania State U. P., University Park, 1973, pp. 247–270.
45. R. H. Miller, in *Research Needs Related to Recycling of Urban Wastewater on Land.* Proceedings of Workshop, Institute for Research on Land and Water Resources, Pennsylvania State University, University Park, 1974, pp. 115–139.

46. C. P. Gerba, C. Wallis, and J. L. Melnick, *J. Irrig. Drain. Div.*, **101,** 157 (1975).
47. N. A. Clarke and S. L. Chang, *J. Am. Water Works Assoc.*, **51,** 1299 (1959).
48. J. E. Vogt, Proc. 1961, *Symp. Groundwater Contamination*, U. S. Public Health Service, technical report No. W61-5, 1961.
49. W. N. Mack, L. Yue-Shoung, and D. B. Coohon, *Health Serv. Rep.*, **87,** 271 (1972).
50. R. Fujioka and P. C. Loh, Abstracts of American Society for Microbiology Annual Meeting, 1974.
51. F. M. Wellings, A. L. Lewis, C. W. Mountain, in J. F. Malina, Jr., and B. P. Sagik, Eds., *Virus Survival in Water and Wastewater Systems*, Texas U. P., Austin, 1974, pp. 253–260.
52. F. M. Wellings, A. L. Lewis, C. W. Mountain, and L. V. Pierce, *Appl. Microbiol.*, **29,** 751 (1975).
53. F. C. McMichael and J. E. McKee, *Research on Wastewater Reclamation at Whittier Narrows*, California Water Quality Control Board, Sacramento, publication No. 33, 1965.
54. R. G. Gilbert, R. C. Rice, H. Bouwer, C. P. Gerba, C. Wallis, and J. L. Melnick, *Science*, **192,** 1004 (1976).
55. J. C. Merrel, Jr. and P. C. Ward, *J. Am. Water Works Assoc.*, **60,** 145 (1968).
56. C. R. O'Melia and D. K. Crapps, *J. Am. Water Works Assoc.*, **56,** 1326 (1964).
57. R. W. Filmer, M. Felton, Jr., and T. Yamamote, in V. L. Snoeyink, Ed., *Virus and Water Quality: Occurrence and Control*, 13th Water Quality Conference, University of Illinois, Urbana–Champaign, 1971, pp. 75–101.
58. W. A. Drewry, in *Landspreading of Muncipal Effluent and Sludge in Florida*, Proceedings of 1973 Workshop, University of Florida, Gainesville, 1973.
59. W. A. Drewry and R. Eliassen, *J. Water Pollut. Control Fed.*, **40,** R257 (1968).
60. K. M. Green, "Sand Filtration for Virus Purification of Septic Tank Effluents," Ph.D. thesis, University of Wisconsin, Madison, 1976.
61. R. Eliassen, W. Ryan, W. A. Drewry, P. Kruger, and G. Tchobanoglous, *Final Report to the Commission on Environmental Hygiene, Armed Forces Expidemiology Board*, Stanford University, Calif., p. 137 (1967).
62. J. C. Lance, C. P. Gerba, and J. L. Melnick, *Appl. Environ. Microbiol.*, **32,** 520 (1976).
63. S. M. Duboise, B. E. Moore, and B. P. Sagik, *Appl. Environ. Microbiol.*, **31,** 536 (1976).
64. D. H. Hori, N. C. Burbank, Jr., R. H. F. Young, L. S. Lau, and H. W. Klemmer, Water Resources Research Center, University of Hawaii, Honolulu, technical report No. 36, 1970.
65. R. E. Reece, "Virus Sorption on Natural Soils," M.S. thesis, University of Arkansas, 1967.
66. R. H. F. Young and N. C. Burbank, Jr., *J. Am. Water Works Assoc.*, **65,** 598 (1973).
67. E. Lefler and Y. Kott, in J. F. Malina, Jr., and B. P. Sagik, Eds., *Virus Survival in Water and Wastewater Systems*, U. P., Austin, 1974, pp. 84–91.
68. G. Bitton, P. R. Scheuerman, G. E. Gifford, and A. R. Overman, in H. T. Odum and K. C. Ewel, Eds., *Cypress Wetlands* (in preparation).
69. G. Bitton, N. Masterson, and G. E. Gifford, *J. Environ. Qual.*, **5,** 370 (1976).
70. S. R. Farrah, S. M. Goyal, C. P. Gerba, C. Wallis and P. T. B. Shaffer, *Water Res.*, **10,** 897 (1976).

71. K. B. McDonald, "The Transport of Water-Borne Viruses in Soil," Ph.D. thesis, University of Guelph, Ontario, Canada, 1971.
72. R. B. Reneau, Jr. and D. E. Pettry, *J. Environ. Qual.*, **4,** 41 (1975).
73. K. M. Green and D. O. Cliver, *Proceedings of National Home Sewage Disposal Symposium* Chicago, Illinois, American Society of Agricultural Engineers, 1974, pp. 137–143.
74. M. L. Peterson, *Am. J. Publ. Health*, **64,** 912 (1974).
75. R. S. Engelbrecht, National Technical Information Service, Springfield, Va, report No. PB234589, 1973.
76. M. D. Sobsey, C. Wallis, and J. L. Melnick, *Appl. Microbiol.*, **30,** 565 (1975).
77. A. L. Novello, "Poliovirus Survival in Landfill Leachates and Migration through Soil Columns," M.S. thesis, University of Cincinnati, Ohio, 1974.
78. A. R. Swoboda and G. W. Thomas, *J. Agric. Food Chem.*, **16,** 923 (1968).
79. O. J. Sproul, in V. L. Snoeyink, Ed., *Virus and Water Quality: Occurrence and Control*, 13th Water Quality Conference, University of Illinois, Urbana-Champaign, 1971, pp. 159–169.
80. O. J. Sproul, in G. Berg, H. L. Bodily, E. H. Lennette, J. L. Melnick, and T. G. Metcalf, Eds., *Viruses in Water*, American Public Health Association, Washington, D. C., 1974, pp. 167–179.
81. B. H. Dieterich, "The Behavior of Bacterial Viruses in Contact with Ordinary and Uniform Filter Sand," M.S. thesis, Harvard University, Cambridge, Mass., 1953.
82. G. Berg, R. B. Dean, and D. R. Dahling, *J. Am. Water Works Assoc.*, **60,** 193 (1968).
83. G. G. Robeck, N. A. Clarke and K. A. Dostal, *J. Am. Water Works Assoc.*, **54,** 1275 (1962).
84. P. Amirhor and R. S. Engelbrecht, *J. Am. Water Works Assoc.*, **67,** 187 (1975).
85. T. S. Brown, J. R. Malina, Jr., and B. D. Moore, *J. Am. Water Works Assoc.*, **66,** 735 (1974).
86. M. Chaudhuri, P. Amirhor, and R. S. Engelbrecht, *J. Environ. Eng. Div.*, **100,** 937 (1974).
87. American Water Works Association, *Water Quality and Treatment*, McGraw-Hill, New York, 1971.
88. S. E. Thayer and O. J. Sproul, *J. Am. Water Works Assoc.*, **58,** 1063 (1966).
89. D. R. Wentworth, R. T. Thorup, and O. J. Sproul, *J. Am. Water Works Assoc.*, **60,** 939 (1968).
90. A. P. Black and R. F. Christman, *J. Am. Water Works Assoc.*, **53,** 737 (1961).
91. T. E. Larson and A M Buswell, *Ind. Eng. Chem.*, **32,** 132 (1940).
92. G. A. Lo Grippo, *Proc. Soc. Exp. Biol. Med.*, **74,** 208 (1950).
93. R. H. Muller, *Proc. Soc. Exp. Biol. Med.*, **73,** 239 (1950).
94. R. H. Muller and H. M. Rose, *Proc. Soc. Exp. Biol. Med.*, **80,** 27 (1952).
95. T. Puck and B. Sagik, *J. Exp. Med.*, **97,** 807 (1953).
96. W. A. Drewry, *Virus Movement in Groundwater Systems*, Arkansas University Water Resources Research, Fayetteville, publication No. 4, 1969.
97. D. H. Foster, R. S. Engelbrecht, R. S. Snoeyink, and V. L. Snoeyink, *Environ. Sci. Technol.*, **11,** 55 (1977).
98. H. J. Carlson, G. M. Ridemour, and C. F. McKahnn, *Am. J. Pub. Health*, **32,** 1256 (1942).

99. C. P. Gerba, M. D. Sobsey, C. Wallis, and J. L. Melnick, *Environ. Sci. Technol.*, **9,** 727 (1975).

100. J. R. Neefe, J. B. Baty, J. G. Reinhold, and J. Stokes, Jr., *Am. J. Pub. Health*, **37,** 365 (1947).

101. O. J. Sproul, M. Warner, L. R. LaRochelle, and D. R. Brunner, *Proceedings of Fourth International Conference on Water Pollution Research*, Prague, Pergamon, London, 1969.

102. J. T. Cookson, Jr., *Environ. Sci. Technol.*, **1,** 157 (1967).

103. J. T. Cookson, Jr., *J. Am. Water Works Assoc.*, **61,** 52 (1969).

104. J. T. Cookson, Jr., *Environ. Sci. Technol.*, **4,** 128 (1970).

105. J. T. Cookson, Jr. and W. J. North, *Environ. Sci. Technol.*, **1,** 46 (1967).

106. E. Kellenberger, A. Bolle, E. Boy de Latour, R. H. Epstein, N. C. Franklin, N. K. Jerne, A. Reale-Schafati, and J. Sechant, *Virology*, **26,** 419 (1965).

107. P. P. Oza and M. Chaudhuri, *Water Res.*, **9,** 707 (1975).

108. P. P. Oza and M. Chaudhuri, *J. Environ. Eng. Div.*, **102,** 1255 (1976).

109. N. Buras, *Water Res.*, **10,** 295 (1976).

110. W. F. Hill, Jr., E. W. Akin, and W. H. Benton, *Water Res.*, **5,** 967 (1971).

111. E. Katzenelson, in G. Berg, H. L. Bodily, E. H. Lennette, J. L. Melnick, and T. G. Metcalf, Eds., *Viruses in Water*, American Public Health Association, Washington, D. C., 1974, pp. 152–164.

112. E. Lund, in F. M. D'Itri, Ed., *Wastewater Renovation and Reuse*, Marcel Dekker, New York, 1977, pp. 421–452.

113. M. D. Sobsey, in G. Berg, H. L. Bodily, E. H. Lennette, J. L. Melnick, and T. G. Metcalf, Eds., *Viruses in Water*, American Public Health Association, Washington, D. C., 1974, pp. 89–127.

114. J. L. Melnick, J. Emmons, E. M. Opton, and J. H. Coffey, *Am. J. Hyg.*, **59,** 185 (1954).

115. B. Fattal, E. Katznelson, and H. I. Shuval, in J. F. Malina, Jr. and B. P. Sagik, Eds., *Virus Survival in Water and Wastewater Systems*, Texas U. P., Austin, 1974, pp. 19–30.

116. O. C. Liu, D. A. Brashear, H. R. Seraichekas, J. A. Barnick, and T. G. Metcalf, *Appl. Microb.*, **21,** 405 (1971).

117. D. O. Cliver, in G. Berg, Ed., *Transmission of Viruses by the Water Route*, Wiley-Interscience, New York, 1967.

118. H. I. Shuval, S. Cymbalista, B. Fattal, and N. Goldblum, in G. Berg, Ed., *Transmission of Viruses by the Water Route*, Wiley-Interscience, New York, 1967, pp. 45–55.

119. J. Grindrod and D. O. Cliver, *Arch. Ges. Virusforsch*, **31,** 315 (1970).

120. E. Lund and C. E. Hedstrom, *Am. J. Epidemiol.*, **84,** 287 (1966).

121. H. I. Shuval, B. Fattal, S. Cymbalista, and N. Goldblum, *Water Res.*, **3,** 225 (1969).

122. D. O. Cliver and J. Yeatman, *Appl. Microbiol.*, **13,** 387 (1965).

123. S. H. Rubestein, H. G. Orback, N. Shuber, and E. King, *J. Am. Water Works Assoc.*, **63,** 301 (1971).

124. S. H. Rubestein, J. Fenters, H. Orbach, N. Shuber, J. Reed, and E. Molloy, *J. Am. Water Works Assoc.*, **65,** 200 (1973).

125. H. Gaertner, in G. Berg, Ed., *Transmission of Viruses by the Water Route*, Wiley-Interscience, New York, 1967.

126. B. A. Sweet, R. D. Ellender, and J. K. L. Leong, *Dev. Ind. Microbiol.*, **15,** 143 (1974).

127. D. O. Cliver, *Biotechnol. Bioeng.*, **10,** 877 (1968).
128. T. W. Mix, *Dev. Ind. Microb.*, **15,** 136 (1974).
129. C. Wallis, M. Henderson, and J. L. Melnick, *Appl. Microbiol.*, **23,** 476 (1972).
130. V. C. Rao and N. A. Labzoffsky, *Can. J. Microbiol.*, **15,** 399 (1969).
131. C. Wallis and J. L. Melnick, *J. Virol.*, **1,** 472 (1967).
132. M. Moore, P. P. Ludovici, and W. S. Jeter, *J. Water Pollut. Control Fed.*, **42,** R21 (1970).
133. S. R. Farrah, S. M. Goyal, C. P. Gerba, C. Wallis, and P. T. B. Shaffer, *Water Res.*, **10,** 897 (1970).
134. S. R. Farrah, C. P. Gerba, C. Wallis, and J. L. Melnick, *Appl. Environ. Microbiol.*, **31,** 221 (1976).
135. W. F. Hill, Jr., E. W. Akin, W. H. Benton, C. J. Mayhew, and W. Jakubowski, *Appl. Microb.*, **27,** 1177 (1974).
136. W. F. Hill, Jr., W. Jakubowski, E. W. Akin, and N. A. Clarke, *Appl. Environ. Microbiol.*, **31,** 254 (1976).
137. W. Jakubowski, J. C. Hoff, N. C. Anthony, and W. F. Hill, Jr., *Appl. Microbiol.*, **28,** 501 (1974).
138. M. D. Sobsey, C. Wallis, M. Henderson, and J. L. Melnick, *Appl. Microbiol.*, **26,** 529 (1973).
139. C. Wallis, A. Homma, and J. L. Melnick, *J. Am. Water Works Assoc.*, **64,** 189 (1972).
140. W. F. Hill, Jr., E. W. Akin, W. H. Benton, C. J. Mayhew, and T. G. Metcalf, *Appl. Microbiol.*, **27,** 506 (1974).
141. H. A. Fields and T. G. Metcalf, *Water Res.*, **9,** 357 (1975).
142. C. Wallis, C. H. Stagg, S. R. Farrah, and J. L. Melnick, in L. B. Baldwin, J. M. Davidson, and J. F. Gerber, Eds., *Virus Aspects of Applying Municipal Wastes to Land*, Florida U. P., Gainesville, 1976, pp. 37–44.
143. F. M. Wellings, A. L. Lewis, and C. W. Mountain, in L. B. Baldwin, J. M. Davidson, and J. F. Gerber, Eds., *Virus Aspects of Applying Municipal Wastes to Land*, Florida U. P., Gainesville, 1979, pp. 45–51.
144. M. D. Sobsey, C. Wallis, and J. L. Melnick, *Appl. Microbiol.*, **28,** 232 (1974).
145. J. H. Johnson, J. E. Fields, and W. A. Darlington, *Nature* (*Lond.*), **213,** 665 (1967).
146. C. Wallis, J. L. Melnick, and J. E. Fields, *Appl. Microbiol.*, **21,** 703 (1971).
147. M. Chaudhury and R. S. Engelbrecht, *J. Am. Water Works Assoc.*, **62,** 563 (1972).
148. R. R. Thorup, F. P. Nixon, D. F. Wentworth, and O. J. Sproul, *J. Am. Water Works Assoc.*, **62,** 97 (1970).
149. J. T. Cookson, *Dev. Ind. Microbiol.*, **15,** 160 (1974).
150. C. P. A. Priesing, *Ind. Eng. Chem.*, **54,** 38 (1962).
151. C. Wallis, J. L. Melnick, and J. E. Fields, *Water Res.*, **4,** 787 (1970).
152. C. Wallis, S. Grinstein, J. L. Melnick, and J. E. Fields, *Appl. Microbiol.*, **18,** 1007 (1969).
153. C. Wallis and J. L. Melnick, *Am. J. Epidemiol.*, **85,** 459 (1967).
154. C. Wallis and J. L. Melnick, in G. Berg, Ed., *Transmission of Viruses by the Water Route*, Wiley-Interscience, New York, 1967.
155. G. Bitton and R. Mitchell, *Water Res.*, **8,** 549 (1974).

156. G. Bitton, O. Pancorbo, and G. E. Gifford, *Water Res.*, **10,** 973 (1976).
157. T. G. Metcalf, Water Resources Research Center, University of New Hampshire, report No. 10, Durham, 1974.
158. G. Berg, in S. H. Jenkins, Eds., *Progress in Water Technology*, Pergamon, New York, 1973.
159. F. E. Hamblet, W. F. Hill, E. W. Akin, and W. H. Benton, *Am. J. Epidemiol.*, **89,** 562 (1969).
160. J. C. Hoff and R. C. Becker, *Am. J. Epidemiol.*, **90,** 53 (1969).
161. S. A. Schaub and B. P. Sagik, *Appl. Microbiol.*, **30,** 212 (1975).
162. S. A. Schaub, C. A. Sorber and G. W. Taylor, in J. F. Malina and B. P. Sagik, Eds., *Virus Survival in Water and Wastewater Systems*, Texas U. P., Austin, 1974, pp. 71–83.
163. V. P. Shirobokov, *Acta Virol.*, **12,** 185 (1968).
164. E. W. Akin, W. H. Benton, and W. F. Hill, Jr., in V. L. Snoeyink, Ed., *Virus and Water Quality: Occurrence and Control*, 13th Water Quality Conference University of Illinois, Urbana-Champaign, 1973, pp. 59–74.
165. E. W. Akin, W. F. Hill, Jr., and N. A. Clarke, in A. L. H. Gameson, Ed., *Discharge of Sewage from Sea Outfalls*, Pergamon, London, 1975, pp. 227–235.
166. G. Bitton, in R. Mitchell, Ed., *Water Pollution Microbiology*, Vol. 2, Wiley-Interscience, New York, 1978, 273–299.
167. J. F. Brisou, *Rev. Hyg. Med. Soc.*, **13,** 359 (1965).
168. R. Mitchell and H. W. Jannasch, *Environ. Sci. Technol.*, **3,** 941 (1969).
169. G. Bitton and R. Mitchell, *Water Res.*, **8,** 227 (1974).
170. C. P. Gerba and G. E. Schaiberger, *J. Water Pollut. Control Fed.*, **47,** 93 (1975).
171. G. D. Boardman and O. J. Sproul, 48th Annual Conference on Water Pollution Control Federation, Miami Beach, Fla., 1975.
172. C. P. Gerba, S. M. Goyal, E. M. Smith, and J. L. Melnick, *Marine Pollut. Bull.*, **8,** 279 (1977).
173. S. C. Rittenberg, T. Mittwer, and D. Ivler, *Limnol. Oceanog.*, **3,** 101 (1958).
174. D. J. Van Donsel and E. E. Geldreich, *Water Res.*, **5,** 1079 (1971).
175. T. G. Metcalf, C. Wallis and J. L. Melnick, in J. F. Malina, Jr. and B. P. Sagik, Eds., *Virus Survival in Water and Wastewater Systems*, Texas U. P., Austin, 1974, pp. 57–70.
176. S. DeFlora, G. P. DeRenzi, and G. Baldolati, *Appl. Microbiol.*, **30,** 472 (1975).
177. C. P. Gerba and J. S. McLeod, *Appl. Environ. Microbiol.*, **32,** 114 (1976).
178. Y. Myamoto, *Virology*, **7,** 250 (1959).
179. Y. Myamoto, *Virology*, **9,** 290 (1959).
180. T. H. Thung and J. Dijkstra, *Tijdschr. Plantezichten*, **64,** 411 (1958).
181. H. P. H. Van der Want, *Proceedings of Conference on Plant Virus Diseases*, Brussels, 1951.
182. G. Berg, in J. F. Malina, Jr. and B. P. Sagik, Eds., *Virus Survival in Water and Wastewater Systems*, Texas U. P., Austin, 1974, pp. xii–xvii.
183. R. L. Culp, in J. F. Malina, Jr. and B. P. Sagik, Eds., *Virus Survival in Water and Wastewater Systems*, Texas U. P., Austin, 1974, pp. 158–165.
184. C. H. Stagg, C. Wallis, C. H. Ward, and C. P. Gerba, *Progr. Water Technol.* (in press).
185. R. Floyd, J. D. Johnson, and D. G. Sharp, *Appl. Environ. Microbiol.*, **31,** 298 (1976).

CHAPTER **11**

Attachment of Microorganisms to Living and Detrital Surfaces in Freshwater Systems

HANS W. PAERL
Institute of Marine Sciences, University of North Carolina, Morehead City

CONTENTS

11.1 INTRODUCTION

Various surfaces serve to support attached microorganisms among diverse freshwater ecosystems. Prior to microscopic observations, chemical determinations, and the application of radiotracer techniques directly implicating microorganisms, man was well aware of the effects of microbial growth on solid surfaces in streams, lakes, and estuaries. Fouling of boat hulls and submersed structures such as piers, dams, and bridges by slime-forming microorganisms has been a persistent problem ever since the use of inland waters for transport. The formation of nuisance slimes on stream and lake bottoms due to either naturally abundant or eutrophication-linked stocks of attached microbes has been a historic problem, especially in densely populated regions. Certain habitats, including thermal springs, streams rich in dissolved and particulate nutrients, and swamps having high suspended sediment loads are often rich in attached microflora, as witnessed by the slippery appearance of rocks or large particles, as well as luxuriant growths trailing from solid surfaces.

Largely due to the advent and practical applications of the light microscope, we became more aware of the important roles of surfaces as microbial habitats in inland waters. Leeuwenhoek's original observations included mention of the close relationships between bacteria ("animalcules") and plant and animal tissues [1]. During the late nineteenth and early twentieth century mention there reports on the presence of bacteria on particles, wood, concrete, and glass surfaces, as well as on algal cells during microscopic examinations of fresh samples [2–5]. Such observations essentially were repeated without elaboration well into the twentieth century. Not until the studies of Henrici [6,7], Waksman and Carey [8], and Zobell and Anderson [9] were some of the functional aspects of microbial attachment discussed and tested. During this era it was shown that surfaces rapidly attracted and stimulated the growth of several aquatic bacterial species and that the ensuing attached growth had profound effects on the structure as well as chemical characteristics of these surfaces. During an early study of bacterial growth in containers it was shown that the amount of bacterial growth on the areas of container walls was inversely proportional to the size of the container [8,10]; essentially, this study was a forerunner to the later "bottle-effect" experiments illustrating the likelihood of greatly accelerating microbial growth by performing experiments in small containers [11,12]. Since then we have become increasingly aware of the importance and function of high surface:volume ratios responsible for stimulating bacterial growth [13].

Initial observations on the functional importance of surface habitats have been supplemented by the application of various staining and enumeration techniques [6,14,15]. A flurry of studies in Europe and North America has revealed that, in certain aquatic environments, significant fractions of the total bacterial biomass can be associated with surfaces and particles [16–22]. Several studies have followed the sequential recruitment of free-floating bacteria by artificial and natural solid substrates, often showing strong and irreversible attraction [13,23]. Taxonomic surveys have identified and categorized genera that are especially predominant on surfaces [16,24]. These genera constitute a vast group, including Gram-negative and Gram-positive bacteria and hyphomicrobia, which possess stalks and fibrillar excretions used for attachment, as well as numerous fungal, blue–green algal (cyanophycean or cyanobacterial), and eucaryotic algal genera that form adhesive pads, extracellular slime, and fibrils aiding adhesion [25–28] (see Chapter 4). Various adhesion mechanisms and appendages have been illustrated by the use of transmission and scanning electron microscopy (TEM and SEM, respectively) [29–34].

Studies on microbial colonization of both natural and artificial particulates and surfaces indicated that the presence of surfaces alone, regardless of nutrient content, could stimulate microbial growth [9,13,35]. The original studies by Waksman and Carey [8,36] and Heukelekian and Hiller [37] indicated that the stimulation of bacterial growth by particles was probably due to both available surface area and available nutrient content. Both factors acted independently of each other. Indeed, more recent studies indicate that chemically inert particles can often stimulate microbial growth, especially in waters containing low dissolved nutrient concentrations [13,38,39]. In such systems attached microorganisms may benefit from nutrient concentrating mechanisms and favorable pH, O_2, and CO_2 gradients present on particle surfaces [40].

In a majority of cases freshwater particles and surfaces represent both attachment sites and nutrient sources. Such substrates as decaying plant and animal matter, soil and mineral particles, and living plants and animals represent concentrated nutrient sources in the dilute aquatic environment. Hence it is not surprising that diverse bacterial genera possess the chemotactic faculties to sense and, subsequently, adsorb to these substrates [41]. Chemotaxis appears to be effective means of locating relatively concentrated nutrient sources in dilute aquatic environments [42]. Specific genera exhibit positive chemotaxis toward growth-enhancing substrates, and others display negative chemotaxis in response to inhibitory or toxic substances [43]. An example of chemotaxis initiating bacterial attachment to a nutrient-rich surface can be shown in the case of bacterial associations with the filamentous N_2-fixing cyanobacterial genera *Anabaena* and *Aphanizomenon* in lakes. Bacteria show preferential attachment to the N_2-fixing cell (heterocyst) in the filament and are often found lodged in the junction between a heterocyst and neighboring vegetative cell (44) (Fig. 11.1). When observed microscopically, natural bacteria added to axenic *Anabaena* (also derived from the same lake) swim, using polar flagellation, toward the *Anabaena* filaments. After several encounters with the filaments, a highly significant ($p < 0.01$) number of the bacteria associated with the filaments settled on the heterocyst–vegetative cell junctions, making this area a permanent attachment site. This site was shown to be a source of

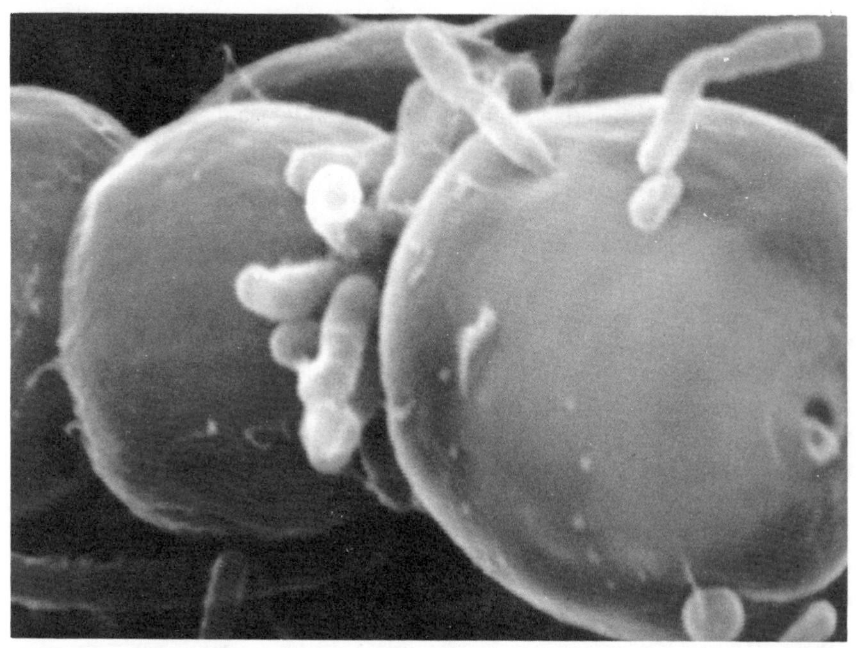

Figure 11.1 Bacteria concentrated in junction of heterocyst (N_2-fixing cell) and vegetative cell of cyanobacterium *Anabaena circinalis*. Bacteria prefer junctions for attachment, presumably because they are often sites for cyanobacterial excretions (×13,375).

Anabaena excretion products, which the bacteria appeared to utilize directly [45]. When tested for chemotactic abilities, these bacteria show positive chemotaxis toward organic compounds previously reported as being *Anabaena* excretory products [44].

Chemotaxis may be an important factor dictating some specific bacterial–particle interactions in fresh water. There are numerous other instances, however, where nonspecific attachment dominates, leading to diverse microflora inhabiting surfaces [20,46,47]. Furthermore, nonmotile bacteria (incapable of a rapid chemotactic response), fungi, and algae often form a dominant part of the attached microflora [48,49]. It appears that other factors, such as ionic and electrostatic characteristics of cells [50], the excretion of adhesive webs and appendages [31], and cell sizes and shapes, also play important roles in determining the likelihood of attachment [51]. Finally, particle composition and the presence of adequate attachment sites are of crucial importance in determining the qualitative and quantitative aspects of microbial attachment [13,52].

11.2 SUBSTRATES FOR MICROBIAL ATTACHMENT IN FRESH WATER

Suspended, benthic, and littoral surfaces (including all surfaces, interfaces, and particles) suitable for microbial attachment are usually of diverse origin and composition among aquatic ecosystems. The organic carbon content of suspended particles in lakes, for example, can vary from as little as <1 to near 50% of the total dry weight [53,54]. This percentage is usually dictated by catchment characteristics. For example, glacier-fed lakes often contain appreciable percentages of inorganic particulates (commonly called *glacial flour*), originating from freshly eroded mineral sediments, scoured from the head of the glacier. In these lakes both water and particle replenishment originate at similar sites. Large, deep tectonic and volcanic lakes, which tend to be oligotrophic and are less influenced by inorganic sediment inflow from surrounding catchments, usually contain a more heterogenous mixture of organic and inorganic particles. In Lake Tahoe (California–Nevada), for example, approximately 30–75% of the particulate matter in open waters is normally of organic composition [55,56]. Accordingly, 15–30% of the particulate dry weight is organic carbon in this ultraoligotrophic lake. In contrast, eutrophic lakes can often yield 25–50% of the particulate dry weight as organic carbon [57]. Inorganic sediment loads, unless accelerated by man, usually form a low percentage of the input in these lakes.

Inorganic particles include basic rock-derived materials, such as feldspars, limestone, mica, quartz, pumice (in volcanic lakes), and metal-containing rocks, such as pyrites, and a variety of clays. All these materials are capable of supporting attached microflora. Organic particles can be roughly divided among plant and animal origins, although aggregates of both, also including mineral components, are common [58–60]. In limnological and oceanographical terms, particles containing a nonliving organic and inorganic core and often supporting microbial growth are referred to as *detritus* [61]. The most common plant-linked particles are live phytoplankton or remains of phytoplankton [62], often containing a mineral component [31]. In lakes supporting submersed rooted macrophytes, attached algae (periphyton), littoral water plants or free-floating surface plants such as *Lemna* or *Azolla*, these plants and their detritus may form a significant fraction of the particulate load [58,63,64]. Additional minor particle sources can originate from decaying trees, shrubs, and ground cover washed or blown in from the catchment [65–67]. Bacteria or fungi may form microcolonies or clumps that can act as particles for successional growths of microorganisms. Such clumps are encountered at times in both lakes and streams [28,68].

Copepod, cladoceran, and rotifer zooplankton and their carapaces and other detrital remains are likely to be the most important animal sources of particles in lakes. This general conclusion may not apply to all systems, however, and is dependent on the degrees of dominance by these grazing plankters. At times, remains of fish (following fish kills or spawning runs), benthic invertebrates, and protozoans could constitute potentially important particle sources. Recently, Ferrante and Parker [34] have shown zooplankton fecal pellets to be an important site for bacterial attachment in Lake Michigan (Fig. 11.2).

Virtually all of these particle sources are also present in streams, but the degrees of importance vary somewhat. For example, fast flowing streams are often depauperate in phytoplankton and zooplankton. Instead, a bulk of the algal and animal production resides on the streambed in the forms of benthic periphyton, a range of invertebrate herbivores and detritivores, insect larvae, and protozoans. Furthermore, streams generally receive larger quantities of leaf litter [66,69], pollen, and soil runoff [28] than do lakes. In deciduous forests, leaf litter alone can provide a bulk of the yearly particu-

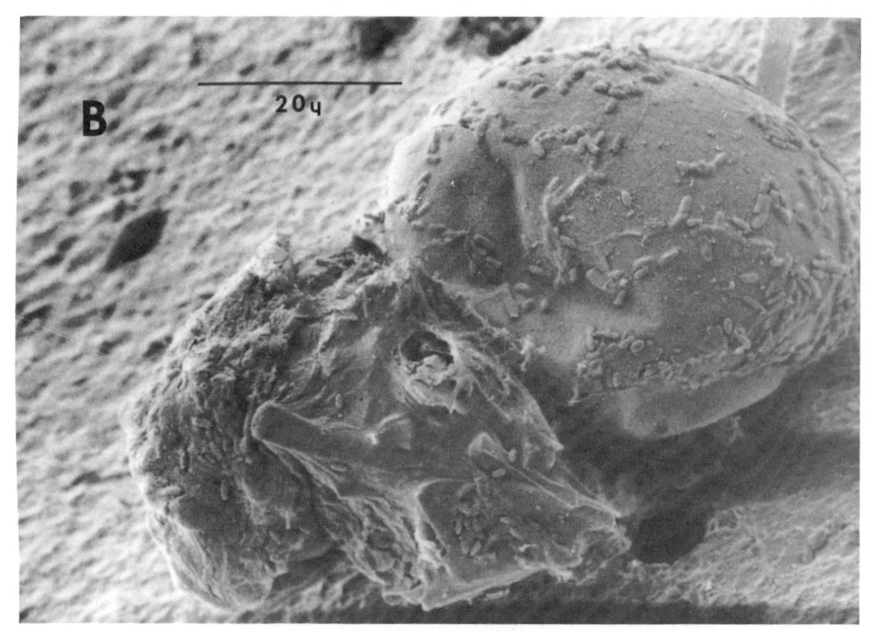

Figure 11.2 Low-magnification SEM of intact zooplankton fecal pellet populated by bacteria. Pellet was freshly sampled from Lake Michigan and immediately fixed. Courtesy of J. G. Ferrante, Argonne National Laboratories, Argonne, Illinois (×1475).

late and dissolved organic carbon input [67]. Bacterial and fungal attachment to leaf litter is extensive [70].

11.3 OBSERVATIONS OF MICROBIAL ATTACHMENT IN FRESH WATER

Electron-microscopic examinations of natural attached freshwater microorganisms are not as extensive as those on laboratory organisms [47, 71–73]. Based on current evidence, however, there appear to be marked similarities between the two situations in the morphologies of extracellular excretions, including polysaccharide pads [29], microfibrils [74], and anchoring appendages [73,75]. Although some quantitative differences exist between laboratory and natural populations regarding extracellular production of mucilage in response to nutritional status, the basic attachment mechanisms appear similar.

Certain bacteria exhibit subtle structural connections between themselves and the substrate. An example of such attachment can be obtained from TEM and SEM observations of bacteria attached to the heterocysts and vegetative cells of the cyanobacterium, *Anabaena* (Fig. 11.3). When observed by SEM, no obvious structural attachment mechanisms were evident. High-magnification TEM observations, however, reveal that deposits of fine fibrous material can be lodged between the bacterium and substrate in some cases. This material can be stained with ruthenium salts, thus indicating that it contains acidic polysaccharides. In the case of bacterial attachment to vegetative cells of *Anabaena*, close associations of fibrils from cyanobacterial and attached bacteria are evident (Fig. 11.4). Again, the positive reaction to ruthenium salts suggests that both sets of fibrils are rich in acidic polysaccharides [27]. The preceding examples appear to be in general agreement with a model of interacting polymeric microfibrils discussed by Costerton et al. [76] where respective polysaccharide chains are depicted to be cross-linked by means of charged residual glycocalyx strands. This mode of attachment may also be similar to the formation of acidic polysaccharide pads found between bacteria and attachment surfaces by Fletcher and Floodgate [27].

Attachment by virtue of these pads is common in fresh water. In addition to cyanobacterial and algal–bacterial interactions, bacterial–sediment and bacterial–detrital interactions also appear to employ fibrillar structures and mucilagenous pads, which are often only resolved by TEM [77]. These pads are highly effective in maintaining the bacteria in an attached state during periods of high turbulence and shear stress. Geesey et al. [77] have shown the use of padlike connections to be one of the main mechanisms that bacteria use to attach to rocks in fast-flowing streams.

In a more obvious fashion, several periphytic filamentous green algae excrete polysaccharide material from distal cells (holdfast cells), allowing these cells to serve as anchors on rocks in lakes and streams [78–80]. Common genera in this group are *Cladophora*, *Oedogonium*, and *Ulothrix*. In the case of algae the deposition of adhesive pads is much more extensive than for bacteria. Attached green algae and cyanobacteria excrete similar fibrillar pad material [75,81]. The colonial nitrogen-fixing cyanobacterium *Nostoc* is commonly found bonded to stream and lake bottoms by a thick layer of mucilage covering the entire colony. Other filamentous cyanobacterial genera, including *Anabaena*, *Oscillatoria*, and *Stigonema*, are commonly found gliding on rock and plant surfaces, using mucilagenous excretions. Benthic and periphytic diatom genera are anchored firmly by

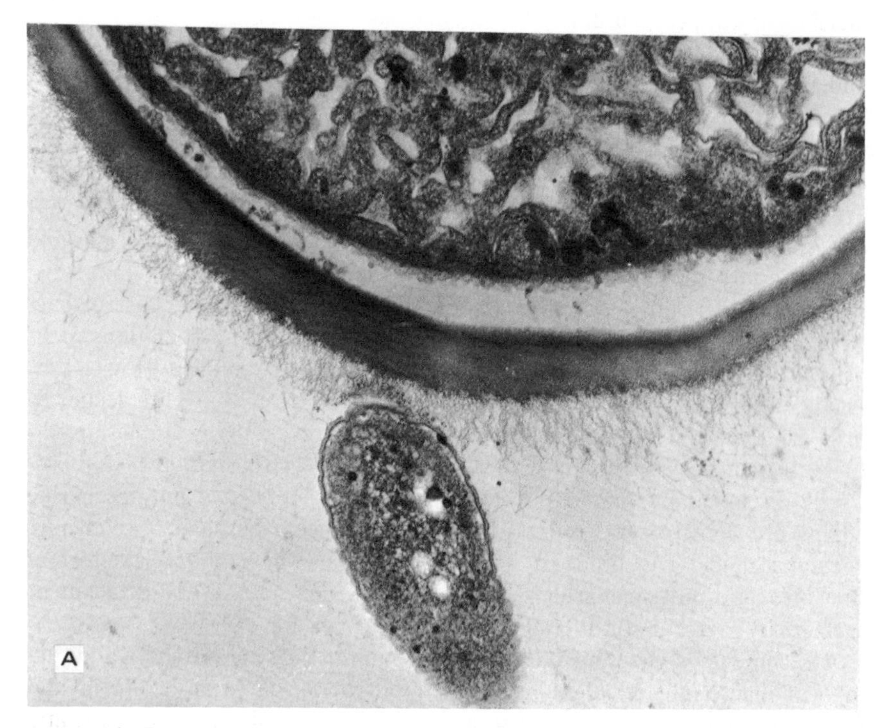

Figure 11.3 Transmission (3A) and scanning (3B) electron micrographs of bacteria attached to heterocyst of *Anabaena* sp. Note fibrillar nature of heterocyst outer coating, which is resolved by TEM but not SEM. Transmission electron micrographs courtesy of A. M. Massalski, Canada Centre for Inland Waters, Burlington, Ontario (TEM, ×40,500; SEM, ×12,625).

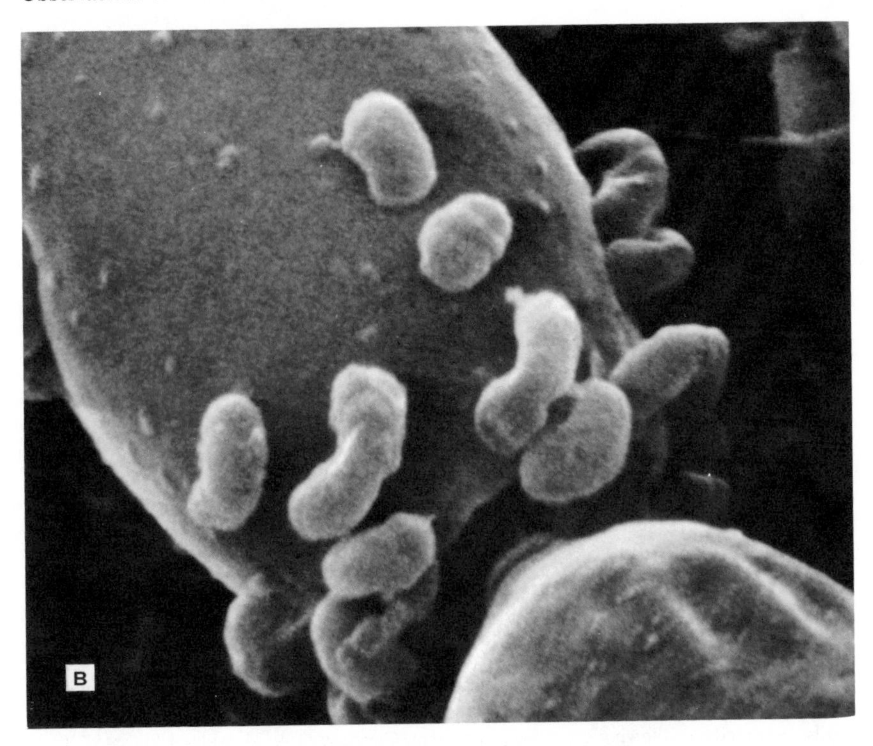

proteinaceous stalks, which are excreted through pores and channels located in the siliceous cell wall or frustule (Fig. 11.5).

Freshwater bacteria use also specialized appendages, reminiscent of pili, as attachment anchors. One example can be seen in the case of relatively large (2–5-μm) rod-shaped bacteria attached to periphytic algae in a fast-flowing alpine stream (Fig. 11.6). These bacteria utilized threadlike projections that were shown to be effective attachment anchors over a wide range of vacuum filtration pressures intended to dislodge the cells. Such anchors are distinctly different from microfibrillar pads in that threads are at least an order of magnitude thicker than fibrils and radiate from cells in all directions. They are also more easily resolved by SEM. The use of both threadlike and fibrillar structures for attachment appears to be widespread as well among aquatic bacteria but has not been observed as frequently as extensive pad formation [47].

In this regard it should be noted that the distinction between threadlike and fibrillar structures is tenuous at best. Normally on initial attachment a few thin fibrils are observed. Following attachment, continued production of polymers normally leads to the formation of more obvious pads [27]. It is

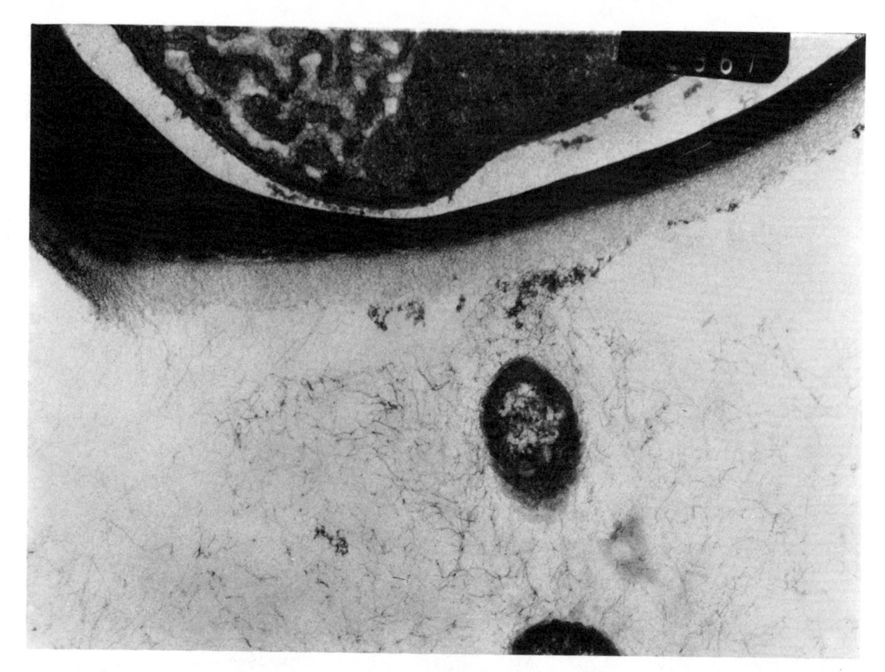

Figure 11.4 Transmission electron micrograph of fibrillar linkage between bacteria and *Anabaena* sp. heterocysts. At times both cyanobacteria and bacteria excrete copious amounts of fibrils and mucilage, allowing extensive attachment. Courtesy of A. M. Massalski (×45,750).

thus possible that threadlike structures observed by SEM may be aggregates of fibrils or enhanced production of polymers surrounding fibrillar cores. This raises the possibility that both TEM and SEM observations may deal with similar substances, but that adequate resolution with SEM can only be achieved when extensive production of polymers is occurring. For an in-depth review of the morphological and chemical nature of surface components potentially involved in freshwater attachment, see Chapter 4.

The extent of extracellular fibril, pad, or web formation in both freefloating and attached bacteria may vary somewhat according to nutritive status and general trophic conditions. For example, bacteria attached to inorganic (mineral) sources of detritus in oligotrophic Lake Tahoe showed only slight web formation, whereas bacteria attached to detritus rich in organic carbon (recently killed phytoplankton) showed more extensive fibrillar networks [60]. The magnitude of web formation in free-floating bacteria also appears to reflect the trophic states of lakes. In eutrophic lakes, especially those supporting high dissolved and particulate organic carbon levels, a significant fraction of free-floating bacteria (at times approaching 50%) routinely show associated fibrillar networks (Fig. 11.7), morphologically identical to those of attached bacteria. In oligotrophic lakes such network formation is normally greatly reduced among freefloating bacteria. However, when bacteria are attached to nutritive substances in these lakes, their fibrillar networks can often be extensive.

The excretion and formation of webs and appendages no doubt serve other functions in addition to attachment. It has been observed that both algal and bacterial extracellular excretions accumulate growth-limiting nutrients [82]. Algal excretory products have been implicated as binding and chelating agents for important trace elements, especially iron [83,84]. The possible physiological roles of extracellular fibrils and appendages in terms of iron uptake and transport have been discussed by Leppard and Colvin [85]. It has been suggested that these excretions serve to protect cells from viral attack, high ambient oxygen concentrations, and toxic substances or hydrolytic enzymes [86]. Thus there may be multiple benefits to the excretion of fibrillar adhesive structures by microorganisms in aquatic environments.

Figure 11.5 Scanning electron micrograph of periphytic diatom *Gomphonema* and its proteinaceous attachment stalk (arrow). This freshly fixed sample was taken from an alpine stream near Lake Tahoe, California (×5125).

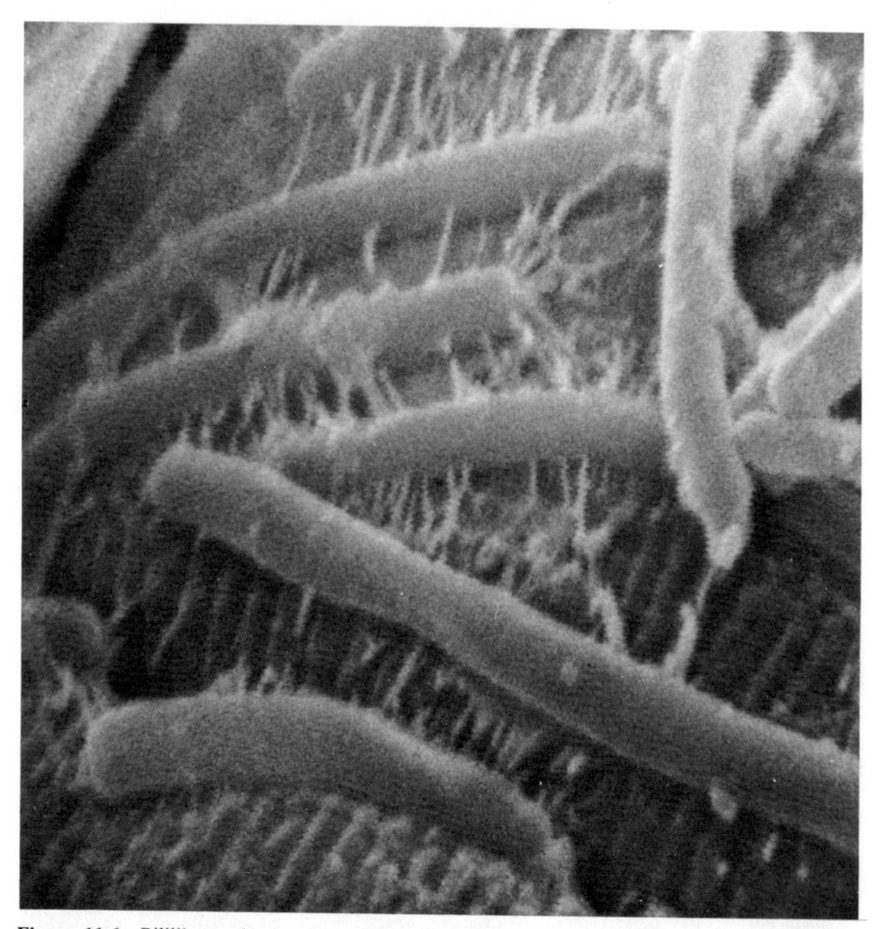

Figure 11.6 Pililike anchoring appendages found on bacteria tightly attached to periphytic algae in fast-flowing stream. Extremely high filtration pressures as well as mild sonication did not allow dislodging of the bacteria (×15,750).

A specific group of rod-shaped aquatic bacteria, the caulobacters, form cellular extensions, or stalks, enclosed by the cell wall [87]. The stalk is a short (~0.5-μm) extension terminating in an adhesive disk, the holdfast, which serves for attachment. *Caulobacter* is a common inhabitant of decaying organic matter in the aerobic epilimnion of many lakes. It can often be found in radiate colonies termed *rosettes*, where groups of cells are attached to each other at the holdfast regions of stalks. Other bacteria, including the iron bacteria *Gallionella* and *Sphaerotilus*, which are found in ponds containing abundant reduced iron salts, form stalks through the extracellular deposition of ferric hydroxide. As a result, large colonies of these bacteria are easily distinguished in ponds by their crusty flocculent appearance [88].

The *Hyphomicrobium* group, a set of prosthecate bacteria, are pear-shaped cells having long filiform prosthecea, somewhat similar to those of *Caulobacter*. These bacteria are highly adapted to low nutrient aquatic conditions and are oxidizers of simple organic compounds. Attachment of these bacteria is normally via the base of the mother cell [89]. Prosthecate-like bacteria are often found entwined with and attached to algal cells (Fig. 11.8). In all likelihood such bacteria utilize simple organics and excretia from host algae. Their ability to do so has been followed by autoradiography. Autoradiography revealed uptake of simple sugars and organic acids by these bacteria in Lake Tahoe water samples [90]. In general, little attention has been paid to the possible ecological importance of prosthecate bacteria in lakes and streams. Likewise, the numerical and functional significance of starshaped forms resembling these bacteria, often encountered in fresh water,

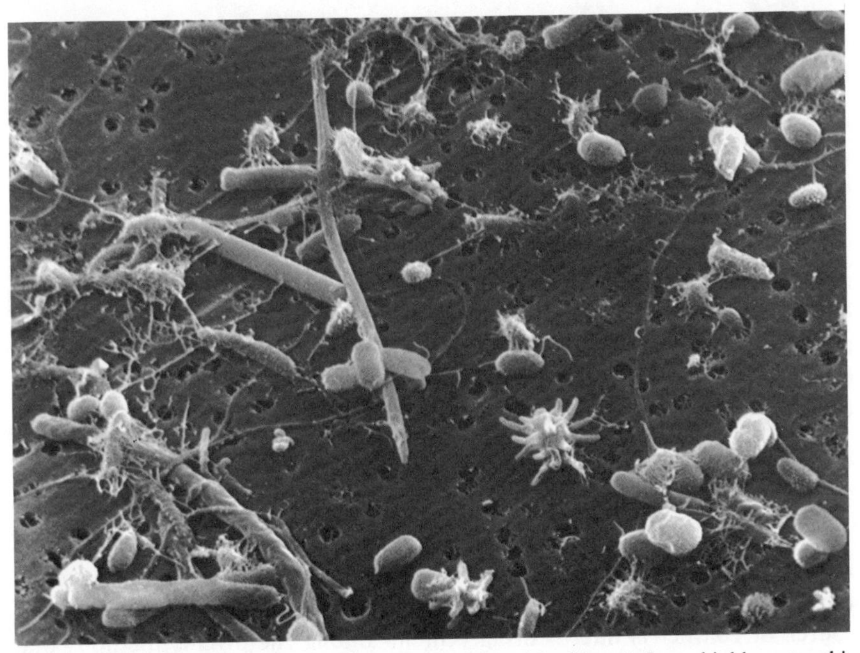

Figure 11.7 Assortment of free-living bacteria freshly filtered from highly eutrophic Thompson Lake, Ontario. Bacteria viewed on surface of a Nuclepore filter. Although attached bacteria often reveal fibrillar and pililike appendages, these free-living cells also possess similar structures. These appendages may serve to anchor bacteria when nutritive particle is encountered.

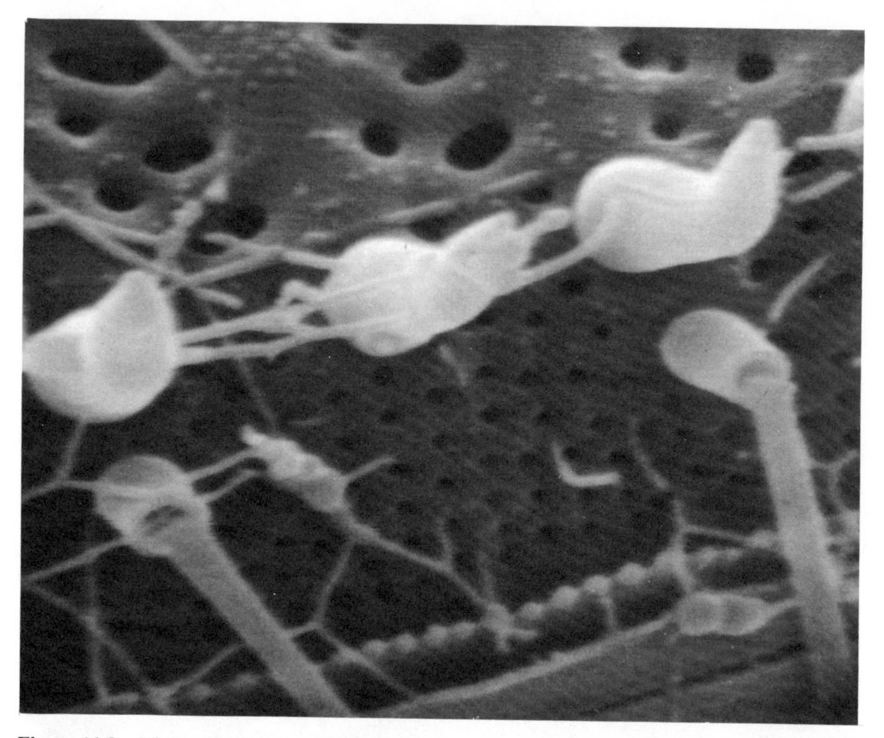

Figure 11.8 High-magnification SEM view of frustule of diatom *Stephanodiscus astrea*. Frustules are often infested with possible prosthecate bacteria. Such algal–bacterial associations are common during mid and late summer months in Lake Tahoe, California.

remains obscure [89,91]. Although elaborate prosthecate networks can be observed in this group, attachment can at times be rare. It has been suggested that prosthecea may also function as a buoyancy aid, maintaining clumps of cells near the air–water interface, to avoid sinking into the anoxic hypolimnia of eutrophic lakes [24].

11.4 QUANTITATIVE ASPECTS OF BACTERIAL ATTACHMENT

At present, conflicting data exist in both freshwater and marine systems on the quantitative significance of attached versus free-floating bacteria. One source of conflict is due to disagreement among the diverse methods that have been used in detecting and enumerating aquatic bacteria. Prior to the use of fluorescent DNA and RNA stains [92–94] and fluorescent antibody staining [95], a wide variety of conventional bacteriological stains were applied to aquatic studies [96,97]. A range of conflicting ratios of free-floating to attached bacteria resulted, and claims and counterclaims as to the specificity of respective methods were too numerous to mention here. In summary, no single method clearly revealed viable bacterial biomass without staining detrital material, or vice versa. To worsen the problem, when investigators attempted to utilize each other's techniques, poor parallel results were obtained—further evidence that universal interpretation of bacterial biomass based on conventional staining techniques was tenuous at best.

The application of fluorescent antibody and nucleic-acid staining has been a significant step forward in the enumeration of natural bacterial populations, especially attached forms. Fixation and initial preparation of samples for counting is relatively simple and can be accomplished in the field [98]. The separation of bacterial cells from detritus using UV fluorescent dyes is far superior to conventional techniques employing a transmitted or phase-contrast light microscope. The method, however, is not entirely free of interpretational problems. High background detrital fluorescence, or a lack of observable bacterial fluorescence can occur unpredictably. In addition, viable bacteria can fluoresce in more than one color. However, in a majority of cases experienced, counters familiar with the waters under investigation are aware of these anomalies. In addition to fluorescent staining, SEM can be used for enumeration purposes [99]. Although representing a significant improvement in resolution over light microscopes, the use of SEM is tedious and expensive on a routine basis. Scanning electron microscopy is, however, a useful tool in obtaining initial ratios of free-floating versus attached bacteria in natural waters, as well as providing high resolution views of detritus and surfaces as microbial habitats. The use of electronic image analyzers for discriminating both attached and free-floating bacteria is currently under investigation and may yield additional methodology for bacterial enumeration.

Another approach to investigating the relative importance of attached versus free-floating bacteria is through size-fractionation experiments using membrane filters. Nuclepore® filters, which are polycarbonate nonfibrous filters having precise uniform pores, are ideally suited for this purpose [99]. Water samples can be filtered through filters of various pore sizes followed by determination of biomass parameters such as pigment, adenosine triphosphate (ATP), muramic acid, and protein analyses. One interpretational problem with the subsequent analyses is that they are not specific for bacterial biomass. In fact, at present we do not possess a biochemical method that qualitatively or quantitatively separates algal from bacterial biomass. Cellular constituents used as measurements of biomass exist to various extents in both groups of microorganisms. Although pigment analyses offer the greatest hope for separating algal from bacterial biomass, pigment content per amount of algal biomass often shows sizable fluctuations in response to environmental gradients.

Nuclepore filtration has been used for the size fractionation of microbial assimilation of radioactively labeled organic substrates [100,101]. At trace concentrations many dissolved organic compounds are assimilated exclusively by bacteria. This can be shown by autoradiography of size-fractionation samples incubated with either ^{14}C- or ^{3}H-labelled organics [28]. In certain cases, however, algal incorporation of these compounds may occur [101], so routine parallel autoradiographic examination is advisable at all times. Substrates known to be assimilated by a wide variety of heterotrophic bacteria are preferred and include simple sugars such as glucose and galactose; organic acids such as acetate, malate, citrate; and a combination of amino acids (labeled *protein hydrolysate*). Separation of free-floating from attached rod-shaped or coccal bacteria can be accomplished by using a range of filter pore sizes. Our experiments usually include 0.2 μm (trapping all microorganisms), 1 μm, 2 μm, and 3 μm (trapping bacteria attached to algal cells and detrital particles) and 5 μm and 8 μm (trapping only large particles). Results from available work on size fractionation of microbial organic carbon uptake in a variety of inland ecosystems are given in Table 11.1. In addition to the problems of agreement between different enumeration methods on frequencies of attached versus free-floating bacteria, there is disagreement as to the overall importance of bacterial attachment in aquatic ecosystems in general [13,102]. The results in Table 11.1 show that the incidence of attached versus free-floating bacteria and their activities can vary dramatically from one system to another and between different areas or depths within one system. In general, near shore sections of lakes, particularly those lakes having high particle content, the highest fractions of bacterial activity are associated with particles that are larger than bacteria in size. Midlake and open ocean waters, which are usually more devoid of particles, tend to have a major bulk of bacterial activity present in free-floating state when no suitable substrates are present. The relationships of trophic states to incidence of attachment are not as clear-cut. Highly eutrophic lakes tend to support a significant percentage of bacterial activity and cells on particles, but the free-floating populations often show periodically explosive blooms leaving the attached populations as a relatively small fraction. Conversely, oligotrophic systems, possibly due to a lack of suitable attachment sites, normally develop relatively high free-floating:attached ratios. But periodic algal blooms can at times attract bacteria to large-sized particles (presumably algal cells in a majority of cases), increasing the significance of the attached bacterial fraction.

In summary, it is evident that blanket conclusions concerning the importance or insignificance of bacterial attachment are dangerous and probably inapplicable to a majority of aquatic ecosystems. To a limited extent, we can within reason predict that pelagic off-shore systems will favor high free-floating: attached bacterial ratios and that near-shore environments should yield significantly higher attached microbial populations. However, extensive size fractionation and enumeration work will need to be undertaken in defining these ratios in the vast array of inland waters supporting diverse particle forming and degrading processes.

11.5 ECOLOGICAL SIGNIFICANCE OF BACTERIAL ATTACHMENT IN FRESH WATER

Although numerical and descriptive evaluations of the importance of bacterial attachment have been numerous, the functional significance of this subject has been examined and evaluated in only a handful of studies. One reason for this disparity is methodologically related; it is difficult to separate, either spatially or temporally, the activities of attached from free floating bacteria under either laboratory or natural conditions. Rarely are all bacteria in a sample either associated with particles or in a free-floating state.

The surfaces of particles are dynamic areas supporting strong physicochemical gradients and rapid turnover of nutrients caused by laminar flow in moving waters as well as the metabolic activities of attached microbes. Respective rates and quantities transported and cycled by these processes are extremely difficult to measure. A great majority of the physical, chemical, and biological activities associated with freshwater particles occur in ultrathin layers that have been termed *microzones* [103], of <1 to 10 μm width, which border particles. Included in this zone is the diffuse double layer [104] of ions, where a strong ionic gradient, in terms of cation and anion accumulation, exists. The placement of a colloidal or bacteria-sized object in the layer causes rearrangement of the distribution of ionic charges [105]. In addition, the metabolic activities of microorganisms, in combination with the chemical environment bordering particles, impart a unique interplay of variables in each freshwater system. Even within one ecosystem, it is easy to envision the contrasting microzones present in the oxygenated epilimnion versus deoxygenated hypolimnion during summer stratification. Strong oxygen gradients exist between these water masses. The degree to which decomposition, and thus oxygen consumption, of certain organic particles (by attached microorganisms) can proceed is another factor deter-

Table 11.1 Summary of size fractions responsible for the incorporation of ^{14}C- and ^{3}H-labeled organic substances in natural waters[a]

Lake or stream, location and trophic status	Organic substrate added	Total incorporation in size fractions			
		$<1\ \mu m$	$1\ \mu m < 3\ \mu m$	$3\ \mu m < 5\ \mu m$	$>5\ \mu m$
Lake Tahoe, California—ultraoligotrophic					
Pelagic waters 20*m*	^{3}H-glucose	75	9	8	8
	^{14}C-acetate	81	7	3	9
	^{3}H-glycine	79	10	2	9
Turbid near shore waters—surface	^{14}C-acetate	55	12	11	22
Lake Taupo, New Zealand—oligotrophic					
Pelagic waters 20*m*	^{3}H-glucose	91	3	3	3
	^{3}H-serine	87	4	7	2
	^{3}H-acetate	90	3	2	5
	^{14}C-protein hydrolysate	85	5	4	6
Turbid near shore waters	^{3}H-glucose	60	11	13	16
Surface	^{3}H-acetate	51	12	14	23

Lake Rotorua, New Zealand—mesotrophic					
Pelagic waters 3*m*	^{3}H-glucose	72	11	3	14
	^{3}H-serine	81	9	2	8
	^{14}C-acetate	71	9	8	12
	^{14}C-protein hydrolysate	86	4	1	9
Lake St. George, Ontario—mesotrophic					
Midlake 4*m*	^{3}H-glucose	91	1	2	6
	^{3}H-serine	82	2	5	11
	^{3}H-acetate	92	2	2	4
Lake Rotongaio, New Zealand—eutrophic					
Midlake 2*m*	^{3}H-glucose	79	2	3	16
	^{3}H-serine	81	4	2	13
	^{14}C-acetate	75	3	4	18
	^{14}C-protein hydrolysate	72	5	2	21
Midlake 10*m*	^{3}H-glucose	61	6	3	30
	^{3}H-serine	54	4	5	37
	^{14}C-acetate	62	3	8	27
	^{14}C-protein hydrolysate	67	2	9	22

[a] All values are corrected for isotope adsorption to filters and particulates. The total incorporation includes all incorporation into particles of at least 0.2 μm in size.

mining the set of physicochemical factors making up the microzone. In short, the multiplicity of interacting variables in microzones is often highly variable in a lake or stream ecosystem as a whole but may be reasonably consistent in specific horizontal layers during thermally stratified periods.

Physical and chemical measurements of microzones are prohibitive without significant manipulation of a natural sample (i.e., concentration by filtration, artificial oxygenation and deoxygenation, and disruption of detrital structures). Large detrital aggregates are open to microprobe analyses, but only on a macro ($>10\mu m$) scale, thereby overlooking microzone gradients. Despite these technical limitations, there are at present useful, albeit generalistic and simplistic, means of determining some physiochemical gradients associated with particles. The use of redox indicators such as methylene blue, rezazurin, and tetrazolium salts in natural lake-water samples appears to be of value in describing microbial habitats bordering particles. In particular, tetrazolium salts should be of potential use as indicators of reduced or oxidized microzones. These salts form either colored or clear solutions when present in oxidized form. When exposed to reduced conditions, the salts form colored precipitates, which are deposited directly in the reduced microzone and can be observed with a light microscope [44,106]. A range of tetrazolium salts is commercially available, each one forming its respective colored formazan precipitate at a specific redox potential. For example, we [44] have used nitro blue tetrazolium (NBT; $E_o = 50$ mV) and 2,3,5-triphenyl tetrazolium chloride (TTC; $E_o = 490$ mV) as indicators of weak or strong reduced conditions in regions of bacterial attachment bordering cyanobacterial filaments. In a similar fashion, NBT was employed as an indicator of reduced microzones surrounding particles in fully oxygenated epilimnia of some New Zealand lakes. The reduced conditions may be due to bacterial activity associated with particles, for formalin-treated and heat-sterilized lake waters failed to show similar reduction. These findings were of interest in conjunction with chemical data from Lake Rotorua, one of the lakes where NBT experiments were conducted. Chemical data indicated that denitrification may have been occurring in the epilimnion [107]. Since ambient oxygen concentrations reached saturation in this section of the water column, it was initially difficult to reconcile chemical with physical evidence since denitrification is an obligate anaerobic process. Given the presence of reduced microzones inside and bordering particles, as shown by NBT deposition, it became clear that ambient oxygen measurements were not *a priori* realistic indicators of denitrification potential and that denitrifiers may well have been residing in microzones associated with these particles.

A similar example can be obtained from sediment-related studies. It was discovered that surface sediments in several eutrophic New Zealand lakes revealed maximum acetylene reduction (N_2 fixation) rates immediately following turnover of the water column [108]. This seemed paradoxical since the entire water column, including the region bordering the surface sediments, becomes highly oxygenated at turnover. Oxygen has repeatedly been shown to be strong inhibitor of N_2 fixation in a wide variety of microorganisms, especially bacteria [109]. Microscopic examination showed bacteria, as opposed to cyanobacteria, to be present in these sediments. In particular, bacteria attached to the filamentous gliding bacterium *Beggiatoa* were abundant. On addition of TTC, which is only converted to its insoluble formazan form under highly reduced conditions, it was found that bacteria with *Beggiatoa* caused such conditions. The fact that TTC deposition readily occurred would indicate that these bacteria formed microzones reduced sufficiently to allow N_2 fixation to proceed. Such conditions are similar to those in the highly reduced heterocysts of the N_2-fixing cyanobacterium *Anabaena* [44,106]. At present it is not known whether the bacteria associated with *Beggiatoa* or *Beggiatoa* itself is responsible for the observed N_2 fixation.

Bacteria associated with clumps of *Oscillatoria*, *Aphanizomenon*, and *Anabaena* are also capable of forming reduced microzones in O_2 supersaturated waters. The presence of such bacteria appears to enhance the N_2-fixing potential of these cyanobacteria [44].

Independent investigations have shown that many algae and the cyanobacteria show enchanced growth rates in the presence of associated bacteria. Several reasons for this enhancement have been given, and although disagreement exists as to the most important, all explanations focus on exchange mechanisms present in microzone regions where the associations occur. Lange [110,111] has indicated that CO_2 replenishment by heterotrophic bacteria associated with photosynthetic algal cells is responsible for the enhancement. Others, including Provasoli and Carlucci [112], have stressed the importance of growth factor (vitamins and growth-enhancing metabolities) exchange in microbial associations. It is likely that inorganic nutrients representing decomposition or end products from one partner and growth factors for another (NH_4^{+1}, PO_4^{-3}, and NO_3^{-1}) may be exchanged. Finally, O_2 removal by bacteria associated with O_2- sensitive photosynthetic microorganisms during freshwater blooms appears to be an important functional aspect of algal–bacterial associations [44].

In oligotrophic lakes, which usually contain extremely low levels of dissolved nutrients, an attached existence may be essential for successful growth among certain bacteria. Nitrifying bacteria may be one example. In Lake Tahoe nitrification appears to take place in the entire water column, but ammonium, even in the nonphotosynthetic (aphotic) deep waters, is undetectable throughout much of the year [113]. As a consequence of this environmental constraint, nitrifying bacteria appear to have formed close association with decomposing organic detritus, which undoubtedly represents a main source of reduced nitrogen available for nitrification [113]. Because of their close association with particulate matter, nitrifying bacteria can be concentrated from Lake Tahoe water by filtering particulate matter greater than 3 μm. This is further evidence that a significant proportion of the nitrifying community is attached. As is the case with soil experiments [114], aquatic nitrification is usually stimulated by the addition of particles, especially those containing organic nitrogen.

Due to surface charges associated with particles, nutrients found in low supply can be attracted to and concentrated by particles. Such a concentrating mechanism would provide relatively nutrient-rich microzones in aquatic environments. Experimental evidence appears to support this hypothesis. For example, the addition of clay minerals, specifically bentonite, allophane, and montmorillonite, as well as carbonates, results in a concentration of organic compounds [13,115], phosphate [116], and iron [117] from lakewaters. These clays can also concentrate bacteria [39], whose growth in natural waters is often reliant on the nutrients mentioned earlier. Mineral particles and organic detritus in Lake Tahoe [28], as well as Lawrence Lake, Michigan [118], are known to concentrate a range of organic compounds. In particular, amino acids, organic acids, and other polar organic compounds adsorb to freshwater particles, presumably leaving a concentrated source of nutrients in microzones where bacterial attachment can occur. It would be advantageous for bacteria to exist in such an environment, considering the growth-limiting dissolved nutrient concentrations characteristic of a bulk of the world's freshwater systems.

Bacterial degradation of organic particles and their adsorbed nutrients is of obvious importance to aquatic food chains, since known decomposition products, including NH_4^{+1}, PO_4^{-3}, CO_2, and a range of reduced trace elements are essential algal growth nutrients. In the absence of aerobic and anaerobic decomposition of such particulates, a vast amount of potential nutrients would remain locked up in the sediments of lakes or exit via outflows. In addition, most mineral cycles in lakes, including the nitrogen, sulfur, phosphorus, and carbon cycles, would eventually exhibit huge accumulations of material arrested in a variety of oxidation states.

An important result of bacterial growth on particle surfaces that is often overlooked in ecological assessments is the role that associated bacteria play as a food source. Bacteria appear to be important harvesting organisms of dissolved organic carbon (DOC) compounds in lakes. It has been shown that attached bacteria can derive their carbon sources from the dissolved pool [28]. Measured mineralization of a variety of added radioactive organic substrates is often less than half the amount of substrates incorporated [119]. This indicates that significant fractions of the DOC pool incorporated are converted to nonrespired, structural components of the bacterial community. In light of the relatively large reservoir of DOC present in most lakes [the dissolved organic:particulate organic carbon (POC) ratios in freshwater average on the order of 10], this uptake and conversion route represents an important source of POC formation. Unfortunately, we know little about the composition of DOC pools in lakes, that is, what percentage of the DOC pool is easily metabolized (reactive) versus unmetabolized (refractile) by bacteria. We do know, however, that significant proportions of algal photosynthate are excreted at times into the aquatic environment [120–122] and, in effect, represent a loss from the POC to the DOC pool. There is evidence that bacteria readily assimilate these products, thereby converting "lost" DOC back to POC [101,123]. Much of this recycling may occur in algal–bacterial associations, which are very common during cyanobacterial and chlorophycean blooms [124]. There is little reason to doubt that bacteria attached to organic and inorganic detritus cannot assimilate such DOC sources as well. In a functional sense, bacterial assimilation of DOC represents a form of harvesting of fixed carbon that might otherwise be lost from lakes and streams via outflows.

Zooplankton grazers, particularly cladoceran and copepod crustacea, rotifers and protozoans, are known to utilize detrital matter [125,126]. Detritus supporting bacteria would enhance their nutritional status, since it is known that bacteria alone can sustain these grazers [54]. Some numerically important copepod zooplankton appear to be restricted to filtering particles larger than 2–3 μm in length, thereby eliminating many free-floating bacteria from their diet. Bacteria present on detritus and algal cells larger than 3 μm would be grazed preferentially by this zooplankton group. Benthic detritivores would no doubt profit from bacterial conditioning of detritus as it settles down the water column. The presence of bacteria, in addition to grazable free-floating bacteria, would represent a means by which DOC not directly consumed by animal grazers can be harvested as a filterable POC food source. Bacterial heterotrophic activity, therefore, appears to be an important means of optimizing the transfer of photosynthetically fixed carbon to grazer and detritivore links in aquatic food chains.

11.6 CONCLUDING REMARKS

Diverse groups of microorganisms are found attached to nonliving and living particles in freshwater. The means and mechanisms of microbial attachment also appear diverse, although this subject deserves further study on ultrastructural and macromolecular levels. Judging by contrasting data, the incidence of microbial attachment appears to vary dramatically, not only within one freshwater system but also over a range of systems covering a trophic scale. Hence it is ill advised to make general conclusions regarding the frequency of attached versus free-floating microorganisms. Spatial and temporal variability of soluble and particulate nutrient states, even within one freshwater system, is a dominant factor in determining the ratios of total versus free-floating microorganisms.

The importance of bacterial particle formation and modification has been grossly underestimated in freshwater environments. Light-microscopic observations of detrital material often do not reveal the complex networks of extracellular fibrils, mucilage, and cellular remains attributable to previous and current bacterial growth periods. Much of the cellular carbon that is laid down by bacteria is extracellular and is either transparent or beyond the resolution of the light microscope. Thus biomass enumeration by cell counts converted to cell volumes often represents an underestimate of the true bacterial POC production. In some cases the adhesive structural modifications and appendages that bacteria produce represent a bulk of the biomass of these organisms in fresh water. Hence it is important that more attention be paid to strategies that aquatic attached microorganisms employ to optimize their growth. The trophic and ecological implications of these strategies appear to be key elements in freshwater food chains and nutrient cycles.

11.7 ACKNOWLEDGMENTS

Much of the work discussed was conducted at the following facilities: Facility for Advanced Instrumentation, University of California, Davis; Freshwater Section, DSIR Taupo, New Zealand; and Canada Centre for Inland Waters, Burlington, Ontario. I would like to thank the many coworkers at these institutes for their help, advice, and critiques. In particular, I am grateful to Mr. J. Pangborn and Drs. C. R. Goldman, E. White, J. Gibb, and A. F. Carlucci for their enthusiastic support of my previous research efforts.

11.8 REFERENCES

1. C. Dobell, *Antony van Leeuwenhoek and His "Little Animals,"* Staples Press, London, 1932.
2. M. W. Beijerinck, *Proc. Acad. Sci. Amst.* **3,** 352 (1900).
3. S. Winogradsky, *Anales Inst. Pasteur, Paris.* **39,** 299 (1925).

4. H. J. Conn, *New York Agric. Exp. Sta. Tech. Bull.* **64,** (1918).
5. W. Bulloch, *The History of Bacteriology*, Oxford U. P., New York, 1960, p. 25.
6. A. J. Henrici, *J. Bacteriol.*, **25,** 277 (1933).
7. A. J. Henrici and E. McCoy, *Transact. Wisc. Acad. Sci. Arts Lett.*, **31,** 323 (1938).
8. S. A. Waksman and C. L. Carey, *J. Bacteriol.*, **29,** 531 (1935).
9. C. E. Zobell and D. Q. Anderson, *Biol. Bull.*, **71,** 324 (1936).
10. B. Lloyd and J. Roy, *Tech. College, Glasgow*, **4,** 137 (1937).
11. R. A. Vollenweider, *A Manual on Methods for Measuring Primary Production in Aquatic Environments*, IBP Handbook 12, Blackwell Scientific Publications, Oxford, 1969.
12. A. Nauwerck, *Symbol. Bot. Upsalien.*, **17,** 163 (1963).
13. H. W. Jannasch and P. H. Pritchard, in U. Melchiorri Santolini and J. W. Hopton, Eds., *Detritus and Its Role in Aquatic Ecosystems*, proceedings of an IBP–UNESCO Symposium; *Mem. Ist. Ital. Idrobiol.*, **29** (suppl.), 289 (1972).
14. A. G. Rodina, in *Methods in Aquatic Microbiology*, R. R. Colwell and M. S. Zambruski, Eds., University Park Press, Baltimore, 1972.
15. Y. I. Sorokin and H. Kadota, Eds., *Techniques for the Assessment of Microbial Production and Decomposition in Fresh Waters*, IBP Program Handbook 23, Blackwell Scientific Publications, Oxford, 1972.
16. A. T. Henrici and D. E. Johnson, *J. Bacteriol.*, **30,** 61 (1935).
17. A. G. Rodina, *Limnol. Oceanogr.*, **8,** 388 (1963).
18. A. G. Rodina, *Verh. Int. Ver. Limnol.*, **16,** 1513 (1967).
19. H. W. Jannasch, *J. Gen. Microbiol.*, **18,** 609 (1958).
20. J. Oláh, *Mem. Ist. Ital. Idrobiol.*, **29** (suppl.), 385 (1972).
21. H. W. Paerl and C. R. Goldman, *Limnol. Oceanogr.*, **17,** 145 (1972).
22. B. L. Madsen, *Mem. Ist. Ital. Idrobiol.*, **29,** (suppl.), 385 (1972).
23. K. C. Marshall, R. Stout, and R. Mitchell, *Can. J. Microbiol.*, **17,** 1413 (1971).
24. R. Stanier, M. Doudoroff, and E. A. Adelberg, *The Microbial World*, 3rd ed., Prentice-Hall, Englewood Cliffs, N. J., 1970.
25. T. M. Skerman, *N. Z. Sci. Technol. Ser. B*, **38,** 44 (1956).
26. A. O. Christie and G. D. Floodgate, *Nature* (*Lond.*), **212,** 308 (1966).
27. M. Fletcher and G. D. Floodgate, *J. Gen. Microbiol.*, **74,** 325 (1973).
28. H. W. Paerl, *Limnol. Oceanogr.*, **19,** 966 (1974).
29. G. D. Floodgate, *Mem. Ist. Ital. Idrobiol.*, **29,** (suppl.), 309 (1972).
30. J. S. Waid, in T. Rosswall, Ed., *Modern Methods in the Study of Microbial Ecology*, Swedish Natural Science Research Council, bulletin No. 17, 1973, p. 103.
31. H. W. Paerl, *Science*, **180,** 496 (1973).
32. K. C. Marshall, in A. D. McLaren, J. J. Skujins, and M. Dekker, Eds., *Soil Biochemistry*, Vol. II, Chapter 14, 1972.
33. D. I. Nikitin, in T. Rosswall, Ed., *Modern Methods in the Study of Microbial Ecology*, Swedish Natural Science Research Council, bulletin No. 17, 1973, p. 85.
34. J. G. Ferrante and J. A. Parker, *Limnol. Oceanogr.*, **22,** 221 (1977).
35. P. Hirsch and S. H. Pankratz, *Z. Allg. Mikrobiol.* **10,** 589 (1970).
36. S. A. Waksman and C. L. Carey, *J. Bacteriol.*, **29,** 545 (1935).

37. H. Heukelekian and A. Hiller, *J. Bacteriol.*, **40,** 546 (1940).
38. H. W. Paerl and C. R. Goldman, *Mem. Ist. Ital. Idrobiol.*, **29** (suppl.), 129 (1972).
39. G. Stotzky, in T. Rosswall, Ed., *Modern Methods in the Study of Microbial Ecology*, Swedish Natural Science Research Council, bulletin No. 17, 1973, p. 17.
40. G. Stotzky, *Transact. N. Y. Acad. Sci.*, **30,** 11 (1967).
41. J. Adler, *Science*, **166,** 1588 (1969).
42. S. Fogel, I. Chet, and R. Mitchell, *Chemoreception in Marine Bacteria*, U.S. Office of Naval Research, technical report No. 2, Contract No. 00014-67-A-0]98-6626, 1971.
43. J. Adler, *Sci. Am.*, **234,** 40 (1976).
44. H. W. Paerl and P. E. Kellar, *J. Phycol.* (in press).
45. H. W. Paerl, *Limnol. Oceanogr.* (in press).
46. H. C. Jones, I. L. Roth, and W. M. Saunders, III, *J. Bacteriol.*, **99,** 316 (1969).
47. H. W. Paerl, *Microbial. Ecol.*, **2,** 73 (1975).
48. W. J. Weibe and L. R. Pomeroy, *Mem. Ist. Ital. Idrobiol.*, **29,** (suppl.), 325 (1972).
49. J. McN. Sieburth, *Microbial Seascapes*, University Park Press, Baltimore, 1975.
50. G. Stotzky, *CRC Crit. Rev. Microbiol.*, **2,** 59 (1972).
51. H. W. Jannasch, *Verh. Interna. Verein. Limnol. Microbiol.*, **17,** 25 (1969).
52. H. W. Jannasch, *Naturwissenschaften*, **41,** 42 (1954).
53. E. A. Birge and C. Juday, *Ecol. Monogr.*, **4,** 440 (1934).
54. R. G. Wetzel, *Limnology*, Saunders, Philadelphia, 1975.
55. O. Holm-Hansen, C. R. Goldman, R. G. Richards, and P. M. Williams, *Limnol. Oceanogr.*, **21,** 548 (1976).
56. H. W. Paerl and C. R. Goldman, unpublished results.
57. E. A. Birge and C. Juday, *Bull. U.S. Bur. Fish.*, **42,** 185 (1926).
58. E. P. Odum and A. A. de la Cruz, in G. H. Lauff, Ed., *Estuaries*, AAAS publication, 1967, pp. 383–388.
59. G. A. Riley, *Adv. Mar. Biol.*, **8,** 1 (1970).
60. H. W. Paerl, R. D. Thomson, and C. R. Goldman, *Verh. Internat. Verein. Limnol.*, **19,** 826 (1975).
61. T. R. Parsons and J. D. H. Strickland, *Science*, **136,** 313 (1962).
62. H. L. Golterman, *Mem. Ist. Ital. Idrobiol.*, **29,** (suppl.), 84 (1972).
63. E. P. Odum and A. A. de la Cruz, *Am. Inst. Biol. Sci. Bull*, **13,** 39 (1963).
64. T. Fenchel, *Limnol. Oceanogr.*, **15,** 14 (1970).
65. K. W. Cummins, M. J. Klug, R. G. Wetzel, R. C. Peterson, K. F. Suberkropp, B. A. Manny, J. C. Wuycheck, and F. O. Hoard, *Bioscience*, **22,** 719 (1974).
66. N. K. Kaushik and H. G. N. Hynes, *Arch. Hydrobiol.*, **68,** 465 (1971).
67. D. L. Lush and H. B. N. Hynes, *Limnol. Oceanogr.*, **18,** 968 (1973).
68. H. W. Paerl and M. A. Perkins, unpublished results.
69. M. A. Lock and H. B. N. Hynes, *Limnol. Oceanogr.*, **21,** 436 (1976).
70. R. L. Todd, K. Cromack, and R. M. Knutson, in T. Rosswall, Ed., *Modern Methods in the Study of Microbial Ecology*, Swedish National Science Research Council, bulletin No. 17, 1973, p. 109.
71. R. L. Todd and T. J. Kerr, *Appl. Microbiol.*, **23,** 1160 (1972).

72. W. A. Corpe, in *Adhesion in Biological Systems*, R. Manly, Ed., Academic, New York, 1970, pp. 73–85.
73. H. W. Paerl and A. M. Massalski, in preparation.
74. G. G. Leppard and S. Ramamoorthy, *Can. J. Bot.*, **53,** 1729 (1975).
75. H. W. Paerl and D. R. S. Lean, *J. Fish. Res. Bd. Can.*, **33,** 2805 (1976).
76. J. W. Costerton, G. G. Geesey, and K. J. Cheng, *Sci. Am.* **238,** 86 (1978).
77. G. G. Geesey, J. W. Costerton, and K. J. Cheng, *Limnol. Oceanogr.* (in press).
78. A. Sladeckova, *Verh. Internat. Verein. Limnol.*, **16,** 753 (1966).
79. R. G. Wetzel, Internat. Rev. Ges. Hydrobiol., **49,** 1 (1964).
80. F. E. Round, in D. F. Jackson, Ed., *Algae and Man*, Plenum, New York, 1964, pp. 138–184.
81. P. S. Meadows and J. G. Anderson, *J. Mar. Biol. Assoc. U.K.*, **48,** 161 (1968).
82. G. G. Leppard, *Science*, **185,** 1066 (1974).
83. C. E. Lankford, *CRC Crit. Rev. Microbiol.*, **2,** 273 (1973).
84. T. P. Murphy, D. R. S. Lean, and C. Nalewajko, *Science*, **192,** 900 (1976).
85. G. G. Leppard and J. R. Colvin, *J. Cell Biol.*, **53,** 695 (1972).
86. B. A. Whitton, in N. G. Carr and B. A. Whitton, Eds., *The Biology of Blue-Green Algae*, Blackwell Scientific Publications, Oxford, 1973.
87. J. S. Poindexter, *Bacteriol. Rev.*, **28,** 231 (1964).
88. E. G. Mulder, *J. Appl. Bacteriol.*, **27,** 151 (1964).
89. K. C. Marshall and R. H. Cruickshank, *Arch. Mikrobiol.*, **91,** 29 (1973).
90. H. W. Paerl, unpublished results.
91. J. T. Staley, *J. Bacteriol.*, **95.** 1921 (1968).
92. S. Strugger, *Fluoreszenz Mikroskopie und Mikrobiologie*, Schaper-Verlag, Hannover, Germany, 1949.
93. M. A. Darken, *Appl. Microbiol.* **10,** 387 (1962).
94. E. F. J. Wood and C. H. Oppenheimer, *Zeits, Allg. Mikrobiol.*, **2,** 164 (1962).
95. E. L. Schmidt, in T. Rosswall, Ed., *Modern Methods in the Study of Microbial Ecology*, Swedish Natural Science Research Council, bulletin No. 17, 1973, p. 67.
96. S. I. Kuznetzov, *Die Rolle der Mikroorganismen in Stroffkreislauf der Seen*, VEB Deutsch. Verlag Wissenschaften, Berlin, 1959.
97. A. G. Rodina, *Metody Vodnoi Mikrobiologii*, Izdatel Stuno Nauka, Moscow, 1965.
98. J. E. Hobbie, R. J. Daley, and S. Jasper, *Appl. Environ. Microbiol.*, **33,** 1225 (1977).
99. H. W. Paerl and S. L. Shimp, *Limnol. Oceanogr.*, **18,** 802 (1973).
100. T. Berman, *Marine Biol.* **33,** 215 (1975).
101. H. W. Paerl, *Proc. Roy. Soc. N. Z.* (in press).
102. H. W. Jannasch and G. E. Jones, *Limnol. Oceanogr.*, **4,** 128 (1959).
103. D. I. Nikitin, in Thomas Rosswall, Ed., *Modern Methods in the Study of Microbial Ecology*, Swedish Natural Science Research Council, bulletin No. 17, 1973, p. 357.
104. L. Wiklander, in F. E. Bear, Ed., *Chemistry of the Soil*, Van Nostrand, New York, 1964.
105. S. Ramamoorthy and G. G. Leppard, *Naturwissenschaften* **63,** 579 (1976).
106. P. Fay S. A. Kulasooriya, *Arch. Mikrobiol.*, **87,** 341 (1972).
107. E. White, unpublished results.

108. A. L. Mackenzie, unpublished results.

109. W. D. P. Stewart, *Ambio*, **6,** 166 (1977).

110. W. Lange, *Nature*, **215,** 1277 (1967).

111. W. Lange, *Can. J. Microbiol.*, **17,** 304 (1971).

112. L. Provasoli and A. F. Carlucci, in W. D. P. Steward, Ed., *Algal Physiology and Biochemistry*, California U. P., Berkeley, 1974, pp. 741–787.

113. H. W. Paerl, R. C. Richards, R. L. Leonard, and C. R. Goldman, *Limnol. Oceanogr.*, **20,** 1 (1974).

114. M. Alexander, *Microbial Ecology*, Wiley, New York, 1971.

115. P. G. Wetzel, P. H. Rich, M. C. Miller, and H. L. Allen, *Mem. Ist. Ital. Idrobiol.*, **29,** (suppl.), 185 (1972).

116. E. White and G. W. Payne, unpublished results.

117. H. W. Paerl and T. P. Murphy, unpublished results.

118. R. G. Wetzel and H. L. Allen, in Z. Kajak and A. Hillbricht-Ilkowska, Eds., *Productivity Problems of Freshwaters*, PWN, Polish Scientific Publishers, Warsaw, 1970, pp. 333–347.

119. J. E. Hobbie and C. C. Crawford, *Limnol. Oceanogr.*, **14,** 528 (1969).

120. C. E. Fogg, *Proc. Roy. Soc.* (*Lond.*) *B*, **139,** 372 (1952).

121. C. E. Fogg and D. F. Westlake, *Verh. Int. Verin. Limnol.*, **12,** 219 (1955).

122. A. E. Walsby, *Br. Phycol. J.*, **9,** 371 (1974).

123. C. Nalewajko, T. G. Dunstall, and H. Shear, *J. Phycol.*, **12,** 1 (1976).

124. D. E. Caldwell, *CRC Critical Rev. Microbiol.*, **305** (1977).

125. G. W. Saunders, *Mem. Ist. Ital. Idrobiol.*, **29,** (suppl.), 261 (1972).

126. Y. I. Sorokin, *Mem. Ist. Ital. Idrobiol.*, **18,** (suppl.), 187 (1965).

CHAPTER **12**

Some Techniques Involved in Study of Adsorption of Microorganisms to Surfaces

J. W. COSTERTON
Biology Department, University of Calgary, Calgary, Alberta, Canada

CONTENTS

12.1 INTRODUCTION

The foregoing chapters have established that many microorganisms have the ability to adsorb onto surfaces and that the resultant adsorbed microbial populations are found in a wide variety of ecosystems. Perhaps it must now be determined how important these absorbed microbial populations are in numerical and functional terms. Our ability to determine this depends on the development of techniques for the enumeration and the recovery of the bacterial cells that comprise these adsorbed populations. Startling insights have resulted from some studies in which these techniques have been used. For instance, 1 cm^2 of submerged surface in an alpine stream bears a bacterial population numerically equivalent to the bacteria in a whole liter of the flowing water, which means that adsorbed bacteria actually comprise the predominant microbial population in these streams [1]. Furthermore, adsorbed bacteria carry out specific digestive functions in the bovine rumen [2] and taxonomically distinct adsorbed bacteria are entirely responsible for the production of an enzyme (urease) that is essential for the normal functioning of this organ [3].

Generally, the study of an adsorbed microbial population begins with the microscopic demonstration of its existence and its distribution on the surface in question. These studies, especially if they involve electron microscopy, may also yield information on the variety of morphologic types, the predominant bacterial cell-wall type (Gram-negative in aquatic systems), the secondary order of the adsorbed population (layering of morphologic types), and the predominant adsorption mechanism. Next, the bacteria in the adsorbed microbial population must be enumerated, and this is best accomplished by direct methods (e.g., epifluorescence) or by the measurement of a specific unique bacterial component and extrapolation to bacterial biomass [4]. Taxonomic identification and the determination of the physiologic capabilities of the bacteria that comprise an adsorbed population must await their recovery from the surface in pure cultures. Finally, the isolation of adsorbed organisms allows the production of specific antibodies that can be used to enumerate a particular species and to map its distribution on the surface under study. These steps could be summarized as the demonstration, enumeration, isolation, and final mapping of an adsorbed microbial population.

12.2 TECHNIQUES FOR STUDYING NATURAL ADSORBED MICROBIAL POPULATIONS

The demonstration of an adsorbed microbial population is best accomplished by some form of microscopic examination, and since each type of microscopy has unique advantages and disadvantages, a combination of microscopic techniques is almost always necessary.

12.2.1 Light Microscopy

The use of phase microscopy to examine living bacterial populations adsorbed onto surfaces has the advantages of avoiding the loss of cells during fixation and processing and of providing a topographic "face" view of large areas of colonized surface. Coverslips or other optically suitable substrates are immersed in natural systems or in bacterial suspensions [5] under a variety of ionic conditions, and both the initial reversible adsorption and the subsequent irreversible adsorption of bacteria can be distinguished by their relative sensitivity to rinsing [5]. The disadvantages of this method are the distinctly limited number of optically suitable substrates and the low resolution, which is insufficient to reveal adhesion mechanisms.

12.2.2 Scanning Electron Microscopy

This technique shares many advantages with light microscopy in that it offers a "face" view of surfaces and easily relates the distribution of bacteria to the topography of the surface in question. In this way adsorbed bacteria can be related to specific cells in a tissue surface (Fig. 12.1) or to topographic features of an inert surface. Like light microscopy, it suffers from its inability to detect the bacterial colonization of interior surfaces and from limitations of resolution. For instance, the rumen bacteria that adsorb to forage materials and effect their digestion do so predominantly in the interior parenchymous tissue (Figs. 12.2 and 12.3*a*). Scanning electron microscopy fails to reveal this important process and, in the intact leaf, shows only those few bacteria on the forage leaf surface (Fig. 12.3*b*). Because of the limited resolution of SEM, adsorbed bacteria may fail to show any of the fine extracellular structures that mediate their adsorption to the surface (Fig. 12.4), whereas the superior resolution of transmission electron microscopy (TEM), coupled with the use of specific stains, reveals these important structures in fine detail (Fig. 12.5).

Improvements in preparation techniques for SEM have minimized the drying-related collapse of tissues and of adsorbed bacteria by the use of critical-point drying [6] and eliminated the "burying" of fine structure during metal deposition by the use of heavy-metal impregnation techniques [7]. Colonized subtrates, therefore, can be fixed with aldehyde fixatives for stabilization, impregnated with large amounts of osmium by the thiocarbohydrazide techique for conductivity, and then dried by the critical point method to avoid drying damage. These technical advances have allowed us to resolve structures very close to the limits of resolution of the SEM (Fig. 12.6). The SEM preparative protocol used in our laboratory is detailed in Appendix A.

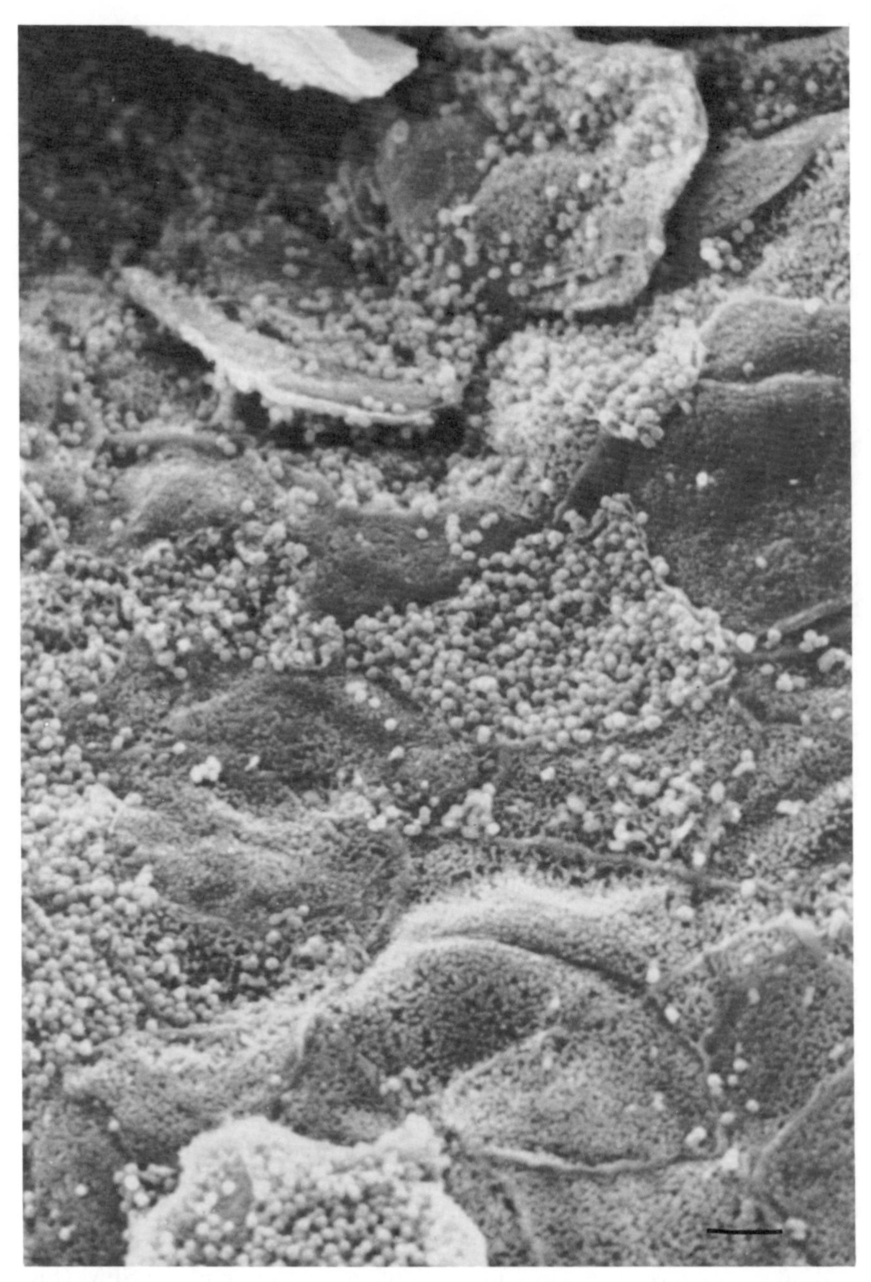

Figure 12.1 Scanning electron micrograph of surface of stratified squamous epithelium of bovine rumen showing heavy bacterial colonization of only some of the tissue cells whose outlines are clearly visible. Note extensive bacterial colonization of cells underlying partly detached tissue cells at top of micrograph (bar = 5 μm).

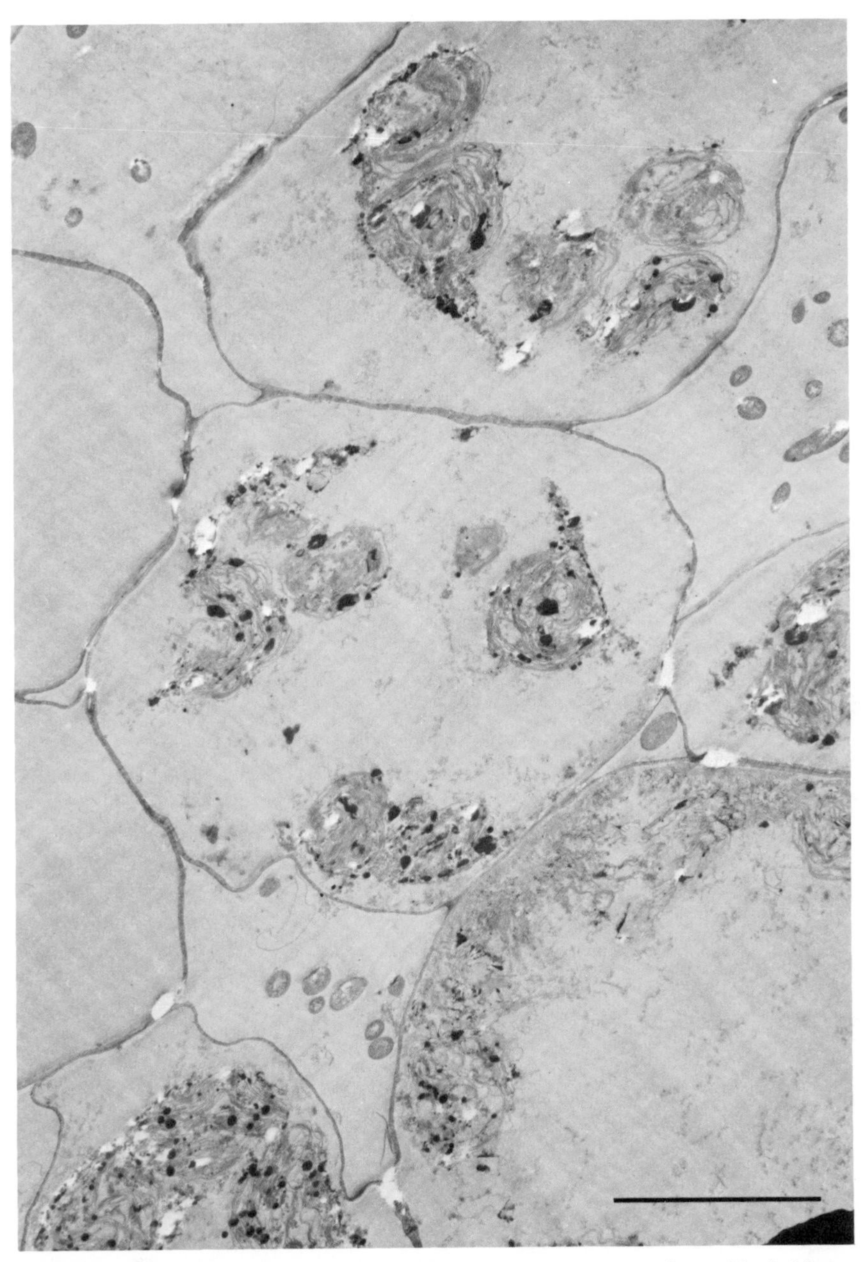

Figure 12.2 Transmission electron micrograph of inner parenchymous tissue of leaf of forage plant incubated with bacteria from bovine rumen. Note extensive colonization of intercellular spaces by bacteria that tend to form discrete microcolonies, whereas intracellular spaces are devoid of bacteria and identifiable by their content of damaged chloroplasts (bar = 5 μm).

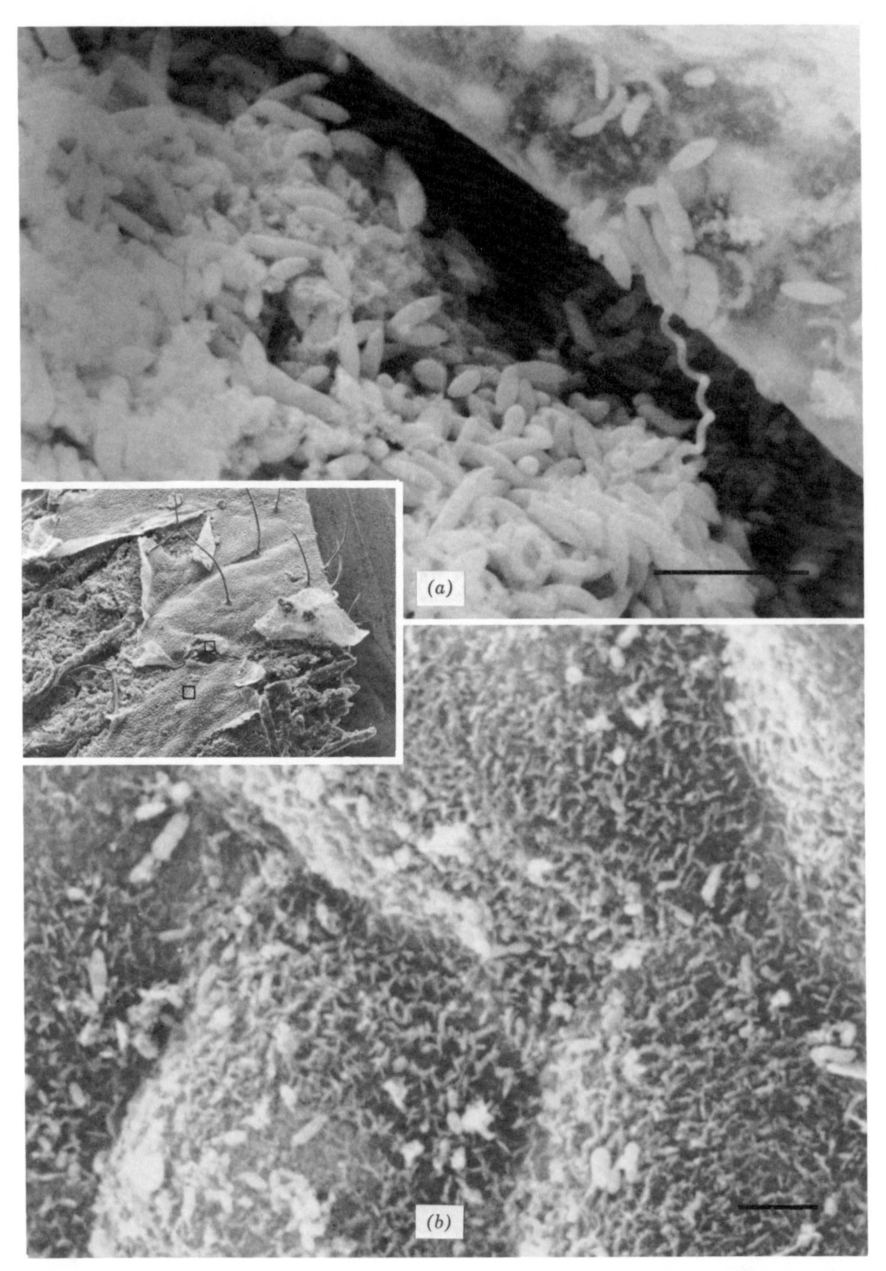

Figure 12.3 Scanning electron micrograph of leaf of forage plant incubated with rumen bacteria (inset shows leaf at low power): (A) presence of very large heterogenous bacterial population in parenchymous tissue where epidermis is damaged; (B) intact epidermal surface, partly colonized by relatively small bacterial population [bars in (A) and (B) = 5 μm].

Figure 12.4 Scanning electron micrograph of trichome-forming bacterium adsorbed to epithelial surface of dorsal part of bovine tongue. Relationship of these bacteria to tissue cells is well represented, but fine bacterial fibrils that mediate this attachment are not resolved (bar = 5 μm).

12.2.3 Transmission Electron Microscopy

This technique can be used to good effect when grids are immersed in water, colonized by bacteria, and examined by negative staining [5], but TEM of sections of embedded material yields very different morphological data in that we normally obtain a slice of the tissue or substrate at a right angle to the colonized surface, and we can derive very little information concerning the distribution of bacteria on the surface in question. On the other hand, bacterial colonization of the interior of tissues is detected by this method (Fig. 12.2), in contrast to SEM, and the greater resolving power of TEM allows the resolution of fine structures that are involved in adhesion to the surface (Fig. 12.5) and to each other (Fig. 12.7). Because TEM provides a sectional image at right angles to the colonized surface it is especially useful in detecting secondary organization in adsorbed bacterial populations, as in Fig. 12.8, which shows a layer of cellulolytic bacteria that are digesting a cellulose substrate and are overlaid by a layer of cells of a different bacterium with which a physiological cooperation is suggested. This morphological expression of a simple consortium, a cooperative physiological unit described in physiological terms by R. S. Wolfe and others, could not be detected by SEM, and this underlines the usefulness of TEM in the

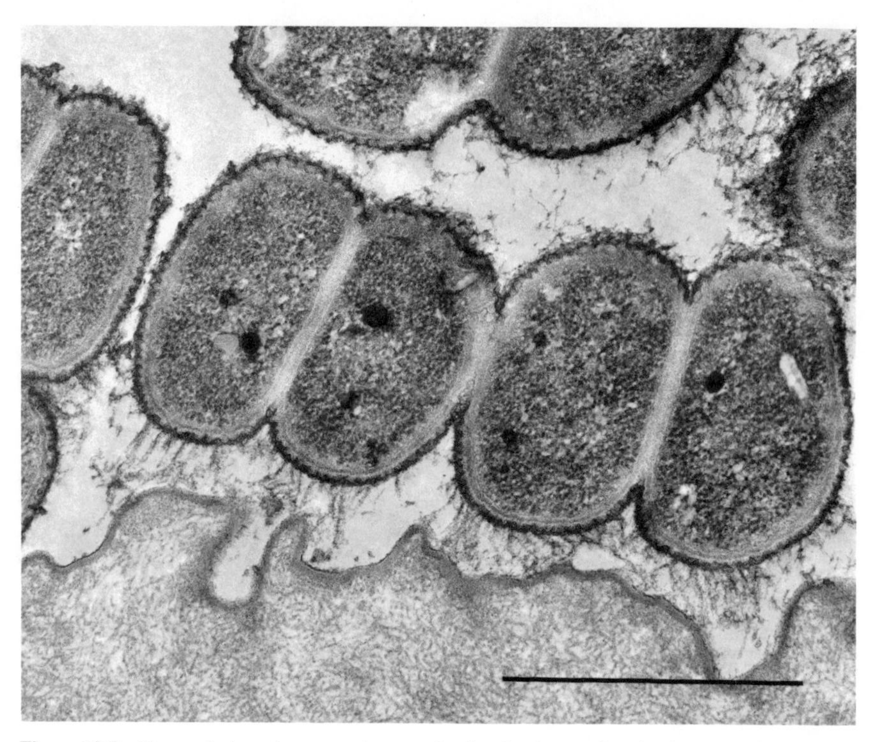

Figure 12.5 Transmission electron micrograph of ruthenium red-stained preparation of same region of bovine tongue showing same type of adsorbed bacteria. Note resolution of fine bacterial fibers that clearly mediate adhesion of these organisms to this tissue (bar = 1 μm).

detection of the relationship of adsorbed bacteria with the surface and with each other.

The preparative steps for TEM include the use of electron-dense stains, and some of these stains are sufficiently specific to yield a histochemical characterization of certain structures. For example, ruthenium red is specific for polyanions such as polysaccharides [8], and most extracellular carbohydrate fibers are specifically stained by this reagent (Figs. 12.5, 12.7, 12.8, 12.10, 12.12, and 12.13). These extracellular carbohydrate fibers, which are very important in the adhesion of microbial populations [9], can be stabilized against collapse and condensation during dehydration by reaction with specific antibodies [10,11] or specific lectins [12] that cross-link individual fibers and preserve the extracellular fibrous mass in its true configuration (Fig. 12.9).

Modern preparative techniques for TEM involve the use of aldehyde fixatives that are delivered, in an ionic milieu as close as possible to natural fluids, by a nondisruptive method (perfusion in some cases). The aldehyde

Figure 12.6 High-magnification SEM of trichomes from pure culture of organism seen in Figs. 13.4and 13.5. Note that critical point drying followed by thiocarbohydrazide treatment (Appendix A) preserves bacterial fibers seen on only one side of these specialized cells (bar = 1 μm).

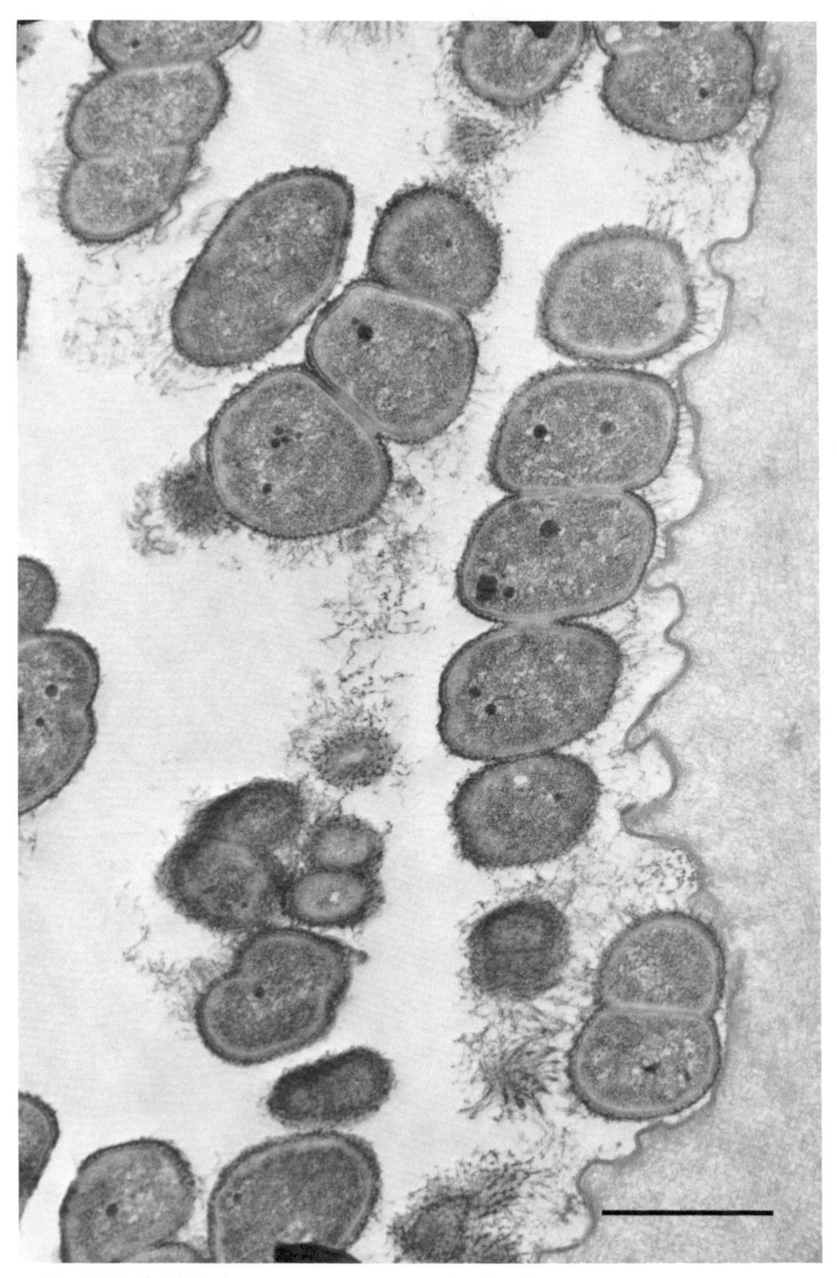

Figure 12.7 Transmission electron micrograph of ruthenium red-stained preparation of trichome-forming bacterium adhering to cells of bovine tongue and forming multilayered adherent population by further bacterium–bacterium attachment (bar = 1 μm).

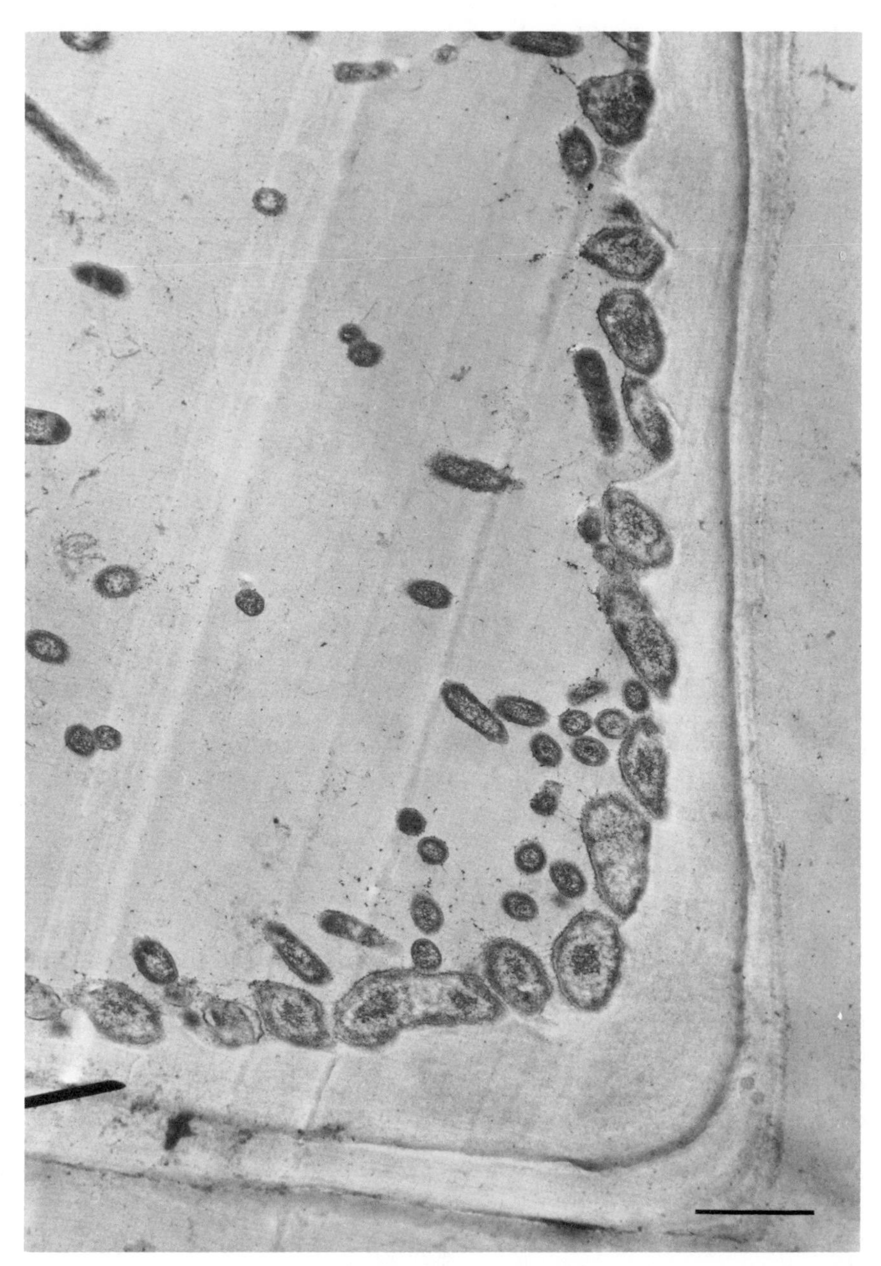

Figure 12.8 Transmission electron micrograph of partly digested L-shaped fragment of plant cell wall. Note that this cell wall is colonized by comparatively large cells of one bacterial type and that these cells appear to have digested "pits" in cellulose into which they have entered. Note that, even though thousands of species are present in this mixed rumen bacterial population, very thin cells of another bacterial type have formed a second layer in this adherent population. Thus the "secondary" structure of an adherent bacterial population is revealed by transmission electron microscopy (bar = 1 μm).

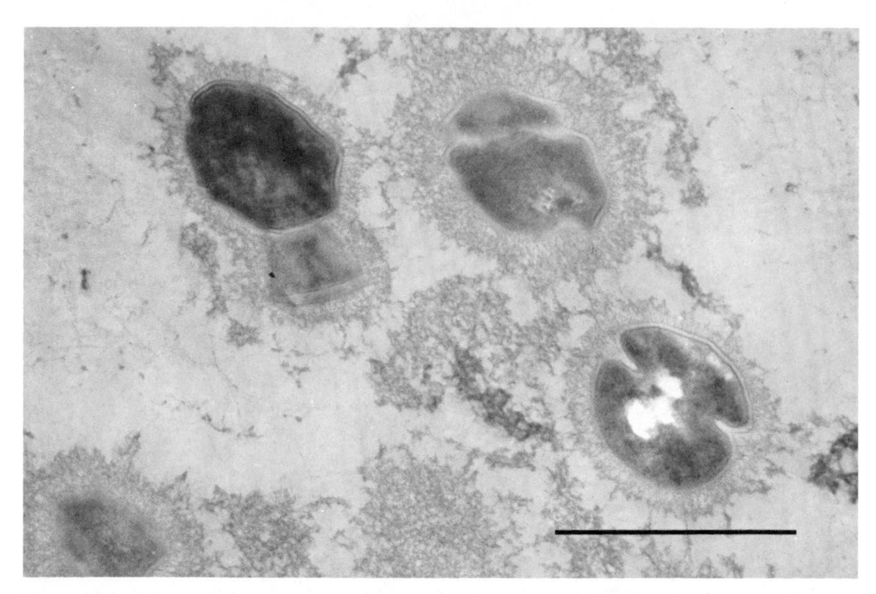

Figure 12.9 Transmission electron micrograph of streptococcal cells whose extracellular fibrils have been stabilized by reaction with specific antibodies. Note even distribution and fine fibrous texture of this extracellular "glycocalyx" in comparison with ruthenium red-stained preparations of cells whose fibrillar "coats" have not been stabilized with antibodies (bar = 1 μm).

fixative can be mixed with a specific stain (ruthenium red) that, because of its solubility, must be maintained in all subsequent aqueous washes, and general electron density is obtained by staining with osmium compounds. Where the substrate is a tissue or is penetrable by the embedding resin, the colonized surface is simply included in the hardened plastic block, but populations adsorbed to metallic and other hard surfaces can also be embedded by polymerizing an embedding plastic around the fixed cells and then shearing the metal away from the hardened plastic [13] by sharp temperature changes (dipping in liquid nitrogen). The TEM preparative protocol used in our laboratory is detailed in Appendix B.

12.2.4 Enumeration

The quantitation of bacterial populations has been approached by many indirect methods because plate counts have often proved to be unreliable due to bacterial clumping [4], and direct counting methods have proved to be very laborious. Indirect chemical methods have measured ATP [14], muramic acid [15], poly-β-hydroxybutyrate [15], DNA [16], and lipopolysaccharide (LPS) [17]. The limulus lysate assay for LPS is the only chemical parameter that has been successfully related to bacterial biomass and bacterial numbers in a significant number of environments [17], and we must bear in mind that it only detects Gram-negative bacteria [18]. The enumeration of bacteria is discussed in detail in a recent publication of the American Society for Testing and Measurement (ASTM) [4].

The aposition of adsorbed bacteria to opaque solid surfaces poses special problems in both direct and indirect methods of enumeration, and, except for rare instances in which adsorbed bacteria have been directly counted on rock surfaces by epifluorescence, the adherent microbial slime layer normally must be removed by scrapping [1,19] so that the bacteria can be enumerated by epifluorescence or by the LPS assay. Although bacteria are clearly lost in this procedure by breakage and by incomplete removal, the data are useful if we remember that the counts represent a minimum estimate of the adsorbed population.

Modern techniques for bacterial enumeration by epifluorescence involve the disruption of large clumps by mechanical agitation, the recovery of the separated bacterial cells by filtration, and the staining of these cells with a fluorescent dye and their enumeration by epifluorescence microscopy [20]. The protocol for the enumeration of adsorbed bacteria by epifluorescence microscopy as used in our laboratory is detailed in Appendix C.

12.2.5 Isolation of Component Bacterial Strains

Isolation is made somewhat complex by the different degrees of adhesion of the individual strains of bacteria that make up adsorbed microbial populations. A convenient index for the definition of adsorbed bacteria is continued adhesion through agitation in several changes of a sterile fluid resembling the natural fluid (e.g., Bryant and Burkey's solution for rumen bacteria [21]) at a specific intensity for a specific time. This procedure removes adventitious organisms and leaves a truly adherent population, some of whose component cells are so deeply embedded in the tissue [22] as to be resistant to removal by swabbing or washing (Fig. 12.10). To recover all these adsorbed bacteria, we have homogenized the tissue and isolated the aerobic bacteria by plating and the anaerobic bacteria by the modified roll-tube method [23]. This method can be used with all soft tissues and substrates, but scraping followed by homogenization is necessary for the recovery of adsorbed bacteria from metal and rock surfaces. Since it has been established that masses of adsorbed bacteria only 10 cells thick can produce anaerobic conditions on a colonized surface [24], any study of adsorbed bacteria should bear in mind that anaerobes may comprise an important part of these populations. Once pure cultures of the isolates have

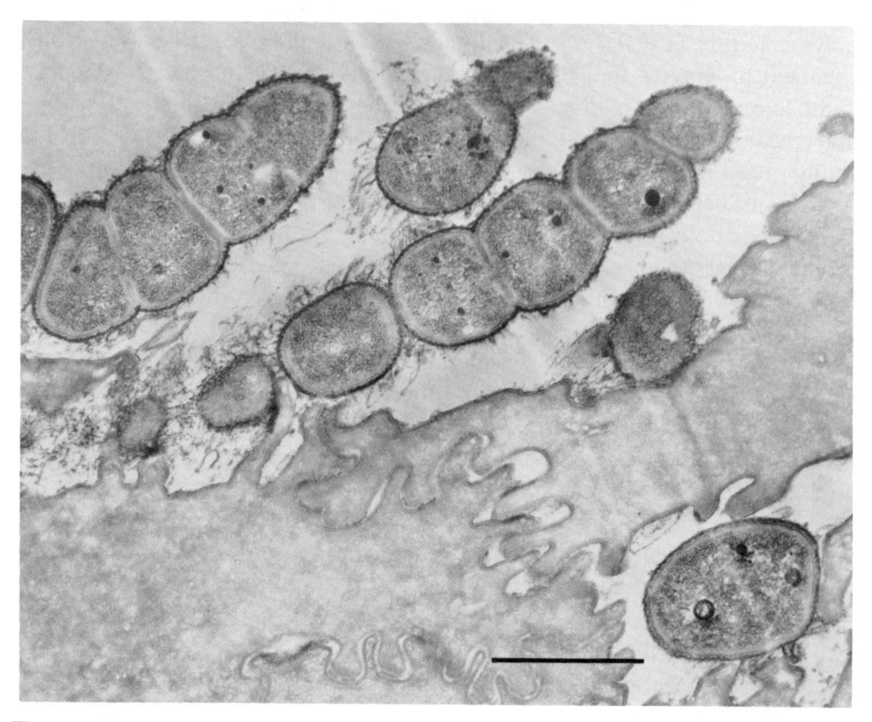

Figure 12.10 Transmission electron micrograph of trichome-forming bacterium that has colonized both bovine tongue epithelial-cell surface and cryptlike intercellular spaces of this tissue. (bar = 1 μm).

been established, physiologic studies can determine their metabolic abilities and reveal their role in the ecosystem. Urease-producing facultative anaerobes, for example, were found on the rumen wall (Cheng and Costerton, unpublished results), and we could conclude that these organisms are involved in oxygen scavenging and the conversion of urea to ammonia.

12.2.6 Mapping the Distribution of Specific Bacteria

Mapping requires a method for the microscopic identification of the strain in question, and this has been accomplished using immunofluorescence. When bacteria are recovered in pure culture from a natural population, they can be used to produce specific antibodies, which can then be coupled to fluorescent compounds and used to identify cells of the species in mixed natural populations [25]. This very elegant technique has been used in several ecosystems [26], but usually to quantitate a specific organism within zones of soil or aquatic systems [27] and only rarely to study adsorbed microbial populations [28]. A complete understanding of adsorbed microbial populations obviously requires the development of techniques such as immunofluorescence at the level of light microscopy and the use of ferritin–antibody complexes at the elctron-microscope level [29], so that the distribution of cells of a particular species can be mapped in relation to the surface to which they are adsorbed.

12.3 TECHNIQUES FOR STUDYING MICROBIAL COLONIZATION PROCESS

In our attempts to understand the very complex processes by which adsorbed bacterial populations are built up on a surface, we tend to revert to the fundamentals of our scientific training and begin, as it were, with a "clean slate." But, paradoxically, our success will depend on the degree to which we realize that there is no such thing as a "clean" surface (see Chapter 3). For instance, dry metal surfaces are composed of oxides and organic contaminants from the air, and all surfaces in aquatic environments attract and bind dissolved ions and organic molecules [30]. Tissue surfaces are complicated by serum factors, including antibodies and complement, and even tissue-culture cells are coated by a complex layer of organic compounds. Thus we must seek to understand the actual chemical milieu at the surface in question and to determine the effect of this milieu on microbial adsorption and on the nutrition and physiology of the resultant adsorbed populations. Since we cannot effectively remove contaminants from a surface, except in high vaccum by the use of "plasma scrubbing," we can best understand the effects of ions and molecules by testing a range of concentrations of each for its effects on microbial adsorption [31].

Once we realize that even a chemically inert surface comprises a zone of concentration of ions and organic molecules, the basic bacterial adsorption experiment suggests itself. A sterile surface is exposed to a population of bacteria in a variety of ionic and molecular milieus, and the rate at which bacteria adsorb to the surface is quantitated by the microscopic and chemical techniques discussed previously in connection with the examination of natural adsorbed populations. We have examined the adsorption of cells of *Salmonella dublin* to tissue-culture cells of the "intestine 407" line using epifluorescence microscopy (Irvin and Costerton, unpublished), and our method is detailed in Appendix D. The adsorption of pathogenic and saprophytic bacteria to tissues and tissue cultures has been examined in literally hundreds of systems [32–36]; however, although these experiments are of interest, they are very difficult to relate to the pathogenic or saprophytic colonization of tissues in intact animals and plants where the milieus are conditioned and defense mechanisms are operative.

When we make a morphological study of the adsorption of bacteria onto a surface, SEM yields valuable data on the distribution of the adherent bacteria and on their relationship to topographical features of the tissue (Figs. 12.1, 12.3, and 12.11), and TEM resolves the fine anionic extracellular fibers that mediate attachment in the early stages of adsorption (Figs. 12.5, 12.7, 12.10, and 12.12). As an adsorbed bacterial population develops on the previously sterile surface, TEM may be utilized to document the development of a matrix of extracellular fibers (Fig. 12.13) and of the resultant microcolonies of bacterial cells of the same strain that make up a newly adherent population.

12.4 TECHNIQUES FOR STUDYING BACTERIAL COLONIZATION OF SPECIFIC SUBSTRATES

Because tissue surfaces and inert surfaces in natural environments display disturbing amounts of heterogeneity (Figs. 12.1 and 12.11), the kinetics of the adsorption of bacteria and the effects of various milieus are often difficult to assess. The simplest specific bacterial adsorption systems described to date are those in which certain bacteria in ruminant digestive systems adhere to and digest the particular insoluble digestive components that constitute their preferred nutrient [2]. These systems include amylase-producing bacteria that adhere to starch granules [37] and cellulase-producing bacteria that adhere to cellulose fibers [38] or to plant cell walls [39] when these insoluble nutrients are introduced into rumen fluid containing a mixture of bacteria. This adsorption is very rapid, and most of the cells in pure cultures of certain species can be removed from culture fluid by their adhesion to an insoluble nutrient (Fig. 12.14), just as most of the cells of certain species of bacteria can be separated from the general mixed population of rumen bacteria by this adhesion to cellulose or starch.

The use of critical-point drying (Appendix A) followed by shadowing, and examination by TEM [38] has been very useful in the determination of the spatial relationship of adsorbed bacteria to their specific insoluble substrate.

12.5 CONCLUDING REMARKS

The study of adsorbed bacterial populations is important because of numerical dominance in some cases and of special physiologic functions in other cases, and it has necessitated the development of a whole battery of special techniques. Initial description of the population is usually accomplished by light microscopy or SEM, which allows the observer to

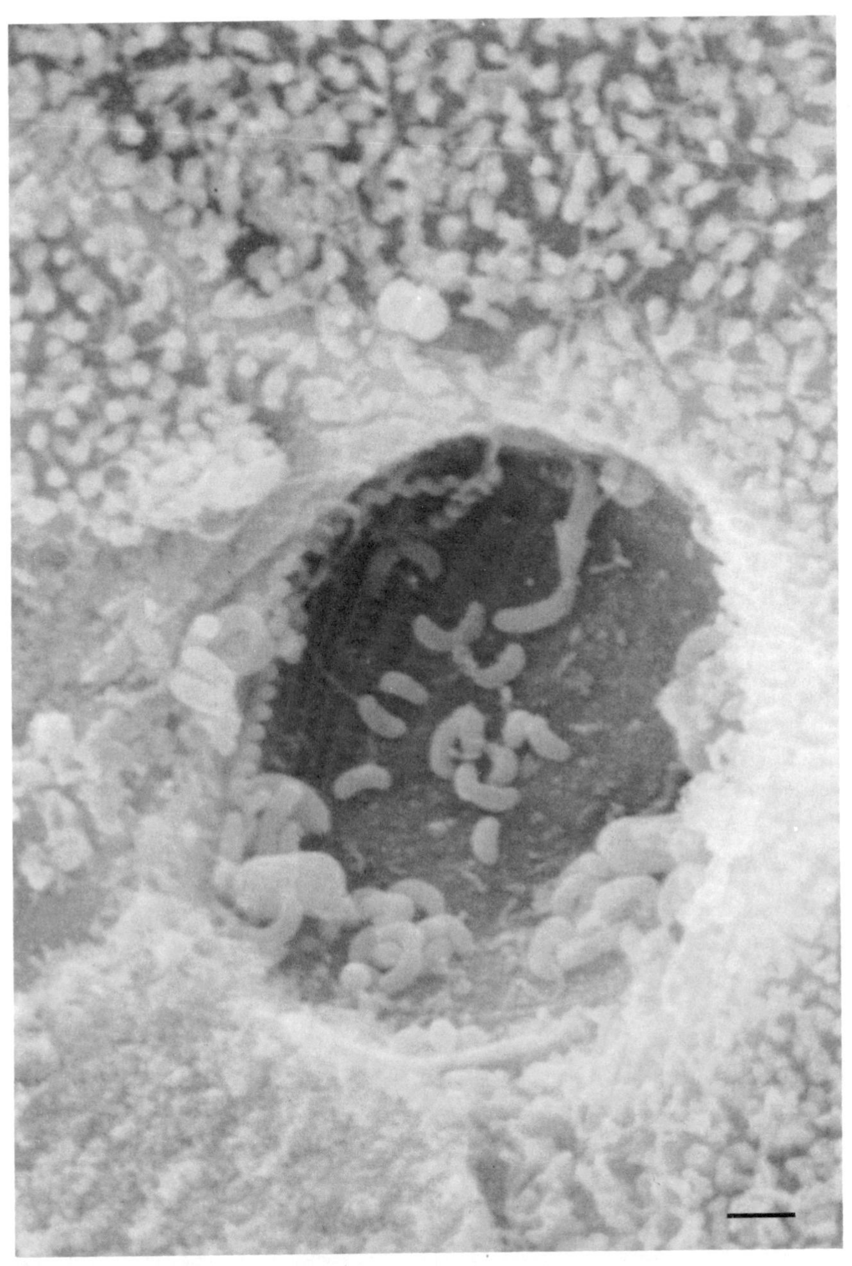

Figure 12.11 Scanning electron micrograph of "pit" between cells of epithelium of bovine rumen showing colonization of pit floor by curved rod-shaped cells and occurrence of spiral organisms on pit walls (bar = 1 μm).

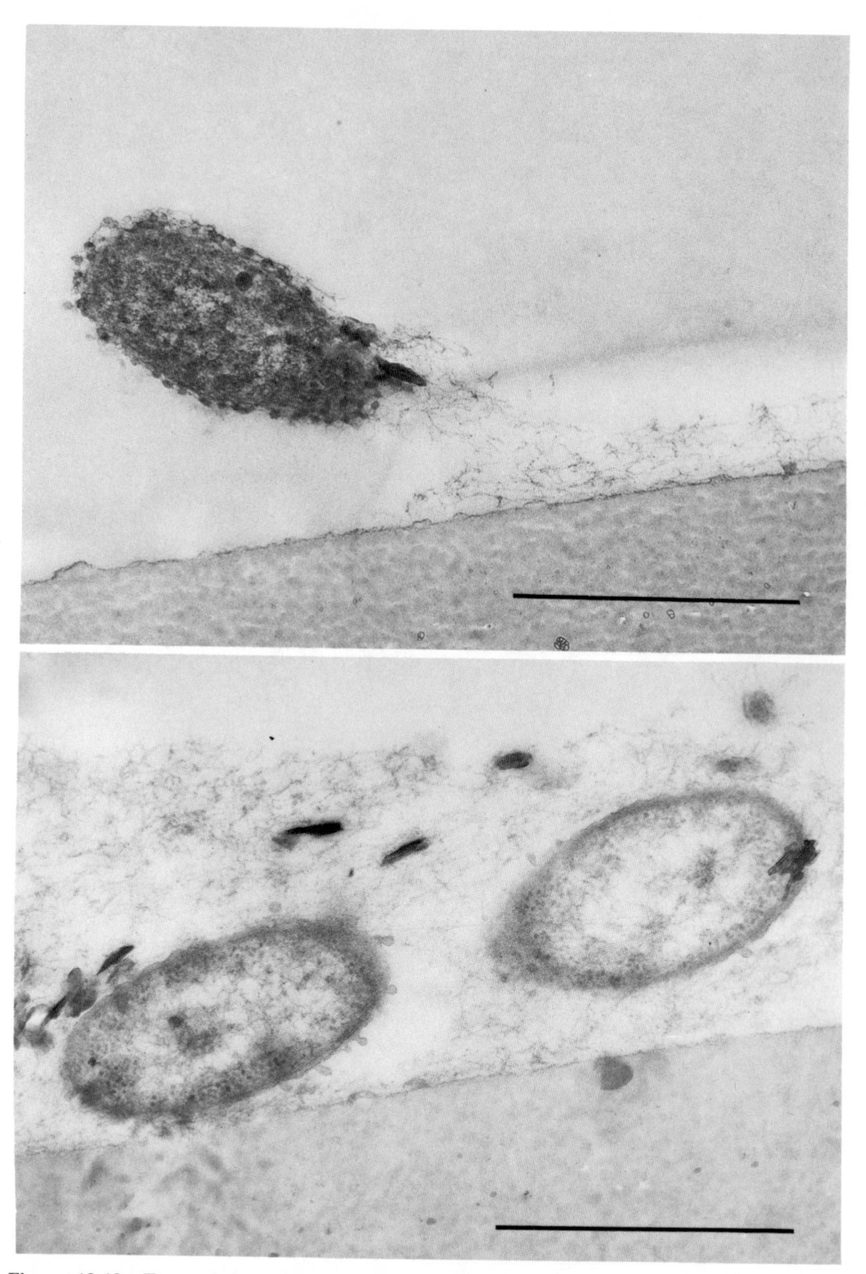

Figure 12.12 Transmission electron micrograph of ruthenium red-stained preparation of bacterium that has initiated colonization of plastic disk suspended in alpine stream. Preparation is made by reembedding exposed disc [40]. Note well-stained extracellular fibrils that clearly mediate attachment of this cell to this surface (bar = 1 μm).

Figure 12.13 Micrograph from same experiment in which bacteria colonizing this plastic surface are enclosed in matrix of extracellular fibers that has trapped some small clay platelets (bar = 1 μm).

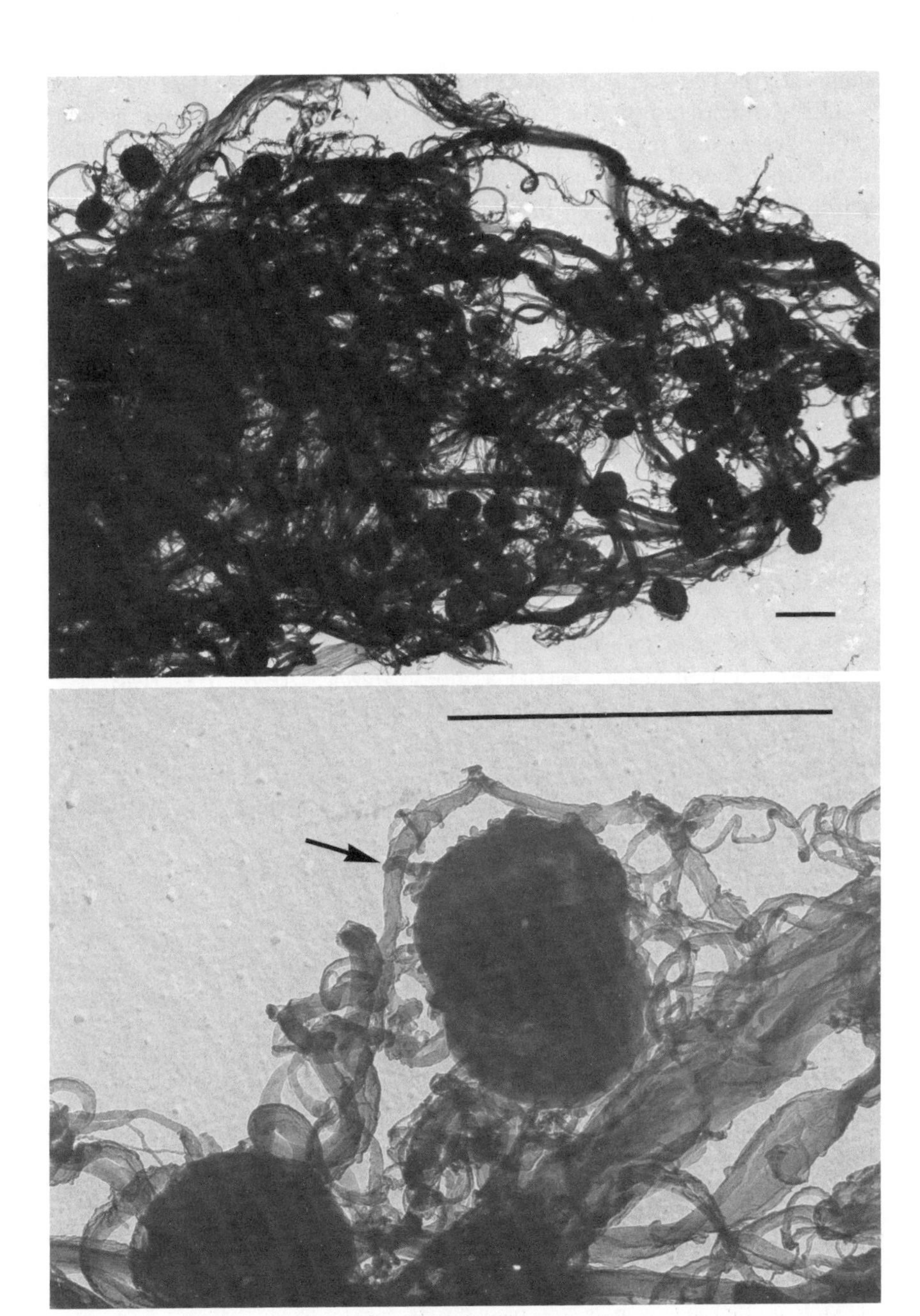

Figure 12.14 Transmission electron micrograph of critical-point-dried preparation of cells from pure culture of cellulolytic rumen bacterium *Ruminococcus albus* previously allowed to adhere to sterile cellulose fibers. Note condensed bacterial fibrils (arrow) coiled around cellulose fibers and appearing to mediate this adsorption [bars in (A) and (B) = 1 μm].

enumerate the bacteria and relate them to the topography of the surface and by TEM, which resolves the extracellular fibrous matrix that may mediate both adhesion and microcolony formation. A complete characterization of an adsorbed population requires the isolation of component species and their mapping by immunofluorescence. An understanding of the development of adherent bacterial populations is best approached by careful studies of the kinetics of adsorption of individual species to a sterile surface, in a variety of milieux.

12.6 REFERENCES

1. G. G. Geesey, R. Mutch, J. W. Costerton, and R. B. Green, *Limnol. Oceanogr.*, **23,** 1214 (1978).
2. K.-J. Cheng, D. E. Akin, and J. W. Costerton, *Fed. Proc.*, **36,** 193 (1977).
3. K.-J. Cheng, R. P. McCowan, and J. W. Costerton, *Am. J. Clin. Nutr.*, **32,** 139 (1979).
4. J. W. Costerton and R. R. Colwell, Eds., *Native Aquatic Bacteria: Enumeration, Activity and Ecology*. ASTM Press, Philadelphia, 1979.
5. .K. C. Marshall, R. Stout, and R. Mitchell, *J. Gen. Microbiol.*, **68,** 337 (1971).
6. A. L. Cohen, D. P. Marlow, and G. E. Garner, *J. Microscop.* (*Paris*), **7,** 331 (1968).
7. L. E. Malick and R. B. Wilson, *Stain Technol.*, **50,** 265 (1975).
8. J. H. Luft, *Anat. Rec.*, **171,** 347 (1971).
9. J. W. Costerton, G. G. Geesey, and K.-J. Cheng, *Sci. Am.* **238,** 86 (1978).
10. E. B. Mackie, K. W. Brown, J. Lam, and J. W. Costerton, *J. Bacteriol.*, **138,** 609 (1979).
11. M. E. Bayer, and H. Thurow, *J. Bacteriol.*, **130,** 911 (1977).
12. D. C. Birdsell, R. J. Doyle, and M. Morgenstein, *J. Bacteriol.*, **121,** 726 (1975).
13. R. D. P. Eaton, E. Meerovitch, and J. W. Costerton, *Transact. Roy. Soc. Trop Med. Hyg.* **63,** 678 (1969).
14. W. J. Wiebe and K. Bancroft, *Proc. Nat. Acad. Sci.* (*USA*), **72,** 2112 (1975).
15. D. C. White, R. J. Bobbie, J. S. Herron, J. D. King, and S. J. Morrison, in J. W. Costerton and R. R. Colwell Eds., *Native Aquatic Bacteria: Enumeration, Activity and Ecology*, ASTM Press, Philadelphia, 1979.
16. J. E. Hobbie, O. Holm-Hansen, T. T. Packard, L. R. Pomeroy, R. W. Sheldon, J. P. Thomas, and W. J. Wiebe, *Limnol. Oceanogr.*, **17,** 544 (1972).
17. S. W. Watson, T. J. Novitsky, H. G. Quinby, and F. W. Valois, *Appl. Environ. Microbiol.*, **33,** 940 (1977).
18. J. W. Costerton, J. M. Ingram, and K.-J. Cheng, *Bacteriol. Rev.*, **38,** 87 (1974).
19. G. G. Gessey, W. T. Richardson, H. G. Yeomans, R. T. Irvin, and J. W. Costerton, *Can. J. Microbiol.*, **23,** 1733 (1978).
20. R. Zimmerman and L.-A. Meyer Riel, *Kiel Meeresforsch.*, **30,** 24 (1974).
21. M. P. Bryant and L. A. Burkey, *J. Dairy Sci.*, **36,** 205 (1953).
22. R. P. McCowan, K.-J. Cheng, C. B. M. Bailey, and J. W. Costerton, *Appl. Environ. Microbiol.*, **35,** 149 (1978).

23. M. P. Bryant, *Am. J. Clin. Nutr.*, **27,** 1313 (1972).
24. W. A. Coulter, and C. Russell, *J. Appl. Bacteriol.*, **40,** 73 (1976).
25. C. B. Fliermans and E. L. Schmidt, *Appl. Microbiol.*, **28,** 676 (1975).
26. P. M. Stanley, M. A. Gage, and E. L. Schmidt, in J. W, Costerton and R. R. Colwell, Eds., *Native Aquatic Bacteria: Enumeration, Activity and Ecology*, ASTM Press, Philadelphia 1979.
27. E. L. Schmidt, *Soil Science*, **118,** 141 (1974).
28. E. L. Schmidt, J. A. Biesbrock, B. B. Bohlool, and D. H. Marx, *Can. J. Microbiol.*, **29,** 137 (1974).
29. T. J. MacAlister, R. T. Irvin, and J. W. Costerton, *J. Bacteriol.*, **130,** 318 (1977).
30. A. T. Henrici, *J. Bacteriol.*, **25,** 277 (1933).
31. J. W. Costerton and G. G. Geesey, *Proceedings of Fourth International Symposium on Contamination Control*, 1979.
32. R. Aly, H. I. Shinefield, W. G. Strauss, and H. I. Maibach, *Infect. Immun.*, **17,** 546 (1977).
33. H. U. Bertschinger, H. W. Moon, and S. C. Whipp, *Infect. Immun.*, **5,** 595 (1972).
34. J. E. Heckels, B. Blackett, J. S. Everson, and M. E. Ward, *J. Gen. Microbiol.* **96,** 359 (1976).
35. G. W. Jones, G. D. Abrams, and R. Freter, *Infect. Immun.*, **14,** 232 (1976).
36. Y. Osada, M. Nakajo, T. Une, and H. Ogowa, *Jap. J. Microbiol.*, **16,** 525 (1972).
37. L. J. Hamlin and R. E. Hungate, *J. Bacteriol.*, **72,** 548 (1955).
38. H. Patterson, R. Irvin, J. W. Costerton, and K.-J. Cheng, *J. Bacteriol.*, **122,** 278 (1975).
39. D. E. Akin and H. E. Amos, *Appl. Microbiol.*, **29,** 692 (1975).

APPENDIX **A**

Preparation of Colonized Tissue Surfaces for Scanning Electron Microscopy

References

1. A. L. Cohen, D. P. Marlow, and G. E. Garner, "A Rapid Critical Point Using Fluorocarbon ("Freons") as Intermediate and Transitional Fluids," *J. Microsc.* (*Oxford*), **7**, 331–342 (1968).
2. R. P. McCowan, K.-J. Cheng, C. B. M. Bailey, and J. W. Costerton, 1978. "Adhesion of Bacteria to Epithelial Cell Surfaces within the Reticulo-rumen of Cattle," *Appl. Environ. Microbiol.*, **35**, 149–155 (1978).
3. Linda E. Malick and B. W. Wilson, "Modified Thiocarbohydrazide Procedure for Scanning Electron Microscopy: Routine Use for Normal, Pathological, or Experimental Tissues," *Stain Technol.*, **50**, (4), 265–269 (1975).

Rationale

Controlled washing allows us to distinguish adventitious from adherent bacteria and exposure to thiocarbohydrazide (TCH) and osmium tetroxide imparts sufficient atomic density and conductivity to allow optimal examination by SEM. The colonized face of the tissue block is oriented in reference to a suture placed in the opposite face. The tissue is dehydrated in alcohol and then replaced with Freon. The tissue is then "dried" by the critical-point method, which minimizes drying forces and preserves the tissue without collapse for examination by SEM.

Procedure

1. Tissues are washed in a reproducible manner by perfusion or by agitation in a fluid resembling their normal milieu (0.85% NaCl in the case of rumen tissue).

2. The tissue is prefixed by perfusion or immersion in a prefixative solution of 0.5% glutaraldehyde in 0.1*M* cacodylate buffer (pH 7.2) for 30 min.
3. Tissue blocks are trimmed to 1 cm^2 after prefixation and immersed in a fixation solution of 5% glutaraldehyde in 0.1*M* cacodylate buffer (pH 7.2) at room temperature for 2 hr.
4. After fixation the tissue blocks are washed five times (30 min per wash with constant agitation of the wash solutions) in cacodylate buffer to remove the glutaraldehyde and the exposed side is marked by the use of a suture on the opposite side.
5. Postfixation is carried out by immersing the tissues in a solution of 2% osmium tetroxide in cacodylate buffer for 2 hr. The tissue blocks are then washed five times as described in step 4 using cacodylate buffer.
   a. After washing, the tissue blocks are treated with a 1% TCH solution (which is prepared by stirring in distilled water for approximately 10 min to dissolve as much TCH as possible). Then the whole solution is filtered through a Buchner's funnel using suction for 20–30 min. This step is followed by washing the tissue six times with distilled water within a period of 10–15 min.
   b. The tissue blocks are treated with 1% osmium tetroxide in distilled water for 2–3 hr followed by six washes with distilled water within a period of 10–15 min.
   c. The 1% TCH solution treatment for 20–30 min is repeated once again and followed by six washes with distilled water within a period of 10–15 min.
   d. the treatment with 1% osmium tetroxide is also repeated once more followed by six washes within a 10–15 min period.

(Note: Steps 5a–d can be omitted if gold-palladium coating is preferred [Ref. 3].)

6. Dehydration is accomplished in a graded ethanol series made up with water of 10%, 20%, 30%, 50%, 70%, 90%, 95%, 100%, and 100%. Each dehydration step is 30 min with constant agitation. This is followed by another dehydration series using 10%, 20%, 30%, 50%, 70%, 90%, 100% and 100% Freon-113 using absolute ethanol as the mixing solvent. Each step in this series is also 30 min with constant agitation.
7. The tissue blocks are then further dehydrated by critical-point drying in Freon-13 to preserve the surface from drying damage.
8. After all the dehydration steps, the tissue blocks are glued onto stubs using a silver paint as the glue and kept inside tightly sealed containers with silica gel until they are examined by SEM.

APPENDIX **B**

Preparation of Colonized Tissue Surfaces for Transmission Electron Microscopy

References

1. R. P. McCowan, K.-J. Cheng, C. B. M. Bailey, and J. W. Costerton, "Adhesion of Bacteria to Epithelial Cell Surfaces within the Reticulo-rumen of Cattle," *Appl. Environ. Microbiol.*, **35**, 149–155 (1978).
2. E. S. Reynold, "The Use of Lead Citrate at High pH as Electron-opaque Stain in Electron Microscopy, *J. Cell Biol.*, **17**, 208–242 (1963).

Rationale

Controlled washing allows separation of adventitious from truly adsorbed bacteria, and exposure to an anion-specific stain (ruthenium red) at the time of fixation imparts electron contrast to the polyanionic extracellular structures that enclose adsorbed bacterial populations. The previously exposed side of a tissue block must be marked for future identification so that sections can be cut at a right angle to this colonized surface, and the examination of these sections yields excellent morphological data on the organization of the adsorbed bacterial population and on the fine structure of cell envelopes and cell surface structures.

Procedure

1. Tissues are washed in a reproducible manner by perfusion or by agitation in a fluid resembling their normal milieu (0.85% NaCl in the case of rumen tissue).
2. The tissue is prefixed by perfusion or immersion in a prefixative solution of 0.5% glutaraldehyde (purchased as 70% glutaraldehyde under argon

and opened <1 hr before use) in 0.1*M* cacodylate buffer (pH 7.2) containing 0.15% ruthenium red at room temperature for 30 min to a maximum of 2 hr.

3. After prefixation, tissue blocks are trimmed to 2 cm^2 and immersed in a fixation solution of 5% glutaraldehyde in 0.1*M* cacodylate buffer (pH 7.2) containing 0.05% ruthenium red at room temperature for 2 hr.
4. After fixation the tissue blocks are washed five times (10–15 min per wash) in cacodylate buffer containing 0.05% ruthenium red to remove the glutaraldehyde, and the previously exposed side is marked by shaping the block or by the use of a suture.
5. Postfixation is carried out using 2% osmium tetroxide in cacodylate buffer containing 0.05% ruthenium red, and this step is followed by five washes (1 hr per wash) in cacodylate buffer containing 0.05% of the stain.
6. Dehydration is accomplished in a graded acetone series in which the 30% and 50% steps contain ruthenium red (0.05%) in the aqueous buffer component, but the 70% and 90% acetone series are made up with distilled water. The initial steps are carried out for 30 min, but the 90% step is shortened to 10 min to minimize loss of stain and is followed by two washes in freshly distilled 100% acetone and two washes in propylene oxide.
7. The tissue blocks are embedded in Vestopal after soaking in a graded (¼, ½, and ¾) series of propylene oxide–resin mixtures and are then positioned in flat rubber embedding trays so that the colonized surface lies at a ±90° angle to the long axis of the resin block.
8. Sections cut from these tissues are stained with uranyl acetate and lead citrate by Reynolds's method [Ref. 2] and are supported with carbon before examination by TEM.

APPENDIX C

Preparation of Sessile Populations for Epifluorescence Microscopy

Reference

The method is a modification of that of Meyer-Reil [R. Zimmerman and L.-A. Meyer-Reil, "A New Method for Fluorescence Staining of Bacterial Populations," *Kiel Meeresforsch.*, **30**, 24–27 (1974)].

Rationale

Epifluorescence microscopy allows the direct counting of bacterial cells and thus avoids the problems that cell clumping poses in indirect methods (e.g., plate counts). The cells are removed from the colonized surface by scraping or by homogenization, depending on the nature of the substrate, and are recovered on a flat filter (Nucleopore) and stained with fluorescent dye. Both orange and green bacteria are counted in these stained preparations.

Procedure

1. Two adjacent 4-cm^2 areas of sessile microbial populations are taken from each source.
2. Each sample is transferred to a sterile vial containing 0.5% glutaraldehyde (purchased as 70% glutaraldehyde under argon and opened no more than 1 hr before use) fixative in 0.1*M* cacodylate buffer (pH 7).
3. Samples are blended in sterile 150-ml Waring blender cups for 30–45 sec.
4. Aliquots (1–5 ml) of the sample are then filtered through 25-mm and 0.2-μm Nucleopore filters.
5. The filter membranes containing trapped cells are placed on a filter support prewetted with filter sterilized isopropyl alcohol.

6. A small vacuum is applied to the filter (1–2 in. of mercury), and 1 ml of sterile 0.01% acridine orange (prepared in 0.1*M* phosphate buffer, pH 7.5) is added to the filter.
7. After 2 min a strong vacuum (7–10 in. of mercury) is applied to the filter.
8. The filter is immediately destained with 1.5 ml of isopropyl alcohol and air dried for 10 min.
9. A quarter section of the filter is mounted on a microscope slide using immersion oil as a mounting medium.
10. Cells are counted with a Zeiss Standard 16 microscope (1000× magnification) equipped with epifluorescence. The filter combinations consisted of: excitation, KP 490 and LP 455; beam splitter, LP 510; and barrier filter, LP 520.

APPENDIX D

A Method for Examination of Bacterial Adsorption to Eucaryotic Cells

References

The adhesion assay utilized has been adapted from the following sources:

1. C. L. Hartley, C. M. Robbins, and M. H. Richmond, "Quantitative Assessment of Bacterial Adhesion to Eukaryotic Cells of Human Origin," *Appl. Bacteriol.*, **24,** 91–97 (1978).
2. R. Zimmerman and L. A. Meyer-Reil, "A New Method for Fluorescence Staining of Bacterial Populations on Membrane Filters," *Kiel Meeresforsch.*, **30,** 24–27 (1974).

Rationale

This adhesion assay is reasonably versatile and can be utilized for examination of kinetic parameters of adhesion or morphologic aspects of adhesion. A nonconfluent monolayer of human cells, washed free of cellular debris, is exposed to a bacterial suspension, washed, prefixed, and fixed with glutaraldehyde and prepared for fluorescence microscopy, TEM or SEM.

Procedure

1. Intestinal epithelial cells (ATCC 6) are grown in Eagles media with Hank's balanced salts, 2m*M* l-glutamine, and 10% dialyzed calf serum. Tissue cells are split and grown in either Lab Teck II slides or Lux Multiplates at 37°C, 95% relative humidity, and 5% CO_2 for 3 days (cells are not yet a confluent monolayer).
2. The tissue cells are washed three times with Eagles basal medium containing 2m*M* l-glutamine (BME).
3. Bacterial cells are grown overnight at 37°C in brain–heart infusion (BHI) broth from a single colony isolate. The overnight culture is used

to inoculate fresh BHI broth, and the cells are grown for 2–3 hr at 37°C or until the cell density reaches approximately 10^9 cells ml^{-1}.

4. A bacterial suspension is prepared by diluting the BHI broth culture with BME until cell density is $\sim 10^7$ ml^{-1}.
5. Aspirate BME from tissue-culture monolayer and replace with the bacterial suspension. Incubate for 30 min at room temperature.
6. Filter 1 ml of bacterial suspension using a 0.2-μm Nucleopore filter. Wash with 20 ml of H_2O. Place filter in a small petri dish on a Whatman No. 1 filter presoaked in 5% glutaraldehyde.
7. After 30 min of incubation, aspirate the bacterial suspension and wash three times with BME.
8. Add BME and $\frac{1}{10}$ volume of 5% glutaraldehyde in BME and allow to incubate for 20 min at room temperature. Aspirate and replace with 5% glutaraldehyde in BME and allow to fix for 1.5 hr at room temperature. Wash cells three times with BME. Cells may be prepared for TEM or SEM rather than light microscopy.
9. Perform acridine orange staining by rinsing fixed and washed cells with H_2O. Cover cells with 1 mg ml^{-1} of acridine orange in 50m*M* phosphate buffer (pH 7) and allow to incubate for 2 min at room temperature. Rinse with H_2O and cover cells with isopropanol of 30 sec, rinse with H_2O, and allow to air dry. The Nucleopore filter is stained in a filter apparatus in a similar manner.
10. Examine the stained cells by epifluorescence using the blue excitation. The results are easily photographed using high-speed Ektachrome film.

Index